B + J
5/2/84
52.50
✓

Benchmark Papers
in Geology

Series Editor: Rhodes W. Fairbridge
Columbia University

RIVER MORPHOLOGY / Stanley A. Schumm
SLOPE MORPHOLOGY / Stanley A. Schumm and M. Paul Mosley
SPITS AND BARS / Maurice L. Schwartz
BARRIER ISLANDS / Maurice L. Schwartz
ENVIRONMENTAL GEOMORPHOLOGY AND LANDSCAPE CONSERVATION, VOLUME I: Prior to 1900 / VOLUME II: Urban Areas / VOLUME III: Non-Urban Regions / Donald R. Coates
TEKTITES / Virgil E. Barnes and Mildred A. Barnes
GEOCHRONOLOGY: Radiometric Dating of Rocks and Minerals / C. T. Harper
MARINE EVAPORITES: Origins, Diagenesis, and Geochemistry / Douglas W. Kirkland and Robert Evans
GLACIAL ISOSTASY / John T. Andrews
GLACIAL DEPOSITS / Richard P. Goldthwait
PHILOSOPHY OF GEOHISTORY: 1785–1970 / Claude C. Albritton, Jr.
GEOCHEMISTRY OF GERMANIUM / Jon N. Weber
GEOCHEMISTRY AND THE ORIGIN OF LIFE / Keith A. Kvenvolden
GEOCHEMISTRY OF WATER / Yasushi Kitano
GEOCHEMISTRY OF IRON / Henry Lepp
GEOCHEMISTRY OF BORON / C. T. Walker
SEDIMENTARY ROCKS: Concepts and History / Albert V. Carozzi
METAMORPHISM AND PLATE TECTONIC REGIMES / W. G. Ernst
SUBDUCTION ZONE METAMORPHISM / W. G. Ernst
PLAYAS AND DRIED LAKES: Occurrence and Development / James T. Neal
PLANATION SURFACES: Peneplains, Pediplains, and Etchplains / George Adams
SUBMARINE CANYONS AND DEEP-SEA FANS: Modern and Ancient / J. H. McD. Whitaker
ENVIRONMENTAL GEOLOGY / Frederick Betz, Jr.
LOESS: Lithology and Genesis / Ian J. Smalley
PERIGLACIAL DEPOSITS / Cuchlaine A. M. King
LANDFORMS AND GEOMORPHOLOGY: Concepts and History / Cuchlaine A. M. King
METALLOGENY AND GLOBAL TECTONICS / Wilfred Walker
HOLOCENE TIDAL SEDIMENTATION / George deVries Klein
PALEOBIOGEOGRAPHY / Charles A. Ross
MECHANICS OF THRUST FAULTS AND DÉCOLLEMENT / Barry Voight
WEST INDIES ISLAND ARCS / Peter H. Mattson
CRYSTAL FORM AND STRUCTURE / Cecil J. Schneer
METEORITE CRATERS / G. J. H. McCall
AIR PHOTOGRAPHY AND COASTAL PROBLEMS / Mohamed T. El-Ashry
DIAGENESIS OF DEEP-SEA BIOGENIC SEDIMENTS / Gerrit J. van der Lingen
DRAINAGE BASIC MORPHOLOGY / Stanley A. Schumm
SEA WATER: Cycles of the Major Elements / J. I. Drever
MINERAL DEPOSITS AND PLATE TECTONICS / J. B. Wright
ANCIENT CONTINENTAL DEPOSITS / Franklyn B. Van Houten
BEACH PROCESSES AND COASTAL HYDRODYNAMICS / John S. Fisher and Robert Dolan
PALYNOLOGY, PART I: Spores and Pollen / Marjorie D. Muir and William A. S. Sarjeant
PALYNOLOGY, PART II: Dinoflagellates, Acritarchs, and Other Microfossils / Marjorie D. Muir and William A. S. Sarjeant

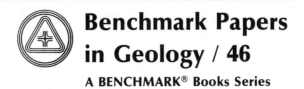

Benchmark Papers in Geology / 46
A BENCHMARK® Books Series

PALYNOLOGY, PART I
Spores and Pollen

Edited by

MARJORIE D. MUIR
Imperial College of Science and Technology

William A. S. SARJEANT
University of Saskatchewan, Saskatoon, Canada

Stroudsburg, Pennsylvania

Wingate College Library

Copyright © 1977 by **Dowden, Hutchinson & Ross, Inc.**
Benchmark Papers in Geology, Volume 46
Library of Congress Catalog Card Number: 77-24110
ISBN: 0-87933-306-5

All rights reserved. No part of this book covered by the copyrights hereon may be reproduced or transmitted in any form or by any means—graphic, electronic, or mechanical, including photocopying, recording, taping, or information storage and retrieval systems—without written permission of the publisher.

79 78 77 1 2 3 4 5
Manufactured in the United States of America.

LIBRARY OF CONGRESS CATALOGING IN PUBLICATION DATA

Main entry under title:
Palynology.
   (Benchmark papers in geology; 46 / 47)
   Includes indexes.
   CONTENTS: 1. Spores and pollen; 2. Dinoflagellates, acritarchs, and other microfossils
   1. Palynology—Addresses, essays, lectures. I. Muir, Marjorie D. II. Sarjeant, William Antony S.
QE993.P32   561'.13'08   77-24110
ISBN 0-87933-306-5 (1)   0-87933-285-9 (2)

SERIES EDITOR'S FOREWORD

The philosophy behind the "Benchmark Papers in Geology" is one of collection, sifting, and rediffusion. Scientific literature today is so vast, so dispersed, and, in the case of old papers, so inaccessible for readers not in the immediate neighborhood of major libraries that much valuable information has been ignored by default. It has become just so difficult, or so time consuming, to search out the key papers in any basic area of research that one can hardly blame a busy man for skimping on some of his "homework."

This series of volumes has been devised, therefore, to make a practical contribution to this critical problem. The geologist, perhaps even more than any other scientist, often suffers from twin difficulties—isolation from central library resources and immensely diffused sources of material. New colleges and industrial libraries simply cannot afford to purchase complete runs of all the world's earth science literature. Specialists simply cannot locate reprints or copies of all their principal reference materials. So it is that we are now making a concerted effort to gather into single volumes the critical material needed to reconstruct the background of any and every major topic of our discipline.

We are interpreting "geology" in its broadest sense: the fundamental science of the planet Earth, its materials, its history, and its dynamics. Because of training and experience in "earthy" materials, we also take in astrogeology, the corresponding aspect of the planetary sciences. Besides the classical core disciplines such as mineralogy, petrology, structure, geomorphology, paleontology, and stratigraphy, we embrace the newer fields of geophysics and geochemistry, applied also to oceanography, geochronology, and paleoecology. We recognize the work of the mining geologists, the petroleum geologists, the hydrologists, the engineering and environmental geologists. Each specialist needs his working library. We are endeavoring to make his task a little easier.

Series Editor's Foreword

  Each volume in the series contains an Introduction prepared by a specialist (the volume editor)—a "state of the art" opening or a summary of the object and content of the volume. The articles, usually some twenty to fifty reproduced either in their entirety or in significant extracts, are selected in an attempt to cover the field, from the key papers of the last century to fairly recent work. Where the original works are in foreign languages, we have endeavored to locate or commission translations. Geologists, because of their global subject, are often acutely aware of the oneness of our world. The selections cannot, therefore, be restricted to any one country, and whenever possible an attempt is made to scan the world literature.

  To each article, or group of kindred articles, some sort of "highlight commentary" is usually supplied by the volume editor. This commentary should serve to bring that article into historical perspective and to emphasize its particular role in the growth of the field. References, or citations, wherever possible, will be reproduced in their entirety—for by this means the observant reader can assess the background material available to that particular author, or, if he wishes, he, too, can double check the earlier sources.

  A "benchmark," in surveyor's terminology, is an established point on the ground, recorded on our maps. It is usually anything that is a vantage point, from a modest hill to a mountain peak. From the historical viewpoint, these benchmarks are the bricks of our scientific edifice.

RHODES W. FAIRBRIDGE

PREFACE

For the average person, spores and pollen are probably brought to mind only when, as airborne irritants, they provoke summertime outbreaks of sneezing by hay-fever sufferers. For the botanist, they are simply the means by which plants propagate themselves or are induced to propagate by horticulturalists; a minute component of the whole plant and occupying a proportionately small amount of attention.

For the archaeologist and for the geologist, however, these microscopical objects have become of great, and indeed often of crucial, importance as a means for the dating of soils, sediments, and human occupancy sites and for the interpretation of the environment and climate of past times.

During the last two decades in particular, it has become increasingly evident that other microfossils seen in microscopical preparations alongside the spores and pollen are also well worthy of attention and—especially in pre-Quaternary sediments—quite often of even greater value in making such interpretations.

The term "palynology" was formulated by Hyde and Williams (1944) to identify this branch of scientific endeavor. Its stem is the Greek word πδλυζω, "to strew, or sprinkle, flour or dust," appropriately enough, since pollen and spores are indeed strewn around by plants and are a prominent component of dust. Originally the word applied specifically to the study of pollen and spores; but the concept came progressively to be enlarged to incorporate other microfossil groups. Indeed, the definition of J.W. Funkhouser (1959) embraced not only spores, pollen, a variety of unicellular and multicellular algal groups, fungal elements, and the microscopic remains of higher plants—all of which are composed of organic substances—but also calcareous and siliceous microfossils such as coccoliths, diatoms, and even radiolaria! Not many specialists want to stretch the boundaries of the subject quite so far; a reasonable present-day definition might be

Preface

> Palynology is that branch of micropaleontology which is concerned with the study of microfossils having a wall composed of an organic substance—a compound of carbon, hydrogen, nitrogen and oxygen, not capable of dissolution in hydrochloric or hydrofluoric acids.

The fossils with which the palynologist is concerned are indeed often called "acid-insoluble," but this is a wholly inappropriate term, since they speedily dissolve in other acids (fuming nitric, sulphuric, even chromic). Some have been termed "organic-shelled," but, though their walls are indeed made up of organic compounds, this term too is an inexact one since calcareous or siliceous shells secreted by organisms are equally organic shells. A better general term might be "palynomorph," but this has tended to be used for pollen and spores rather than for the other components of assemblages.

That no single term can conveniently be applied to all the components of the assemblages seen under the microscope by the palynologist may seem surprising, but it becomes less so when the diversity of objects in assemblages is recognized. In addition to spores and pollen, these include planktonic zoospores of Prasinophycean algae, resting or reproductive cysts of dinoflagellates, linings of early chambers of foraminifera, clustered cells of colonial algae, fragmentary or entire jaw apparatuses of worms, fungal elements, and a variety of objects of problematical character. (Cuticular fragments and carbonized wood are also present in virtually all post-Lower Silurian preparations, but are not sufficiently informative to attract much study.) So diverse indeed are these other groups that we decided to make them the subject of a complete Benchmark volume entitled *Palynology, Part II: Dinoflagellates, Acritarchs, and Other Microfossils*, published simultaneously with this. In the present volume, attention is concentrated on pollen and spores, though there is inevitably some overlap in coverage.

MARJORIE D. MUIR
WILLIAM A. S. SARJEANT

DEDICATION

It will be clear to the reader from the selection of papers in this volume that one of the most far-seeing thinkers in stratigraphic palynology is Jan Muller, who has contributed greatly to our understanding both of the principles of stratigraphy as applied to pollen and spores, and also to our knowledge of the sedimentary controls which affect spore assemblages. We would like to dedicate this volume to that most talented palynologist, Jan Muller.

CONTENTS

Series Editor's Foreword	v
Preface	vii
Dedication	ix
Contents by Author	xv
Introduction	1

PART I: SPORE AND POLLEN GRAIN WALLS—APPLICATION TO GEOCHEMISTRY

Editors' Comments on Papers 1 Through 5 — 4

1. BROOKS, J., and G. SHAW: Chemical Structure of the Exine of Pollen Walls and a New Function for Carotenoids in Nature — 8
   Nature **219**(5153):532–533 (1968)

2. ROWLEY, J. R., and D. SOUTHWORTH: Deposition of Sporopollenin on Lamellae of Unit Membrane Dimensions — 13
   Nature **213**(5077):703–704 (1967)

3. SENGUPTA, S., and J. R. ROWLEY: Re-Exposure of Tapes at High Temperature and Pressure in the *Lycopodium clavatum* Spore Exine — 17
   Grana **14**:143–151 (1974)

4. GUTJAHR, C. C. M.: Carbonization Measurements of Pollen-Grains and Spores and Their Application — 26
   Leidse Geol. Meded. **38**:1–10, 13–29 (1966)

5. GIJZEL, P. van: Autofluorescence of Fossil Pollen and Spores with Special Reference to Age Determination and Coalification — 53
   Leidse Geol. Meded. **40**:263–268, 276–293, 300–312, 314–317 (1967)

PART II: SPORES AND POLLEN GRAINS—CLASSIFICATION AND STRATIGRAPHIC APPLICATION

Editors' Comments on Papers 6 Through 10 — 96

6. KUYL, O. S., J. MULLER, and H. Th. WATERBOLK: The Application of Palynology to Oil Geology with Special Reference to Western Venezuela — 103
   Geologie en Mijnbouw **17**(3), New Ser.:49–75 (1955)

Contents

7 SCHOPF, J. M., L. R. WILSON, and R. BENTALL: An Annotated Synopsis of Paleozoic Fossil Spores and the Definition of Generic Groups — 130
Illinois Geol. Survey Rep. Inv. **91**:7–10, 61 (1944)

8 POTONIÉ, R., and G. O. W. KREMP: Die *Sporae dispersae* des Ruhrkarbons. Teil I — 136
Palaeontographica **98B**:1–28 (1955)

9 TRAVERSE, A.: Pollen Annalysis of the Brandon Lignite of Vermont — 167
U.S. Bur. Mines Rept. Inv. **5151**:7–10, 81–90 (1955)

10 GERMERAAD, J. H., C. A. HOPPING, and J. MULLER: Palynology of Tertiary Sediments from Tropical Areas — 184
Rev. Palaeobotany and Palynology **6**:189–192, 203–206, 221–222, 230–248, 263–268, 346–348

PART III: REWORKING AND OTHER STRATIGRAPHIC PROBLEMS

Editors' Comments on Papers 11, 12, and 13 — 222

11 WILSON, L. R.: Recycling, Stratigraphic Leakage, and Faulty Techniques in Palynology — 224
Grana Palynologica **5**(3):425–436 (1964)

12 TURNAU, E.: The Age of Coal Fragments from the Cretaceous Deposits in the Outer Carpathians, Determined on Microspores — 236
Acad. Polonaise Sci. Bull., Ser. Sci. Géol. et Géog. **10**(2):85–90 (1962)

13 HAMILTON, L. H., R. HELBY, and G. H. TAYLOR: The Occurrence and Significance of Triassic Coal in the Volcanic Necks Near Sydney — 242
Royal Soc. New South Wales Jour. and Proc. **102**, Pts. 3 & 4:169–171 (1970)

PART IV: MEGASPORES

Editors' Comments on Papers 14 and 15 — 246

14 DIJKSTRA, S. J.: Carboniferous Megaspores in Tertiary and Quaternary Deposits of S.E. England — 248
Annals and Mag. Nat. History **3**, Ser. 12:865–877 (Oct. 1950)

15 HUGHES, N. F.: Wealden Plant Microfossils — 267
Geol. Mag. **92**(3):201–217 (1955)

PART V: DISTRIBUTION OF SPORES AND POLLEN IN SEDIMENTS

Editors' Comments on Papers 16 Through 19 — 288

16 MULLER, J.: Palynology of Recent Orinoco Delta and Shelf Sediments: Reports of the Orinoco Shelf Expedition; Volume 5 — 291
Micropaleontology **5**(1):1–32 (1959)

Contents

17 NEVES, R.: Upper Carboniferous Plant Spore Assemblages from the *Gastrioceras subcrenatum* Horizon, North Staffordshire   323
    Geol. Mag. **95**(1):1–3, 12–18 (1958)

18 CHALONER, W. G.: The Carboniferous Upland Flora   333
    Geol. Mag. **95**(3):261–262 (1958)

19 SMITH, A. H. V.: The Palaeoecology of Carboniferous Peats Based on the Miospores and Petrography of Bituminous Coals   334
    Yorkshire Geol. Soc. Proc. **33**(4):423, 428–439, 446–465 (1962)

Author Citation Index   369
Subject Index   375
About the Editors   383

CONTENTS BY AUTHOR

Bentall, R., 130
Brooks, J., 8
Chaloner, W. G., 333
Dijkstra, S. J., 248
Germeraad, J. H., 184
Gijzel, P. van, 53
Gutjahr, C. C. M., 26
Hamilton, L. H., 242
Helby, R., 242
Hopping, C. A., 184
Hughes, N. F., 267
Kremp, G. O. W., 136
Kuyl, O. S., 103
Muller, J., 103, 184, 291

Neves, R., 323
Potonié, R., 136
Rowley, J. R., 13, 17
Schopf, J. M., 130
Sengupta, S., 17
Shaw, G., 8
Smith, A. H. V., 334
Southworth, D., 13
Taylor, G. H., 242
Traverse, A., 167
Turnau, E., 236
Waterbolk, H. Th., 103
Wilson, L. R:, 130, 224

INTRODUCTION

Although spores and pollen of living plants have been studied as microscopical objects for a very great number of years, investigation of *fossil* spores and pollen grains, with which this volume is concerned, began more recently. The earliest studies of fossil spores were associated with nineteenth century work on coal thin sections. These contain sections through both megaspores and microspores, and two types of coal are named "crassidurain" or "tenuidurain" depending upon whether they contain thick or thin walled spores. However, in order to understand the three-dimensional geometry of the spores, laborious reconstructions had to be made from many thin sections; and while it was possible to use the results stratigraphically, the sample size was far too small and obviously inadequate for stratigraphic work.

The development by Lennart von Post in 1916 of the new science of pollen analysis by extraction and concentration of pollen grains from Pleistocene peat deposits offered the possibility of using pollen stratigraphically. In the 1920s and 1930s, pollen analysis was greatly refined by means of new preparation methods such as acetolysis which made identification simpler. Pollen analysis developed into a sophisticated method of correlating scattered deposits of Holocene and Pleistocene ages, using changes in tree pollen assemblages. These changes are on a very fine time scale and are sensitive indicators of climatic fluctuations. Pollen analysis or Pleistocene palynology embraces a vast amount of literature, and cannot be considered in detail in this volume which is concerned with pre-Quaternary palynology. A number of references to classic papers are given at the end of Part II, and these will serve to introduce the reader to this field.

About the time that von Post was developing pollen analysis methodology, the great German organic chemist, Zetzsche, assisted by the

Introduction

young palaeobotanist Robert Potonié, began his far-reaching studies on the organic geochemistry of coal and fossil plant substances. He found that the material of which the spore walls were composed, and which he called *sporopollenin*, was resistant to all forms of chemical treatment except for strong oxidation. Controlled oxidation released the spores and pollen grains from the coals and made them translucent, so that they could be studied under the microscope. This was, in many ways, a greater breakthrough than von Post's, because correlation of the Coal Measures of Europe and North America had been a major geological problem for many years. The virtual absence of marine microfossils, the difficulties of using plant macrofossils, and the lack of closely spaced universal marker horizons (such as marine bands or tonsteins) make close correlation difficult, and in mining operations, rapid and accurate seam identification is essential. The use of spores and pollen extracted from coal seams offered the possibility of close correlation for the first time. Raistrick, in Britain, Potonié and his coworkers in Germany, Naumova and others, in the USSR, and Schopf, in the United States, were quick to seize this opportunity: the present day accuracy of Coal Measures correlation is derived from their pioneering efforts.

In the late 1930s, it was realized that spores and pollen grains could be extracted from nearly every kind of sedimentary rock, excepting only those that had suffered severe oxidation. A number of oil companies became interested enough to set up small palynological laboratories, although few of these became operational until after the Second World War. Spores and pollen grains are ideally suited to oil exploration since they can be recovered from drill core, sidewall core, or cuttings in sufficient numbers to date the samples. They can also be found in all facies, unlike, for example, foraminiferida and ostracoda, and can be used for dating continental as well as marine successions. However, they are not totally facies independent and they can be used as quite sensitive indicators of palaeoenvironments.

Finally, in the late 1950s, it was discovered that the colors of spores and pollen grains darkened with increasing temperature and that this color change was irreversible. The color of the spores indicates the former maximum palaeotemperature of the sediment and can be used to estimate the hydrocarbon potential of the sample.

In this present volume, the development of ideas in pre-Quaternary palynology from the end of the Second World War will be outlined by a selection of important papers. Although many earlier papers in palynology are considered classic, most of these refer to only one geological horizon, or to only one locality, or are descriptive papers of new spore types. Because of limitations of space in this volume, all such papers have been omitted, and only those papers that, in our opinion, contain new ideas with far-reaching effects have been considered for inclusion in this volume.

Part I

SPORE AND POLLEN GRAIN WALLS – APPLICATION TO GEOCHEMISTRY

Editors' Comments on Papers 1 Through 5

1   BROOKS and SHAW
       Chemical Structure of the Exine of Pollen Walls and a New Function for Carotenoids in Nature

2   ROWLEY and SOUTHWORTH
       Deposition of Sporopollenin on Lamellae of Unit Membrane Dimensions

3   SENGUPTA and ROWLEY
       Re-Exposure of Tapes at High Temperature and Pressure in the Lycopodium clavatum *Spore Exine*

4   GUTJAHR
       Excerpts from *Carbonization Measurements of Pollen-Grains and Spores and Their Applications*

5   GIJZEL
       Excerpts from *Autofluorescence of Fossil Pollen and Spores with Special Reference to Age Determination and Coalification*

Spores and pollen grains represent the sexual stage in the land plant life cycle. They contain the gametophytes which fertilize each other to create the new sporophyte generation. In the reproductive organs, one mother cell undergoes meiotic division and forms four daughter cells; these usually remain together in *tetrads* during the early stages of development. In primitive land plants, male and female gametophytes originate from spores which are morphologically identical. Such plants are described as being *homosporous* and the spores they produce are called *isospores*. In rather more advanced plants, the female spore is much larger than the male. This condition is termed *heterospory*; the male and female organs are *microspores* and *megaspores* respectively. In still more advanced plants, the megaspore is retained within the parent plant to form a *seed*; the microspore which fertilizes the seed is defined as a *pollen grain*. Microspores, isospores, and megaspores all germinate proximally through the *tetrad scar*. Pollen grains germinate distally or equatorially. Nevertheless it is often difficult to distinguish between these two types of germination from the morphology of fossil spores. It is therefore convenient to use a term that encompasses them all: *small spores, miospores,* and *polospores* have all been proposed to avoid having to use the long-winded term "spores and pollen" each time. In this volume, we shall use "miospores," unless either "microspores" or "pollen" is specifically intended.

Only the wall, or exine, of the miospore or megaspore resists degradation in the sediment and is capable of being fossilized (Elsik, 1971: Havinga, 1971). The chemical nature of the material of which the exine is composed is intractable to normal methods of chemical analysis. Zetsche (1932) coined the terms "sporonin" for spore wall material, and "pollenin" for the material in pollen walls. The elemental composition of these materials, which are chemically indistinguishable and are now referred to as "sporopollenin," is variable, but a formula of $C_{90}H_{140}O_{18}$ approximates to the composition of most sporopollenins. It is known to be one of a class of biological polymers; analyses using techniques such as ozonolysis have indicated that it contains hydroxyl groups, phenols, and fatty acids, among others.

The first breakthrough in our understanding of the nature of sporopollenin came from a series of experiments carried out by Brooks and Shaw (Paper 1). They discovered that at early stages of pollen development in *Lilium*, although the pollen grain was already morphologically identifiable, no sporopollenin was present in the exine. At a later stage after the maturation of the pollen grain, their exines did contain sporopollenin. Its appearance was a rapid process which could be correlated with a vast increase in the amount of the yellow pigment β-carotene in the anther. After the deposition of the sporopollenin wall, the amount of β-carotene decreased again. Brooks and Shaw deduced that sporopollenin might be an oxidative polymer of β-carotene, and using this as a precursor, proceeded to synthesize material which is chemically indistinguishable from sporopollenin. Since then, several different types of sporopollenin have been synthesized, using various carotenoid and carotenoid esters as sporopollenin precursors (Shaw, 1971).

Miospore exines are multi-layered, but the two most easily distinguishable layers are the endexine (= nexine) and ektexine (= sexine). These two layers have different origins: the endexine is laid down by the daughter cell, whereas the ektexine is deposited on the spore by the tapetal cells of the reproductive organs (Heslop Harrison, 1971: Echlin, 1971). The sporopollenin of the endexine is deposited on discrete lamellae in the spore wall (Rowley and Southworth, Paper 2). The number and arrangement of these lamellae can vary according to the type of parent plant (see various papers in Ferguson and Muller, 1976); their chemical nature is not well understood, but it appears to be a polysaccharide of some kind. Sengupta and Rowley (Paper 3) have demonstrated that if spore and pollen grain walls are experimentally heated at various temperatures (up to 450° C), the lamellae are freed from the spore wall, and become active sites for deposition of new material. The experimental work carried out by Sengupta and Rowley was designed to substantiate results reported by Gutjahr (Paper 4), who observed a gradual darkening of exine color with increasing temperature. Gutjahr was able to correlate this with rank of coal, which is itself a measure of the degree of thermal metamorphism undergone by organic-matter-rich sediments. One of the major controls in the evolution of petroleum is the degree of thermal metamorphism undergone by its source rock. Crudely speaking, if sedimentary organic matter is yellow to light brown, then oil can be expected; if it is dark brown, gas can be predicted; and if it is black, no hydrocarbons can be expected at all (Staplin, 1969; McIntyre, 1972). Exine color is now used by most exploration laboratories as a rough test for determining hydrocarbon potential.

Very subtle chemical changes in the exine can be detected by the use of the fluroescence technique, first applied by van Gijzel (Paper 5). Using a UV source, spores are first irradiated and the color of the emitted light is then measured. The spectra produced are characteristic of age, depth of burial, thermal metamorphism, oxidation, and a number of other factors. This method is extremely sensitive and is now being applied to other aspects of organic geochemistry and botany (van Gijzel, 1973).

REFERENCES

Echlin, P. (1971). Production of sporopollenin by the tapetum. In: J. Brooks et al. (eds.), *Sporopollenin*. London and New York: Academic Press, pp. 220-249.

Elsik, W. C. (1971). Microbiological degradation of sporopollenin. In: J. Brooks et al., (eds.), *Sporopollenin*. London and New York: Academic Press, pp. 470-509.

Ferguson, I. K., and Muller, J. (1976). *The Evolutionary Significance of the Exine*. London and New York: Academic Press.

Havinga, A. J. (1971). An experimental investigation into the decay of pollen and spores in various soil types. In: J. Brooks et al. (eds.), *Sporopollenin* London and New York: Academic Press, pp. 446-478.

Heslop Harrison, J. (1971). Sporopollenin in the biological context. In: J. Brooks et al. (eds.), *Sporopollenin*. London and New York: Academic Press, pp. 1-30.

McIntyre, D.J. (1972). Effect of experimental metamorphism on pollen in a lignite. *Geoscience and Man* 4:111-117.

Shaw, G. (1971). The chemistry of sporopollenin. In: J. Brooks et al. (eds.), *Sporopollenin* London and New York: Academic Press, pp. 305-348.

Staplin, F. L. (1969). Sedimentary organic matter, organic metamorphism, and oil and gas occurrence. *Bull. Canadian Petrol. Geol.* 17:47-66.

van Gijzel, P. (1973). Polychromatic Uv-fluorescence microphotometry of fresh and fossil plant substances and the identification of dispersed organic matter in rocks. Extrait Colloque Internationale: "Petrographie de la matiere organique des sediments, relations avec la paleotemperature et le potential petrolier." C.N.R.S., Paris, September 15-17, 1973.

Zetszche, F. (1932). *Sporopollenine.* In: G. Klein (ed.), *Handbuch der Pflanzen Analyse.* Vienna.

ADDITIONAL REFERENCES

Brooks, J.; Grant, P. R.; Muir, M. D.; Shaw, G.; and van Gijzel, P. (Eds.). (1971). *Sporopollenin.* London and New York: Academic Press.

Correia, M. (1969). Contribution à la recherche de zones favorables à la genèse du pétrole par l'observation microscopique de la matière organique figurée. *Revue Inst. Fr. Pétrôle* **24**:1417-1454.

Heslop Harrison, J. (1971). *Pollen: Development and Physiology.* London: Butterworths.

Chemical Structure of the Exine of Pollen Walls and a New Function for Carotenoids in Nature

J. Brooks
G. Shaw

School of Chemistry,
University of Bradford.

RECENT communications[1,2] have confirmed and extended the observation of Zetzsche et al.[3] that there is a close chemical similarity between the nitrogen-free walls of a wide variety of pollen and spores. In particular the walls of *Lycopodium clavatum* and *Pinus silvestris*, which were studied in most detail, were found to consist of (*a*) an almost pure cellulose intine (10–15 per cent by weight of the wall) which retained the original shape of the wall after removal of the exine (sporopollenin of Zetzsche) by oxidation; (*b*) an ill-defined "xylan" fraction (10 per cent); (*c*) a fraction regarded as lignin-like (10–15 per cent) because the walls produced phenolic acids when fused with potassium hydroxide; (*d*) a major fraction (55–65 per cent, the exine) which is very resistant to most chemical reagents but readily oxidized to a mixture of mono and dicarboxylic acids containing eighteen carbon atoms or less.

The extreme resistance of the pollen exine to chemical degradation other than oxidation was a severely limiting factor in attempts to extend the chemical studies. We decided therefore that additional useful information about the chemical structure of the exine might be best obtained by following the course of formation of pollen exine in a particular plant and correlating this development with any parallel development of chemical substances in the anthers. It has been known for some time that exine material is formed in the tapetal cells, probably during the early stages of meiosis[4], and correlation of development of the pollen wall with anther or bud size is roughly possible in certain *Lilium* species[5]. Anthers from buds of 100 *Lilium henryii* plants were removed at intervals, their lengths measured, portions examined microscopically and the remainder extracted with solvents and the extracts examined for chemical constituents. Some of the plants were allowed to develop to maturity, the pollen was collected and the walls were isolated by the general methods described earlier[1-3]; chemically they were very similar to those from other plants.

It soon became apparent that the formation of exine material in *L. henryii* was accompanied by a parallel formation of carotenoids. No carotenoids were detected in anthers less than 1·1 cm long and this corresponded to the presence of sporogenous tissue only (anthers < 0·9 cm long) or to only tetrads without exine (anthers < 1·1 cm long); with anther lengths greater than 1·1 cm exine was being increasingly deposited and corresponded to increas-

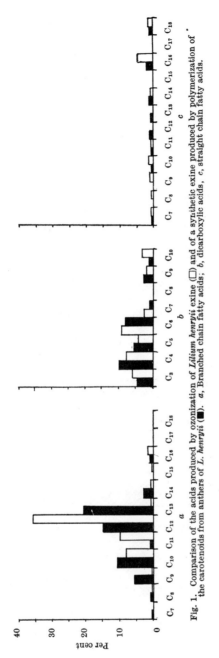

Fig. 1. Comparison of the acids produced by ozonization of *Lilium henryii* exine (□) and of a synthetic exine produced by polymerization of the carotenoids from anthers of *L. henryii* (■). *a*, Branched chain fatty acids; *b*, dicarboxylic acids; *c*, straight chain fatty acids.

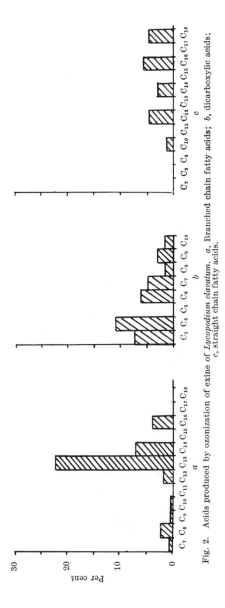

Fig. 2. Acids produced by ozonization of exine of *Lycopodium clavatum*. *a*, Branched chain fatty acids; *b*, dicarboxylic acids; *c*, straight chain fatty acids.

Table 1

Material	Molecular formula†	Per cent acids produced by ozonization		
		Branched chain mono-acids	Straight chain mono-acids	Dicarboxylic acids
L. henryii Pollen wall*	$C_{90}H_{142}O_{36}$	59·3	8·4	32·9
L. henryii Carotenoid/carotenoid esters polymer	$C_{90}H_{148}O_{38}$	57·3	6·5	36·4
L. henryii carotenoid polymer	$C_{90}H_{110}O_{33}$	64·1	3·4	33·4
Lycopodium clavatum Spore wall*	$C_{90}H_{144}O_{27}$	38·8	21·4	37·7
β-carotene polymer	$C_{90}H_{130}O_{30}$	61·3	1·9	35·0
Vitamin A-palmitate polymer	$C_{90}H_{150}O_{13}$	28·3	47·6	25·8

See Figures for more detail.

* The figures are for walls from which cellulose has been removed.

† Molecular formulae are arbitrarily recorded on a C_{90} basis to facilitate comparison with earlier references 1–3

ing formation of carotenoids. Extraction of the carotenoids from the anthers revealed that they consisted of a mixture of free carotenoid and carotenoid esters (ratio 2·2 : 1). The mixture was saponified and the fatty acids were examined (as methyl esters) by gas–liquid chromatography. They contained 90 per cent straight chain acids, C_{16} (80 per cent), C_{18} (6·6 per cent), C_7–C_{13} (3·4 per cent), and 10 per cent branched chain acids, C_{16} (4·6 per cent), C_{14} (2·7 per cent), C_{11} (1·2 per cent) and C_7–C_{14} (1·5 per cent). Further work to establish the structure of the carotenoids is under way.

The presence of carotenoids in anthers is known[6] but their possible function as precursors of pollen exine has so far been unsuspected. We have examined the polymerization of both the free carotenoids and total carotenoids and carotenoid esters, from the anther extracts of *L. henryii* with trace amounts of a boron trifluoride catalyst in methylene dichloride solution in the presence of oxygen. As model compounds we have similarly examined the polymerization of β-carotene and vitamin A palmitate. In all cases insoluble polymers were formed which contained substantial amounts of oxygen and which had molecular formulae very similar to those of the pollen exine (Table 1). Each of the synthetic polymers was degraded with ozone[1,2] under the same conditions used for the degradation of pollen exine and the resultant mixture of branched, non-branched fatty acids and dicarboxylic acids was examined (as methyl esters) by gas–liquid chromatography. The results (see Figures) show the remarkable similarity, both qualitative and quantitative, between the degradation products derived from the synthetic polymers and those from the pollen exine of *L. henryii*. Analogous results are included for the sporopollenin of *L. clavatum*, and a very similar spectrum of compounds is obtained although the greater proportion of straight chain acids would indicate a higher ratio of carotenoid esters to carotenoids in the anthers. Potash fusion of either *L. henryii* pollen exine, or the synthetic polymer derived from the total-carotenoids of the anther extracts gave in each case *p*-hydroxy-benzoic acid as the principal phenolic acid.

Our results suggest that the pollen exine is formed by an oxidative polymerization of the mixture of carotenoids and carotenoid esters contained in the anther material. The total wall is now seen to consist of a cellulose intine covered by a polymerized carotenoid exine with probably a small amount of some other material acting as a cement between the two. Because phenolic acids are clearly obtainable from polymerized carotenoids there is now perhaps no need to postulate lignin-like material in the sporopollenin.

We thank Professor J. Heslop-Harrison for his advice on material and use of anther and bud measurements. We also thank Professor E. Lees and Mr K. Evans for growing the *L. henryii*.

Received April 26; revised May 20, 1968.

[1] Shaw, G., and Yeadon, A., *Grana Palynologica*, **5** (2), 247 (1964).
[2] Shaw, G., and Yeadon, A., *J. Chem. Soc.* (C), 16 (1966).
[3] Zetzsche, F., Kalt, P., Liechti, J., and Ziegler, E., *J. Prakt. Chem.*, **148**, 267 (1937).
[4] Heslop-Harrison, J., *Nature*, **195**, 1069 (1962).
[5] Heslop-Harrison, J., and Mackenzie, A., *J. Cell Sci.*, **2**, 387 (1967).
[6] Karrer, P., and Oswald, A., *Helv. Chim. Acta*, **18**, 1303 (1935).

Copyright © 1967 by Macmillan Journals Ltd.

Reprinted from *Nature* **213**(5077):703-704 (1967)

Deposition of Sporopollenin on Lamellae of Unit Membrane Dimensions

J. R. ROWLEY

Botany Department,
University of Massachusetts,
Amherst, Massachusetts.

D. SOUTHWORTH

Botany Department,
University of California,
Berkeley.

THERE are several reports of a lamellar inner part of the exine of pollen grains[1-8]. The number of species examined is small, but the sampling of plant families is wide enough to conclude that the deposition of sporopollenin in some kind of lamellar form is probably characteristic for at least the inner unornamented part of the exine of pollen grains.

In microspores of *Anthurium* sp., fixed in formalin[9] and stained with uranyl acetate and lead citrate[10], we have direct evidence for the formation of some of the exine on lamellae of unit membrane dimensions. Sporopollenin accumulates on such membranes, often asymmetrically, to form the buttressing lamellae of the germinal apertures (Fig. 1). These buttressing lamellae grow to 0·3–0·4μ in cross-sectional height, but they retain the line of low density characteristic of the unit membrane until near pollen maturity. In mature pollen only a staining discontinuity at the relative position of the uppermost line of low density around the apertures remains as an indication of the mode of origin (Fig. 2).

Fig. 1. Section of one side of a germinal aperture region in a microspore of *Anthurium*. The accumulation of sporopollenin on some of the unit membrane-like lamellae is asymmetric (arrow). Lines similar to the characteristic region of low density in the above lamellae are traceable (double-headed arrow) in all of the buttressing parts of the exine around the germinal aperture (P) including the one that is outermost. (× c. 19,000.)

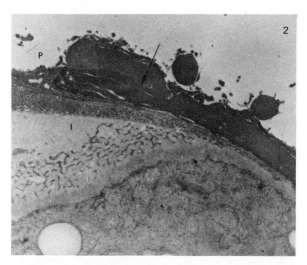

Fig. 2. Medial section of one side of a germinal aperture in a mature pollen grain of *Anthurium*. A staining discontinuity (arrow) can be traced in the outer part of the exine beside the pore (*P*). Strands of cytoplasm in the intine (*I*) are bounded by plasma membrane. (× *c*. 15,000.)

In the spores of the liverwort *Scapania nemorosa* most of the spore wall has a laminar or unit membrane origin (Fig. 3). The wall of immature spores of *Scapania* is composed of a layer of low density overlain by acetolysis-resistant material (sporopollenin). The resistant components consist of about five lamellae having some characteristics of unit membranes surmounted by stacks of disks bounded by a membrane (Fig. 3). The piles of disks are irregularly shaped and spaced and appear similar to the ornamentations on spores of mosses reported by von Wettstein[11] and McClymont and Larson[12]. Working with *Geothallus*, Doyle[13] determined that two of the four wall layers of the spores gave reactions which indicated the presence of waxy compounds on a carbohydrate framework. The inner layer of the above two was composed of a series of lamellae and both layers gave histochemical reactions indicative of sporopollenin.

Heslop-Harrison[6,7] and Larson and Lewis[4] have presented micrographs showing endoplasmic reticulum oriented parallel to the plasma membrane in regions of presumptive germinal apertures. These writers have strongly suggested that the position of elements of the endoplasmic reticulum influences the deposition of the exine. We do not know the source of the membranes, but lines of low density are about 40 Å in cross-sectional height and highly regular as would be expected of a paracrystalline molecular system such as the unit membrane. The minimal dimensions for the two dense outer areas measured

Lamellae of Unit Membrane Dimensions

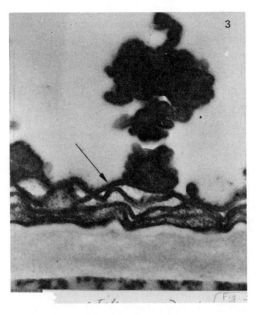

Fig. 3. Section from an immature *Scapania nemorosa* spore wall. About five unit membrane-like lamellae (arrow) form the exine. The piles of dense material that constitute the ornamentation over the lamellae may also form on membranes but the amount of sporopollenin already present obscures that information. At spore maturity the membraneous origin of all parts of the wall is obscured. (× *c.* 50,000.)

60 or 100–120 Å. The 60 Å dense bands may be a single strand of sporopollenin similar to Rowley's[14] description of stranded sporopollenin in the exine of *Poa annua* microspores or to the description of Afzelius[1] of aligned granules. The dense parts that were measured at 100–120 Å in thickness may consist of two layers of sporopollenin strands.

Afzelius[15–17] found lines of unit membrane dimensions in mature exines of *Acmopyle* and *Sequoia* pollen and of *Lycopodium* spores. In *Nuphar*, Rowley[18] observed low density lines of 35–40 Å width aligned both parallel and perpendicular to the ornamented surface of the exine. In *Nuphar* the low density lines extend through the entire length of 0·5μ spinules.

The presence of unit membranes around isolated agglomerations of sporopollenin (Ubisch bodies), which are found at considerable distances from microspores, is further evidence of a universal mode of sporopollenin deposition on unit membranes. While working with Professor G. Erdtman on pollen of *Populis tremula*, we observed membrane-bound Ubisch bodies located between tapetal cells.

This work was supported by a grant from the U.S. National Science Foundation. Part of the work was done while one of us (J. R. R.) was a John Simon Guggenheim fellow.

[1] Afzelius, B., *Bot. Not.*, **108**, 141 (1955).
[2] Ehrlich, H. G., *Exp. Cell Res.*, **115**, 463 (1958).
[3] Larson, D. A., and Lewis, C. W., *Amer. J. Bot.*, **48**, 934 (1961).
[4] Larson, D. A., and Lewis, C. W., *Grana Palyn.*, **3**, 21 (1962).
[5] Larson, D. A., Skvarla, J. J., and Lewis, C. W., *Pollen et Spores*, **4**, 233 (1962).
[6] Heslop-Harrison, J., *Grana Palyn.*, **4**, 7 (1963).
[7] Heslop-Harrison, J., in *Pollen Physiology and Fertilization* (edit. by Linskens, H. F.), 39 (North-Holland, Amsterdam, 1964).
[8] Larson, D. A., *Grana Palyn.*, **5**, 265 (1964).
[9] Pease, D. C., *Histological Techniques for Electron Microscopy*, 52 (Academic Press, New York, 1964).
[10] Reynolds, E. S., *J. Cell Biol.*, **17**, 208 (1963).
[11] Wettstein, D. von, *Exp. Cell Res.*, **12**, 427 (1957).
[12] McClymont, J. W., and Larson, D. A., *Amer. J. Bot.*, **51**, 195 (1964).
[13] Doyle, W. T., *Univ. California Pubs. Bot.*, **33**, 185 (1962).
[14] Rowley, J. R., *Science*, **137**, 526 (1962).
[15] Afzelius, B., Erdtman, G., and Sjöstrand, F. S., *Svensk Bot. Tidskr.*, **48**, 155 (1954).
[16] Afzelius, B., *Grana Palyn.*, **1**, 20 (1956).
[17] Afzelius, B., *Grana Palyn.*, **6**, 435 (1966).
[18] Rowley, J. R. *Rev.*, *Palaeobot. Palynol.* (in the press).

Copyright © 1974 by the Almquist & Wiksell Periodical Co.

Reprinted from *Grana* 14:143-151 (1974)

RE-EXPOSURE OF TAPES AT HIGH TEMPERATURE AND PRESSURE IN THE *LYCOPODIUM CLAVATUM* SPORE EXINE

Sukla Sengupta and John R. Rowley

(Received November 11, 1974)

ABSTRACT

Sengupta, S. (Dept. of Geology, Imperial College of Science and Technology, London, Great Britain) and Rowley, J. R. (Palynological Laboratory, Wallenberglaboratoriet, Stockholm, Sweden). *Re-exposure of tapes at high temperature and pressure in the Lycopodium clavatum spore exine.* Grana 14: 143-151, 1974.—Lamellations are visualizable through the staining commonly used in transmission electron microscopy during exine formation on *Lycopodium* and other spores, and the nexine of pollen grains. The lamellations so exposed consist of dark tapes at either side of an unstained (white) line. Neither tapes nor white lines are visualizable in the exine of mature spores of *Lycopodium*. The continued presence of lamellations having tape-white line spacing has been demonstrated with inorganic tracers in the nexine of pollen in which lamellations otherwise appeared to be absent. Through high contrast staining methods for TEM we have observed lamellations in the residual exine following heat treatment (350°C) of mature spores of *Lycopodium clavatum*. The surface of these residual exines was etched by treatment with hot 2-aminoethanol and filaments were observed to protrude from the etched surfaces. The residual exine stained darkly. *Lycopodium* spores heated to 300°C at 1 kb pressure had long filaments exposed at the surface of the residual exine (sporopollenin). Sections of the pellet remaining after heat and pressure treatment also included bundles of closely parallel filaments and masses of isolated filaments. The filaments were stained while the exine residue, assumed to include sporopollenin, was not. Isolated filaments produce a stable metachromasia with toluidine blue indicating the presence of many closely spaced sites of negative charge. The staining of intact exines with basic dyes may result from anionic sites on filaments embedded within the exine rather than being due to sporopollenin. The results of our experiments indicate that the filaments are more resistant to heat and pressure and 2-aminoethanol than is sporopollenin. We propose that the trilamellar elements commonly called tapes and white lines might be composed of two filaments bridged by polybasic molecules.

INTRODUCTION

The first descriptions of lamellae in exines resulted from work by Afzelius (Gullvåg) et al. (1954) on spores of *Lycopodium*. Since then lamellae in the exines of spores and in the inner zone of the exine (nexine) of pollen grains have been reported by many observers, e.g., Angold (1957), Argue (1972), Christensen & Horner (1974), Dickinson & Heslop-Harrison (1968), Dunbar (1973), Godwin et al. (1967), Larson & Lewis (1962), Le Thomas & Lugardon (1972), Lugardon (1971), Nabli (1971), Oltmann (1974), Rowley & Southworth (1967), Southworth (1966), and many others. The presence of lamellae in *Lycopodium clavatum* spores is considered by Gullvåg (1966) and Pettitt (1966, 1971) with regard to their ultrastructural and morphological studies.

Early in the ontogeny of pollen grains and in limited sites of mature pollen the lamellations of the nexine are observed as triple-layered structures similar in appearance to unit membranes. Larson & Lewis (1962, figs. 8 and 10) seem to have first called attention to these trilamellar structures in pollen grains. As exine accumulation progresses the two outer layers commonly become indistinguishable from the sporopollenin which accumulates on these tripartite lamellations so that only the middle non-staining "white" lamellation is observed on micrographs. Under these circumstances the unstainable lamellations are commonly referred to as *white lines*. The morphology of the ontogenetic change from discrete triple-layered filaments to white lines embedded in homogeneous exine subdivisions is similar in pollen (Godwin et al., 1967) and spores, e.g., *Lycopodium* and *Selaginella* (Pettitt, 1971).

Still later in exine maturation white lines completely disappear from exine regions in which they have been prevalent. This alteration is convincingly illustrated in the comparable exine sections shown by Godwin et al. (1967, Pl. II, fig. D and Pl. III, fig. C) for *Ipomoea* and Rowley (1967, figs. 5 and 6) for *Anthurium*. Visualizability of white lines is also lost in spores of *Lycopodium* and of ferns toward exine maturation (Lugardon, 1971, *Ophioglossum, Osmunda, Blechnum*). The white lines present in great abundance on spores of liverworts are, on the other hand, often observable on mature spores. Some of our most extensive descriptions and sequential observations of white lines during spore ontogeny have been made on liverworts (Horner et al., 1966, *Riccardia*; Heckman, 1970, Jungermanniales; Denizot, 1971 a–c, Sphaerocarpales and Marchantiales). Heckman (1970) referred to the white line centered trilamellar structures as *slips* to indicate their three-dimensional form and Denizot (1971 a, fig. 17) determined that the tripartite lamellations in spores of *Conocephalum* which were observable as slips or sheets of great lateral dimensions, were locally modified to tubules having a white central core.

At the International Palynological Conference at Utrecht, where some of the first data on white lines during microspore ontogeny was reported (e.g. Godwin et al., 1967; Angold, 1967) and discussed (Southworth, 1966) these tripartite structures were referred to as tapes. We will use the word tape although appreciating that these lamellations may be slips or sheets which are subject to modification, e.g., the tubules of Denizot (1971 a). The origin(s), composition(s) and role(s) of tapes in spore and pollen development is one of the grand problems in palynology and cytology. Without implying any homology it may be noted that white lines and tapes, similar in appearance to those on spores and pollen grains, are observed in wall and matrical regions of many cells. They are seen, for example, in the root of *Lemna minor* (Fagerlind & Massalski, 1974), seeds of peas and beans (Mollenhauer & Totten, 1971), the regenerating wall of isolated tomato fruit protoplasts (Pojnar et al., 1967) and the cell walls of bacteria (e.g., de Petris, 1967; Wang et al., 1970).

The experiments we report concern conditions which have led to the re-exposure of tapes in mature spores of *Lycopodium*.

MATERIAL AND METHODS

Fresh *Lycopodium clavatum* spores were heated at 350°C under atmospheric pressure in a nichrome wire resistance furnace for 100 hours and for the same period in a modified triaxial deformation apparatus at 300°C under 1 kb pressure in a mixture of spores, silica sand and sea water in the proportion of 1:4:5 (Sengupta, 1974).

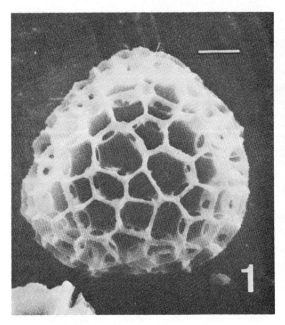
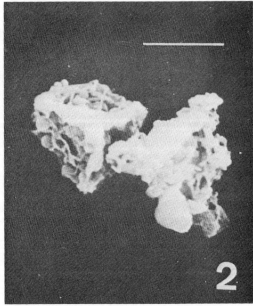

Fig. 1. Distal surface of a fresh *Lycopodium clavatum* spore. SEM. Magnification marker line ca. 5 µm.

Fig. 2. Spores heated to 350°C at atmospheric pressure are greatly reduced in size. Spore at left shows trilete mark of proximal surface and right spore the residual reticulum on the distal surface. SEM. Marker line ca. 5 µm.

The spores that had been heated to 350°C were divided into two portions. The first portion was treated with 2-aminoethanol for an hour with a gradual increase in temperature from 90° to 140°C. The spores were then washed in water and stained twice with osmium tetroxide using a divalent hydrizide for application of osmium to the original binding sites between the first and second exposure to osmium. The osmium tetroxide (0.1%) was dissolved in water that was deionized, then glass distilled (final pH 6) and held on the treated spores overnight. The spores were washed with water, placed in 0.2% thiocarbohydrazide (TCH) dissolved in 20% acetic acid (Thiéry, 1967) for 30 min, washed in water, and then placed in 0.1% osmium tetroxide for 2 hours. Dehydration was in an acetone series followed by propylene oxide as an intermediate solvent for Epon-Araldite embedding in Mollenhauer's (1964) mixture No. 1.

The second portion of the spores heated to 350°C and spores heated to 300°C under 1 kb pressure were stained, dehydrated, and embedded as above without 2-aminoethanol treatment. Sections were cut using a diamond knife and picked up on gold grids. The sections were stained by immersing the grids (to stain both cut surfaces) in TCH, 10% acetic acid, deionized-glass distilled water, and 1% silver proteinate (Thiéry, 1967). Sections were examined and micrographed using a Zeiss EM-9S transmission electron microscope with a 50 μm objective aperture.

Scanning electron micrographs were made using both the Mark IIA and S600 Cambridge "Stereoscan" microscopes (Sengupta, 1974). Toluidine blue stain for light microscopy consisted of 0.05% toluidine blue in 0.2 M citrate buffer at pH 4.4 (Feder & O'Brien, 1968) and 0.5% toluidine blue in water to which was added 0.5% borax (final pH 5).

RESULTS

Lycopodium spores heated to 350°C for 100 hours at atmospheric pressure are reduced in overall size (Fig. 2) to about one fourth the size of the fresh spores (Fig. 1). At this high temperature most of the exine ridges are lost or reduced although the non-ornamented part of the exine (nexine) and the triradiate mark survive without obvious change.

TEM observations of the thin sections of *Lycopodium* spores exposed to heat, heat followed by 2-aminoethanol, and heat plus pressure are described under the following three subheadings. All material illustrated was stained with OsO_4–TCH–OsO_4 prior to embedding and after embedding sections were stained with TCH–SP.

1. *Spores heated to 350°C at atmospheric pressure and not treated with hot 2-aminoethanol*

The section in Fig. 3 reveals dark lamellations or lines which have a minimal thickness of 15 nm, and grade up in width to 100 nm or more, like twisted ribbons (Fig. 3, arrowhead). Regions of minimal thickness are darker with sharp boundaries between stained profile and background; when wider than 40–60 nm they are less dark and difficult to distinguish from the background. This phenomenon accounts for the difficulty in tracing lamellations in Fig. 3 which are apparently continuous over relatively great distances. Assuming that the lamellations are stained throughout the thickness of the section, then the widest records could represent an oblique profile across the entire 80–100 nm estimated thickness of the section. On one side the lamellations appear to be continuous with the medium dark material forming the bulk of the sectioned nexine, which we consider to be residual sporopollenin, and on the other side by the less dark material (Fig. 3, short arrow) limited in extent. Faint striations are present throughout the section and appear in a few sites to have the trilamellar arrangement of two dark lines with an unstained space in between (Fig. 3, long arrows) characteristic of tapes and white lines.

Numerous holes are present in the section. This spongy effect may be the result of volatilization of the exine during heat treatment; holes are frequent in the less dark material adjacent to dark lamellations. Holes are common in the bases of exine ridges in fresh spores.

2. *Spores heated to 350°C at atmospheric pressure and then treated with hot 2-aminoethanol*

Surviving ridges and nexine surfaces are deeply etched. In sections the etched areas are saw-toothed or step-like in outline (Fig. 4). Filaments having similar minimal and maximal dimensions as the dark lamellations in Fig. 3 (subheading No. 1) are observed to protrude from these etched areas (Fig. 4, short arrows). Except for many generally parallel and unstained lines (Fig. 4, long arrows) the residual exine

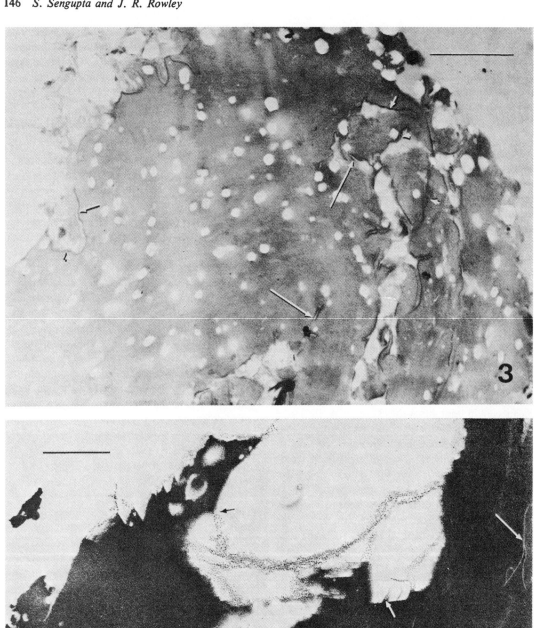

is darkly stained (Fig. 4). Enough of the spore wall remained so that it could be recognized as *Lycopodium*.

There are circular holes in the residual exine (Fig. 4) although they are not so numerous or random in distribution as they were prior to treatment with hot 2-aminoethanol (Fig. 3).

3. *Spores heated to 300°C at elevated (1 kb) pressure*

The following three kinds of formed objects were observed: (*a*) Nonstaining spongy material with moderately stained filaments perpendicular to parallel to its surface as in Fig. 5, (*b*) dark stained bundles of filaments as in Fig. 5 and (*c*) isolated loose masses of moderately stained filaments in association with little or no nonstaining spongy material (Fig. 6).

In some few cases the spongy material had morphological features of the spores of *Lycopodium* (see section on light microscopy) and we assume that it is a residue of the exine, probably sporopollenin. There are few silver granules, as a result of the stain sequence OsO_4–TCH–OsO_4–TCH–SP, on the exine residue. Many of the silver granules present are aligned in rows which near the surface of the spongy residue are continuous with exposed filaments (Fig. 5, long arrows). The surface of exine or sporopollenin remnants is laced by filaments. Filaments occur singly, as loose pairs (Fig. 5, short arrow) and commonly as loops although only one loop is included in Fig. 5.

The dark staining bundles contain filaments which are closely paired and black material having a crystalline outline in transverse section (Fig. 5*c*). The latter material volatilizes in the electron beam. The ends of closely paired filaments can be seen at the edges of dark staining bundles (Fig. 5).

Filaments occurring more or less completely isolated from residual sporopollenin (Fig. 6) have lateral dimensions similar to those cited under subheading No. 1. The greatest widths observed in isolated filaments (e.g., Fig. 6, long arrows) are ca. 150 nm and minimal widths are ca. 15 nm. The shortest filaments, which are considered to be filaments cut transversely, are ca. 120–150 nm (Fig. 6, short arrows). Under our conditions for extraction, filaments ca. 15 nm in thickness by ca. 150 nm in width and several micrometers in length are left after elimination of the volatile parts of exine of the *Lycopodium* spores.

Light microscope observations with the metachromic dye toluidine blue

Spores heated to 350°C at atmospheric pressure and the remains of spores heated to 300°C at elevated pressure stain a bright red in toluidine blue, indicating the presence of numerous closely spaced free anionic sites. After treatment at elevated temperature and pressure the residue of most spores is unrecognizable as *Lycopodium* in sections for TEM (see subheading No. 3). Mounted for light microscopy the triradiate scar is prominently seen on the largest remnants, leaving no doubt that they are the remains of spores. Masses of loops, considered to be isolated filaments similar to those in Fig. 6, also stain bright red in toluidine blue. Fresh mature spores of *Lycopodium* stain a blue color in toluidine blue, indicative of large molecules with regularly arranged and relatively widely spaced anionic sites.

DISCUSSION

We find that filaments about 15 nm in thickness by 150 nm in width and several μm in length can be isolated from exines of *Lycopodium* spores. These filaments resist treatment which removed all or most other components of the *Lycopodium* spore exine, including sporopollenin. Following treatment which eliminated most sporopollenin, exposed filaments produce a stable metachromasia with the basic dye toluidine blue indicating the presence of a large number of closely spaced sites of negative charge.

According to the concept of Michaelis (Pearse,

Figs. 3–6. Prior to embedding *Lycopodium* spore remnants were stained with OsO_4, followed by the hydrazide thiocarbohydrazide (TCH), and then reexposed to OsO_4. Sections were treated with TCH, then silver proteinate, and examined by TEM.

Fig. 3. Dark lines are seen in the residual exine of spores heated to 350°C at atmospheric pressure. In some regions these dark lines are variable in width, like twisted filaments (arrow heads). They are seen in greatest contrast where they (short arrows) border nonstaining areas. In a few instances the nonstaining areas appear as white lines between dark filaments (long arrow). Faint striations run throughout the section. Marker line ca. 1 μm.

Fig. 4. The surface of the spore exine is etched by treatment in hot 2-aminoethanol after being heated to 350°C at atmospheric pressure. Stained filaments (short arrows) are seen to protrude from step- and tooth-shaped etched regions. Long arrows indicate lightly stained lines. The residue exine (sporopollenin) is darkly stained following 2-aminoethanol treatment. Marker line ca. 1 μm.

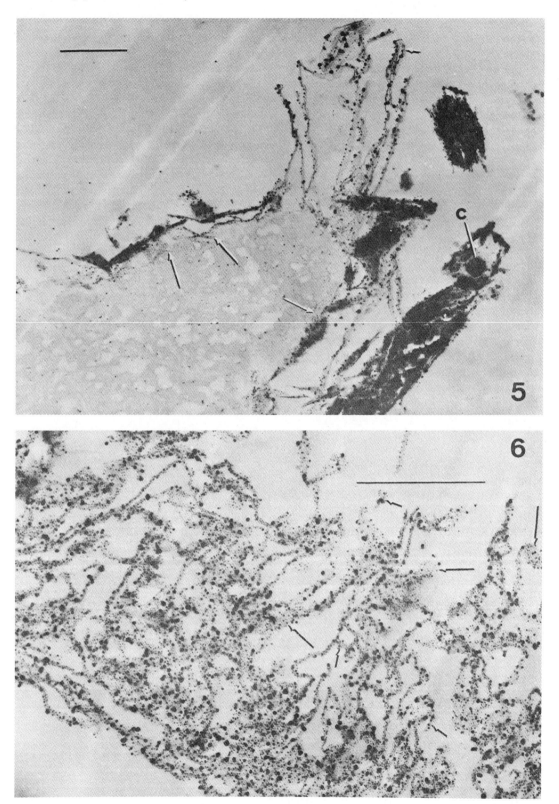

22

1961) polymerization of the substrate, with which the dye combines, induces polymerization of the dye, and hence metachromasia. The monomeric form of toluidine blue is blue and the polymeric form is red. In examples given by Pearse (1961, p. 150) of the minimum surface density of negative charges on the substrate necessary for metachromasia to take place he notes that in hyaluronic acid where the distance between negative charges is about 10 Å, no metachromasia occurs—toluidine blue is blue or purple. In pectic acid the intercharge distance is about 5 Å and this is enough to produce a weak metachromasia. Polysaccharides with intercharge spacing reduced to less than 4 Å give a strong and stable metachromasia; toluidine blue is red or pink.

In the following discussion we will present an argument for the tapes and white lines observed in young spores of *Lycopodium*, as well as in other spores and in pollen grain walls, being composed of two filaments. Close contact between adjacent filaments would be prevented by negative charges. A white line would result when adjacent filaments were held in close proximity by molecules with positive charge sites at either end.

We suggest that the white line phenomenon is the result of polyelectrolyte interaction with the surfaces of tapes. Experiments with polyelectrolytes on biological surfaces have largely involved the cell surface; however, the structure of many cell membranes and intracellular reticulations can be attributed to forces operating between oppositely charged polyions. These and subsequent data on polyelectrolytes and their biological interactions derive from Katchalsky's (1964) summary of concepts in this area of colloid chemistry and major results obtained by various workers of special relevance to biologists.

Katchalsky (1964) and a collaborator (Dr M. Sela) demonstrated many years ago that minute quantities of basic polyelectrolytes cause strong agglutination of bacterial, plant and animal cells. The agglutination is preceded by an adsorption of the polyions to the cell surface. The amounts of polybase adsorbed to negatively charged cell surfaces are several times larger than those required to cover the surface with a flat monolayer, indicating that the polyions are at least partially perpendicular to the cell surface. The fact that agglutination takes place between negatively charged cells indicates that no close contacts between the cells exist; instead macromolecular bridges are formed between one cell surface and the other.

The white line is likely to be such a space with macromolecular bridges between the negatively charged surfaces of two tapes. At high magnifications bridges are observed across white lines (Erdtman, 1969, Pl. 107). The apparent spacing of bridges across white lines is about 10 nm; however, since polybase molecules are adsorbed with large numbers of segments in direct contact with the surface, the spacing of bridges would not indicate the frequency of negative charges on tapes. The strong metachromasia with toluidine blue at *Lycopodium* "spore" surfaces known to have many free or partially free "tapes" (Fig. 5) indicates an intercharge spacing of <5 Å.

The colloid chemist finds that an adsorbed polybase layer is often impossible to remove even by washing with pure solvent. This is consistent with observations of white lines in pollen and spore walls following a great variety of preparational methods. During pollen and spore ontogenesis white lines do disappear or become limited to germinal apertural regions. In *Epilobium* white lines are seen at local sites of translocation whereas before and after this local transfer activity white lines are not visualizable (Rowley, 1971, 1973 *a, b*). The results of experiments with lanthanum nitrate as a tracer indicate that tapes continue to be present after neither tapes nor white lines are demonstrable through ordinary staining methods (Rowley & Flynn, 1971).

Many plant products such as the pectic acids are typical polyelectrolytes and the polyelectrolyte nature of polyacid and polybase polypeptides has been emphasized by Katchalsky (1964).

The structure and the arrangement of the lamellae in the exine are described in many reports but the chemical and physical nature of these lamellations has not been reported in any detail. Dickinson &

Fig. 5. Filaments are exposed at the surface of residual exine (sporopollenin) following heating to 300°C at 1 kb. pressure. The filaments occur as loops, single or loosely double elements (short arrow), and, at the right, in darkly stained bundles. These bundles of filaments include black material (c) which is sublimed by the electron beam. The residual sporopollenin is spongy in appearance and unstained except for small granules which are in rows (long arrows) near the surface of the residual sporopollenin and continuous with filaments. Marker line ca. 1 μm.

Fig. 6. The residue of spore walls after heating to 300°C at elevated pressure contains masses of isolated filaments. Their greatest width is ca. 150 nm (long arrows) and their minimal thickness is ca. 15 nm. Short segments (short arrows) may be cross sections of filaments. Marker line ca. 1 μm.

Heslop–Harrison (1968) reported as a result of electron microscopic staining tests that they may be lipoprotein in nature. Filaments isolated from *Epilobium* and *Artemesia* pollen exines through treatment with hot 2-aminoethanol resist 90% phenol in water (after the method of Weidel et al., 1960) and protolytic enzymes suggesting a lipopolysaccharide component for these filaments (Rowley, 1973a; Rowley, Dahl & Sengupta, unpublished). These filaments react in the same way with toluidine blue as exposed filaments of *Lycopodium* spores indicating the presence of free anionic sites. Southworth (1973, 1974) has considered the reactions of the basic cytochemical stains toluidine blue and azure blue B with regard to the proposal of Brooks & Shaw (1968) that sporopollenin is an oxidative polymer of carotenoids and carotenoid esters. She points out that while the carotenoid ester hypothesis provides a source of unsaturated bonds to react with osmium tetroxide and to absorb ultraviolet light it fails to explain reactions with basic dyes which indicate regularly arranged anionic sites.

In our spores treated at 300°C at 1 kb pressure the residual sporopollenin is unstained while the free filaments continue to be reactive. They are stained by osmium tetroxide and/or bind 2 divalent hydrazide (TCH) or silver proteinate. The exposed filaments are stained by basic dyes. We conclude that it is the filaments (tapes), with regularly arranged anionic sites, which provide the basis for stainability of the *Lycopodium* spore exine with basic dyes rather than sporopollenin.

It is generally accepted that the sporopollenin is the most resistant material in the spore and pollen walls but from our experiment it is found that the sporopollenin partly disintegrates in 2-aminoethanol treatment and also breaks down or is volatilized under elevated pressure and temperature whereas the "tapes" withstand such treatments. Sengupta (1973, 1974) found that gas is released from spores heated to 300°C under pressure. We find tapes to be exposed from exines of *Lycopodium* spores apparently due to the loss of sporopollenin through a combination of physical distortion and volatilization at elevated temperature and pressure. The results of our experiments show the tapes to be highly resistant, perhaps more so than sporopollenin.

ACKNOWLEDGEMENTS

The first writer wishes to thank British Petroleum Company Ltd. for financial assistance, especially for the travel and other expenses while she worked at the Palynological Laboratory in Stockholm and for permission to publish this work. We are grateful to the Swedish Natural Science Research Council who supported the final stage of this work. Our thanks go to Dr Marjorie Muir of Imperial College, London, for going through the manuscript, and to Elisabeth Grafström and Barbro Dahlberg of the Palynological Laboratory for their expert assistance.

REFERENCES

Afzelius, B. M., Erdtman, G. & Sjöstrand, F. S. 1954. On the fine structure of the outer part of the spore wall of Lycopodium clavatum as revealed by the electron microscope. — Svensk Bot. Tidskr. 48: 151–161.

Angold, R. E. 1967. The ontogeny and fine structure of the pollen grain of Endymion non-scriptus. — Rev. Palaeobot. Palynol. 3: 205–212.

Argue, C. L. 1972. Pollen of the Alismataceae and Butomaceae. Development of the nexine in Sagittaria lancifolia L. — Pollen Spores 14: 5–16.

Brooks, J. & Shaw, G. 1968. Chemical structure of the exine of pollen walls and a new function for carotenoids in nature. — Nature 219: 523–524.

Christensen, J. E. & Horner, H. T., Jr. 1974. Pollen pore development and its spatial orientation during microsporogenesis in the grass Sorghum bicolor. — Am. J. Bot. 61: 604–623.

Denizot, J. 1971a. Sur la presence d'ensembles exiniques élémentaires dans le sporoderme de quelques Marchantiales et Sphaerocarpales. — C. R. Acad. Sc. Paris 272: 2166–2169.

— 1971b. Recherches sur l'origine des ensembles exiniques élémentaires du sporoderme de quelques Marchantiales et Sphaerocarpales. — C. R. Acad. Sc. Paris 272: 2305–2308.

— 1971c. Recherches sur les formations callosiques au cours de la sporogenèse de quelques Marchantiales et Sphaerocarpales. — C. R. Acad. Sc. Paris 272: 2679–2722.

Dickinson, H. G. & Heslop-Harrison, J. 1968. Common mode of deposition for the sporopollenin of sexine and nexine. — Nature 220: 926–927.

Dunbar, A. E. 1973. Pollen ontogeny in some species of Campanulaceae. A study by electron microscopy. — Bot. Notiser 126: 277–315.

Erdtman, G. 1969. Handbook of Palynology. — Munksgaard, Copenhagen.

Fagerlind, F. & Massalski, A. 1974. The development of cell walls and intercellulars in the root of Lemna minor L. — Svensk Bot. Tidskr. 68: 64–93.

Feder, N. & O'Brien, T. P. 1968. Plant microtechniques. Some principles and new methods. — Am. J. Bot. 55: 123–142.

Godwin, H., Echlin, P. & Chapman, B. 1967. The development of the pollen wall in Ipomoea purpurea (L.) Roth. — Rev. Palaeobot. Palynol. 3: 181–196.

Gullvåg, B. M. 1966. The fine structure of some gymnosperm pollen walls. — Grana Palynol. 6: 435–475.

Heckman, C. A. 1970. Spore wall structure in the Jungermanniales. — Grana 10: 109–119.

Horner, H. T., Jr, Lersten, N. R. & Bowen, C. C. 1966 Spore development in the liverwort Riccardia pinguis. — Am. J. Bot. 53: 1048–1064.

Katchalsky, A. 1964. Polyelectrolytes and their biological interactions. — Biophys. J. *4* (Suppl.): 9–42.

Larson, D. & Lewis, C. W. 1962, Pollen wall development in Parkinsonia aculeata. — Grana Palynol. *3:* 21–27.

Le Thomas & Lugardon, B. 1972. Sur la structure fine des tétrades de deux Annonacées (Asteranthe asterias et Hexalobus monopetalus). — C. R. Acad. Sc. Paris *275:* 1749–1752.

Lugardon, B. 1971. Contribution a la connaissance de la morphogenèse et de la structure des parois sporales chez les filicinées isosporées. — Thesis, l'Université Paul Sabatier de Toulouse.

Mollenhauer, H. H. 1964. Plastic embedding mixtures for use in electron microscopy. — Stain Technol. *39:* 111–115.

Mollenhauer, H. H. & Totten, C. 1971. Studies on Seeds. II. Origin and degradation of lipid vesicles in pea and bean cotyledons. — J. Cell Biol. *48:* 395–405.

Nabli, M. A. 1971. Ultrastructure de l'endexine et de la tryphine chez quelques espèces du genre Teucrium L. (Labiées). — C. R. Acad. Sc. Paris *273:* 2075–2078.

Oltmann, O. 1974. Licht- und Elektronenmikroskopische Untersuchungen zur Sporogenese von Lophocolea heterophylla (Schrad.) Dum. I. Die Entwicklung von der Sporenmutterzelle bis zum Tetradenstadium. 1974. — Pollen Spores *16:* 5–25.

Pearse, A. E. G. 1961. Histochemistry. — Little, Brown & Co., Boston.

Petris, S. de 1967. Ultrastructure of the cell wall of Escherichia coli and chemical nature of its constituent layers. — J. Ultrastruct. Res. *19:* 45–83.

Pettitt, J. M. 1966. Exine structure in some fossil and recent spores and pollen as revealed by light and electron microscopy. — Bull. British Museum (Natural History), Geology *13:* 223–257.

— 1971. Some ultrastructural aspects of sporoderm formation in Pteridophytes. — *In* Pollen and Spore Morphology/Plant Taxonomy IV (ed. G. Erdtman and P. Sorsa), pp. 227–251. Almqvist & Wiksell, Stockholm. 227 pp.

Pojnar, E., Willison, J. H. M. & Cocking, E. C. 1967. Cell-wall regeneration by isolated tomato-fruit protoplasts. — Protoplasma *64:* 460–480.

Rowley, J. R. 1967. Fibrils, microtubules and lamellae in pollen grains. — Rev. Palaeobot. Palynol. *3:* 213–226.

— 1971. Resolution of channels in the exine by translocation of colloidal iron. *In* 19th Ann. Proc. Electron Microscopy Soc. Amer. (ed. C. J. Arceneaux). — Boston, Mass., 1971.

— 1973 a. Translocation through the pollen wall. Abstr. Scand. Society for Electron Microscopy, Umeå, June 1973.

— 1973 b. Dynamic changes in pollen wall morphology associable with translocation. — J. Cell Biol. *59* (2, pt. 2) 295a (1973).

Rowley, J. & Flynn, J. 1971. Migration of lanthanum through the pollen wall. — Cytobiologie *3:* 1–12.

Rowley, J. R. & Southworth, D. 1967. Deposition of sporopollenin on lamellae of unit membrane dimensions. — Nature *213:* 703–704.

Sengupta, S. 1973. The effects of temperature and pressure on Lycopodium clavatum spores. — *In* Actes du 6ème Congress International de Geochimie Organique. Technip. Paris, pp. 305–306.

— 1974. Size reduction and structural change in Lycopodium clavatum spores, produced by temperature and pressure changes. — *In* Proceedings of the "Scanning Electron Microscopy/1974" (ed. O. Johari & I. Corvin).

Southworth, D. 1966. Ultrastructure of Gerbera jamesonii pollen. — Grana palynol. *6:* 324–337.

— 1973. Cytochemical reactivity of pollen walls. — Journ. Histochemistry & Cytochem. *21:* 73–80.

— 1974. Solubility of pollen exines. — Am. J. Bot. *61:* 36–44.

Thiery, J.-P. 1967. Mise en évidence des polysaccharides sur coupes fines in microscope technique. — J. Microscopie *6:* 987–1018.

Wang, W. S., Korczynski, M. S. & Lundgren, D. G. 1970. Cell envelope of an iron-oxidizing bacterium: Studies of lipo-polysaccharide and peptidoglycan. — J. Bacteriol. *104:* 556–565.

Weidel, W., Frank, H. & Martin, H. H. 1960. The rigid layer of the cell wall of Escherichia coli strain B. — J. Gen. Microbiol. *22:* 158–166.

Sukla Sengupta
Imperial College
Royal School of Mines
Geology Department
London SW7 2BP
Great Britain

John R. Rowley
Palynological Laboratory
Wallenberglaboratoriet
Stockholm University
S-104 05 Stockholm 50
Sweden

4

Copyright © 1966 by Geologische en Mineralogisch Instituut, Nederland

Reprinted from pp. 1-10 and 13-29 of *Leidse Geol. Meded.* **38**:1-29 (1966)

CARBONIZATION MEASUREMENTS OF POLLEN-GRAINS AND SPORES AND THEIR APPLICATION

C. C. M. Gutjahr

SUMMARY

This study presents the equipment and technique used in the author's palynological carbonization studies. The equipment consists mainly of modified Zeiss microphotographic equipment. A photoelectric cell fitted with a field restrictor is used to measure the light absorption of a spore or pollen type. The light-absorption value is considered to be a measure of carbonization, and values of a standard spore or pollen type are used to construct a carbonization line for an individual well.

Isocarbonization lines can be constructed in sections to show local or regional carbonization. Maps, block diagrams, panel diagrams, etc., can be constructed with either isocarbonization or carbonization contours.

Carbonization is a low-grade metamorphic process; an understanding of this process and its effects can be useful to the petroleum industry. The primary advantage of the described technique is that the degree of carbonization of pollen-grains and spores is a measure of low-grade (organic) metamorphism in areas or intervals devoid of coal. In addition the technique is cheaper and faster than the usual chemical coal analyses. Samples normally taken in the petroleum industry, such as cores, sidewall cores, and cuttings, can be used in palynological carbonization studies.

Palynological carbonization studies may be undertaken not only in strata in which palynological time correlations are possible but also in areas or sections in which alteration is too great for routine palynological time correlation. The carbonization method is unsuccessful in extremely carbonized zones, where alteration makes standard types unrecognizable, or, of course, in strata devoid of pollen-grains and spores.

The equipment and technique described here yield reasonably accurate and reproducible data that may shed some light on the temperature history of the organic material, the sediments, and the associated hydrocarbons.

TABLE OF CONTENTS

Introduction 1
   Terminology 2
   Acknowledgments 3

CHAPTER I Previous investigations 4
   1. Coal investigations 4
   2. Changes in the exines of pollen-grains and spores during carbonization 6
   3. Possible cause of changes in the light absorption of pollen-grain and spore exines during carbonization 6

CHAPTER II Causes of carbonization 7
   1. Introduction 7
   2. Temperature 7
   3. Time 7
   4. Pressure 8

CHAPTER III Equipment for carbonization measurements of pollen-grains, spores, and other plant fragments 11
   1. Description of equipment 11
     a) Standard equipment 11
     b) Special equipment and modifications . . . 11

CHAPTER IV Procedures 13
   1. Standard chemical preparation technique . . 13
   2. Selection of standard pollen-grain or spore . . 13

   3. Pre-operational procedures 14
     a) Four-stage amplifier adjustments 14
     b) Location of the sensitive spot 14
   4. Operational procedures 14
     a) Operational procedures for standard Zeiss microscope with modified photographic equipment 14
     b) Operational procedures for standard Zeiss microscope with modified equipment . . . 14

CHAPTER V Recording, presentation and evaluation of data 15
   1. Recording and presentation of data 15
   2. Evaluation of data 15

CHAPTER VI Examples of carbonization measurements 17
   1. Coal samples 17
   2. Sand-shale samples 17

CHAPTER VII Applications 25
   1. Correlation between carbonization and other sediment properties 25
   2. Palynological application 26
   3. Oil industry application 26

Bibliography 27

Pollen-grains and spores in coals from different ranks, studied under the microscope (transmitted light) after maceration methods or in thin sections, gradually change in color from yellow to light brown to dark brown-black. Changes in color of pollen-grains and spores from other sedimentary rocks, such as shales, sandstones, etc., are similar. The term carbonization is used in this study to denote this alteration process.

The chief interest and importance of carbonization investigations of pollen-grains and spores is that the carbonization measurements of pollen-grains and spores are indices of the level of organic metamorphism. In order to avoid confusion with the definition of metamorphism accepted by hard-rock geologists, the term organic metamorphism is suggested in palynological carbonization studies.

The main objective of this investigation is the quantitative determination of the level of organic metamorphism in sediments. The knowledge of the level of organic metamorphism in sediments is important for several reasons. In the first place, a reasonable to optimum preservation of pollen-grains and spores is a necessity for palynological correlation and/or ecological studies. Strongly to intensely carbonized pollen-grains and spores are opaque and show no structure. They are, therefore, difficult to recognize and palynological correlations and/or ecological investigations in areas or intervals in which they occur are difficult and often impossible. Because severe carbonization is a limiting factor in carrying out palynological or ecological studies, it is obvious that the delineation of severely carbonized areas or intervals is important to the palynologist and to the exploration effort in general.

In the second place, the carbonization degree of pollen-grains and spores is an index of the level of organic metamorphism to which also other organic matter (other than pollen-grains and spores) in the sediment has been subjected including oil. It is probable that important economical oil accumulations will diminish progressively as the intensity of organic metamorphism alteration increases.

Another reason why the determination of the level of organic metamorphism may be of importance is the possible correlation between sediment properties and the carbonization of pollen-grains and spores.

In order to determine the level of organic metamorphism, two methods have been developed. The simplest and fastest of the two methods to measure the degree of carbonization of pollen-grains and spores is the subjective evaluation of the color of the pollen-grains and spores by their appearance under the microscope (transmitted light). Several attempts were made to standardize the visual color determination. Reference slides of pollen-grains and spores of different ranks were compared with pollen-grains and spores to be measured. Color charts and color sticks were devised as standards of carbonization values. These attempts were unsatisfactory for detailed investigations because the inherent disadvantage of using color standards. The difficulty is the visual comparison of subtle color differences.

Color changes of pollen-grains and spores with increasing low-grade metamorphism are accompanied by a decrease of translucency to final opacity as observed with transmitted light. As a result of this observation, photoelectric measurements of the translucency of pollen-grains and spores have been made in order to express the carbonization degree of pollen-grains and spores objectively.

Terminology

Most of the following definitions of technical terms, used in this study, have been compiled from *Glossary of Geology and Related Sciences* (American Geological Institute, 1960), *Pollenmorfologiske definitioner og typer* Inversen and Troels-Smith, 1950) and the *International Handbook of Coal Petrography* (International Committee for Coal Petrography 1963).

British thermal unit: Abbreviated B.t.u. A unit of heat which is 1/180th part of that required to raise the temperature of one pound of water from 32 °F to 212 °F at sea level. Usually considered as that amount of heat required to raise the temperature of one pound of water from 63 °F to 64 °F.

Carbon ratio: The ratio of the fixed carbon in any coal to the fixed carbon plus the volatile hydrocarbons; expressed in percentages.

Carbon ratio theory: The theory that in any area the gravity of oil varies inversely as the carbon ratio of the coal. As temperatures and pressures increase, the percentage of the fixed carbon in coal increases, the rank of the coals rises and the oils become lighter. Increase in metamorphic processes results in elimination of volatile constituents from coal but increases the lighter and more volatile hydrocarbons in oil.

Caving: Caving of rock fragments from the wall of the borehole.

Cuttings: The fragmental rock samples broken or torn from the penetrated rock during the course of drilling.

Fixed Carbon: In the case of coal, coke, and bituminous materials, the solid residue other than ash, obtained by destructive distillation, determined by definite prescribed methods (A.S.T.M. D121-30).

Mud: Various fluids and emulsions, collectively termed, used in rotary drilling.

Organic metamorphism: In order to avoid confusion with the definition of metamorphism accepted by hard-rock petrologists, the term organic metamorphism is suggested. This alteration process of organic material, including pollen-grains and spores, alters under much less severe conditions than most inorganic material.

Proximate analyses: In the case of coal and coke, the determination by prescribed methods of moisture, volatile matter, fixed carbon (by difference), and ash (A.S.T.M. D121-30).

Psilate (smooth): A pollen-grain or spore sculpture type without sculpture elements. Depressions are absent or smaller than one micron.

Rank: Describes the stage of coalification attained by a given coal.

Spore: An asexual reproductive structure, commonly unicellular and usually produced in sporangia.

Trilete: Possessing a triradiate tetrad scar.

Vitrinite: The major maceral, or micropetrological unit of vitrain. It occurs as microscopic lenticels in some coals and is of such consistency as to suggest that it is of the same nature as vitrain.

Volatile matter: Those products, exclusive of moisture, given off by a material as gas and vapor, determined by definite prescribed methods, which may vary according to the nature of the material. In the case of coal and coke the standard methods are A.S.T.M. designation, D 271 and A.S.T.M. D 121-30.

ACKNOWLEDGMENTS

The author is greatly indebted to Shell Development Company and Shell Oil Company for permission to formulate the results of the carbonization studies. The advice and critisism of Professor A. Brouwer of the University of Leiden, The Netherlands, are gratefully acknowledged. The author is much indebted to Mr. B. O. Prescott of the Geological Instrumentation, Shell Development, Houston, Texas, for designing the field restrictor and for technical advice. The author's sincere thanks are also due to Dr. M. Teichmüller, Geologische Landesamt Krefeld, German Federal Republic, to Mr. E. Walker of the Bureau of Mines, Pittsburgh, Pennsylvania, U.S.A., and to Shell Oil Company (Houston Area) for providing samples. The author wishes to express his appreciation for the permission from the American Geological Institute, Washinton, D.C., U.S.A., and the Associated Technical Services Inc., East Orange, New Jersey, U.S.A., to use their literature data. The author is much indebted to colleagues and assistants of the Shell Development Company in Houston, Texas, for valuable help, discussions and stimulating interest. Finally, the author is greatly indebted to Mrs. J. Schmidt for typing the manuscript.

CHAPTER I

PREVIOUS INVESTIGATIONS

1. COAL INVESTIGATIONS

In his volume on the geology of Pennsylvania, Rogers (1858) noted a consistent regional change in character of the Pennsylvanian coals in the Appalachian basin. The volatile bituminous coals in the slightly deformed rocks of the Pittsburgh coal district were shown to grade eastward into anthracite coals in the strongly faulted and folded rocks of eastern Pennsylvania. Rogers (1865) later demonstrated that a correlation existed between coal composition, oil and gas occurrences, and the degree of metamorphism of the associated strata. He was the first to recognize the "carbon ratio" theory, later elaborated upon by David White (1915).

In the first formal statement of the theory, White in 1915 described the relationship between the limits of known commercial oil and gas occurrences and regional variations in the intensity of rock metamorphism as indicated by the percentage of "fixed carbon" in the coals. In addition, he suggested that the carbon ratio value marking the "deadline" for oil occurrences probably lay between 66 and 75 percent of fixed carbon (White, 1925). White stressed the importance of dynamic thrust as the cause of coal metamorphism, pointing out that normal faults and folds might serve to diminish locally the intensity of metamorphism. He further suggested that differences in the plant constituents of which the coals were made might account for wide differences in carbon ratio values despite an identical metamorphic history of the associated strata. The last suggestion is still valid today.

Since White's original paper, many attempts have been made to evaluate the carbon ratio theory, particularly in Oklahoma (Fuller, 1920), Arkansas Ozarks (Croneis, 1927), north Texas (Fuller, 1919), West Virginia (Reger, 1921; White, 1925), southwest Virginia (Eby, 1923), Rocky Mountains (Dobbin, 1929), western Canada (Jones, 1928; Hume, 1927), northern Alabama (Semmes, 1921), New Mexico (Storm, 1924), Illinois (Moulton, 1925), Kentucky (Russel, 1925), eastern Kentucky (McFarlan, 1926; White, 1925), and Pennsylvania, Maryland, Virginia, and eastern Ohio (White, 1925). One of the best surveys of the history and development of the carbon ratio theory was prepared by Thom (1934). Several causes of variation of individual coal analyses are as follows:

(a) use of nonstandard coal sampling methods
(b) nonstandard analytical work,
(c) differences in coal-forming vegetation at time of burial,
(d) use of "dry, ash free" as compared with "as received, ash free" forms of analyses (Thom, 1934).

If coal carbon ratios are used, these factors must be considered very carefully.

Now standard procedures are common in North America and Europe. The low rank coals are classified (A.S.T.M.= American Society of Testing Materials) according to their calorific values (or British Thermal Units), the higher rank coals are classified according to their fixed carbon or volatile matter (figure 1).

In order to eliminate the effect of the difference in vegetational matter on either the fixed carbon values or the volatile matter and therefore on the rank determination of the coal, M. and R. Teichmüller (1958), M. Teichmüller (1958) and Ammosov (1961) are analyzing the vitrinite of the coal which is a more precise technique than bulk coal analyses.

A recent paper by M. and R. Teichmüller (1958) discusses the carbon ratio theory in general, the relationships between the chemical and physical characteristics of the coals, and the occurrence of oil and gas in Lower Saxony, West Germany. The study indicates that the degree of coal metamorphism in an area can be estimated by reflected-light observations of the vitrinite of the coal. M. and R. Teichmüller (1958) state that all oil fields in the Lower Saxony basin are restricted to the edge of the basin, which contains low-grade coal, whereas in the highly carbonized center of the basin, only gas has been encountered.

Recent papers from Russia and Germany indicate that carbon ratios of coals are being used to evaluate oil potentialities in several areas. A paper by Vyshermirskii (1958) illustrates how the Russians are utilizing the carbon ratios of coals to block out areas in Siberia which are favorable for oil exploration. Vyshermirskii (1958) notes that there is a direct relationship between the carbon ratios of coals and the degree of metamorphism of the sediments in all regions studied. According to Vyshermirskii, this method of forecasting oil bearing capability has long been known, but "in practice of the Soviet oil men, however, this method did not gain wide use. This is probably due to the fact that up until recently the main volume of oil exploration was conducted in regions almost devoid of coal outcrops."

As a result of extensive work by Ammosov (1961), it has been determined that all main oil fields in the U.S.S.R. are associated with sediments in which the coal carbon ratio is not higher than 60, with isolated oil fields in the fixed carbon range 60—65 (fig. 1).

Since in many basins of the world coals are either rare or absent, the carbon ratio method was of limited use

1 Calorific value of the moist, mineral-matter-free coal.
2 On a ash-free-dry basis.

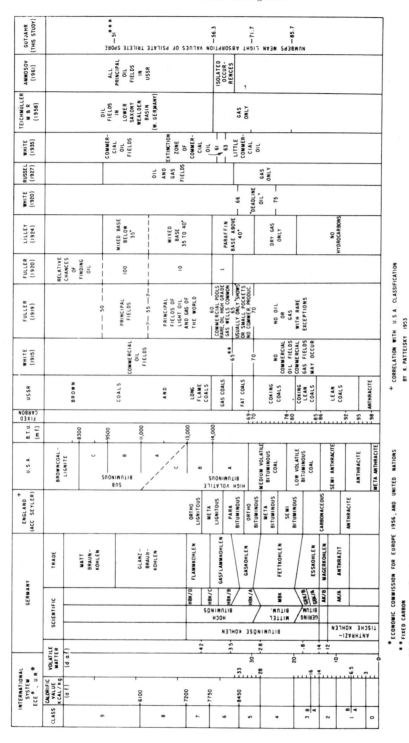

Fig. 1. Relationship between rank of plant remains and the occurrence of commercial oil and/or gas in reservoirs.

for determining the incipient metamorphism of the sediments. World-wide palynological studies have, in general, shown relative abundant pollen-grains and/or spores and other plant matter in the sediments of many basins. If pollen-grains or spores could be used in sediments to express the level of organic metamorphism, then basins with sporadic coal occurrences or devoid of coals could be investigated.

2. CHANGES IN THE EXINES OF POLLEN-GRAINS AND SPORES DURING CARBONIZATION

The pollen-grains and spores in coal beds, studied both isolated by maceration and by thin section, change in color with increasing coal rank from yellow over light brown to dark brown and finally to black. In the range brown coal to high-volatile B bituminous coal pollen-grains and/or spores are in general yellow. From high-volatile A bituminous coal to low-volatile bituminous coal pollen-grains and/or spores change in color from yellow over brown to black. In the past these qualitative changes of pollen-grains and spores have been reported by several authors (Bode, 1928; Potonie, 1930; Kirchheimer, 1934; Ergol'skaya, 1939, 1947; Schopf. 1948; Böttcher and M. and R. Teichmüller, 1949; M. and R. Teichmüller, 1950; Val'ts, 1952; Bogolyubova, 1956; M. Teichmüller, 1958; Ammosov, 1961; Wilson, 1961; Schopf, 1962). No spore exines have been found in coals ranking higher than "Magerkohlen" (equivalent to semi-anthracite, figure 1) (Bode, 1928; Potonie, 1930). This has been caused by chemical changes which transformed pollen-grains and spores into totally carbonized particles Microspores on polished sections from coals with vitrinite containing 20 to 22 percent volatile matter become invisible (Teichmüller, M., 1958). It is of interest in connection with the observations of M. Teichmüller to note that a similar change of pollen-grains and spores has been noticed by Schopf (1948) in thin sections in coals within the range of 23 to 25 percent volatile matter. The pollen-grains and spores from other sedimentary rocks, such as shales and sandstones, change in similar color sequence as is observed in coals, indicating their low-grade metamorphism. It seems that strata with levels of organic metamorphism equivalent to semi-anthracite, anthracite and meta-anthracite are unsuitable for palynological time correlation or ecological investigations. Pollen-grains and spores in these levels are completely carbonized. Sediments with levels of organic metamorphism equivalent to the lower part of medium volatile bituminous coal and of low volatile bituminous coal may contain pollen-grains and spores less suitable for palynological studies, because of their dark brown to black color. Early recognition of the above facts may help the palynologist in the selection of favorable areas for palynological studies.

3. POSSIBLE CAUSE OF CHANGES IN THE LIGHT ABSORPTION OF POLLEN-GRAIN AND SPORE EXINES DURING CARBONIZATION

It is interesting to speculate why light-reflection and light-absorption values of plant material (respectively by reflected and transmitted light) change during the coalification process.

The reflection index and the absorption coefficient of coals depend on the chemical structure of the coal. Also the reflection capability of vitrinite under reflected light (van Krevelen, 1953; H. Stach and M. Teichmüller, 1953) and the absorption of transmitted light by pollen-grains and spores (Gutjahr: this study) depend on the chemical structure of the constituent under consideration. The absorption capability of plant material for light, according to the electromagnetic light theory, depends on the number of electrons with free oscillation, present in the plant material. The oscillation of transmitted light are dampened by the oscillations of the electrons of the plant material and cause light absorption (M. and R. Teichmüller, 1954).

The following statements are taken from R. W. Terhune (1964, p. 39).

"When an ordinary light beam passes from air into a transparent material the most obvious phenomenon is that the light beam slows down. The amount of bending of the light beam, according to geometric optics, depends on the ratio between the index of refraction of air and that of the transparent material. There is therefore an interaction between light and matter. In physical optics, this interaction is viewed as occurring between the oscillating electric field of the light beam and the electric charges in the atoms or molecules in the material. When the beam passes through a material, the oscillating electric field moves the electron cloud surrounding the positively charged nucleus first in one direction, then in the other. The net effect is that each particle becomes a miniature oscillating dipole radiating energy at precisely the same frequency as the incoming wave.

Each dipole acts as a separate source and the scattered radiation from them is called Rayleigh scattering. In the forward direction, both the incident and the re-radiated fields are present. The total field is, therefore, the vector sum of their amplitudes. Although the two fields are phase coherent, the re-radiated field is delayed 90 degrees in time phase from the incident field. The result of this vector addition is a wave with lower velocity than that of the incident wave."

CHAPTER II

CAUSES OF CARBONIZATION

1. INTRODUCTION

From a physical-chemical point of view, the coalification process can be subdivided into three stages (Huck and Karweil, 1962). In the beginning of the coalification process complex chemical reactions are caused by fungi and bacteria. In the second stage (the formation from peat to high volatile bituminous coals, figure 1) the bacterial and chemical changes decrease in favor of a mechanical compression of sediments, which causes expulsion of water. The last stage (the formation from high volatile bituminous coal to anthracite, figure 1) is a combination of physical and chemical changes. The chemical metamorphism of coals is characterized by an increasing aromatization of humic substances. During this process molecule groups of nonaromatic character rich in hydrogen and oxygen are released. At the same time the remaining molecules of aromatic character are growing together (M. and R. Teichmüller, 1954). Among the factors causing carbonization reported in literature are temperature, time, pressure, and radio-activity. If the main cause of carbonization can be found, severely carbonized zones can be predicted. A knowledge of the cause of carbonization therefore is of importance.

2. TEMPERATURE

Recent *Quercus robur* (oak) pollen-grains on a glass slide have been subjected to a temperature controlled hot plate in the open air. The author does not suggest that a similar process will occur in the subsurface of the earth but warns that pollen-grains and spores should not be heated above 100 °C if carbonization measurements are being conducted. The *Quercus* pollen-grains have been heated at 100 °C, 150 °C, and 200 °C for approximately one week (figures 2 and 3). Slight changes in carbonization could be observed if *Quercus robur* pollen-grains were heated at 100 °C for a week. However, if they were heated at 150 °C and 200 °C, important carbonization changes could be observed (figure 2 and 3). A plot of mean carbonization (light absorption) values against time (figure 2) show that for each of the temperatures the rate of carbonization change decreases rapidly and that after some 50 hours of temperature exposure a level is reached and only slight or no further carbonization changes will occur. If cuttings samples, cores, or sidewall cores from wells or Recent samples are used for pollen-grain and spore carbonization studies, temperatures of less than 100 °C are recommended if samples are dried, because temperatures of 100 °C, 150 °C and 200 °C carbonize pollen-grains and spores in a relatively short time interval. This experiment shows that temperature can carbonize pollen-grains.

An example of slight temperature exposure in nature during a considerable time interval is the Lower Carboniferous brown coal in the basin of Moscow. This deposit has never been buried deeply and it has been exposed to low temperatures only. Examples of high temperature exposure in nature are the severe carbonization zones near small or large igneous bodies. Examples are the Pliocene brown coals of Palembang, Sumatra, Indonesia (Mukherjee, 1955), which by contact metamorphism are partly changed into anthracite, and the organic metamorphic zones around the Bramscher massif, German Federal Republic (M. and R. Teichmüller, 1958). These and other studies suggest that temperature is the predominant cause of carbonization.

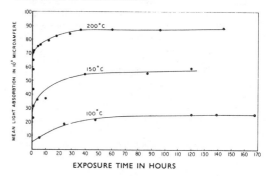

Fig. 2. Mean light absorption values of *Quercus robur* pollen-grains as a function of exposure time and temperature.

3. TIME

The effect of exposure time of low temperature on pollen-grains and spores is not important. The experiment discussed in Section 2 of Chapter II shows this clearly. One of the best examples in nature of the effect of exposure time of lower temperature on organic matter is the Lower Carboniferous brown coal of Moscow. Apparently time alone could not carbonize this carbonaceous material beyond the brown coal rank (M. and R. Teichmüller, 1954). The effect of exposure time of higher temperatures on pollen-grains is of importance. Slight carbonization changes in *Quercus robur* (oak pollen-grains) could be obtained at 100 °C for a week (figure 2), but drastic carbonization changes were observed at higher temperatures (viz. 150 °C and 200 °C) in a relatively short time interval. Huck and Karweil (1955) have published a diagram (figure 4) showing the relationship between carbonization, time, and temperature. The diagram indicates that impor-

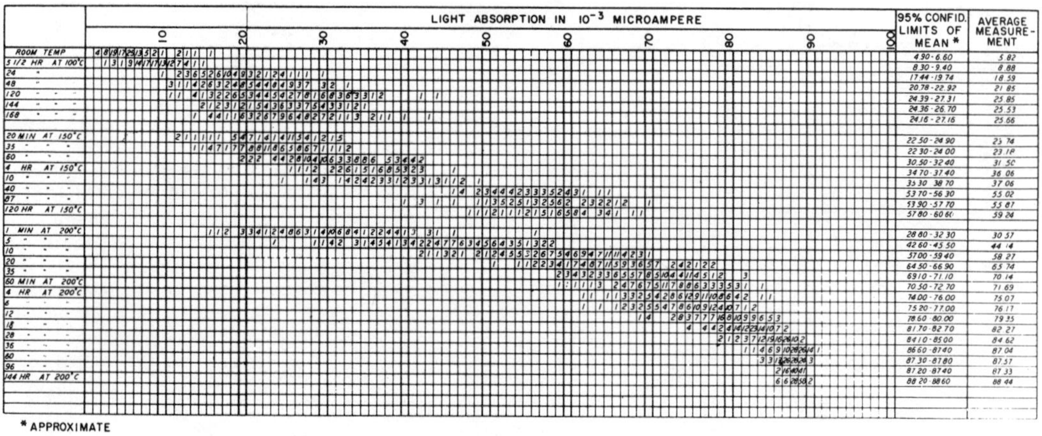

Fig. 3. Light absorption data sheet of *Quercus robur* (oak) showing the relation between light absorption, temperature and exposure time.

tant changes in carbonization can be obtained if carbonaceous matter is exposed longer at relatively higher temperatures.

4. PRESSURE

White (1925) emphasized the importance of "dynamic thrust" as the cause of coal metamorphism. However, Lesley (1879) had earlier proposed that the regional change in the rank of coal in Pennsylvania was due to the higher rank coal having been buried more deeply. He pointed out that the difference in the amount of the original overburden could easily explain why "disturbed" Belgian coals are not anthracites and why Arkansas anthracites are not disturbed.
M. and R. Teichmüller (1954) mentioned a brown coal in a folded belt in Pakistan which was greatly deformed by dynamic thrusting but was only slightly altered in rank.
The western part of the Prokop'evsk-Kisel area (Kuznetsk basin, Russia) is characterized by steeply folded and faulted coal measures (I. I. Ammosov, 1961). These coals are less coalified than the coals in the less disturbed parts of the basin.
In the Westphalian and the French-Belgian coal fields, coal beds are richer in volatile matter (less carbonized) on the highly folded and faulted south side of the coal basin than they are on the north side where the strata are much less disturbed (Reeves, 1928). The above examples show that pressure alone apparently could not increase the rank of the coal.
Various coal samples were subjected (Hoffman, 1936) for four days to a pressure of 18,000 kg per square centimeter. Hoffman noted an average decrease in volatile matter by 1.5 %, indicating very slight carbonization.
According to Hoffman (1936) and Bergius (1913) coalification by pressure is successful only if accompanied by relative high temperature. This was also observed by Gropp and Bode (1932) who noticed a decrease of volatile matter of "immature" coals when subjected to 300 °C and 1800 atm. Gropp and Bode (1932) considered the high temperatures in the synclines of deep basins as the main factor in the carbonization of coal, while pressure only hinders the formation of carbonization products.
Trifonow and Toschew (1940) compressed nine different coal samples by applying a pressure of 10,000 kg per square centimeter and only slight increase of total volatile matter was noticed.
From a thermodynamic viewpoint (Fuchs, 1941), rising pressure in a condensed heterogeneous system will cause phenomena making for decreasing volume, e.g., boiling points will be raised, melting points will be lowered and simple polymerization will be favored. Hence only small changes of proximate analysis, but no significant change of ultimate composition or rank of coal may result from compression.
Also, van Krevelen (1957) mentioned that static pressure has no effect on the chemical properties of the coal since it is inconsistent with thermodynamics. Compression may, conceivably, cause a rise of temperature. Fuchs (1941) calculated that, assuming a maximum compression of 10,000 kg per square centimeter, a starting temperature of 25 °C, a specific heat of 0,3 cal per gram coal, a volume expansivity α of $0.2 \cdot 10^{-4}$ and a specific gravity of 1.3, a temperature increase of 3.5 °C could be expected which is obviously negligible for coalification (carbonization).
Fuchs (1953) and Huck and Karweil (1953) have stated that pressure has a great influence on structural alteration of sediments but that pressure alone cannot change the chemical nature of the coals. Huck, Karweil, and Pateisky (1962) used a heatable pressure devise in order to study the coalification process. They concluded that static pressure delays coalification and does not accelerate the coalification process as is often assumed.

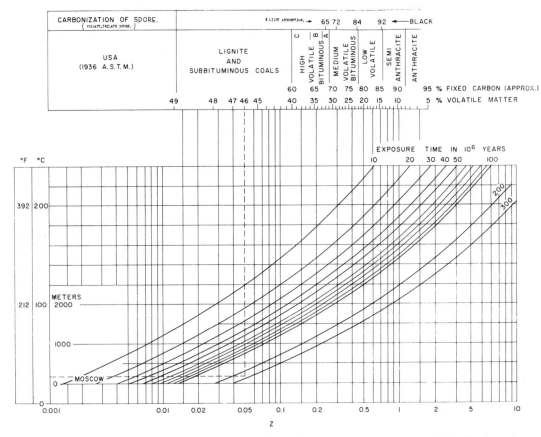

Fig. 4. Relation between carbonization, temperature and exposure time (Karweil 1955, with additions by the author).

Concerning the dynamic pressure in relation to coalification, the same authors mentioned that the influence of dynamic pressure in nature cannot be ignored, but that it is rarely of importance.

E. Lewis (1936) found that "cokes" formed from a coal, when heated under diminished pressure, lie on straight lines (RS, TM, and AB of figure 5a). When, however, coals were heated under pressure, a series of "cokes" was produced (see curved lines JKP and NOP of figure 5b) whose compositions more nearly approximated to the compositions of natural coals (figure 5b). The composition of natural coals is indicated in figure 5a and 5b as a gray band, the so-called coal band.

Regarding the change in composition of a coal under influence of temperature and pressure, Lewis' (1936) study shows that the effect of pressure produces residues that are closer to the coal band than the residues obtained under the action of heat alone.

Carbonization studies of a lignite in pressure vessels by J. Rogers, R. P. Suggate and J. O. Elphick (1962) show (figure 6) that at 1000 atmospheres and at 170 °C and above there is an initial rapid change, whereas up to 100 °C the change in calorific value or in coalification is slow in terms of time of these experiments. It is interesting to note that at 170 °C there were greater changes in calorific value or carbonization at 10 atmospheres pressure than at 1000 atmospheres. This observation is in agreement with the observation of Huck, Karweil and Pateisky (1962) viz. that static pressure retards coalification. Rogers, Suggate and Elphick (1962) furthermore found that there is a marked point of inflexion at about 170 °C in the the calorific value versus termperature (figure 6), obtained at 1000 atmospheres pressure over a period of seven days, the inflexion point at 10 atmospheres and seven days however, is only at about 120 °C. It seems that carbonization is slow at low temperatures until a particular temperature (threshold T) is obtained after which rank increases more rapidly with increasing temperatures. The higher the pressure, the higher this threshold temperature will be (Rogers, Suggate, Elphick, 1962).

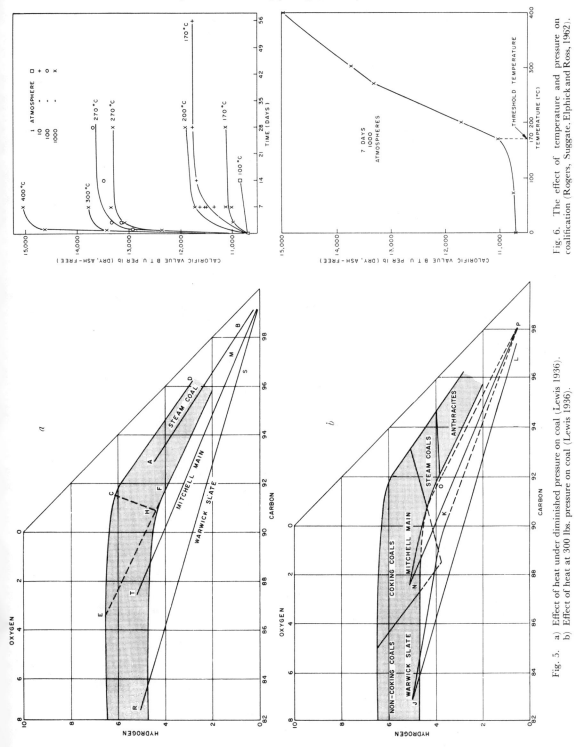

Fig. 6. The effect of temperature and pressure on coalification (Rogers, Suggate, Elphick and Ross, 1962).

Fig. 5. a) Effect of heat under diminished pressure on coal (Lewis 1936).
b) Effect of heat at 300 lbs. pressure on coal (Lewis 1936).

[*Editors' Note:* Chapter III on equipment has been omitted.]

CHAPTER IV

PROCEDURES

1. STANDARD CHEMICAL PREPARATION TECHNIQUE

Spores and pollen-grains are affected to a greater or lesser extent by certain chemicals, and a standard treatment insures that variables will not be introduced by laboratory preparation methods.

Mechanical treatments will not affect the absorption characteristics of pollen-grains and spores. However, prolonged ultrasonic vibrations may crack or fragment specimens which then cannot be measured accurately.

The use of oxidants have been avoided (nitric acid, Schultz's solution, acetolysis, etc.) in the following standard preparation procedure (for calcareous shales, Steps 1-14, for shale, dirty sandstones, and silt, Steps 3-14). The following steps have been used in this investigation:

(1) pulverize sample, and add a solution of 20 percent hydrochloric acid; boil four minutes,
(2) centrifuge, decant, and add water to residue; centrifuge again,
(3) add hydrofluoric acid (52 percent) to residue; boil five minutes,
(4) centrifuge, decant, and add water to residue; centrifuge again,
(5) boil residue with hydrochloric acid (20 percent) for five minutes,
(6) centrifuge, decant, and add water to residue; centrifuge again,
(7) boil residue for 30 seconds in potassium hydroxide (5 percent),
(8) centrifuge, decant, and add water to residue; centrifuge again,
(9) put residue in aqueous suspension in test tube,
(10) insert in an ultrasonic generator,
(11) centrifuge and decant, normal or short centrifuging,
(12) add alcohol, shake, and centrifuge,
(13) separate with heavy liquid using bromoform — alcohol solution (s.g. 2.0),
(14) put float from Step (13) in vial.

It is probable that slight modification of the above standard treatment may yield comparable results. However, for greater confidence in comparing carbonization data obtained in different laboratories, it is suggested that the above preparation procedures be adopted.

2. SELECTION OF STANDARD POLLEN-GRAIN OR SPORE

One of the most important considerations in undertaking light-absorption value measurements is the selection of the type or "species" of the pollen-grain or spore to be measured. It must be recognized that different types have different light-absorption value because of original differences in wall thicknesses and possible original differences in chemical composition. At any given degree of low-grade metamorphism, there will be pollen-grains and spores with different light-absorption values. With increasing metamorphism these absorption differences become less pronounced. It is important to select a type that will give low absorption values when slightly carbonized. If a type with high initial light-absorption values is chosen, the measurable effect of light-absorption alteration is drastically reduced (figure 11).

The standard type should be relatively large, because an area of the exine should be selected which is free of apertures, occasional folds, etc. The exine should be reasonably uniform in thickness, relatively free of ornamentation, preferably smooth (psilate), and generally without cracks, folds, or perforations. The type should be easily recognizable and distinctive. The vertical range and relative abundance of the types available for selection should be considered. It

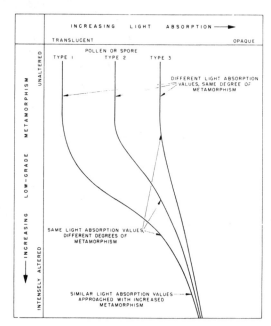

Fig. 11. A schematic comparison of the effect of increasing low-grade metamorphism on the light absorption of three different pollen-grains or spores.

is preferable that the type chosen be relatively abundant throughout the strata to be studied.

If the chosen type becomes relatively rare or disappears in the samples investigated, another type may be selected. The light-absorption values of these different types can be equated in zones in which both occur.

3. PRE-OPERATIONAL PROCEDURES

a. *Four-stage amplifier adjustments* (figure 9)

(a) The indicator of the microampere scale (standard scale) may have to be mechanically adjusted to *0* by adjusting the screw near the base of the dial. The stage selector switch (middle knob) must be on *0* during this procedure. Usually this must be done when the equipment is received from the supplier and at regular intervals, but it is not a part of the operational procedure.

The transformer and the four-stage amplifier are switched on to warm up. It should be noted that longer life and stability can be expected if the four-stage amplifier is turned on Monday morning and is turned off only for the week-end.

(b) The four-stage amplifier is zeroed with the "Nullpunkt" knob on the *0* of the standard scale with the stage selector switch on *1*.

The amplifier is compensated with the "Kompensation" knob on the *0* of the standard scale with the stage selector switch on *1000*. In both adjustments no light has passed to the photocell. The zero and compensation should be checked at regular intervals each day.

b. *Location of the sensitive spot*

The field restrictor confines the field to be measured and defines a sensitive spot or area, approximately six microns in diameter (with the oculars described), with an oil immersion objective. The exact position of this spot is slightly different for each microscope and restrictor equipped photocell.

The sensitive spot is located by the procedure outlined below. The amplifier is zeroed and compensated, and a small (∼ 5 micron) opaque black particle is located under oil immersion in a field that is relatively free of other debris.

The beam-splitting slide is pulled to permit light to the photocell, and the stage selector switch is set on *1000*. The black particle is moved almost out of view, and the lower diaphragm of the microscope is adjusted to obtain *0* light absorption on the reversed scale.

The black particle is moved across the field of view by multiple traverses until the maximum light absorption reading is obtained. The position of the sensitive spot in the field is the position of the black particle at this maximum light absorption reading (maximum reading on reversed scale). The position of the sensitive spot in the field is located for future reference on the scale in the ocular. The accurate location of the sensitive spot in the field is fundamental and should be checked occasionally. Once the location of the sensitive spot is established, neither the field restrictor in housing nor the ocular with reference scale should be moved.

It is suggested that the ocular then be fixed (taped or more permanently fixed) to the body of the inclined ocular tube, because if it and the included scale are rotated after positioning, the reference point to the sensitive spot will be lost. The beam-splitting slide can be moved to shut light from the photocell, and measurement of pollen-grains and spores can be started.

4. OPERATIONAL PROCEDURES

a. *Operational procedures for standard Zeiss microscope with modified photographic equipment*

(a) The pollen-grain or the spore to be measured under oil immersion is offset so that the sensitive spot is located in a fully illuminated vacant area, free of debris.

(b) The beam-splitting slide of basic body FL II is pulled to position 3 (completely out) to permit light on the photocell. The lower microscope diaphragm is adjusted so that the light produces a reading of *0* light absorption on the reversed scale of the four-stage amplifier when set on its most sensitive scale (*1000* on the stage selector switch).

The technique of adjusting the light available to the photocell to produce a scale reading of *0* light absorption (zero on reversed scale) before each pollen-grain and spore light-absorption measurement, eliminates several variables (e.g., differential thickness of microscope slide, cover glasses and variations in mounting media). The light source is kept constant by the Sola constant voltage transformer throughout these operations.

(c) The pollen-grain or the spore is moved so that the sensitive spot is located in the area to be measured. Folds, cracks, thicker and thinner areas in the exine, and debris above and below the pollen-grains or the spores must be avoided.

(d) The light absorption value is read on the reversed scale of the four-stage amplifier meter and is recorded.

(e) The beam-splitting slide is returned to position 1 or 2 so that no light is cast on the photocell. Steps (a) through (e) are repeated for each measurement. The amplifier remains on during operations.

b. *Operational procedures for Zeiss photomicroscope with modified equipment*

The light is excluded from the photocell at positions 1, 2 and 3 of the built-in beam-splitting slide. Normally position 2 is used for observations, and position 3 which masks the area to a 35-mm frame size with cross hairs, is used to locate the sensitive spot.

(a) Same as Step (a) above, except that for consistent magnification the ocular of the microscope is always set on the same magnification factor.

(b) Same as Step (b) above, except for beam-splitting slide position, which is pulled to position 4 to pass light vertically to the photocell.

(c) Same as Step (c) above, except that the beam-

splitting slide must be returned to position 3 for observation and moved back to position 4 for measurement.
(d) Same as Step (d) above.
(e) The beam-splitting slide is returned to position 1,

2 or 3 so that no light is passing through the photocell.
Steps (a) through (e) are repeated for each measurement, with proper position of the beam-splitting slide. The amplifier remains on during operations.

CHAPTER V

RECORDING, PRESENTATION AND EVALUATION OF DATA

1. RECORDING AND PRESENTATION OF DATA

The numerical values read on the reversed scale of the four-stage amplifier are termed light-absorption values expressed in millimicroamperes. Several methods can be devised to record and present graphically these absorption values of measured specimens of the same type.

The absorption values of individual specimens are recorded as they are encountered during traverses of the slide. It is desirable to measure approximately 40 specimens of the same type per sample. With increasing low-grade metamorphism less variation in light-absorption occurs, and fewer measurements per sample are sufficient.

The arithmetical mean or the median light-absorption value can be calculated for each sample depending on the distribution of the light-absorption values. To determine the reliability of the arithmetical mean of each sample, the approximate 95 percent confidence interval can be calculated by standard methods.

From these data a *carbonization diagram* is then plotted, with light-absorption values, expressed in millimicroamperes, as the horizontal scale and depth as the vertical scale (figure 12). For each individual sample a frequency curve is constructed with its base at the sample depth and its height controlled by an arbitrary but uniform vertical scale. These curves have a characteristic "saw blade" pattern. Each point on the curves represents the number of specimens with a given light absorption value. The average light absorption value for each sample is plotted at the base of the frequency curve. These points are connected in succeeding samples to show a line representing the trend of the values called the *carbonization* line. This line is drawn for each well or section and is used to estimate the depth at which certain average light absorption and, thus, carbonization values are encountered.

Carbonization lines for different wells in sections, panel diagrams, etc., can be used to show the carbonization trends in compressed form. In section lines connecting the same light absorption values can be constructed; these are referred to as *isocarbonization lines* (figure 13a). The intervals between succeeding isocarbonization lines, here referred to as a *carbonization zone* (figure 13a), with a specific light absorption range, can be color-coded in sections in order to aid the visual presentation. If carbonization measurements of an adequate number of wells in a certain area are available, several types of contour maps can be drawn. One type shows the depth of a specific carbonization value (or a specific light absorption value). These contours are termed *isocarbonization contours* (figure 13b). Another type shows the carbonization values of a specific datum, which can be either a geological unit or a specific depth. These contours are termed *carbonization contours* (figure 13c). The type of contour map(s) to be made depends upon the purpose of the investigation. Block diagrams, panel diagrams, etc., can be constructed with either type of contour map.

2. EVALUATION OF DATA

The data obtained by the method discussed in Chapter IV and Chapter V section 1, are numerical expressions of the light absorption of pollen-grains and spores. Empirical data have repeatedly shown that these values change with changing carbonization of coals.

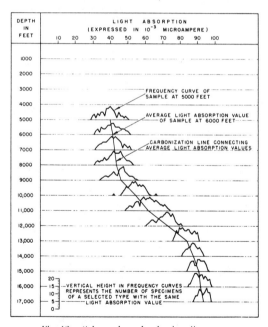

Fig. 12. Schematic carbonization diagram.

As carbonization increases during the coalification process, the spores and pollen-grains become darker and absorb more light. The absorption values of a specific type in succeeding samples in a well are therefore considered to be a measure of the carbonization stage, or "rank", of the strata.

Carbonization is a low-grade metamorphic process which alters organic material, and the absorption values are therefore considered as indicators of this organic metamorphic process.

Carbonization begins initially with slightly altered (translucent) plant material and continues toward intensely altered (opaque) plant material. The term "carbonized" has been applied in palynological studies to spores, pollen-grains, and other plant materials with various meanings. In general, the term ha, been used to indicate that either the spores and pollen-grains or the associated plant debris, or both, have been so altered that the isolation and identification of spores and pollen-grains is impossible. In some cases the term has been used when black plant debris is present, with or without spores and pollen-grains. In general most plant material undergoes more rapid carbonization alteration than the associated pollen-grains and spores. These usages of the term carbonized are unfortunate, because unwarranted conclusions may be drawn. According to our usage, virtually all fossil plant material is carbonized to some extent. Such terms as "completely carbonized" or "totally carbonized" are proposed in this study to indicate that pollen-grains, spores and other plant debris are black (opaque).

The contamination of a sample by spores and pollen-grains which have had a different carbonization history may alter the general aspect of that sample. For example, contamination in any manner by Recent pollen-grains and spores which are less carbonized. mens to a sample. Also, caving usually introduces pollen-grains and spores which are less carbonized. Reworking of sediments, on the other hand, usually introduces types that are more carbonized than those which were deposited in an unaltered state at the time the strata were deposited. Furthermore, the pollen-grains and spores introduced by the drilling mud on and in the cores, sidewall cores, or cuttings may be derived from penetrated rocks, from mud additives, or from Recent plants. These mud-introduced pollen-grains and spores may be either more or less carbonized than the pollen-grains and spores from clean rock samples.

In our experience to date these various types of contamination usually do not severely limit carbonization studies. In order to introduce serious errors, the sample must be contaminated by considerable numbers of a type identical with the the standard pollen-grain or spore type measured. Precautions taken in normal routine palynological studies are sufficient for satisfactory carbonization studies.

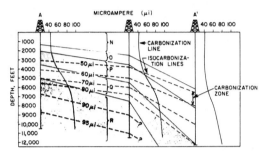

Fig. 13a. Schematic cross section showing carbonization lines, isocarbonization lines, and carbonization zones.

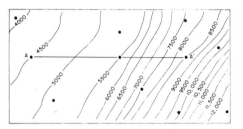

Fig. 13b. Schematic map with isocarbonization contours showing the depth of a specific carbonization line (65 µi).

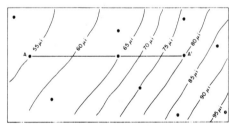

Fig. 13c. Schematic map with carbonization countours showing the values of a specific datum (P-Q contact).

CHAPTER VI

EXAMPLES OF CARBONIZATION MEASUREMENTS

1. COAL SAMPLES

Spore and plant fragment carbonization measurements of humic coal samples have been made. The coal samples were provided by Dr. M. Teichmüller, Geologisches Landesamt, German Federal Republic, Europe and by Mr. F. E. Walker, Bureau of Mines, Pittsburgh, Pennsylvania, U.S.A. (fig. 14, pp. 000, 000).

Light-absorption measurements of plant fragments in the insoluble residue with a specific gravity of 2.00 and less have been made for humic coal of different rank ranging from lignite to meta-anthracite (figure 15). Although varying in thickness and tissue type, plant fragments can, by their opacity, give an approximate measure of carbonization rank. In the low-rank coals analyzed (lignite to high-volatile B bituminous coal), the plant fragments colors range from yellow to brown to black (the light-absorption values range from 35 to 93 $\times$ 10^{-3} microampere). In the range high-volatile A bituminous coal to semi-anthracite, the plant fragments vary from dark brown to black. In the anthracite and meta-anthracite ranks, regardless of the thickness and type of plant fragments, all plant fragments are opaque, indicated by light-absorption values of 91 $\times$ 10^{-3} microampere and higher. It is interesting to note that a shift in the light absorption pattern of plant fragments occur between high-volatile B bituminous coal and high-volatile A bituminous coal. The light-absorption values of plant fragments in the high-volatile B bituminous coal range from 60 to 93 $\times$ 10^{-3} microampere, whereas in the high-volatile A bituminous coal the range has been narrowed down to 85 to 93 $\times$ 10^{-3} microampere.

The carbonization of the psilate trilete spores is hardly noticeable in the range lignite to high-volatile B bituminous coal. The carbonization rate increases in the range high-volatile B bituminous coal to low-volatile bituminous coal. At and beyond the low-volatile bituminous coal rank pollen-grains and spores become totally opaque.

2. SAND-SHALE SAMPLES

The subsurface carbonization trend of a specific psilate trilete spore has been recorded in two sections perpendicular to the strike in the Texas Gulf Coast (figures 17 and 18). In the first section (figure 17) south of Houston, Texas, cuttings samples of eight wells have been analyzed, while in the second section thirteen wells have been analyzed. At specific depths of each well light-absorption measurements of specimens of a psilate trilete spore were made.

The results of these measurements have been plotted on data sheets (figure 19 and 20). For each sample analyzed the carbonization mean and the approximate 95 % confidence limit of this mean have been calculated and plotted on the data sheets (figures 19, 20 and 21).

The carbonization means then have been drawn at specific depths in the wells on the sections (figures 17 and 18).

Isocarbonization lines connecting equal carbonization values in the wells were constructed.

Two obvious observations can be made while studying the carbonization pattern in the two Texas Gulf Coast sections. First of all, there is a clear trend of increasing carbonization with depth. Hilt (in Raistrick and Marshall, 1939), after analyzing coals of the Pas de Calais field, northern France, came to the conclusion that in a vertical sequence at any point in the coal field, the rank of the coals increased with depth, a principle commonly known as Hilt's law. Hilt's law seems to be true for the two sections analyzed in the Texas Gulf Coast.

The second observation is that isocarbonization zones cross time lines. Similar observations have been made by Levenshteyn (1962) in the Donets-basin (figure 22) Russia, and by M. and R. Teichmüller (1958) in the "Bochumer Groszmulde des Ruhrkarbons" east of Bochum, German Federal Republic.

In the Texas Gulf Coast the isocarbonization lines dip

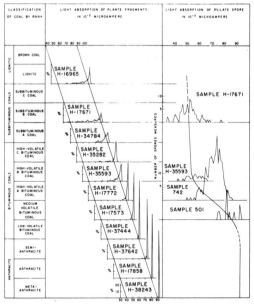

Fig. 15. Light-absorption measurements of plant fragments and psilate spores in humic coals of different ranks.

	Sample	Mine	State	County	Country	Strata
Lignite	H-16965	Kincaid	North Dakota	Barke	U.S.A.	Fort Union
Sub-bituminous B coal	H-17671	Barnham No. 3 coal crop fire project	New Mexico	San Juan	U.S.A.	Fort Union, Mesa Verde
Sub-bituminous A coal	H-34784	Pittsburg and Midway	Colorado	Routt	U.S.A.	Mesa Verde
High-volatile C bituminous coal	H-35282	Orient No. 2	Illinois	Franklin	U.S.A.	Carbondale
High-volatile B bituminous coal	H-35593	Caney Creek	Kentucky	Muhlenberg	U.S.A.	Carbondale
High-volatile A bituminous coal	H-17772	C.C. Conley	West Virginia	Nicholas	U.S.A.	Coalburg in Kanawha group Pottsville Series
Medium-volatile bituminous coal	H 17573	Brubaker	Pennsylvania	Cambria	U.S.A.	Lower Kittanning in Allegheny formation
Low-volatile bituminous coal	H 37444	Pageton	West Virginia	McDowell	U.S.A.	Pottsville
Semi-anthracite	H37642	Glen Burn	Pennsylvania	North Humberland	U.S.A.	Pottsville
Anthracite	H 17858	Greenwood Breaker	Pennsylvania	Schuylhill	U.S.A.	Pottsville
Meta-anthracite	H38243	Granston	Rhode Island	Providence	U.S.A.	Kingston

	Sample	Town	Area	Country	Strata
	742	Wellendorf	Teutoburger Wald	German Federal Republic	Wealden
	501	Reinsen	Wesergebirge	German Federal Republic	Wealden

Fig. 14. Coal data.

Age	Moisture (as received)	Volatile matter (moisture free)	Fixed carbon (moisture free)	Ash (moisture free)	Sulphur (moisture free)	British thermal units (moisture free)
Paleocene	18.2%	42.0%	42.9%	15.1%	1.3%	10560
Cretaceous	8.5%	39.2%	41.9%	18.9%	0.6%	8700
Cretaceous	7.8%	40.9%	49.4%	9.7%	0.6%	
Pennsylvanian	5.4%	36.2%	55.9%	7.9%	1.5%	13110
Pennsylvanian	5.2%	44.2%	49.1%	6.7%	3.5%	13510
Pennsylvanian	1.8%	37.4%	55.6%	7.0%	0.6%	14160
Pennsylvanian	0.6%	21.3%	70.8%	7.9%	1.6%	14430
Pennsylvanian	0.8%	20.8%	74.6%	4.6%		14930
Pennsylvanian	2.2%	8.3%	81.0%	10.7%	0.8%	13480
Pennsylvanian	2.4%	3.4%	88.0%	8.6%	0.5%	13550
Carboniferous	0.3%	2.7%	59.5%	37.4%	0.2%	8510

Age	Volatile matter (in vitrite)	Carbon (in vitrite)	Oxygen (in vitrite)	Hydrogen (in vitrite)	
Cretaceous	31.0%	88.0%	5.8%	6.2%	
Cretaceous	22.0%	92.0%	2.4%	5.6%	

Fig. 14. Coal data (continued).

Fig. 16. Well locations and index.

SECTION 1

1. Davis and Co. Inc. No. 1, John R. McLane
2. Shell No. 2, C. R. Schiurring
3. Shell No. 1, H. M. Kane
4. Shell No. 1, C. W. McDermott, et al.
5. Colorado No. 1, Pryor, et al.
6. Texas No. A-33, Pierce Est.
7. Pure No. 1, Mrs. E. Page
8. Sun No. 1, Pierce Est.
9. Union No. 1, Pinkney
10. Old Ocean No. 1, S. C. Patterson
11. Pan American Old Ocean Unit No. 99
12. Pan American Old Ocean Unit No. 138
13. Pan American Old Ocean Unit No. 77

SECTION 2

14. Gulf No. 1, Katie Ward
15. Pan American No. 1, Sneed
16. Hammond No. 1, Tieman
17. Pan American No. 1, Suderman
18. Phillips No. 2 M. Houston Farms
19. Phillips No. 1 L, Houston Farms
20. Phillips No. 1 Y, Houston Farms
21. Sun No. 1, Houston Farms

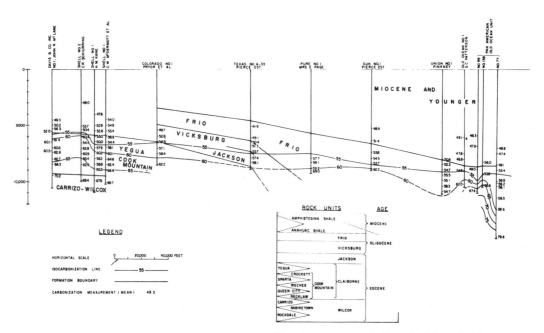

Fig. 17. Variations in the carbonization degree of psilate spores in Section 1, Texas Gulf Coast, U.S.A.

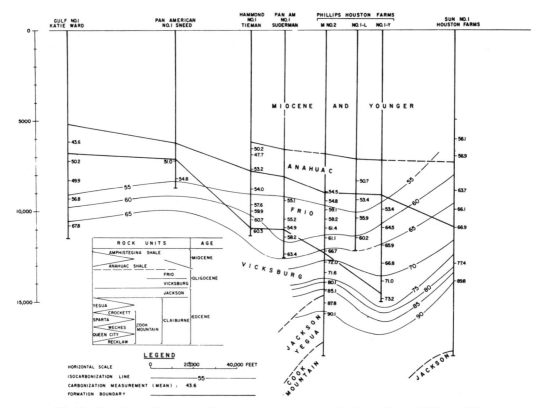

Fig. 18. Variations in the carbonization degree of psilate spores in Section 2, Texas Gulf Coast, U.S.A.

less towards the coast than the time lines. Therefore, rock units like Wilcox, Yegua, Vicksburg, Frio, etc., are less carbonized in their updip areas, and consequently are more carbonized in their downdip directions. If the carbon ratio theory is being applied in the Texas Gulf Coast, oil pools may disappear in the downdip direction of each formation.

Examples (figure 1) showing the relation of commercial oil fields and the fixed carbon percentage of coals, indicate the disappearance of big commercial oil pools on a world-wide basis in sediments containing coal with a fixed carbon of 65—70 and higher values. The 70 % fixed carbon can be correlated approximately with the 72×10^{-3} microampere light absorption [1] of a specific trilete psilate spore (figure 15).

Using above correlation, it is interesting to observe that the Frio Formation in Section 2 (figure 18b) enters the 65—70 fixed carbon values (estimated

---

[1] It must be emphasized that those light-absorption values were obtained with a particular microscope equipped for carbonization studies. If other microscopes are used, calibration is a necessity if results are being compared.

equivalent of $67—72 \times 10^{3}$ microampere light absorption value) in the area Phillips Houston Farms M No. 2 and Sun No. 1, Houston Farms. The Vicksburg Formation reaches similar values updip of the Phillips Houston Farms wells. The Jackson, Yegua, and Cook Mountains Formations in Section 2 (figure 18) are severely carbonized varying from 86×10^{-3} microamperes light absorption to totally opaque. The Miocene and younger, and Anahuac have not reached the values 67 to 72×10^{-3} microampere light absorption in this section (figure 18).

In Section 1 it appears that the Eocene rock units (Carrizo-Wilcox, Cook Mountain, and Yegua) are dipping into the 67 to 72×10^{-3} microamperes light-microamperes light-absorption values updip of well Pure, No. 1 Mrs. E. Page. The Oligocene units (Vicksburg and Frio) will encounter similar values somewhere between the wells Texas, No. A33, Pierce Est and Pure, No. 1, Mrs. E. Page and in the downdip direction of these wells (figure 17). The Miocene and younger strata in this section are lightly carbonized (less than 55×10^{-3} microampere).

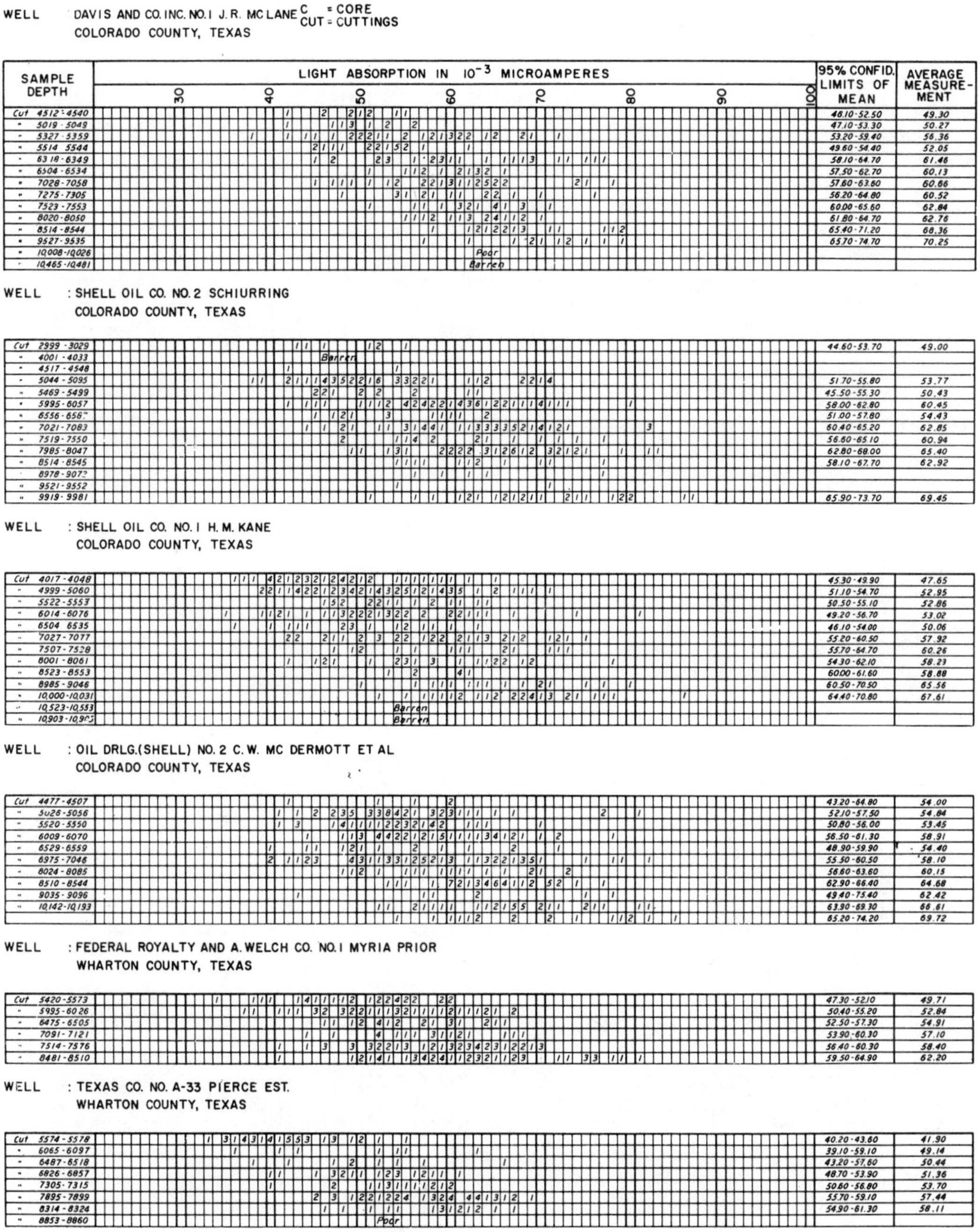

Fig. 19. Light-absorption data of psilate spores from wells in Section 1.

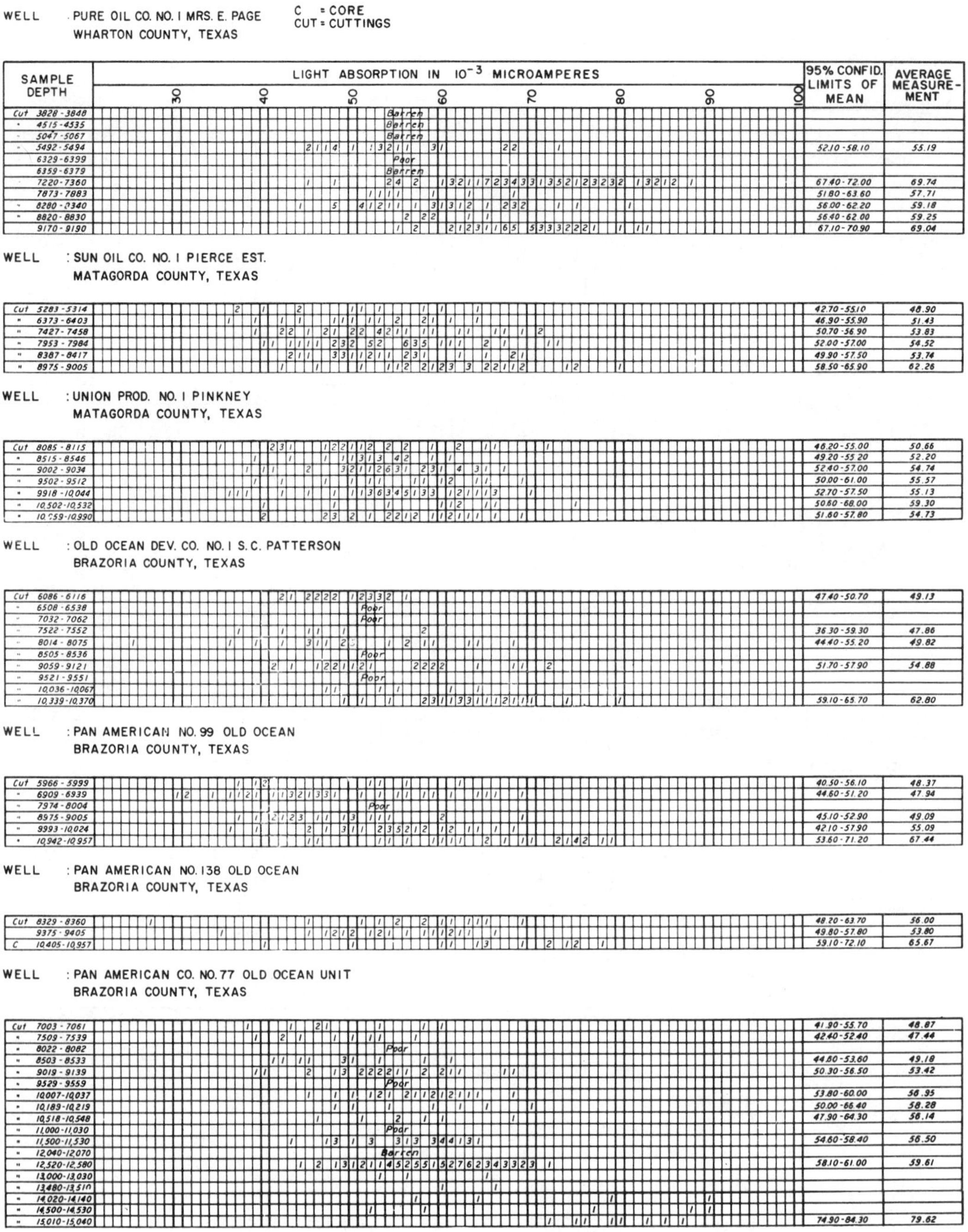

Examples of carbonization measurements

Fig. 20. Light-absorption data of psilate spores from wells in Section 1.

24 Gutjahr: Carbonization of pollen-grains and spores

WELL: GULF OIL CO. NO. 1 KATIE WARD
HARRIS COUNTY, TEXAS

C = CORE
CUT = CUTTINGS

Sample Depth	Light absorption in 10^{-3} microamperes	95% Confid. Limits of Mean	Average Measurement
Cut 6172-6202		39.62-47.62	43.62
" 7243-7280		47.46-53.06	50.26
" 8340-8370		47.13-52.73	49.93
" 9356-9385		53.79-59.99	56.89
" 10,841-10,904		64.25-71.25	67.85

WELL: PAN AMERICAN CO. NO. 4 SNEED
BRAZORIA, TEXAS

Sample Depth		95% Confid.	Average
Cut 5130-5190	Poor		
" 6153-6190	Poor		
" 7225-7255		48.04-54.04	51.04
" 8250-8280		51.46-58.26	54.86

WELL: HAMMOND NO. 1 TIEMAN
BRAZORIA COUNTY, TEXAS

Sample Depth		95% Confid.	Average
Cut 6479-6508		46.15-54.35	50.25
" 6907-6937		44.18-51.38	47.78
" 7740-7770		48.95-57.55	53.25
" 8258-8288	Poor		
" 8807-8837		52.08-55.98	54.03
" 9727-9751		55.62-59.72	57.67
" 10,060-10,090		57.68-62.28	59.98
" 10,551-10,583		58.05-63.45	60.75
" 11,015-11,046		58.07-62.67	60.37

WELL: PAN AMERICAN NO. 1 SUDERMAN
BRAZORIA COUNTY, TEXAS

Sample Depth		95% Confid.	Average
Cut 8990-9020			
" 9470-9500		52.05-58.15	55.10
" 10,490-10,530		52.37-58.17	55.27
" 10,970-11,000		49.41-60.41	54.91
" 11,480-11,510		55.59-60.99	58.29
" 12,440-12,470		61.37-65.57	63.47

WELL: PHILLIPS M-NO. 2 HOUSTON
BRAZORIA COUNTY, TEXAS

Sample Depth		95% Confid.	Average
Cut 8979-9040		52.08-57.08	54.58
" 9474-9536		50.19-59.59	54.89
" 9990-10,040		56.83-61.43	59.13
" 10,524-10,554		54.95-61.55	58.25
" 10,982-11,073		59.54-63.44	61.49
" 11,560-11,620		57.11-65.11	61.11
" 12,260-12,320		62.89-70.69	66.79
" 12,990-13,080		70.33-73.83	72.08
" 13,480-13,540		69.64-73.64	71.64
" 13,990-14,050		76.99-83.39	80.19
" 14,500-14,560		83.17-87.07	85.12
" 15,180-15,240		84.80-90.80	87.80
" 15,780-15,810		87.93-92.93	90.18
" 17,020-17,050	Barren		
" 17,940-17,970	Barren		

WELL: PHILLIPS NO. 1-L HOUSTON FARMS
BRAZORIA COUNTY, TEXAS

Sample Depth		95% Confid.	Average
Cut 8387-8417		42.05-59.45	50.75
" 9460-9490		49.68-57.18	53.43
" 10,472-10,502		53.26-58.66	55.96
" 11,560-11,589		58.15-62.25	60.20

WELL: PHILLIPS NO. 1-Y HOUSTON FARMS
BRAZORIA COUNTY, TEXAS

Sample Depth		95% Confid.	Average
Cut 9970-10,000		50.31-56.51	53.41
" 10,980-11,010		60.72-68.32	64.52
" 11,980-12,010		60.92-70.92	65.92
" 12,970-13,000		63.90-69.70	66.80
" 13,990-14,020		67.79-74.39	71.09
" 14,950-14,980		70.60-75.80	73.20

WELL: SUN OIL CO. NO. 1-HOUSTON FARMS
BRAZORIA COUNTY, TEXAS

Sample Depth		95% Confid.	Average
Cut 6062-6093		50.93-61.43	56.18
" 7021-7052		49.68-64.18	56.93
" 8960-8992		58.20-69.27	63.70
" 10,025-10,057		63.54-68.74	66.14
" 11,028-11,059		60.17-73.67	66.92
" 13,024-13,054		70.91-83.91	77.41
" 14,007-14,039		87.05-92.55	89.80
" 15,013-15,043	Barren		
" 16,015-16,044	Barren		
" 17,024-17,053	Barren		
" 17,960-17,990	Barren		

Fig. 21. Light-absorption data of psilate spores from wells in Section 2.

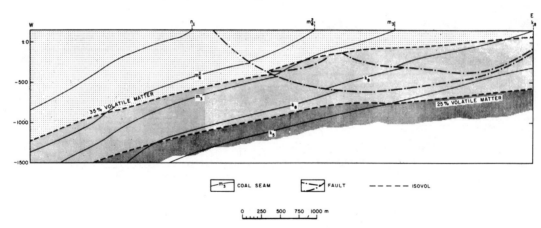

Fig. 22. Variations in degree of metamorphism of coals in the central region of the Donets basin (west of Main anticline,) Russia. (Levenshteyn 1962).

Whether 67 to 72 × 10^{-3} microampere light-absorption values can be used in the Texas Gulf Coast to indicate a transition zone between big commercial oil accumulations and small non-commercial oil accumulations or no oil at all (but condensates and gas only or nothing a all) will be tested by deeper drilling in future, which will provide the additional data necessary to confirm or disprove the abovementioned values.

As more data become available in the future, it may be necessary either to raise or lower the values indicated above. The data so far used are based upon the relation between fixed carbon of coals (and their equivalents expressed in the light absorption of psilate, trilete spores) and the occurrence of commercial and noncommercial or absence of oil pools on a worldwide basis.

CHAPTER VII

APPLICATIONS

1. CORRELATION BETWEEN CARBONIZATION AND OTHER SEDIMENT PROPERTIES

In addition to the correlation between the carbon ratios of coals and the carbonization of pollen-grains, spores, and other plant material, the latter might exhibit a number of other interesting correlations with sediment properties.

A correlation between porosity and carbonization has been noticed by several authors. Weber (in Koslow, 1958) reports a relationship between sandstone porosities and carbon ratios of associated coals in the northern part of the Donets basin of Russia (figure 23). However, no correlation between sandstone porosity and volatile matter of coals in the Ruhr area (West Germany) has been reported by R. Teichmüller (1962).

Russel (1951) reports a correlation between carbon ratio and porosity in sandstone which shows that for loose, porous and permeable sandstone, carbon ratios of associated coals generally fall below 60. For values between 60 and 80, the sandstones are usually very hard and show only local porosity. Very hard quartzitic sandstones and quartzites occur with coals having carbon ratios greater than 80. It should also be recognized, however, that low-porosity sediments can be found in rocks with a low-carbonization value caused by the clay content of the sands, early cementtation, etc.

Peldjakow (in Koslow, 1958) has reported average specific gravities of 2.17-2.35 for rocks occurring in sequences with brown coal and medium volatile bituminous coal in the Donets and Kuznetsk basins and in the Ural region. Characteristically, the average specific gravities of the rocks in the abovementioned areas, intercalated with higher rank low-volatile bituminous coals range between 2.35 and 2.60. The average specific gravities of sandstones associated with high-rank semi-anthracites and anthracites are 2.60—2.75 in the Donetx and Kuznetsk basins and in the Ural region.

Koslow (1958) reports that scismic velocities of dediments occurring with brown coal and lower rank

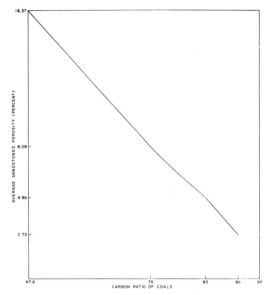

Fig. 23. Relationship between sandstone porosities and carbon ratios of coals in the northern part of the Donets basin, Russia. (Koslow 1958).

semi-anthracites vary from 2600 to 3550 meters per second. Dediments containing high-rank semi-anthracites to anthracites show seismic velocities of 4000 to 5000 meters per second. However, high-pressure shale zones may show moderately to strongly carbonized pollen-grains and spores but have lower seismic velocities and better sandstone porosities than expected for the present depth of burial. We emphasize that low porosities of sandstones, high specific gravities, high seismic velocities of the sediments, etc., can also be due to other factors which do not cause carbonization of pollen-grains and spores. Therefore, sediments with these properties do not invariably have highly carbonized pollen-grains and spores. Furthermore, intensely carbonized pollen-grains and spores may occur in sediments with secondary porosity.

2. PALYNOLOGICAL APPLICATION

A reasonable to optimum preservation of pollen-grains and spores is a necessity for palynological correlations and ecological studies. Strongly to intensely carbonized pollen-grains and spores are opaque and show no structure. They are, therefore, difficult to recognize and palynological correlations and ecological investigations in areas in which they occur are difficult and often impossible. Because severe carbonization is a limiting factor in carrying out palynological correlation or ecological studies, it is obvious that the delineation of severely carbonized areas is important to the palynologist and to the exploration effort in general. In basins in which no palynological studies have been carried out, the rank of the coals might help the palynologist to select favorable areas for palynological standard sections. Samples collected for the purpose of palynological time-correlation or ecological studies ought not to be taken from sediments which level of organic metamorphism is equivalent to semi-anthracite, anthracite or meta-anthracite. Samples taken from sediments which level of organic metamorphism is equivalent to medium-volatile coal and low-volatile coal may contain brown to dark brown spores or pollen-grains. None to slightly carbonized pollen-grains and spores can be expected in sediments with levels of organic metamorphism equivalent to the range lignite to high-volatile A bituminous coal.

3. OIL INDUSTRY APPLICATION

This study suggests that the degree of carbonization of pollen-grains and spores in sedimentary rocks devoid of coals can be used instead of fixed carbon values to determine areas of equal organic metamorphic intensity and to provide a clue to the hydrocarbon potential of these areas. If important commercial oil accumulations are found in slightly to not metamorphosed sediments, and if other severely metamorphosed sediments of essentially the same character are devoid of important oil accumulations, metamorphism may have converted the sediments from a productive to an unproductive condition (Thom, 1934).

Pollen-grains and spores can be used as recorders of incipient rock metamorphism and, as such, can delineate intervals and areas within which certain metamorphic intensities have been reached. Although sharp and definite boundaries to economical oil accumulations do not exist, it is probable that significant economical oil occurrences will decrease progressively as intensity of organic metamorphic alteration increases (Thom, 1934).

BIBLIOGRAPHY

American Geological Institute, 1960. Glossary of Geology and Related Sciences, Second edition 1960, The American Geological Institute.

Ammosov, I. I., 1961. Alteration Stages of Sedimentary Rocks and Paragenetic Relations of Mineral Fuels, Sovetskaya Geol., No. 4. Translation by: Associated Technical Services, Inc., Post Office Box, 271, East Orange, New Jersey, U.S.A.

Baier, C. R., 1938. Chemisch-mikrobiologische Beobachtungen zur Sedimentdiagenese an postglazialen Sedimenten der Eckernförder Bucht (Ostsee), Geol. Rundschau 29, pp. 316—321, Stuttgart 1938.

Baher, C. L., 1928. The Panuco Oil Field, Mexico, Bull. Amer. Assoc. Petroleum Geol., v. 12, no. 4, 395—441.

Barton, D. C., 1934. Natural History of the Gulf Coast Crude Oil, Problems of Petroleum Geology, Amer. Assoc. Petroleum Geol.

Bergius, G., 1913. Die Anwendung hoher Drucke bei chemischen Vorgängen und eine Nachbildung der Entstehungsprozesses der Steinkohle, Halle, W. Knapp.

Bode, H., 1928. Die Mazeration methode in der Kohlenpetrographie, Bergtechnik 21, p. 205.

Bogolyubova, L. I., 1956. "Determining the Intensity of Metamorphism and the Rank of Clarain Coals by the Petrographic Method of Thin Polished Sections", Izvest. Akad. Nauk SSSR Ser. Geol. 1956, no. 7.

Böttcher H. and Teichmüller, M. and R., 1949. Zur Hiltschen Regel in der Bochumer Mulde des Ruhrkarbons, Glückauf, 85, 81—92, Essen 1949.

Campbell, M. R., 1897. Tazewell, Virginia-West Virginia, U.S. Geol. Survey, Geol. Atlas Folio 44, 77 pp. and maps.

Campbell, M. R., 1930. Coal as a Recorder of Incipient Rock Metamorphism, Economic Geology, vol. 25, pp. 675—696.

Croneis, C., 1927. Oil and Gas Possibilities in the Arkansas Ozarks, Bull. Amer. Assoc. Petroleum Geol., v. 11, no. 3, 279—297.

Dobbin, C. E., 1929. Carbon Ratios and Oil Gravities in the Rocky Mountain Region of the United States, Bull. Amer. Assoc. Petroleum Geol., v. 13, no. 10, 1247—1255.

Dorsey, G. E., 1927. The Present Status of the Carbon Ratio Theory, Bull. Amer. Assoc. Petroleum Geol., v. 11, no. 5, 455—465.

Dulhunty, J. A., 1950. Some Effects of Compression on the Physical Properties of Low Rank Coal, J. and Proc. Royal Soc. New South Wales, v. 82, 265—271, Sydney 1950.

—, Hinder, H., and Penrose, R., 1951. Rank Variation in Central Eastern Coalfield of New South Wales, J. and Proc. Royal Soc. New South Wales, v. 84, 99—106, 1951.

Eby, J. B., 1923. The Possibilities of Oil and Gas in South West Virginia as Inferred from Isocarbs, Bull. Amer. Assoc. Petroleum Geol., v. 7, no. 4, 421—246.

Emmons, W. H., 1931. Deformation on Petroliferous Strata, Chapter 10 of Geology of Petroleum, pp. 152—161, second edition McGraw-Hill Book Company Inc., New York, London.

Ergol'skaya, Z. V. 1939. Metamorphism of Coals in the Process of Coalification, Proc. XVII Session of Internat. Geol. Congr., vol. 1, 1939.

— 1947. "Procedures of Petrographic Study of Coal Beds, in Determining the Quality of Coal," Materialy po geol. Zapadn. Sibiri, no. 59. 1947.

Fuchs, W., 1941. Thermodynamics and Coal Formation, American Institute of Mining and Metallurgical Engineers, Techical Publication no. 1333, 1941.

— 1953. Wesentliche Variable in der Systematik und in der Entstehung der Kohlen, Brennstoff-Chemie 34, pp. 161—167.

Fuller, M. L., 1919. Relation of Oil to Carbon Ratios of Pennsylvanian Coals in North Texas, Econ. Geol., v. 14 536—542.

— 1920. Carbon Ratios of Carboniferous Coals of Oklahoma and Their Relation to Petroleum, Econ. Geol., v. 15, 225—235.

Gropp, W. and H. Bode, 1932. Über die Metamorphose der Kohlen und das Problem der Künstlichen Inkohlung, Braunkohle, vol. 31, no. 16, p. 277—284, no. 17, p. 299—302, no. 18, p. 309—313.

Heck, E. T., 1942. Reginoal Metamorphism of Coal in Southeastern Virginia, Bull. Amer. Assoc. Petroleum Geol., v. 27, no. 9, pp. 1184—1227.

Hoffmann, E., 1936. Untersuchungen über Gasbildung und Gas Führung der Steinkohlen des Ruhrgebietes und der Abhängigkeit von Inkohlungsgrad, petrographischer Gefügezusammensetzung und der Einwirkung hoher Drucke, Beiheft zu Angewandte Chemie und die Chemische Fabrik, Nr. 24, pp. 1—12, Berlin 1936.

Huck, G. and J. Karweil, 1953. Physikalisch-chemische Probleme der inkohlung Vortrag 6, Hauptversamml. Dtsch. Ges. Mineralölwiss. Kohlechemie, Goslar 25, September 1953.

—, 1962. Probleme und Ergebnisse der Künstlichen Inkohlung im Bereich der Steinkohlen, Fortschr. Geol. Rheinld. u. Westf. 3, pp. 717—724, February 1962.

— —, and Patteisky, K. 1962. Der Einflusz des Druckes auf die Inkohlung, Brennstoff-Chemie, 43, Essen 1962.

Hume, G. S. 1927. Carbon Ratios of Coal as an Index of Oil and Gas Prospects in Western Canada, Canada Inst. Min. Met. Engr. Bull., pp. 325—345.

International Committee for Coal Petrology, 1963. International Handbook of Coal Petrography. Part I: Nomenclature, Part II: Methods of Analysis. Centre National de la Recherche Scientifique, Paris, Second edition 1963.

Jones, I. W., 1928. Carbon Ratios as an Index of Oil and Gas in Western Canada, Econ. Geol., v. 23, 353—380.

Karweil, J., 1955. Die Metamorphose der Kohlen vom Standpunkt der physikalische Chemie, Deutsche Geol. Gesellschaft, Zeitschrift, v. 107 (1955), p. 132—139.

Kirchheimer, F., 1934. Fossile Sporen und Pollenkörner als Thermometer der Inkohlung, Brennstoff-Chemie, Zeitschrift für Chemie und Chemische Technologie der Brennstoffe und ihre Nebenprodukte B-C, Bd 15, Heft 2, pp. 21—40.

Koslow, W. P., 1958. Die Bedeutung der Regionalmetamorphose der Sedimentgesteine für die Prospektion des Erdols und Gases, Translation from Zeitschrift für Geologie der Sowjet Union, 1958, No. 5, pp. 46—63.

Kreulen, J. W., 1935. Grundzüge der Chemie und Systematik der Kohlen, Amsterdam 1935.

Krevelen, D. W. van, 1953. Physikalische Eigenschaften und Chemische Struktur der Steinkohle, Brennstoff Chemie 34, pp. 167—182.

— and Schuyer, J., 1957. Coal Science, Elsevier Publ. Co., Amsterdam, Holland.

Kuyl, O. S., and Patijn, R. J. H., 1961. Coalification in Relation to Depth of Burial and Geothermic Gradient, Compte Rendu 4ième Congrès de Strat. et Géol. du Carbonifère, (1958), Tome II, pp. 357—365.

Lahee, 1934, in Thom, W. T., 1934. Present Status of Carbon Ratio Theory, Problems of Petroleum Geology, Amer. Assoc. Petroleum Geol.

Iversen, J., and Troel-Smith, J., 1950. Pollenmorfologiske definitioner og typer. Danmarks Geologiske Undersøgelse IV Raekke. Bd 3 Nr. 8; København.

Lesley, J. P., 1879. in McCreath, A. S., 1879. Second Report of Progress, Laboratory of the Survey, 2nd Geol. Survey of Pennsylvania, pp. 155—157.

Levenshteyn, M. L., 1962. Zakonomernosti Metamorfizma Ugley Donetskogo Basseyna, Sovet. Geol., 1962, no. 2, pp. 61—79.

Lewis, E., 1936. The Formation of Coal, with Particular Reference to its Behavior under Heat and Pressure, paper for the Institute of Fuel, March 11, 1936, at Burlington House, Piccadilly, extract in: The Iron and Coal Trades Review, March 13, 1936, p. 501 and 503.

Lilley, E. R., 1924. Coal as an Aid in Oil Exploration Engineering and Mining, Journal Press, v. 117, no. 25, pp. 1009—1012.

McCoy, 1921. A Short Sketch of the Paleogeography and Historical Geology of the Mid-Continent Oil District and its Importance to Petroleum Geology, Amer. Assoc. Petroleum Geol. v. 5, 541—584.

McFarlan, 1926. Carbon Ratio Map of the Eastern Kentucky Coal Field, Kentucky Geol. Survey, Ser. 6 (1926).

Molchanov, I. I., and A. V. Tyzhnov, 1961. Early Metamorphism of Sedimentary Rocks and its Value in Prospecting for Mineral Fuel, Razvedka i Okhrana, Nedr 27, No. 5, 8—15 (1961); Translation by Assoc. Technical Serv., Inc., East Orange, New Jersey.

Moulton, G. F., 1925. Carbon Ratios and Petroleum in Illinois, Geol. Survey Illinois Report Investigations 4.

Mukherjee, A. N., 1935. Ein Beitrag zur Kenntnis der pliozänen Braunkohle des Tandjoeng Kohlenfeldes Palembang, Süd-Sumatra. Dissertation, 52 pages, Würzburg 1935.

Muller, A., and Schwartz, W., 1953. Geomikrobiologische Untersuchungen I, Die mikrobiologischen Verhältnisse in einer spät- und postglazialen Sedimentfolge. Geol., Jb. 67, pp. 195—208, Hannover 1953.

Patteisky, K., and Teichmüller, M., 1960. Inkohlungs-Verlauf, Inkohlungs-Maszstäbe und Klassifikation der Kohlen auf Grund von Vitrit-Analysen, Brennstoff-Chemie No. 3, Bd 41, p. 79—84; No. 4, Bd 41, p. 97—104 No. 5 Bd 41, p. 133—137.

Patijn, R. J. H., 1963. De vorming van aardgas tengevolge van na-inkoling in het Noordoosten van Nederland, Geologie en Mijnbouw, vol. 42, pp. 349—358.

— 1964. Die Entstehung von Erdgas infolge der Nachinkohlung in Nosten der Niederlande. Erdöl und Kohle-Erdgas Petrochemie No. 1, pp. 2—9.

Petrascheck, W., 1953. Die Regel von Hilt, Brennstoff-Chemie 34, pp. 194—196, Essen 1953.

Potonié, R., 1930. Zur Diagenese und Metamorphose der Kohlen. Arbeiten aus dem Institut für Palaeobotanik und Petrographie der Brennsteine, Band 2, Heft 1.

Power S., and F. G. Clapp, 1932. Nature and Origin of Occurrences of Oil, Gas and Bitumen in Igneous and Metamorphic Rocks, Bull. Amer. Assoc. Petroleum Geol., v. 16, No. 8, 718—726.

Raistrick, A., and C. E. Marshall, 1939. The Nature and Origin of Coal and Coal Seams, The English Universities Press Ltd., London.

Reeves, F., 1928. The Carbon Ratio Theory in the Light of Hilt's Law, Bull. Amer. Assoc. Petroleum Geol., v. 12, no. 8, 759—823.

Reger, D. B., 1921. Carbon Ratios of Coals in Western Virginia Oil Fields, Trans. Amer. Inst. Mining and Met. Engr., vol. 65, pp. 522—527.

Rogers, H. D., 1858. Geology of Pennsylvania, J. B. Lippincott and Company, Philadelphia.

— 1860. On the Distribution and Probable Origin of the Petroleum, or Rock Oil of Western Pennsylvania, New York, and Ohio, Proceedings of the Philosophical Society of Glasgow, May 2, 1860.

— 1863. Coal and Petroleum in Harper's New Monhtly Magazine, v. 27, 259—264.

— 1865. On Petroleum, Proc. Phil. Soc. Glasgow, v. 6, 48—60.

Rogers, J., R. P. Suggate, J. O. Elphick and J. B. Ross, 1962. Metamorphism of a Lignite, Nature, Sept. 15, 1962, vol. 195, pp. 1078—1080.

Russel, W. L., 1925. The Relation between the Isocarbs and Oil and Gas Production in Kentucky, Econ. Geol., v. 20, 253.

— 1926. Porosity and Crushing Strength as Indices of Regional Alteration, Bull. Amer. Assoc. of Petroleum Geol., vol. 10, no. 10, pp. 939—952.

— 1927. The Proofs of the Carbon Ratio Theory, Bull. Amer. Assoc. Petroleum Geol., v. 11, no. 9, 877—889.

— 1951. Principles of Petroleum Geology, McGraw-Hill Book Co., Inc., New York.

Schopf, J. M., 1948. Variable Coalification: The Processes Involved in Coal Formation, Econ. Geol., v. 43, no. 3.

Schopf, J. M., 1962. Practical problems and principles in study of plant microfossils. In: Palynology in Oil Exploration. Soc. Econ. Pol. and Min.

Schultze, R., 1948. Die Diagenese von Kohle und Erdöl als thermodynamisches Problem, Vortrag vor dem Ortsverband Harz der Ges. Deutsch. Chemiker, March 5, 1948.

Semmes, D. R., 1921. Oil Possibilities in Northern Alabama, Trans. Amer. Inst. Mining and Met. Engr., v. 65, 176—198.

Seyler, C. A., 1948. The Past and Future of Coal — The Contribution of Petrology, Proc. South Wales Inst., Engin 63, no. 3, pp. 213—243, Cardiff 1948.

Skok, V. I., 1954. "The Intervals of Deepseated Metamorphism of Coals," Izvest. Akad. Nauk, SSSR, Ser. Geol., No. 6, pp. 85—93. English translation by: Assoc. Technical Services, Inc., Post Office Box 271, East Orange, New Jersey, U.S.A.

Stach, H., 1949. Experimentelle Beiträge zur Frage der Brikettierbarkeit von Weich- und Hartkohlen und der Quellung und des Zerfalls von Braun-Kohlenbriketts, Braunkohle, Wärme und Energie 1, pp. 35—44.

Stach, E., 1950. Vulkanische Aschenregen über dem Steinkohlenmoor. Glückauf 86, 41—50, Essen 1950.

Stach, H. and Teichmüller, M., 1953. Zur Chemie und Petrographie der Ionen Austauscher aus Braun-und Steinkohlen, Brennstoff Chemie 34.

Stadnichenko, T., 1934. Progressive Regional Metamorphism of the Lower Kittanning Coal Bed of Western Pennsylvania, Econ. Geol., v. 29, no. 6, pp. 511—543.

Suggate, R. P., 1959. New Zealand Coals, their Geological Setting and its Influence on their Properties, Bull. 134, N.Z. Geol. Survey.

Storm, W., 1924. Carbon Ratios of Cretaceous Coals in New Mexico in their Possible Relation to Oil, Bull. Amer. Assoc. Petroleum Geol., v. 8, pp. 519—524.

Stuart, M., 1927. Carbon Ratios, Trans. Inst. Petrol. Tech., v. 13, pp. 308—310.

Tarr, R. S., 1925. Oil may exist in Southeast Oklahoma, Oil and Gas Journal, v. 24, no. 30, p. 51.

Teichmüller, M. and R., 1950. Das Inkohlungsbild des Niedersächsischen Wealdenbeckens, Zeitschr. Deutsch. Geol. Ges. 100, pp. 498—519, Stuttgart 1950.

— 1954. Die stoffliche und struckturelle metamorphose der Kohlen, Geol. Rundschau, Band 42, 265—296.

Teichmüller, M. and R., 1958. Inkohlungsuntersuchungen und Ihre Nutzanwendung, Geologie en Mijnbouw nr. 2, Nw. Serie 20e Jaargang, pp. 41—66, 2 pl.

Teichmüller, M., 1958. Metamorphisme du charbon et prospection du Pétrole, Rev. Industr. Mineral, Fr., Juillet 1958, pp. 99—113.

Teichmüller, R., 1962. Zusammenfassende Bemerkungen über die Diagenese des Ruhrkarbons und ihre Ursachen., Fortschritte in der Geologie von Rheinland und Westfalen, Band 3, Teil 2, pp. 725—734, Krefeld, February 1962.

Terhune, R. W., 1964. Nonlinear Optics, International Science and Technology, August 1964.

Thom, W. T., 1934. Present Status of the Carbon-Ratio Theory, Problems of Petroleum Geology, Amer. Assoc. Petroleum Geol., pp. 69—95.

Thompson, A. B., 1926. The Significance of Surface Oil Indication, J. Inst. Petroleum Technol., v. 12, 603—622.

Trifonow, I. and G. Toschew, 1940. Änderung der Eigenschaften der Kohlen nach Verpressen unter sehr hohem Druck, Brennstoff-Chem, 21, p. 85, Essen 1940.

Val'ts, I. F., 1952. "Work of the Coal-Petrography Laboratory at the Kuznets-Geologiya Trust, in Servicing Coal Exploration Parties," in Tezisy Dokladov Soveshchaniya Rabotnikov Mineralog. i Petrograf. Lab. Ministerstva Geol. SSSR. Gosgeolizdat (State Publishing House of Geological Literature) 1952.

Vyshermirskii, V. S., 1958. Use of the Method of the Carbon Coefficient in Forecasting Oil-Bearing Capability Petroleum Geol,. v. 2, no. 6-b; an English Translation of the Russian from Geologii Nefti.

Waksman S. A., and K. R. Stevens, 1930. Contribution to the Chemical Composition of Peat: The Role of Micro-Organisms in Peat Formation and Decomposition, Soil Science, v. 28, 315—340.

White, D., 1915. Some Relations in Origin between Coal and Petroleum, J. Washington Acad. Sci. v. 5, 189—212

— 1920. Genetic Problems Affecting Search for New Oil Regions, Trans. Amer. Inst. Mining and Met. Engr. v. 65, 176—198.

— 1925. Progressive Regional Carbonization of Coals, Trans. Amer. Inst. Mining and Met. Engr., v. 71, 282—288.

White, D., 1935. Metamorphism of Organic Sediments and Derived Oils, Bull. Amer. Assoc. Petroleum Geol., v. 19, no. 5, pp. 589—617.

Wilson, J. H., 1926. Lithologic character of Shale as an Index of Metamorphism, Bull. Amer. Assoc. Petroleum Geol., v. 10, no. 6, 625—633.

Wilson, L. R., 1961. Palynological Fossil Response to Low-Grade Metamorphism in the Arkoma Basin, Proc. Seventh Biennial Geol. Symposium, University of Oklahoma, Norman, Oklahoma, March 7—8, 1961.

Woolnough, W. C., 1928. Carbon Ratios as an Index of Oil and Gas, Econ. Geol., v. 23, 809—810.

Zetsche, F., 1932. Untersuchungen uber die Membran der Sporen und Pollen, X: Die Inkohlungstemperatur der Steinkohlen, Helv. Chem. Acta. 1932, p. 675.

Copyright © 1967 by Geologische en Mineralogisch Instituut, Nederland

Reprinted from pp. 263–268, 276–293, 300–312, and 314–317 of
Leidse Geol. Meded. **40**:263-317 (1967)

AUTOFLUORESCENCE OF FOSSIL POLLEN AND SPORES WITH SPECIAL REFERENCE TO AGE DETERMINATION AND COALIFICATION

BY

PIETER VAN GIJZEL

[*Editors' Note:* Photoplate 1, consisting of figures 1 through 9; Plates I through V; Tables 2, 5, 6, and 7; and Figure 19 are not reproduced here.]

ABSTRACT

In the present study the primary fluorescence phenomena of fossil pollen and spores are described. This new method in palynology is based on a large number of fluorescence microscopical observations and spectrophotometrical determinations of palynomorphs from deposits of various type and age. It resulted in three principles: the relationship between fluorescence colour to type or form of pollen and spores (Plate I and figs. 21—22), the change in their fluorescence colour from blue or green to red or brown with increasing geological age (Plate II, III and fig. 24) and a similar colour change with increasing rank of coal of the embedding deposits (fig. 33). These phenomena appear to be in accordance with other fossilization and coalification studies of fossil palynomorphs by various authors.

For the preparation of pollen samples and the microscopical determination of fluorescence colours some special techniques have been adapted or developed.

The discoveries of fluorescence palynology can be applied to various questions, as, for instance, the study of pollen morphology and corrosion susceptibility and the age determination of those deposits, for which conventional pollen analysis fails. Such datings of Cenozoic rocks can be carried out with an accuracy of more than 80 %. As an example a number of age determinations of contaminated sediments is given (Plate V). Besides, fluorescence palynology may be used to determine the rank of coal of palynomorphs in coalified rocks in that part of the coalification series, ranging up to a fixed carbon content of about 70 %.

The explanation of the fluorescence phenomena described, meets still great difficulties, due to the inadequate knowledge of the chemical nature of the walls of fossil pollen and spores. Once again it is proved by this study that fossil palynomorphs are less resistant to fossilization and coalification than has been previously assumed.

CONTENTS

I. Introduction	264
Purpose of the investigation	264
Principles of fluorescence palynology	265
Previous investigations	266
The principle of fluorescence	267
Use of fluorescence microscopy in geology	268
II. Microscopical equipments for fluorescence analysis	268
The Berek fluorescence photometer	268
Principle and equipment	268
Procedure and reproducibility	271
Fluorescence colour and spectral ratio	271
Accuracy of the measurements	271
The UV-microspectrograph	272
Equipment and procedure of the measurements	272
Evaluation of the density spectra	274
Accuracy of the measurements	275
III. Preparation methods	276
Preparation procedures for fluorescence palynology	276
Pulverization of the samples	276
Further treatment	276
Preparation of the slides	277
Influence of the treatment on fluorescence	277
IV. Fluorescence of fresh and subfossil palynomorphs	278
Previous studies	278
Fluorescence colour determinations	278
Applications	281
V. Fluorescence and geological age	281
Introduction	281
Presentation and evaluation of data	282
Geological time-scale	282
Description of fluorescence colours at various ages	282
The second principle of fluorescence palynology	282
Change in fluorescence of vesiculate forms	282
Fluorescence of other forms	286
Fluorescence of acritarchs and dinoflagellates	286
Applications to age determinations	287
Fluorescence analyses for long-distance correlation	287
Other applications	287
Geochronology of the Pleistocene	289
Discussion of results	289
VI. Fluorescence and dating of contaminated sediments	290
The contamination problem	290

Examples of contaminated sediments in literature 291
Examples of fluorescence palynological dating Pleistocene clays in the N. Netherlands and N.W. Germany 293
  a) Previous studies 293
  b) Sources of contamination. 294
Procedure for measuring and evaluation of data 295
Fluorescence analyses. 297
  a) The Lauenburger Ton in N.W. Germany 297
  b) Banded clay of Glindow 297
  c) Peat and clay of Adendorf 298
  d) The "potklei" of the N. Netherlands . . 298
  e) Other Pleistocene clays in the N. Netherlands 298
  f) Clays in the central Netherlands . . . 298
  g) The clay of Terhaagen (N. Belgium) . . 299
Applications to other contamination studies . 299

VII. Fluorescence and geochemical coalification . 300
Introduction 300
Influence of geological time and geochemical coalification on the fluorescence of palynomorphs. 300
  Influence of geological time 300
  Influence of geochemical coalification . . 301
  Influence of corrosion 301
Relationship between fluorescence and rank of coal 303
  Relationship between fluorescence and light absorption 303
  Relationship between fluorescence and rank of coal 304
Conclusions 304

VIII. Fluorescence and chemical character of sporopollenine 305
Introduction 305

Previous chemical analyses 305
  Fresh pollen and spores 305
  Fossil megaspores 306
  Conclusions 306
  Other chemical data 307
Relationship of fluorescence to chemical character 307
Relationship of fluorescence to fossilization . . 308
  Corrosion 308
  Humification 308
  Biochemical coalification 308
Conclusions 309

IX. General conclusions 310

Acknowledgements 312

Résumé, Zusammenfassung 313

References 314

Appendix

Tables of samples nrs. VI, VII, VIII

Plate   I.   Fluorescence spectra of various Holocene forms

Plate   II.   Fluorescence spectra of vesiculate pollen forms of various ages

Plate   III.   Spectral ratio and geological age of vesiculate pollen

Plate   IV.   Fluorescence spectra of acritarchs and dinoflagellates

Plate   V.   Spectral ratio of *Pinus sylvestris* from contaminated sediments

I. INTRODUCTION

PURPOSE OF THE INVESTIGATION

The subject of this study concerns the phenomena of a little known property of fossil pollen and spores: viz. the fluorescence, which is closely related to the physical and chemical nature of their walls, and this can be applied to the study of some palynological problems. The themes developed in this paper are: what happened to these microfossils after their deposition, how does fluorescence change in geological time and how can these changes be applied in palynology?

After burial in the soil, exines and exospores were involved in the sequence of processes of humification, fossilization, biochemical and often of geochemical coalification. The great majority of remains of pollen and spores is very well preserved in deposits under anaerobic conditions and only those with a very weak exine will disappear.

Their resistance to these processes makes possible the investigation of vegetational history and stratigraphy. The pollen-containing deposits may be correlated with the stratigraphy of similar deposits elsewhere, in combination with other geological investigations. In general it may be assumed that the vegetation of a sedimentation area is reflected in the pollen content of a deposit, if the circumstances were favourable for their fossilization. The vegetational changes will then be reflected in the changing pollen composition by which zones in a pollen diagram can be distinguished. Palynological age determinations and correlations are based mainly on these zonations and on the climatological and stratigraphical evidence of the microscopical plant remains.

Unfortunately, it is not always so easy practice as one might think. Every palynologist knows from his own experience that in the diagrams the zonation may be absent or poorly developed, due to various factors.

In the first place, contamination with allochthonous or older pollen grains may occur during sedimentation, which results in an incorrect age determination. The dominance of Tertiary pollen associations in Pleistocene clays in the N. Netherlands is an extreme example of this sort of contamination. The clays then seem to be Tertiary in age owing to this outnumbering of the autochthonous components.

The facies of a sediment is an important factor. In the pollen rain the autochthonous pollen can predominate over that of the surrounding area. Changes in the latter will not be expressed then in the pollen content of the samples. This phenomenon appears repeatedly in lignites, in which the pollen from the dense marsh forests is predominant. Only in sediments deposited in open water, the pollen from vegetations occurring at higher altitudes is present in such quantities, that stratigraphical conclusions can be drawn.

Other factors to be taken in account are climatological differences. The vegetational changes during the Pleistocene of temperate areas, e.g. N.W. Europe, are typical and may be compared with other paleoclimatological data. The Pleistocene vegetational history of Spain (Menéndez Amor and Florschütz, 1962, 1964) for instance and that of Colombia (van der Hammen and Gonzalez, 1964) may also be compared with paleotemperature curves of sea water. Such a comparison still meets with difficulties (van Gijzel et al., 1967). C 14 datings can only be carried out on the uppermost part of the sequence of Pleistocene layers; consequently, a correlation with the Middle and Lower Pleistocene subdivisions of various regions is hardly possible (Chapter VI).

Moreover, pollen and spores may be selected by corrosion, as a result of the oxidation and biochemical activity in the soil during and after the sedimentation. This corrosion mainly occurs in coarse deposits such as sands and humic sandy soils under dry conditions. For this reason the older pollen zones may not be represented in the pollen diagrams of such deposits (Havinga, 1962; Chapter VIII).

Some other problems of age determination, in connection with pollen analyses, arose. For deposits which are younger than 50,000 years and without a distinct palynological character, radiocarbon datings can lead to correlation, but dating and correlation of older deposits still meet with great difficulties (Chapter V). In the present publication these questions are investigated from a new point of view, i.e. the phenomenon of fluorescence of pollen and spores and its changes during geological time.

PRINCIPLES OF FLUORESCENCE PALYNOLOGY

Some years ago, the present author tried to find a means to distinguish autochthonous and secondary pollen for dating contaminated sediments in the N. Netherlands. Fluorescence-microscopical techniques proved to be very useful in solving this problem. Fossil exines observed under UV-light were found to show more or less brilliant fluorescence colours.

In some preliminary notes the outlines of the phenomena were described and the application of fluorescence microscopy to palynology was summarized (van Gijzel, 1961, 1963, 1966). It was then stated that three principles are to be distinguished. (1) Fresh and subfossil pollen grains and spores show various fluorescence colours, dependent on type or species (Chapter IV), (2) differences in colour appear in each pollen type at different ages (Chapter V) and (3) a relation exists between the fluorescence and the rank of coal in coalified rocks (Chapter VII).

All these phenomena are closely related to the chemical character of the pollen and spore walls, which is still poorly understood (Chapter VIII).

It soon became apparent to the author that this technique could be applied extensively in palynology, on the condition, however, that the observed colours are registered objectively by means of a method, which is verified with a standard in order to make analyses reproduceable. A method was required therefore, which was sensitive and accurate enough to enable measurements of such small objects as fluorescing pollen grains (Chapter II).

Besides it appeared that the preparation techniques, used in the laboratory to separate the pollen and spores from other sedimentary constituents, are very important for fluorescence microscopy. The use of acids influences the fluorescence colours harmfully and is therefore to be avoided (Chapter III).

In this study, dealing with the most important aspects of fluorescence palynology, much attention is given to the description of the colours at various geological ages, from which the relations of fluorescence palynology are derived (Chapter IV and V). The applications of this phenomenon to the afore-mentioned palynological and chronostratigraphical problems are extensively investigated (Chapter V—VI), but the use of fluorescence in pollen morphological studies had to be left out of consideration. For a historical review of fluorescence palynology one should refer to van Gijzel (1967 a).

The localities of the rock samples, analysed for this study, are shown in the map of fig. 10. The material is partly registered in the collection of the State Museum for Geology and Mineralogy, Leiden (see Appendix).

In the course of the investigations it became clear that the causes of the observed phenomena could not be easily explained. The complex chemical nature of the sporopollenine, the main constituent of the fossil spore- and pollen-walls, is very difficult to examine and it was only possible to present some hypotheses as to the causes (Chapter VIII).

Nevertheless, it may be possible that this study will be a stimulus to the continued application of fluorescence techniques in palynology and other geological studies, for instance, to micropaleontology (van Gijzel, 1966). But most of all it may be a contribution to a better knowledge of the processes, determining the preservation of pollen and spores, in connection with palynological practice. Still, comparatively little is known about its theoretical basis. Although fluorescence is a fascinating subject, it is not only the more theoretical aspects but the practical applications that should be considered. In the first place fluorescence is geologically and palynologically important. In the present study, therefore, more emphasis is laid on the application of these phenomena than on the theory behind it.

PREVIOUS INVESTIGATIONS

Till now the fluorescence of pollen and spores had seldom been studied, although many investigations on the fluorescence of other plant substances have been published.
The first, who observed fresh pollen under UV-light, between fluorescence and the fossil pollen types, as investigated be the present author, is in clear accordance with a similar relationship among fresh pollen as found by Berger.
Some time later, Asbeck (1955) studied fresh pollen under UV-Light with other purposes in mind. It appeared to him that the exine screens the very sensible chromosomes in the pollen cell against the UV-light of the sun. He noticed that some transparent

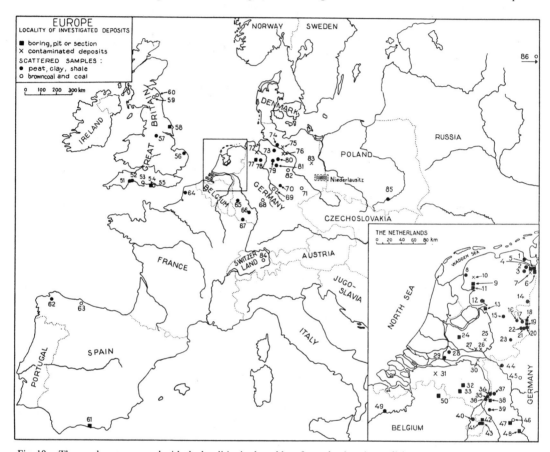

Fig. 10. The numbers correspond with the localities in the tables of samples (see Appendix).

was Berger (1934). He investigated various pollen grains in connection with hay-fever and stated that they show characteristic fluorescence colours, which are more or less constant and typical for every type. However, these analyses concerned living pollen only, which still contains protoplasm, the bright fluorescence of which is dominant over that of the exine. Consequently, the observed colours were of mixed origin and his descriptions cannot be applied to fossil material, as the intine disappears during fossilization and the exine only is fluorescent. The relationship pollen species, showing a white colour in daylight, are fluorescent in ultra-violet radiation and that those of other colours often do not fluoresce. However, neither Berger, nor Asbeck oberved the fluorescence of the pure exine.
Sitte (1960) noticed the differences in fluorescence of the fresh exines of some species and found that a bright blue colour of *Selaginella* spores is changed into yellow-brown by acetolysis. In accordance with this statement is the fact, that the use of acids in the preparation of pollen slides has a similar effect (Chapter III).

As the exine of fresh pollen is fluorescent, one would expect fossil pollen grains to possess the same property. Only a few investigators have observed this phenomenon in fossil remains.

The first study of fossil pollen and spores under the UV-microscope was made by Maier and Wetzel (1958). They discovered the luminescence of some microfossils such as Hystrichosphaeridae (acritarchs) and other organic remains, but found no indication of the fluorescence of pollen and spores. They later concluded that, in general, fossil animal remains show a clear luminescence, while plant substances only possess this property in the fresh state and loose their luminescence when they are fossilized (Wetzel, 1959; Maier, 1959).

As is apparent from the author's present and earlier investigations, there is no doubt that fossil pollen and spores are fluorescent in UV-light (van Gijzel, 1961, 1963), with the exception of those from highly coalified deposits. This phenomenon even occurs in similar deposits and of the same (Tertiary and Mesozoic) ages as those from which the samples of Wetzel and Maier originated. The absence of fluorescence in the objects, which they observed, was probably caused by the preparation of the samples in HF and by oxidation — these two being the greatest enemies of fluorescence — and partly by the use of an embedding medium, being self-fluorescent to such an extent that it prevented the observation of the weak fluorescence of the pollen. Furthermore, one cannot agree with the opinion of Wetzel and Maier that fossil plant and animal remains can be distinguished on the basis of luminescence. The present author has found that dinoflagellates, considered non-luminescent by Wetzel and Maier, show bright green or yellow fluorescence colours. Other fossil plant remains of various origins are fluorescent as well: for instance *Pediastrum*, Fungi, wood, the massulae of *Azolla* species, etc., and even those of pre-Quaternary age (van Gijzel, 1963). The luminescence of fossil Algae, moreover, has been found by Jacob (1961a, 1961b) in some clayey bituminous rocks and by Wolf (1966) in boghead coal. The fluorescence intensities of all these plant remains are in most cases no less than those of the bright green-coloured acritarchs, which should belong in the opinion of Wetzel and Maier to the animal kingdom (see Chapter V).

In the papers of some other investigators, the fluorescence of fossil pollen in lignite samples has been mentioned and illustrated, but without closer examination (see among others: Jacob, 1961a; Eder and Fritsche, 1963).

Shellhorn et al. (1964) were the first to use fluorochrome (acridine orange) in the search for fresh and fossil pollen in soils and deeper layers, but they restricted their paper mainly to a description of the methods used.

THE PRINCIPLE OF FLUORESCENCE

Fluorescence is the property of matter to emit light under the influence of an exciting light (visible or U.V.). The emitted light is, in general, of longer wave-length than the exciting light. If the fluorescence glow persists for an appreciable time after the stimulating rays have been cut off, this afterglow is termed phosphorescence. Both phenomena are termed luminescence. The use of the latter term for the fluorescence phenomena by many authors is incorrect and gives rise to confusion.

The present study concerns fluorescence only, phosphorescence being left out of consideration as it has not been observed in fossil pollen and spores.

The fluorescence emission spectrum of a specimen is the sum of fluorescence spectra of atoms and molecules.

Two kinds of visible fluorescence are distinguishable: *autofluorescence or primary fluorescence*, which is a property of the substance itself and is not activated by staining, and *secondary fluorescence*, exited by microchemical reactions, caused by staining liquids (known as flurochromes) such as eosin, fuchsin, fluorescein and many others, for which various elements, moreover, show a specific affinity.

Autofluorescence may show splendid colours and has appeared to be the most important for age determination. Investigations on the secondary fluorescence of fossil pollen and spores are in progress and will not be dealt with here.

The spectral distribution of fluorescent substances may range from ultra-violet into infra-red (see among others: Goodwin, 1953; Krieg, 1955). Of the absorbed UV-energy, most is dissipated as heat and a part is emitted in the visible region of the spectrum. The fluorescent light of fossil pollen and spores is also polychromatic. The absorbed UV-light is mainly derived from the Hg-band at 365.5 nm and, to a much lesser extent, from the faint bands at 313 and 334 nm (see Chapter II, fig. 13).

Fluorescence in UV-light has been applied on a large scale in chemistry and technical sciences for analysing certain minerals, organic and inorganic chemicals, metal-organic compounds, etc. Many substances emit a typical fluorescence, which provides an effective method for quantitative and qualitative analyses, even in extremely small concentrations.

Fluorescence is, moreover, an important property of substances in living organisms, such as in plants and the human body. The various colours of organs and tissues are usually produced by an assortment of compounds. Different cell structures in plants show characteristic fluorescence colours, depending on their chemical nature, e.g. cuticle whitish, epidermis blue, phloem bundles blue, xylem violet or blue-green, chloroplasts red, etc.

It is remarkable that fossil plant substances often show fluorescence. The majority of the tissues left in Young *Sphagnum* Peat are fluorescent, the cell-walls of *Sphagnum* leaves themselves showing faint blue or green colours (Photoplate I: fig. 2). In older deposits this fluorescence is hardly visible and only the palynomorphs and some other remains continue to fluoresce. Fluorescence is often the key to the isolation and identification of many biochemical constituents by means of special techniques (Goodwin, 1953). The phenomenon of fluorescence has been applied successfully in physiological studies. For the application of the techniques and microscopy of fluorescence, reference should be made to the following studies, dealing with various aspects of the application to investigations of organic matter: Haitinger (1938), Bukatsch (1914), Bräutigam and Grabner (1949), Goodwin (1953), Gottschewski (1954), Clark (1961) and many others.

USE OF FLUORESCENCE MICROSCOPY IN GEOLOGY

In the past the use of fluorescence in earth sciences has been restricted to certain fields only. Przibram (1962) used fluorescence for the identification and the study of the distribution of organic remains in natural inorganic material. It has been applied for instance, in geochemistry, to the fluorescence analysis of hydrocarbon combinations in connection with the migration of petroleum and it has been used in the study of the origin of oil by means of the fluorescence microscopy of bituminous rocks (Jacob, 1961b). Investigations of facies and diagenesis of organic deposits under the UV-microscope were made by Wetzel (1939, 1959, 1962), Maier (1959) and Haberlandt (1942). Overbeck (1964) investigated the fluorescence phenomena of plant fossils and molluscs, in particular those of skeletal substances which are found in animal fossils only.

Sacchi Vialli (1962, 1964) investigated the fluorescence of vertebrate teeth and some other macrofossils and this opened up a new field in the study of palaeobiochemistry.

In mineralogy, luminescence has proved to be a useful property for the detection and identification of minerals and gemstones (see among others: Gleason, 1960). They often show brilliant fluorescence colours: adamite, green; calcite, red, white and yellow; scheelite, bluish white; wollastonite, orange; uranium minerals, green; fluorite, blue etc.

It is, therefore, not surprising that microfossils with tests, consisting of minerals, such as Foraminifera and Bryozoa, appear to be fluorescent as well. Their colours may be analysed by means of the fluorescence photometrical methods, described in the next chapter (van Gijzel, 1963, 1966).

[*Editors' Note:* Section II has been omitted.]

III. PREPARATION METHODS

PREPARATION PROCEDURES FOR FLUORESCENCE PALYNOLOGY

Pulverization of the samples

It is necessary in many cases to pulverize slightly the sample in a mortar before further treatment. The pestle must be handled very carefully to avoid destruction of pollen. But for hardened rocks such as shale, browncoal, coal and limestone, this means of pulverization is insufficient. To disassociate the mineral grains from each other, various treatments can be applied:

The so-called *benzene method* delivers good results for fluorescence studies. Its procedure is as follows:

(1) The sample is placed in a drying apparatus to remove the water from the pores in the rock pieces,

(2) After cooling the sample, a sufficient quantity of liquid is introduced, which possesses a surface tension lower than that of water, as, for instance, xylene or benzene. The tumbler is shaken now and then until the light liquid has filled all pores of the rock.

(3) After the particles have settled down, the liquid is poured off and hot water is added to the sample instantly. Then the water molecules try to replace those of the light liquid from the rock pores, resulting in a strong pressure, which will break up the rock fragments into separate grains (see also Staplin et al., 1960).

(4) The sample is filtrated and dried. This treatment can be repeated until the sample is pulverized completely. The sample is ready now for further treatment.

Another method is *ultrasonic disintegration*. It is used on a large scale for our purpose, for which various equipments exist (Stevens et al., 1960; Dumait, 1962a, 1962b; Gibson, 1963; Streel, 1964). Exines and exosporia can resist without serious damage, ultrasonic vibrations with a frequency up to 20,000 cycles/sec for 10 min. Repeated ultrasonic treatment, however, must be carried out very carefully to avoid destruction of pollen.

Further treatment

We will pass now to a description of preparation of various deposits, such as coal, browncoal, hardrock, peat, clay and sand.

Browncoal and *coal* are treated previously by means of oxidation with 10 % warm H_2O_2 for 10 min. **Do not boil**! Coalified rocks must always be oxidized to break the pollen away from the rock particles. This treatment must be done carefully, in order to prevent oxidation of the pollen grains as much as possible. Therefore, during the treatment a regular control under the microscope is necessary. Oxidation with hydrogen peroxide is preferred rather than a treatment with acids as HNO_3.

After cooling of the sample the peroxide must be removed. Centrifuge and wash the sample twice with aqua destillata. Further treatment is similar to that used for peat and clay, and is used only in case sand or clay are present in the browncoal or coal.

Peat samples are treated in the following way:

(1) Boil the sample for 10 min with 10 % NaOH. Meanwhile some water is added to prevent higher alkaline concentrations.

(2) Sieve the sample through a coarse and a fine sieve with mashes of min. 200 μ. Transfer the filtrate simultaneously to a glass centrifuge tube.

(3) Centrifuge and wash twice with aqua destillata. For peat and browncoal without inorganic material this procedure will suffice.

Clay, sand, diatomite and other sediments with inorganic matter are initially treated with the first three steps of the peat treatment. The removal of minerals must be carried out as follows:

(1) The sample is washed with 96 % alcohol to remove the water. Centrifuge and repeat this washing. If some water is left in the sample when it is brought into a bromoform alcohol solution, the latter becomes an emulsion and will prevent the mixing of the bromoform with the sample.

(2) Add some bromoform, diluted in alcohol, with a s.g. of 2.0 until the tube is filled half. Stir the mass until no clods are left. This takes a long time and can be accelerated by means of ultrasonic vibration of the mixture or by means of a special glass tube shaker, by which the lumps will disintegrate.

(3) Centrifuge rather short (3 min) at low speed (1500 r/min). The minerals will settle down on the bottom of the tube, while the pollen is floating.

(4) Pour off the liquid with the pollen into another centrifuge tube and fill it with alcohol (96 %). Stir the liquid. The specific gravity of the mixture will be reduced so much that the botanical substances settle down on the bottom. Centrifuge at high speed (5000 r/min) and long (10 min). If necessary this separation may be repeated once or more.

(5) Wash several times with alcohol (96 %) and centrifuge to remove the bromoform. Water cannot be added instantly, because emulsion can be formed.

(6) Wash several times with distilled water and centrifuge to remove the alcohol.

(7) Turn the tube upside down to allow the water

to leak out from the sample. Water would cause a drying up of the permanent pollen slides afterwards. Add some drops of pure glycerine (water-free) and stir the mixture.

For the preparation of *hardrocks* after the pulverization of the sample the same procedure is followed.

Preparation of the slides

The slides must be prepared in the usual way with the utmost care. The object-glass and cover-slip are cleaned very carefully, at which no dust may left on the glass (dust is self-fluorescent!). The use of fluorescent media for conservation, as glycerine gelatine, silicone oil, canadabalsem or plastics causes an undesirable illumination of the picture of the UV-microscope. But this picture has to be as dark as possible for the observation and measurement of the fluorescence. Water-free, chemically pure glycerine, being non-fluorescent, has proved to be most suitable. **Do not stain the residue** with saffranine or other liquids, which may cause secondary fluorescence (Haitinger, 1938; Krieg, 1955).

The greatest care and accuracy is required in collecting, preparing and studying of the samples. The cleanliness in a palynological laboratory must be equal to that in a bacteriological one. This maxim, given by Faegri and Iversen (1950) holds the more for the study of fluorescence palynology.

INFLUENCE OF THE TREATMENT ON FLUORESCENCE

For the separation of pollen grains, spores and other plant remains from minerals and humic substances (humic acids) in sediments, numerous methods are used (see Faegri and Iversen, 1950; Brown, 1960; Staplin et al., 1960; and other papers). However, nearly all of them concern treatments with acids or strong oxidizing agents. Although their action on the exines and exosporia, have generally been considered as neglectible, it appeared to the present author that these procedures may have an important influence on the chemical properties of sporopollenine. Mechanical techniques are preferable for this separation. This is in accordance with the experience of Felix (1963), who stated that mechanical disintegration of shales without the use of acids contributes to an improved recovery of pollen, spores and other microfossils.

The use of the well-known acetolysis method of Erdtman (1954, 1960) and similar contact of pollen with acids as HF, HCl, H_2SO_4 and HNO_3 causes a change in fluorescence colour towards the red part of the spectrum. By such treatments the colour of, for instance, *Pinus sylvestris* grains, changes from green to yellow, orange or brown and the total intensity of the fluorescence decreases. Sitte (1960) already noticed that the use of acetolysis at the preparation of fresh *Selaginella* spores changes the fluorescence from blue to brown. As appears to the present author, similar changes occur at many other types of palynomorphs in fresh condition. Besides, it appeared that the activity of acids and other strong reagents works more or less selective; certain types are corroded more than other ones and a large variation in fluorescence occurs for grains of the same type.

Extensive experiments have been made on samples of Holocene clay and peat, which confirm this statement. In comparison with that part of the samples, treated with NaOH only, the boiling with HF for longer than three minutes resulted in a remarkable different pollen composition. Mainly *Pinus sylvestris*, *Tilia* and *Alnus*, being the most resistant, increased in percents. Therefore the author cannot agree with the opinion of Faegri and Iversen (1950) and other investigators, who supposed that organic remains as pollen are not, or only to a small degree, damaged by HF.

Oxidation of material is used at the maceration of browncoal and coal for the preparation of pollen slides (Potonié, 1931; and other papers). It causes a similar change in fluorescence colour, but to a lesser extent if the treatment has been stopped at the right moment, i.e. as the oxygen can attack the grains too strongly. This is in accordance with the oxidation experiments on palynomorphs from browncoal, carried out by Kirchheimer (1933a, 1933b, 1934).

The preparation of browncoal and coal with strong cleansing reagents as NaClO, transforms the fluorescence in an opposite direction; for instance fresh grains of *Alnus*, normally yellow or white fluorescent, are changed hereby towards green or blue. Other types show the same phenomenon. Although this reagent is very useful for the preparation of such rocks, it is unsuitable for fluorescence studies.

Washing the samples insufficiently with water after the oxidation procedure and heating afterwards with NaOH may result in similar blue colours. It is not clear what happens at the violent reaction, occurring when the sample, centrifuged only and still containing much hydrogen peroxide, is brought into NaOH. Possibly a natrification of the sporopollenine takes place.

Although the treatment with acids can be replaced completely by other methods, the oxidation method cannot be avoided in all cases in order to pulverize strongly indurated rocks for the fluorescence study. The influence of alkaline solvents like NaOH, KOH, bromoform and alcohol on the fluorescence of fossil palynomorphs can be neglected. To diminish or eliminate the possible undesirable influence of the solvents used, all treatments have been carried out as much as possible under standardized conditions: equal duration of preparation, concentration of reagents and sequence of treatments.

Very clean pollen slides are obtained by the use of bromoform instead of HF and HCl for the separation of the pollen and other plant remains from the mineral particles, according to the procedure, described by Urban (1961), Felix (1963) and other authors. Then heating with HCl can be avoided, because the lime is removed as well. This method is similar to that used in sediment-petrology before long and is based in principle on the differences in specific gravity between minerals and plant material. In a solution of bromoform in alcohol with a specific gravity of 2.0 or somewhat lower (min. 1.75), the plant substances remain floating. Nearly all the minerals will settle down at the bottom of the centrifuge tube. This treatment is very suitable for fluorescence studies and has been used, if necessary, for all samples.

The mentioned preparation methods have been previously tested by comparative fluorescence analyses of samples of various sediment types and ages. This test was made of the following samples: a young *Sphagnum* peat, a humic Holocene clay, a strongly humified Pleistocene peat, an Eocene and a Paleocene browncoal and a black humic Cretaceous clay. Each sample was split up into equal parts, which were prepared in different ways. They were treated respectively either only with distilled water, or NaOH, or HF and NaOH, or HCl, HF and NaOH, or NaOH, alcohol and bromoform, or H_2O_2 and NaOH, or NaClO and NaOH. It appeared that distilled water, alkaline solvents, alcohol and bromoform may only be used without drawbacks. Chemical treatment must be limited as much as possible. Oxidation, if inevitable, must be carried out very carefully and held under control by testing a drop of the sample from time to time under the microscope. After each treatment the sample must be washed repeatedly with water to remove the suspended and solved compounds, which may cause an adverse effect on the microscopical observations.

IV. FLUORESCENCE OF FRESH AND SUBFOSSIL PALYNOMORPHS

PREVIOUS STUDIES

Berger (1934), Asbeck (1955) and some other allergologists established that various types of fresh pollen grains and spores show different fluorescence colours, dependent on type or form. They are caused by the combined fluorescence of the various substances of the living pollen cell, in which that of the protoplasm is dominant. This is shown clearly, for instance, by fresh exines of *Pinus sylvestris* (Photoplate: fig. 1). The protoplasm, showing a white fluorescence colour, can be more or less removed by treatment of the grains with KOH. The fluorescence colour of the exine, light blue, then becomes visible. In view of the fact that during fossilization the exine appears to be resistant to destruction, it is to be expected that the pollen walls will remain fluorescent.

Observations on the fluorescence of subrecent palynomorphs have been made from thin *Sphagnum* layers, situated some centimeters below the surface of a living peat bog and other peat formations or from young, slightly humified, *Sphagnum* peats. As the protoplasm has disappeared, their fluorescence colours are considerably different from the fresh pollen grains described by Berger and Asbeck. *Pinus sylvestris* pollen, for instance, shows more bluish colours in fresh condition, as against the bluish green or dark green fluorescence of subrecent grains (see Pl. III: uppermost samples and fig. 1 and 4).

It appeared that subrecent palynomorphs show more or less bright fluorescence colours, which are different in the various forms. From these phenomena the first principle of fluorescence palynology has been previously derived (van Gijzel, 1961, 1963). It can be defined more exactly now as follows:

Similar to the differences in fluorescence between fresh pollen and spores the subfossil and fossil palynomorphs show a more or less typical fluorescence colour.

These phenomena will be described below by means of extensive analyses of the various forms.

The terms fossil and subfossil have been used in this study with regard to the state of preservation, being a concept, different from that generally used by geologists. Pollen grains are considered as subfossil or fossil when the protoplasm and other living substances have disappeared; when the pollen walls have been humified the pollen is considered as fossil. Recent and subrecent are terms, which concern geological time of less than 10,000 years.

It appeared furthermore that other plant remains e.g. leaf fragments and cuticles, resin, wood and cork tissues which are preserved and fossilized in the soil, frequently remain fluorescent, even when these are from older Pleistocene peat and clay or Tertiary browncoal.

FLUORESCENCE COLOUR DETERMINATIONS

The fluorescence picture of palynomorphs from the so-called young *Sphagnum* peats is clearly illustrated by the figs. 2—4 of the Photoplate. As examples of such deposits have been chosen the peats occurring at Ekamp[1]) near Winschoten, N. Netherlands (prep. nr. EK 65/L) and at Kloosterhaar in the E. Netherlands (prep. nr. KH/JV/2). The figures show that spores possess blue fluorescence colours, whereas the pollen grains are green, white, yellow, orange or pink. In certain cases they show a rather large variation in fluorescence colour, due to differences in their state of preservation. This variation may even have been caused by autoxidation before burial in the soil (see Chapter VIII). The fluorescence of these objects is

---

[1]) For palynological data see van Gijzel (1967c).

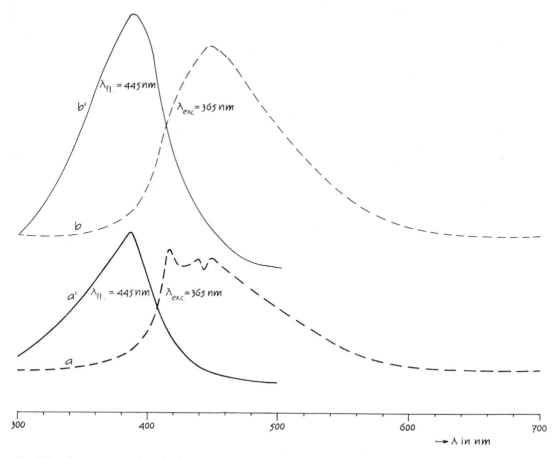

Fig. 19b. Fluorescence and excitation spectra of fresh spores of *Lycopodium*, measured in suspension in aqua bidistillata, by means of a Fluorispec fluorescence spectrophotometer (made by Baird — Atomic Inc., Cambridge, Mass., U.S.A.).

a) fluorescence spectrum, excited at 365 nm; spores treated previously with acetolysis and KOH,

a′) excitation spectrum belonging to the fluorescence at 445 nm; spores treated similar to a,

b) fluorescence spectrum, excited at 365 nm, spores treated with KOH only,

b′) excitation spectrum, belonging to the fluorescence at 445 nm; spores treated similar to b.

The fluorescence colour of these spores is similar to the blue colour of *Filix* spores (fig. 19a).

similar to, or but slightly different from that of subfossil grains from living peat bogs.

In order to describe the fluorescence colours more objectively than is possible by means of the human eye, a number of relative Berek fluorescence spectra have been determined, as summarized in Plate I and some other figures. It appears that spores of Filices and Sphagna are blue in ultra-violet light, while pollen grains of *Pinus sylvestris*, *Abies*, *Carpinus*, Cyperaceae, Chenopodiaceae and Ericaceae are greenish white, or yellowish green in fluorescence; *Alnus* and *Fagus* yellowish white; *Picea*, *Quercus*, *Tilia*, *Ulmus*, *Betula*, *Myrica* and *Corylus* yellow or orange and the Poaceae (Gramineae) orange or pink.

The difference in spectral composition between the blue, green and yellow coloured objects, is formed by a gradual shifting of the main intensity from blue to red, which is expressed by the gradual decrease of the maxima at 474 and 488 nm in favour of those at 601 and 621 nm. The striking difference between the spectra in blue spores and the pollen of the orange or pink Poaceae (Gramineae) gives a clear impression of the range in fluorescence colour to be encountered.

In plate I the spectra have been arranged in a fluorescence series from blue to red. A similar gradual change also occurs in each pollen type with increasing geological age (see next chapter and Plate II). The similarity of the spectra is remarkable, for these always show two distinct maxima. These analyses, carried out by means of the Berek photometer are in accordance with the UV-microspectrographic determinations of some palynomorphs from the Ekamp peat (fig. 19 in Chapter II). These spectra show a more complete picture of the fluorescence emission with four instead of two maxima. When one compares the spectra of a blue fern, a green *Pinus* and an orange *Picea*, respectively, it is clearly observable that the blue and green maxima diminish gradually in favour of the two maxima in the yellow, orange and red part of the spectrum. Both types of spectra have been measured and calculated according to the procedures, described above, in which each line represents the fluorescence spectrum of one specimen.

These phenomena are very important in relation to the problems of the chemical nature of sporopollenine (see Chapter VIII). The walls of fresh and fossil palynomorphs do not appear to be fundamentally different in chemical composition, but only in the proportions of certain omnipresent compounds.

The spectra of Plate I show just like the colour pictures of fig. 2—4, that a certain variation in fluorescence colour exists in the forms studied. The specimens measured were chosen in such a way that the largest variation was included. For technical reasons only part of the spectra could be reproduced in the plate and figures; those which coincided are represented by one curve only. As a result the spectra of *Myrica* (nr. 50), *Betula* (nr. 65), *Corylus* (nr. 41) and *Alnus* (nr. 53) are illustrated in Plate I (left below), while the Cyperaceae (fig. 21) are represented by another curve (nr. 60). The spectrum for the Chenopodiaceae (nr. 55) is an average, derived from three spectra, which could not be separately represented.

The Berek photometrical determinations of a number of grains of the Myricaceae and Betulaceae have been summarized in fig. 22. It appears from these spectra and numerous visual observations that the following differences in fluorescence colour can be distinguished: *Myrica* pollen grains show a smaller variation in colour

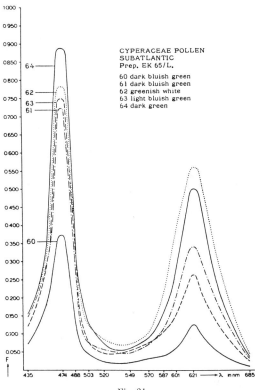

Fig. 21.

than those of *Betula*. On the other hand they differ only slightly from those of *Corylus*, which possess an equal colour intensity as *Myrica* grains. The total fluorescence intensities of *Myrica* and *Corylus* are larger than those of *Betula*, which shows more orange colours, in contrast to the white and yellow of *Myrica* and *Corylus*. *Alnus* pollen shows more white and green colours in comparison with the yellowish shades of *Myrica* and *Corylus*.

These differences become the more distinct by calculations of the spectral ratio according to the method described above. Together with the Q values of other forms measured, they can be arranged in a fluorescence series or colour scale (Table III), which may

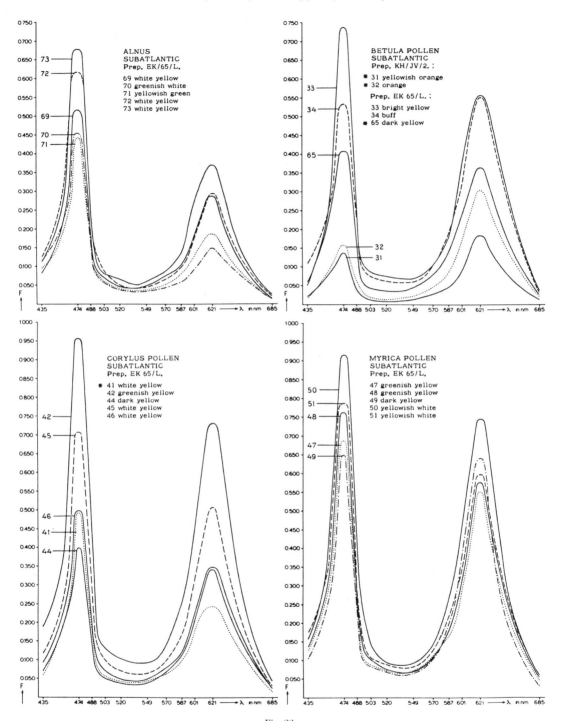

Fig. 22.

be compared with the corrosion susceptibility series of fresh pollen and spores, published by Havinga (1964). It appears that a general agreement exists between both series. Blue and green fluorescing palynomorphs are more resistant to oxidation than the yellow and orange coloured ones, whereas those which are white occupy an intermediate position. Although some differences in the arrangement of both series occur, the general trend appears to be similar. The question arises whether fresh pollen and spores, which have been oxidized by treatments in the laboratory, may be directly compared with those, obtained from young *Sphagnum* peats, which have been fossilized and changed chemically in another way. More ...ensive investigations on fresh and fossil material are ... ssary in order to obtain more data to amplify th fluorescence series. It may be important to make measurements on oxidized pollen material, which has also been treated in other ways.

In the table below those forms of which less than three determinations are available have been left out of consideration, since these types show a rather large variation in fluorescence colour. Moreover, more measurements of the types mentioned are required before a more accurate arrangement of this fluorescence series can be presented.

TABLE III. *Corrosion susceptibility and spectral ...o of some fresh and fossil palynomorphs*

Corrosion susceptibility of fresh pollen and spores, increasing downwards (after Havinga, 1964):	Average Q values of the fluorescence of some palynomorphs from young Sphagnum peats:	
	Sphagna	4.38 (8)
Lycopodium	Filices	4.25 (8)
	Cyperaceae	2.28 (7)
Pinus sylvestris	*Pinus sylvestris*	2.21 (11)
Tilia	*Abies*	1.84 (6)
Alnus	*Alnus*	1.69 (10)
Corylus	*Tilia*	1.67 (5)
Myrica	*Carpinus*	1.65 (5)
Betula	Chenopodiaceae	1.46 (3)
Carpinus	Ericaceae	1.45 (8)
Populus	*Picea*	1.44 (4)
Quercus	*Corylus*	1.34 (5)
Ulmus	*Myrica*	1.24 (8)
Fagus	*Fagus*	1.13 (4)
Fraxinus	Gramineae	1.08 (7)
Acer	*Betula*	0.86 (5)
Salix		

(8) = number of measurements on each form

APPLICATIONS

It is important to point out the practical use of fluorescence microscopy to the analysis of pollen preparations, when these are very poor in pollen. These can be more easily studied under ultraviolet light, in particular when a darkfield microscope condensor is used. The light and coloured grains can be found rapidly in the dark image projected by the microscope, even when the preparations are from older deposits, unless the material from these has been coalified so strongly that the fluorescence is weak or absent. After the detection of an object, the illumination can be easily switched from ultra-violet to normal for identification. Not only small concentrations of pollen grains, but those of other plant remains, e.g. *Pediastrum*, sporangia of *Azolla*, spores of Fungi, specimens of acritarchs (Hystrichosphaeridae) and dinoflagellates can be easily detected and counted in this way.

The importance of the first principle of fluorescence palynology is restricted to pollen morphological studies in which determination of the spectral ratio can be used for distinguishing pollen species. A number of grains of related species have then to be measured in order to be able to give an accurate and significant average Q value for each of them. Care must be taken, however, that no acetolysis or any treatment with acids is used in the preparation of the slides in the laboratory. The number of measurements of the spectral ratio will be somewhat smaller than with fossil material, due to the smaller variation in fluorescence colour of fresh exines.

In conclusion it ought to be pointed out that by means of the ultra-violet microscope the recognition of the various layers composing the exine may be facilitated and yield results important for pollen morphology.

The applications mentioned are under further investigation.

V. FLUORESCENCE AND GEOLOGICAL AGE

INTRODUCTION

From the study of contaminated sediments by means of fluorescence microscopy it appeared that a certain relationship exists between the fluorescence colour and geological age of fossil pollen and spores. This is considered the second principle of fluorescence palynology (van Gijzel, 1961, 1963). In the first stage of the present investigations some experiments were carried out to test the utility of fluorescence microscopy with regard to palynological purposes. The differences in fluorescence colour between Holocene palynomorphs occurring in a young *Sphagnum* peat and those in a Tertiary clay, the residues of which were mixed with each other, gave rise to the assumption that such a relationship exists.

Continued studies of more than one hundred deposits of different ages and of various origins now furnish proof of this supposition and result in many new data.

The results are summarized and discussed in this chapter which deals with the description of the phenomena of fluorescence colour change.

PRESENTATION AND EVALUATION OF DATA

With regard to the interpretation of the figures and diagrams reproduced below, a short explanation is given.
The procedure for measuring and calculating is described in Chapter II. At least eight pollen grains in each slide have been tested. The number depends on the variation of fluorescence colour in each sample. More or less weathered deposits, in which corroded pollen occurs, have been left out of consideration with regard to the calculations of the average fluorescence values per sample. These samples have been indicated in the diagrams with an asterisk.
In the figures and diagrams each line represents the fluorescence spectrum and each dot the spectral ratio of the pollen grain measured. All objects were chosen arbitrarily, unless the preparations contained but few pollen grains.
The colour of the curves and dots, in Plate II and III, resembles the fluorescence colour observed in each pollen grain. The boundaries between the colour groups merge into each other, due to the fact that visual observations of fluorescence colours under UV-microscope are always a question of personal judgement. It appeared to be very difficult to establish visually whether a pollen grain is, for instance, either greenish yellow or yellowish green or white.
After continued investigations it appeared that it is sufficient to determine the spectral ratio Q — a new term which is introduced and defined in Chapter II — instead of the complete fluorescence spectra. This proportional number is represented as the horizontal scale in the standard diagrams. It expresses the proportion of the blue-green and orange-red part of the spectrum of each fluorescent pollen grain measured. The number of the sample (in code) refers to the deposit studied. The Q values of each sample are plotted on a thin horizontal line behind it. The average spectral ratio-values of all samples of a certain geological age are presented in a separate column. For geological and other data of the material, reference should be made to Table VII (see Appendix). The localities of the deposits studied are indicated in fig. 10 (Chapter I).
Geochemical coalification has an important influence on the fluorescence colour of fossil palynomorphs (see Chapter VII). One therefore meets with great difficulties in finding for the standard diagram a sufficient number of non-coalified deposits from Mesozoic and Palaeozoic eras. They may often have been originally buried more or less deep in the subsoil, even when they are found at the present erosion surface. The average spectral ratio-values for ages older than the Cenozoic have therefore to be regarded as preliminary. Finally, for the statistical evaluation of fluorescence data, an attempt has been made to find a statistical model, that should show how much Q determinations are needed per sample. This appeared to be at least 6 to 8 per sample, depending on the variation in fluorescence colour.

GEOLOGICAL TIME-SCALE

The time-scale used in Plate III, demands some explanation. For practical reasons the absolute time boundaries are plotted on a logarithmic scale. The geological ages have been derived from various papers and data.
The absolute pre-Quaternary dates are those according to Holmes (1959). Some years ago, new data became available in the so-called "Geological Society Phanerozoic time-scale 1964", i.e. a list abstracted from various chronostratigraphical papers.[1] They differ not much from those mentioned by Holmes.
The ages for the Upper Pleistocene and Holocene time units are based on generally accepted C 14 dates. The problem of the actual limits of the Lower and Middle Pleistocene glacials and interglacials still constitutes an important problem, to which more attention will be paid below. In the diagram of Plate III those absolute time lines have been chosen, which are in accordance with the highest known values, viz. those given by Eberl (1928). The Pleistocene sequence of glacials and interglacials, introduced by Zagwijn (1957) and other investigators, is used here as no generally accepted nomenclature exists for this epoch. Some informal terms have been used for the subdivision of various Tertiary epochs. These have been printed in italics in Plate III. These are used since for continental deposits (the great majority of the samples studied here), another stratigraphical nomenclature is necessary than that for marine sediments.

DESCRIPTION OF FLUORESCENCE COLOURS AT VARIOUS AGES

The second principle of fluorescence palynology
This can be defined as follows:
With increasing geological age all fossil palynomorphs show a gradual change in fluorescence colour from blue, green or white towards orange, red or brown colours, followed by extinction of the fluorescence.
The gradual change is effected by the decrease of the blue-green maxima of the fluorescence spectra in favour of those in the yellow, orange and red part of the spectrum. The relative spectra, obtained by means of the Berek photometer appeared to be in full accordance with the UV-microspectrographical observations.

Change in fluorescence of vesiculate forms
Vesiculate pollen grains belong to a group of palynomorphs, showing very clearly the change of fluorescence colour with increasing geological age. Recent

---

[1]. Published in: Quart. J. geol. Soc. London, 120, s, p. 260—262.

grains show a blue or green fluorescence. Consequentely, with increasing age nearly the whole visible spectrum is available for the colour change, contrary to other recent palynomorphs, which initially show yellow or orange fluorescence colours. Besides, many vesiculate grains are generally present in humic deposits and these forms furthermore show a very long stratigraphical range.

The relation of fluorescence colour and geological age is closely tied up with the following assumptions: The fresh exines of Tertiary and older *Pinus* species probably showed originally a fluorescence colour similar to that of recent *Pinus sylvestris*. Should this be the case, then *Pinus sylvestris*-like types from those ages may be compared directly with younger *Pinus sylvestris* pollen. Furthermore *Abies* and *Pinus haploxylon*-types from the Cenozoic are similar in fluorescence to *Pinus sylvestris*. Their change of colour during the Mesozoic may consequently be considered a direct continuation of the colour change of *Pinus sylvestris* forms.

The fluorescence of all these forms changes as follows in time. The Holocene is characterized by a predominance of green and white colours, the Pleistocene by white and yellow, the Tertiary by bright yellow and the first distinct appearance of orange, the Mesozoic and Palaeozoic by dark yellow and orange with the continuous appearance of brown, announcing the final extinction of fluorescence. With increasing age the total intensities gradually decrease, but far less regularly than the change in spectral composition. The decrease of intensity is therefore unsuitable for use in palynology.

The colour change phenomenon is clearly illustrated in Plate II by a number of Berek photometrical spectra. The same gradual change appears from the complete UV-microspectrographical spectra in fig. 23. All determinations of the spectral ratio are summarized in the standard diagrams. Plate III gives a general idea and fig. 24 the complete picture of the Pleistocene. The second principle of fluorescence palynology is expressed by two curves: the average Q connection line (striped) and the general tendency line (full). Both curves have been calculated in the Cenozoic for *Pinus sylvestris* and *haploxylon* types only, in older eras for all vesiculate forms measured. Only the Quaternary *Picea* grains differ much in fluorescence colour from the other vesiculate grains. The general tendency is a parabolic regression curve, calculated by means of the so-called arithmetic method of the smallest averages. Its curved course may be evidence as to the chemical nature of the fluorescence phenomena (see Chapter VIII). The area of the range of fluorescence colour groups shows the different variation in spectral ratio for the various deposits studied.

Palynologists, engaged in palaeoclimatological studies, must notice, that the oscillating line, connecting the average Q values, cannot be interpreted like climatic or normal pollen curves. A sudden rise or fall in the connecting line has nothing to do with climatic changes. Similarly the regressive line reflecting the general tendency in fluorescence cannot be interpreted in detail.

Both curves prove the direct relationship between the fluorescence colour and the geological age of the vesiculate forms. At least thirteen fluorescence colour groups can be distinguished in the standard diagrams, viz. recent, subrecent, Holocene, Upper-, Middle- and Lower Pleistocene, Pliocene, Miocene, Oligocene, Eocene with Palaeocene, Upper Mesozoic, Middle with Lower Mesozoic and finally the Palaeozoic. Some of the Holocene, Pleistocene and Upper Tertiary groups may be subdivided as demonstrated for the Pleistocene in fig. 24.

The point in time of complete extinction of fluorescence which is brought about by the geological age only, is still unknown due to the lack of suitable material. It must be situated somewhere in pre-Carboniferous time, because in a number of Carboniferous rocks fluorescent palynomorphs have been found, and in some older deposits fluorescence appeared to be absent.

Only part of the material, collected for this study, has been measured. Selection of samples for fluorescence analysis has been made for practical reasons such as the presence of suitable pollen forms and the degree of coalification.

Geochemical coalification may have strongly influenced the fluorescence of palynomorphs. This fact must be taken into account when one considers Tertiary and pre-Cenozoic strata, composed of either lignite or coal. Mesozoic and Palaeozoic deposits, studied here, are in all probability not coalified to such an extent that the real Q values for these eras must be considered to amount much more. With regard to the Permian and Carboniferous it may be assumed that the Q values are at most 0.10 on the small side, but this is not important for the general trend. More data must become available, however, before it will be possible to establish the lowermost course of the general tendency line exactly.

The results of these investigations may, consequently, be applied to age determinations of deposits in which conventional pollen analysis fails to be reliable (see below). Eighty five deposits have been measured in order to compile a standard diagram for vesiculate forms. Fourteen of these deposits show divergent average Q values, being either too high or too low compared with other samples of the same age. This means that the accuracy of this method amounts to at least 83 %.

A number of sedimentary properties may be responsible for these differences. Values which are too low may be caused by weathering of a sediment during or after its deposition. This can be confirmed by the rather bad state of preservation of the pollen grains and spores. Numerous corroded pollen grains occur, for instance, in the Lateglacial peat of Lutterzand in the E. Netherlands (prep. nr. LTZ/A), which outcrops in the bank of the river Dinkel, some metres above the normal water level in summer. Another example is the Waalian clay at a depth of 97 m in the boring Eindhoven (prep. nr. EH 97), at which depth an old erosion surface has been considered to be present (personal communication by Dr. W. H. Zagwijn).

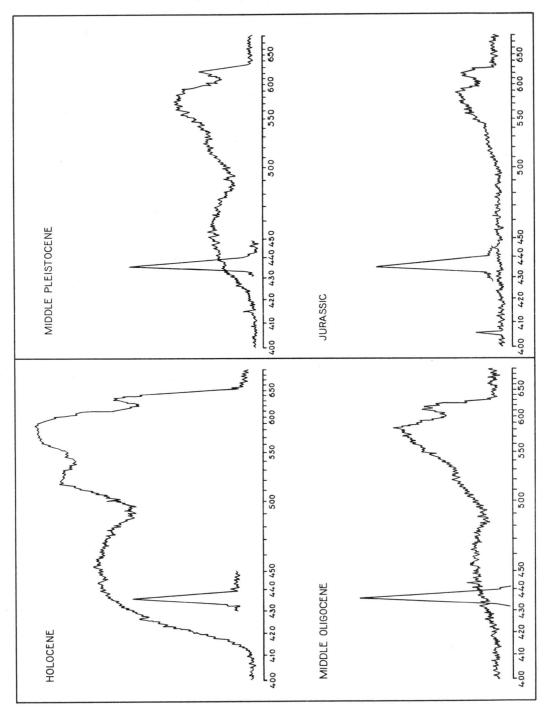

Fig. 23. Fluorescence spectra of some vesiculate pollen grains of various ages, obtained by means of an UV-microspectrograph.

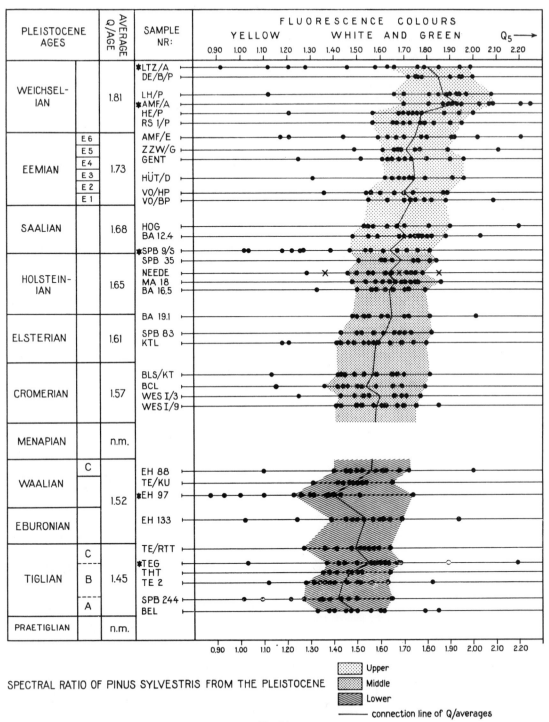

SPECTRAL RATIO OF PINUS SYLVESTRIS FROM THE PLEISTOCENE

Fig. 24.

Corrosion of pollen and spores occurs repeatedly in older deposits as well, for instance in coals (Wilson, 1961, 1964). The chance of meeting with corroded pollen is greater in coarse sediments such as sand and loam than in fine deposits like clay and peat, in particular when they have been situated above the groundwater level for some time. A more intensive air circulation then results in damaged exines.

On the other hand, no differences between the fluorescence colours of pollen exines preserved in either clay or peat or diatomite were observed. It is remarkable, however, that in many cases small differences in fluorescence seem to exist between pollen grains from clay and from browncoal in the same Tertiary areas. More research is necessary in order to establish whether the action of geochemical coalification on clay and browncoal results in different fluorescence of the pollen grains.

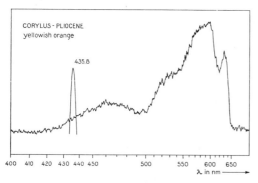

Fig. 25. Fluorescence spectrum of a *Corylus* pollen grain from the Pliocene, obtained by means of an UV-microspectrograph.

The fluorescence values of certain samples which were found to be too high for their geological age, are not so easy to explain. The processes of chemical transformation of the sporopollenine under the influence of geological time apparently have been delayed in certain deposits for unknown reasons.

Fluorescence of other forms

With increasing geological age, other forms of fossil palynomorphs show changes in fluorescence, similar to the colour series of Plates II and III for vesiculate pollen. A number of relative fluorescence spectra of other types from certain geological periods has been published previously (van Gijzel, 1963). UV-microspectrographical analyses demonstrate this phenomenon in more detail. Fig. 25 shows the fluorescence spectrum of a Pliocene *Corylus* pollen grain, which shows much resemblance with that of an orange-yellow *Picea* from the Holocene (fig. 19, orange curve). The spectrum of a Holocene *Corylus* pollen grain is similar to other yellowish white spectra, as, for instance, that of a Middle Pleistocene *Pinus sylvestris* (fig. 23). It appears from a comparison of such spectra, that those pollen grains from young deposits which show yellow or orange fluorescence colours, e.g. *Corylus*, can shift in colour with increasing age, but to a lesser extent than in forms showing green or white colours, e.g. *Pinus* and *Abies*. This is especially true for spores which are dark blue in recent and subrecent material.

Recent and subrecent palynomorphs often show large differences in fluorescence colour, which have been described above (Chapter IV). With increasing geological age these differences diminish gradually and finally disappear almost entirely. In slides from the oldest rocks investigated here, viz. those from the Carboniferous, various forms of palynomorphs still show differences in fluorescence colour.

Wilson (1961, 1964) described the selective susceptibility of fossil spores from coal to corrosion and increase of temperature. With increasing rank of coal, some forms become opaque earlier than others. Where old erosion surfaces are present in coal, the spores appeared to be corroded selectively. The fluorescence phenomena appear to be in accordance with this selective susceptibility.

Photoplate I illustrates the colour change of other forms of palynomorphs, while fig. 5 shows a picture of a Tertiary pollen slide containing grains of *Pinus haploxylon* type and some Myricaceae. They all show fluorescence colours which are less different than those of grains of *Pinus sylvestris* and *Myrica* in the Young Holocene. Tertiary Myricaceae are more yellow and less white than younger ones. In fig. 8 a Palaeocene spore is illustrated, showing a yellow fluorescence colour. It may be assumed that this grain originally possessed a blue fluorescence.

Fluorescence of acritarchs and dinoflagellates

The fluorescence of fossil acritarchs and dinoflagellates was compared to that of palynomorphs. A number of spectra were determined by means of the Berek fluorescence photometry, the results of which are shown in Plate IV. The ages of the deposits studied, range from Middle Pleistocene to the Blue Lias (Jurassic). Some of the slides were also used for measurements of vesiculate pollen. It appears from these spectra that no relation exists between fluorescence colour and geological age of these microfossils. This is in accordance with many other visual observations. The orange, yellow and green colours occur together at nearly every geological age.

A relationship between the fluorescence colour and type or species has not been established up to now, but the possibility cannot be excluded. This could explain the rather large variation in colour even in those specimens of old age, compared with fossil palynomorphs.

The influence of geological time, which is different from that on fossil palynomorphs, may be due to the fact that sporopollenine differs fundamentally in chemical nature from the tests of acritarchs and dinoflagellates, which consist of a chitinous substance. It appears to be much more resistant to the action of coalification and geological time than sporopollenine. In certain cases the fluorescence colour is determined by the state of preservation. Brown and orange coloured microfossils occur simultaneously with more or less corroded palynomorphs as in contaminated deposits. Damage during the transport of the rebedded material may be the cause of a change of fluorescence colour as well.

APPLICATIONS TO AGE DETERMINATIONS
Fluorescence analyses for long-distance correlation

As stated in Chapter I, palynological age determination and correlation sometimes meet with great difficulties. This is the result of various factors, e.g. contamination with rebedded material, the facies of the sediment, climatological differences and corrosion. Fluorescence palynology opens up the possibility of age determinations and correlations, in particular for Cenozoic beds older than 50,000 years. In that era the differences in fluorescence values are larger than those of older eras. It may be successfully applied, under the condition, however, that the pollen grains are well-preserved and are not too strongly coalified. Besides, a large number of measurements on *Pinus sylvestris* grains, or pollen which is comparable in fluorescence colour and stratigraphical range, must be made. The variation in fluorescence colour and the average Q values can be compared with those of standard diagrams such as Plate III and fig. 24.

How fluorescence palynology may be applied to the study of these problems, will now be shown more extensively by some examples. The first is the long-distance correlation of Pleistocene sequences. Those from Padul (Spain), previously investigated by Menéndez Amor and Florschütz (1964), and from the Sabana de Bogota (Colombia, S. America), studied by van der Hammen and Gonzalez (1964), are compared with the Pleistocene pollen stratigraphy of the Netherlands, which has been established by Zagwijn (1961, 1963). The uppermost parts of the climatic curves are in accordance with each other and these correlations are supported by a number of 14 C dates. Difficulties arise, however, in the correlation of the older parts of these diagrams, for which the application of fluorescence palynology appeared to be useful (van Gijzel et al., 1967). For more details on palynological data, interpretations of the diagrams and climatic curves, reference should be made to that and the other papers mentioned.

Long-distance correlations can be effected by comparing fluorescence measurements according to the procedure mentioned above. A large number of spectral ratio determinations of these sequences are compared with those of the standard diagrams, based in many cases on deposits previously studied by Zagwijn. The results are summarized in fig. 26 and Table IV.

In the Colombian sequence, grains of the vesiculate *Podocarpus* were measured, the fluorescence values of which are similar to those of *Pinus sylvestris*. In this table the arithmetic averages of the fluorescence analyses of groups of samples, belonging to certain supposed ages, were calculated. These averages for Q were placed in the table in the same position as those of the standard values. Not only in the standard diagram, but also in the two sequences, the change in fluorescence with increasing geological age is obvious. The divergent averages for some samples and ages may be caused by the action of corrosion, for instance, in the Eemian lake-chalk of Padul, the figures of which were left out of consideration here.

On the basis of fluorescence analyses the following correlation appears to be possible. The similarity between the Weichselian and Saalian glacials of Padul and N.W. Europe is evident. The data from the deeper layers give support to the opinion that a pre-Cromerian interglacial may be present in the Padul section at a depth of about 60—70 m. The correlation with the diagram of the Sabana de Bogota appears to be more complicated. The values of the lowermost part of this sequence are in accordance with the standard values of the Tiglian in Europe. This supports the idea that the Tiglian interglacial was represented here. In the interval between 95 and 125 m, the fluorescence figures suggest the presence of the Cromerian interglacial.

Other applications

Another example of the successful application of fluorescence palynology with reference to age determination is found in the study of contaminated Pleistocene deposits, the results of which are to be discussed in the next Chapter. Furthermore it may be applied, for instance, to the dating of scattered samples from borings or deposits, dredged out of the subsoil and in general to those beds, in which typical pollen associations or zonations in the pollen content are absent. Examples are pollen spectra from clay and peat, occurring in fossil skulls and bones of mammals, which have been dredged out of the river Meuse (the Netherlands) or from other localities where Pleistocene sands are exploited. Such spectra are often contaminated with rebedded material, which even may be of Mesozoic age. Numerous other examples of deposits, of which only a few samples could be obtained, can be easily found.

As stated above, fluorescence palynology can only be applied when the pollen grains are not corroded or coalified too strongly. In such cases the accuracy of fluorescence-palynological dating can amount to more than 80 %.

In addition, some other aspects of this method may be quoted here. As appears from the standard diagram of fig. 24, at least three groups of fluorescence colours can be distinguished in the Pleistocene: a Lower, a Middle and an Upper Pleistocene group. This number may even be extended to six groups: Weichselian, Eemian, Saalian with Holsteinian, Elsterian with Cromerian, Menapian with Waalian and Eburonian and a group of Tiglian with Pretiglian. This method

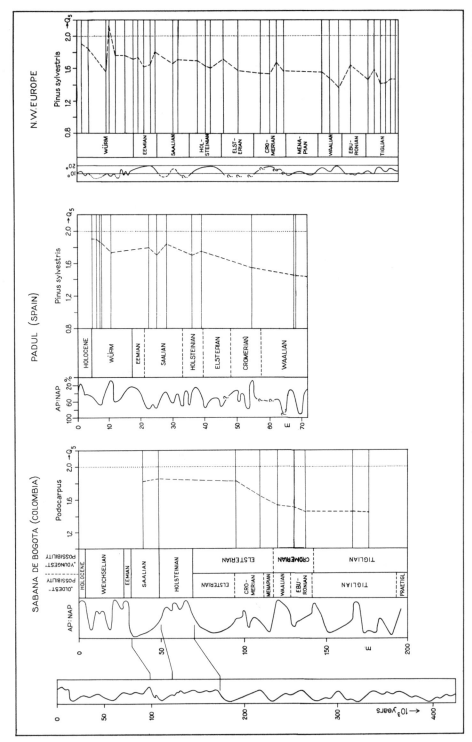

Fig. 26. Average values of the spectral ratio of fluorescent pollen grains used for long-distance correlation of climatic curves (after van Gijzel et al., 1967). The sea water temperature curve at the left is after Emiliani (1966); the stratigraphic subdivisions in the diagrams are after various authors (see text).

thus covers a range in age, between the lower limit of the 14 C method and the Pliocene, for which other absolute age determination methods exist. Some other new geochronological methods fill this gap, but they are still inadequate for determining the age of continental deposits. After some improvements to the procedure, fluorescence palynology may possibly furnish the solution to this problem.

Geochronology of the Pleistocene

It is possible that this method can contribute to the solution of another important problem, viz. the duration of the Pleistocene epoch. With regard to this a great controversy still exists.

Fluorescence data may contribute to the solution of this question. The samples from the Lower Pleistocene have been plotted in the absolute time scale of Plate III, according to the ages given by Eberl. With regard to the general tendency line, the Q averages for the Tiglian are situated somewhat too low. This may indicate that Eberl's figures may be somewhat too low as well. This could mean that the Plio-Pleistocene boundary is to be drawn at one million years or more ago. If so, then some of the oldest glacials and interglacials must have had a somewhat longer duration than assumed by Eberl. It appeared later on, however, that Eberl's opinion about the duration of the Weichselian glacial was too premature, as has been proved by

TABLE IV. *Average values of the spectral ration of fluorescent pleistocene pollen grains* [1]

Age	Northwestern Europe (Pinus sylvestris) average of Q	Padul (Spain) (Pinus sylvestris) depth in m	average of Q	Sabana de Bogota (Podocarpus) prep. no.	average of Q
Weichselian	1.81 (37)	4.5—10.5	1.87 (46)		n.m
Eemian	1.73 (32)	16.5—18.0	1.45 (12)		n.m.
Saalian	1.68 (16)	22.4—28.0	1.70 (34)	220/236	1.75 (12)
Holsteinian	1.65 (26)	35.7—39.2	1.72 (17)		n.m.
Elsterian	1.63 (20)		n.m.		n.m.
Cromerian	1.57 (26)	54.5	1.54 (10)	339/367	1.79 (12)
		67.5—71.2	1.44 (22)		
Menapian	n.m.				n.m.
Waalian	1.52 (42)			389	1.47 (8)
Eburonian				409/420	1.42 (14)
Tiglian	1.45 (46)			483/492	1.43 (14)
Praetiglian	n.m.				n.m.

1. Numbers in brackets indicate the number of objects measured; n.m. = no measurements; the preparation no. of the Sabana de Bogota are from Van der Hammen and Gonzalez (1964), (after Van Gijzel, et al., 1967).

The geochronological data are mainly based on the curve of changes in the radiation of the sun during the last one million years, calculated by Milankovitch and others. Among the numerous interpretations of this curve, Eberl (1928) mentioned the highest value for the lower boundary of the first glacial, situated at 761,000 years. In Plate III this figure has been drawn as the Plio-Pleistocene boundary.

On the other hand, it is remarkable that for this boundary the radio-active age determinations delivered figures of at least one million years (Holmes, 1959) or much more (1.5 m.y. according to the Geological Society Phanerozoic time-scale 1964). Furthermore, during the last few decades palynologists have succeeded in establishing a complete picture of the vegetational history of the Pleistocene. In the author's opinion the possibility that one or more glacials and interglacials should have been overlooked is highly unlikely.

14 C datings. On the other hand, no indication exists that the Lower Pleistocene stages were of longer duration than those in the Upper Pleistocene.

The situation of the Plio-Pleistocene boundary is obviously a very complex problem. Many more fluorescence palynological data are needed, however, to find a synthesis, which is in accordance with both absolute dating figures and palaeoclimatological studies.

DISCUSSION OF RESULTS

In reviewing the fluorescence data with increasing geological age, some critical remarks have to be made.
1. A number of samples studied, show average Q values, diverging too much from the general tendency and the fluorescence figures of the time unit to which they belong. In order to increase the accuracy of this method, it is necessary to carry out a larger number

of determinations on comparable deposits. A study of the causes of the fluorescence phenomena described is needed therefore. This refers in particular to the large variation in fluorescence colour of fossil palynomorphs in certain sediments.

2. As stated above, the average Q values for the Mesozoic and Palaeozoic must be established more exactly by using samples of selected material, which are not or only slightly coalified. Such deposits are not easy to find.

3. The disadvantages of the Berek photometer must be taken into account in the interpretation of the fluorescence data. More accurate figures of the spectral ratio may be obtained when intensities in the green and yellow range of the spectrum can be measured.

4. In spite of these restrictions, fluorescence palynological dating is delivering reliable results. It needs, however, further investigation. Much care must be taken in sampling, preparing and measuring of the material.

VI. FLUORESCENCE AND DATING OF CONTAMINATED SEDIMENTS

THE CONTAMINATION PROBLEM

In the Introduction it was stated that the problem of pollen contamination was the starting point of the present fluorescence investigations. Contaminated deposits represent an old and complex question. It is useful to quote the essential part of earlier attempts at reaching a solution.

Contamination takes place when *autochthonous* pollen grains and spores produced by the vegetation in and around the area of sedimentation are mixed during their deposition with *secondary* or *redeposited* specimens. This reworked material is found in different quantities, varying from the one case with a small amount consisting of a few species to the case in which the allochthonous components may entirely predominate over the autochthonous components.

Palynological dating and correlation are based mainly on the presence of typical pollen assemblages and zonations in pollen diagrams, by which the vegetational and climatological history is reflected. These palynological characteristics may become unclear or even disappear by contamination with either older and and reworked pollen and spores or with contemporaneous material from vegetations situated at a great distance from the site of sedimentation. Even a small amount of such material may prevent the establishment of zonations.

The quintessence of the question is how to distinguish the non-contemporaneous secondary pollen and spores from the autochthonous grains in case that both belong to the same form or species. For instance, in a mixture of Miocene and Pleistocene elements, the Tertiary representatives are for the greater part recognizable and can be left out of consideration. But how does one ascertain which number of such forms as e.g. *Pinus sylvestris*, *Picea*, *Abies*, *Alnus*, *Betula*, *Carpinus*, *Myrica*, belongs to the autochthonous and which to the reworked part of a pollen spectrum?

Thomson (1935) described a contamination from older interglacial layers in Late Glacial deposits.

Iversen (1936) tried to find a solution in his paper "Sekundäres Pollen als Fehlerquelle". He analysed the boulder clay of Egebjerg, Fyn, Denmark, which contains numerous pollen grains of thermophilous trees. A large number of them also occurs as contamination in the overlying Late Glacial deposits. The amount of redeposition could be determined by subtracting the quantitative pollen composition in the boulder clay from that in the younger layers. The remaining pollen spectra corresponded to the autochthonous vegetation and resulted in an uncontaminated diagram.

Iversen's method is, however, very limited in its possibilities. In general it can only be used if the origin of the contamination is known and is situated nearby as was shown above, or if part of a layer has remained uncontaminated and the vegetation has not changed during sedimentation.[1] In that case only is it permitted to assume a constant average pollen content, which may be subtracted from the reworked spectra.

No difficulties arise when the contaminations are so different in age, that the autochthonous components have nothing in common with the secondary species. Such old reworked pollen sometimes occurs in the Pleistocene desits of the river Meuse in the S. Netherlands, in Middle Tertiary marine clays and glauconitic sands in the Netherlands and N.W. Germany and in deep-sea sediments, which are only contaminated with Mesozoic or older material.

In most cases, however, the origin, age and amount of contamination are difficult to estimate with any degree of accuracy. The source may have been removed by erosion and then calculations according to the Iversen method become impossible (see Davis, 1961, and others). Besides, the source deposit itself possesses in general a certain variation in its pollen content and an average pollen content will not be easy to find. The homogeneous dispersion of pollen grains in boulder clay, caused by the kneading and mixing activity of the inland ice, forms a favourable exception.

Furthermore, selection of secondary pollen during erosion and transportation over long distances in river or sea water may occur and the chance of more sources of contamination increases in such a case, delivering

---

[1]. For example in the Upper Saalian or Lower Eemian clay of Amsterdam—Slotermeer in the Netherlands, analysed by Zagwijn (personal communication).

reworked pollen in variable amounts. It will therefore be obvious that by means of conventional methods qualitative comparison at best is permitted and that successful quantitative analyses are favourable exceptions.

Thomson (1952) remarked that rebedded pollen can be expected mainly in all fine-grained clayey sediments, where they sustain destructive alkaline reactions and oxidation. He mentioned the clay of Satzvey in Germany as an example, the lower part of which is without doubt of a Middle Tertiary age. But the clay in the overlying browncoal conglomerate contains numerous Lower Tertiary species, resulting in pollen spectra which are comparable to the Lower Eocene of Helmstedt. Thomson presumed that the well-known "helle Schichten" (light bands), often occurring in lignites and fimminites (lake deposits, consisting mainly of pollen exines) might be the source of reworked material, rebedded in the clay and resulting in a domination of the secondary pollen over the autochthonous pollen content.

Contamination can be caused, however, by long distance airborne transport of fresh pollen as well. Since Erdtman made his trip across the Atlantic (1937), it has been generally accepted that airborne pollen grains can be transported over long distances. Such pollen, especially bisaccate grains as in the conifers, may have their origin far away. Aario (1940), for instance, found in surface samples from the tundra vegetation of Lappland tree pollen, mainly of conifers, but of *Tilia* as well, which must have been transported by the wind over hundreds of kilometers and was finally deposited in the tundra of Northern Europe. Such contamination is not recognizable if it consists of the same pollen species as occurring in the autochthonous vegetation far from the area of sedimentation. In general its influence will be rather small. This aspect of the contamination demonstrates however that it is not easy to define this phenomenon exactly. When does one consider airborne pollen secondary and where does one draw the limits of the area of origin of the autochthonous group? On the other hand, what is the minimum difference in age, necessary to establish whether a pollen diagram shows contamination or not? The question naturally arises whether the occurrence of *Picea* and *Abies* pollen in the Upper Holocene pollen diagrams of the Netherlands has to be considered a contamination either by air or by water transport (see below).

If considerable differences in age may be assumed, the contamination problem can be investigated successfully by the application of fluorescence palynology, on the condition, however, that for each important pollen form various groups of fluorescence colour can be distinguished. Only then it is certain if a grain belongs to the autochthonous part or not. Fluorescence palynology produces, consequently, a direct method of testing autochthony contrary to the indirect method of Iversen. Its procedure and use will be described below and illustrated by some examples of fluorescence analyses of contaminated Pleistocene deposits.

EXAMPLES OF CONTAMINATED SEDIMENTS IN LITERATURE

The increasing importance of the study of contamination phenomena is apparent from the steady increase in the number of publications on this subject. It therefore appears useful to review some of the most important studies, of which those by Iversen (1936) and Thomson (1935, 1952) were already mentioned above.

Pflug (1963) attempted to solve the problem of rebedded pollen at his review of the pollen analysis of salt deposits on the Gulf Coast (U.S.A.), which were carried out earlier by Jux. Pflug examined the original preparations, which show a considerable contamination, among others by means of fluorescence techniques but he kept silence upon the method used. Examples of long distance pollen transport are to be found in some Holocene clays from the Netherlands and N.W. Germany. With the lithological change from peat into brackish clay the values of *Pinus sylvestris* increase rapidly, whilst in many cases *Picea* and *Abies* appear simultaneously in small numbers, though they must have been absent in the Atlantic and Subboreal vegetation of these areas (Overbeck and Schmitz, 1931; Polak, 1936). A similar occurrence has been found by Florschütz and Jonker (1939) in a clay and peat section at Wijk bij Duurstede, The Netherlands. Particles of these clays, deposited by the river Rhine, settled down simultaneously with conifer pollen derived from the forests of the upper Rhine region, where *Picea* and *Abies* composed the most important elements of the vegetation. In all these cases, however, it is uncertain whether the sum total of this pollen had been transported by sea or river water only, because it occurs repeatedly in small percentages in the peat as well.

Much more attention has been paid in the literature to contamination with older reworked pollen. An extreme example is the dominance of Tertiary pollen associations in Pleistocene clays in the N. Netherlands to such an extent (60 % or more), that these clays often seem to be of Tertiary age, as for instance the so-called "potklei" (pottery-clay). The contamination prevented the finding of a palynological correlation of the stratigraphical sequence in this area with that of the Lower and Middle Pleistocene in the S. Netherlands (van Gijzel, 1961, 1963).

In the central part of the Netherlands contamination has been reported for the Middle Pleistocene at Loenermark (Polak et al., 1962), at Zweiersdal and Duno (Teunissen and Florschütz, 1957; Teunissen, 1961) and for the Late Glacial sand and lake-marl of the "Bleeke Meer" (Polak, 1963).

Marine and brackish sediments often appear to be contaminated, as, for instance, the Plio-Pleistocene clays in the harbour pit of Antwerp (Hacquaert, 1960) and the marine Miocene glauconitic sands and the Oligocene "Septarienton" (septaria clay) in the subsoil of N.W. Germany and the Netherlands.

Some examples in other areas may be quoted. Davis (1961) studied the problem of rebedded pollen in Late

Glacial sediments of Taunton, Mass. (U.S.A.) by means of Iversen's method and concluded that the mixed autochthonous and secondary pollen content could not be interpreted successfully. Cushing (1964) found pre-Quaternary microfossils in Late-Wisconsin silty sediments of East-Central Minnesota (U.S.A.). Moreover they contain corroded pollen, derived from inwash of soil from the surrounding slopes. Subtraction appeared here again inadequate to distinguish both groups.

Rebedded remains of lignites and other deposits from the Upper Pliocene were found at Fürstenhagen near Hessisch-Lichtenau in W. Germany by Brosius (1958). They were redeposited in the Lower Pleistocene under periglacial circumstances. It appeared difficult, however, to establish the degree of contamination due to the presence of only a few typical Tertiary species.

Recently a large number of contaminated sediments have been reported from the U.S.S.R. Chiguryayeva and Voronina (1960) described rebedded microspores of various ages (Palaeozoic to Cenozoic) in Upper Pleistocene deposits in the North Caspian area. They tried to distinguish the secondary pollen by staining and other treatments, but the corrections of the spectra obtained, appeared to be unsatisfactory. Ralska-Jasiewiczowa (1960) noticed that Pleistocene layers at Zablocie on the river Bug were partly contaminated with numerous grains from the underlying Tertiary. Arkhipov and Matveyeva (1960) reported the occurrence of much redeposited pollen from the Jurassic and Cretaceous (max. 45 %) and numerous individual grains of Tertiary origin in the Pleistocene pre-Samarova deposits in the Yenisei region of the West Siberian Lowland.

It is remarkable that contamination occurs so often in continental Pleistocene deposits. Various factors may be responsible for this. In the first place the chemical weathering and other destructive processes by which the pollen grains can be damaged, were not as intensive in a cold or temperate climate as in a (sub-)tropical one. Consequently the reworked pollen may have more easily sustained the destruction by erosion and transport during the Pleistocene glaciations, than in warmer epochs. Secondly, the thick and extensive Tertiary browncoal layers contain an abundance of pollen and spores. Thus a great opportunity existed during the Pleistocene — an epoch with strong glacial and river erosion in many regions — for these deposits to become a prey of river or inland-ice erosion and for this reason to become the source of contamination in the succeeding glacials. Many Tertiary browncoals in N.W. Germany appear to have been partly or completely eroded during the Pleistocene and supplied their pollen to the rivers (see below).

The mechanical destruction during river transport influenced the state of preservation of the rebedded pollen to a lesser extent than did the chemical weathering. An analogy is found in the conservation of other rebedded microfossils as for instance Bryozoa in boulder-clay, which due to their small size, better resisted the strong kneading by the inland-ice than the weathering of this clay after its deposition (Veenstra, 1963). Fossil pollen and spores in boulder-clay show a similar resistance with respect to mechanical destruction. They resist the less destructive transport by river water rather easily.

Recently the study of contamination has been enlarged in scope by the development of marine palynology. During the last few years deep-sea sediments have become a new field for palynological investigation, by which pollen diagrams of Holocene and even Pleistocene sediment sequences may be obtained. Pollen and spores have been deposited more or less continuously on the ocean floor, contrary to continental sedimentation where erosion and sea-level changes gave rise to incomplete lithological successions. It might therefore be expected that pollen analysis of deep-sea sediments, where the influence of erosion is to be neglected, might be a means of establishing complete climatic curves for the Quaternary.

In many cases, however, it appeared that these deposits contain a mixture of reworked pollen grains and spores, ranging from the Palaeozoic to the Quaternary (Groot and Groot, 1966; Stanley, 1965, 1966a). This contamination hampers the investigation of climatic curves considerably. Even sediments from the sea floor at small depths contain a large amount of rebedded pollen, as has been found by Zagwijn and Veenstra (1966) in the North Sea, where Mesozoic and Paleozoic miospores often reach more than 25 %. As pointed out by Groot (1964, 1966) it is very important to find a solution to this contamination problem, in order to obtain reliable pollen diagrams and climatic curves. It is also important for the future use and development of marine palynology in connection with other investigations of these sediments, for instance the study of Foraminifera and sea-water temperature determinations, that this problem cleared up.

Stanley (1965, 1966a, 1966b) tried to distinguish the secondary material from the non-reworked grains by means of staining with safranin "O" liquid. He found four groups of contamination, based on the decreasing susceptibility for staining with increasing geological age: Palaeozoic-Lower Mesozoic, Mesozoic, Upper Mesozoic and Lower Tertiary and the group of Quaternary and recent. This statement is in accordance with the results of the first experiments on the contamination problem, made by the present author, from which appeared, however, that the staining method is in fact inadequate, as it does not give any information on contamination within the Cenozoic era. Besides this the limits of Stanley's groups are very vague. Nevertheless it means an important advance in the search for a solution to the contamination problem, which is one of the most difficult questions in palynology.

Fluorescence palynology appears to be a useful expedient for contamination investigations. This will be shown here by fluorescence analyses of a number of strongly contaminated Pleistocene sediments from the N. Netherlands, as for example the above-mentioned "potklei" and older deposits, of which the source of the rebedded material could be established. Some examples of contamination to a minor degree are given in addition.

[*Editors' Note:* The material on examples and applications of fluorescence palynological dating has been omitted.]

VII. FLUORESCENCE AND GEOCHEMICAL COALIFICATION

INTRODUCTION

When a peat layer is progressively buried under sedimentary accumulations, it passes into lignite, which, on its turn, can be transformed into bituminous coal and further into anthracite. During this transformation the chemical composition of the plant substances is fundamentally altered. The carbon content of the remaining plant material relatively increases and other chemical compounds are progressively decomposed. These chemical changes are called coalification. The coalification series is characterized by the sequence peat — lignite — subbituminous coal — bituminous coal — anthracite, in which a general relationship between the successive stages exists. With regard to this series two fundamentally different mechanisms are distinguishable: (1) transformation by biochemical processes which is called *biochemical coalification* and (2) *geochemical coalification* (considered by the International Committee for Coal Petrography as metamorphism, see below).

Biochemical coalification takes place at or nearby the surface in the soil under atmospheric pressure and temperature and ends probably at the soft lignitic stage. It is caused by microorganisms occurring in peat and other humic deposits by which the less resistant plant substances are decomposed. It may have had an important influence on the chemical change of fossil palynomorphs during geological time, which resulted in a change in fluorescence colour of exines and exosporia (see next Chapter).

After a peat is lignified it may become involved in geochemical coalification. The fundamental chemical changes, occurring in the deposit, are the disappearance of nitrogen, a diminution of the oxygen and hydrogen content. The volatile matter yield (% V.M.[1]) and the "fixed carbon" (% F.C.[1]) relatively decrease and increase respectively. The fixed carbon content is used by the present author to express the rank of coal, except for soft browncoals. Rank is the stage reached by plant material after coalification; it represents the degree of coalification.

In fact the F.C. content and other chemical characteristics of coals mentioned do not reflect exclusively the degree of coalification, but also vary somewhat with their original petrographic and organic composition. In order to exclude the influence of variable petrographic compositions, the properties of the maceral vitrinite — in particular its optical reflectivity (e.g. Kötter, 1960; International Committee for Coal Petrography, 1963) — are used in some recent studies on coalification as a dependable parameter of rank (e.g. Patteisky and Teichmüller, 1960).

The degree of coalification has been long known to increase in vertical sections in sedimentary basins ("Hilt's rule"). Investigations of the regional increase in coalification have shown that the rank of coal is controlled mainly by the factors of temperature (i.e. maximum depth of burial and geothermal gradient), duration of exposure to temperature increase (M. and R. Teichmüller, 1954; Karweil, 1956; Patteisky and Teichmüller, 1960), and to a lesser extent by loading pressure, structural pressure and the presence of radioactive material in the buried rocks. For more details regarding the coalification problem and the factors governing its processes, reference should be made to the recent review papers of M. and R. Teichmüller (1954, 1958, 1965, 1966), M. Teichmüller and Thomson (1958), Karweil (1956), van Krevelen (1961), Kröger (1966), Hedemann and Teichmüller (1966) and many others.

Some remarks have to be made on the terminology used in coalification studies. In the literature one may find the term carbonization incorrectly used instead of coalification. Carbonization, however, is the result of rapid heating of coal in the laboratory or industry or by igneous intrusions into coal seams. The term coalification is used here according to the terminology accepted by the International Committee for Coal Petrography (1963).

This Committee and many authors consider coalification as a process of metamorphism. The physical conditions of geochemical coalification are largely similar to those of late diagenesis (epidiagenesis) or "burial metamorphism". A distinct boundary between diagenesis and metamorphism cannot be drawn, but browncoals and low-rank coals may be considered to be diagenetic. The relation between rank of coal and late diagenetic or burial metamorphic alteration in associated sedimentary rocks has been reviewed by M. and R. Teichmüller (1966) and Kisch (1966a, 1966b).

INFLUENCE OF GEOLOGICAL TIME AND GEOCHEMICAL COALIFICATION ON THE FLUORESCENCE OF PALYNOMORPHS

In the preceding chapters of the present study much attention has been paid to the important influence of geological time on the chemical nature of fossil palynomorphs, as appears from the change in their fluorescence colour. It has become known from the work of various authors that the chemical character of exines and exosporia has been influenced also by geochemical coalification (Wilson, 1961, 1964; Gutjahr, 1966). Therefore we will pass now to a discussion of the importance of both time and rank of coal. After a short review of the influence of geological time, the change in fluorescence as a result of geochemical coalification will be discussed.

Influence of geological time

It appeared that the fluorescence of fossil palynomorphs changes gradually with increasing age from

---

[1]) All V.M. and F.C. percentages throughout the text are given on a dry, ashfree basis (d.a.f.).

colours at lower wave-lengths into colours at higher wave-lengths. In Chapter V this phenomenon has been described by means of a large number of fluorescence measurements of vesiculate pollen grains from various ages, ranging from the Paleozoic up to the Recent (Plate III). Fresh pollen grains of *Pinus sylvestris* show light blue or bluish white fluorescence colours. Those from the Holocene are dark or light green or bright white in ultra-violet light. The Pleistocene grains of this type show more light green and yellowish green or greenish yellow colours. In the Tertiary the *Pinus sylvestris*-type is mainly white and yellow in its fluorescence, while orange colours appear regularly. In the Mesozoic and Paleozoic the vesiculate grains show mainly orange and brown colours.

This great change includes almost the entire visible spectrum, except the violet and dark blue region. As the fluorescence is closely related to the chemical character of palynomorphs, it may be assumed that their chemical nature has been considerably changed by geological time. As will be discussed in the next Chapter, this change may be a result of biochemical coalification. From this large scale of fluorescence colour change with increasing age, it may be concluded that time forms the most important factor in the process of chemical alteration of palynomorphs. All samples from the Miocene and younger, mentioned in Plate III and fig. 24, belong to deposits, which have been only slightly buried in the subsoil. Their rank has never reached the lignitic stage. Therefore the influence of temperature on this process may be neglected.

Influence of geochemical coalification

It has become well known that with increasing rank of coal, fossil pollen grains and spores from coal beds or deep sections change in colour in normal light from yellow to light brown and then to dark brown. Their light absorption increases gradually (Gutjahr, 1966). They finally become black as soon as the exines and the exosporia have been totally transformed into coalified particles. In coals with an F.C. content of approx. 75 or 80 % or more, they have become indeterminable.

Some experiments have been carried out in the laboratory in order to reproduce the coalification phenomena of fossil palynomorphs. Kirchheimer (1933a, 1933b, 1935) tested fresh *Lycopodium* spores under high temperature and pressure. Similar experiments on spores from Eocene browncoal resulted in a chemical decomposition of the exosporia. Kirchheimer suggested that the damage which he observed on spores from Eocene browncoals, has been caused in a similar way by coalification. Gutjahr (1966) subjected fresh pollen grains of *Quercus robur* to a temperature test. Heating at various temperatures above 100° C at various exposure times showed that the colour of these pollen grains changes with higher temperatures. In the present author's opinion, however, such experiments have a limited value, as exposure time and natural conditions in the lithosphere can hardly be reproduced in the laboratory.

In order to determine the influence of geochemical coalification, a number of fluorescence determinations of palynomorphs from browncoals and coals of various age have been made. The rank of coal of the samples has been obtained by means of measurements of the optical reflectivity of vitrinite macerals in the same samples. The determination of the reflectivity has been carried out, according to the directions of the International Committee for Coal Petrography, by means of the photo-electrical method, and using the curves, published by Kötter (1960).

Fig. 33 shows a number of fluorescence measurements on palynomorphs of various age and rank [1]). The fluorescence values from Upper Tertiary browncoals have been separately plotted on a somewhat stretched horizontal scale in the left part of the figure, because the Miocene and Eocene browncoals coincide in rank. The reflectivity values of the samples are average values, calculated from at least 25 measurements on vitrinite or collinite. For localities and other data of the samples, reference should be made to Table VIII (Appendix).

The results of these comparative measurements must be interpreted very carefully, due to the presence of scattered samples and the small number of fluorescence measurements and samples. It appears, however, from the general decrease in spectral ratio with increasing reflectivity, that a correlation between fluorescence and reflectivity may be assumed.

It appears that the change in fluorescence with increasing rank runs parallel to the change with increasing age (Plate III). An increase in rank, however, resulted in a much smaller change of fluorescence of palynomorphs in low-rank deposits (from yellow to orange or brown) than that caused by time (from blue or white to yellow). As will be discussed below, time may be a more important factor than rank in the process of chemical alteration of palynomorphs from low-rank deposits.

Influence of corrosion

Another factor that may play a role at the change in fluorescence of palynomorphs must be mentioned, viz. the corrosion. Kirchheimer (1935) and Wilson (1964) already called attention to the importance of corrosion to the study of coalified palynomorphs. It also causes a change in the fluorescence properties — even in material that has not been coalified (see next Chapter) — probably as a result of erosion of the browncoal or coal on ancient erosional surfaces, which were covered afterwards by younger sediments. A considerable number of corroded palynomorphs is present in two samples mentioned in fig. 33. The Eocene browncoal from Hampstead, Great Britain (sample VI/B), shows some corrosion, observable

---

[1]) Reflectivity measurements kindly communicated by Shell Exploration and Production Laboratory, Rijswijk (Z.-H.), The Netherlands.

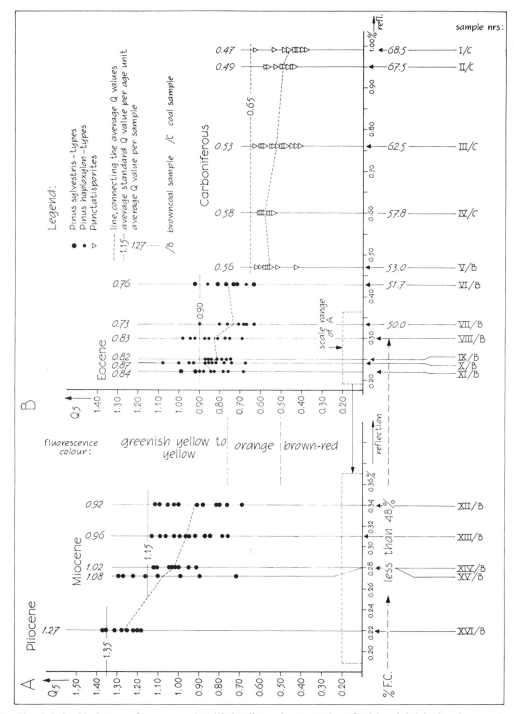

Fig. 33. Relationship between fluorescence of coalified pollen and spores to the reflectivity of vitrinite from browncoal and coal of various geological ages. A: Upper Tertiary browncoal B: Lower Tertiary and Carboniferous browncoal and coal.

from the brown colours of the pollen grains both in normal and ultra-violet light, and macroscopically visible by the browncoal as well. Of this and another sample, viz. the coal from Lippe, W. Germany (sample II/C), only the highest Q values have been drawn in fig. 33, and have been taken into account at the calculation of the spectral ratio averages.

RELATIONSHIP BETWEEN FLUORESCENCE AND RANK OF COAL

After the search for a correlation between fluorescence and optical reflectivity that has been mentioned above, it should be useful to investigate the relation between fluorescence and light absorption, in order to obtain a complete picture of the correlation between fluorescence and rank of coal.

Relationship between fluorescence and light absorption

The increase in absorption of light by fossil palynomorphs with increasing rank has been investigated by Gutjahr (1966). He used equipment consisting of a photocell, fixed on the microscope body and connected with a separate four-stage amplifier. Gutjahr suggested that a correlation exists between the degree of coalification and the light absorption of palynomorphs. The light absorption values have therefore been considered to be a measure of the rank of coal. The light absorption of palynomorphs appeared to increase with increasing depth of burial of the sediments, probably as a result of temperature increase.

Judging from the results of a small number of comparative measurements of fluorescence and light absorption (fig. 34), it seems that fluorescence and light absorption of palynomorphs may be correlated, whereby the fluorescence shifts towards the red part of the spectrum with increasing light absorption. The fluorescence measurements are plotted in fig. 34 on the vertical scale; the absorption values [1]) are given horizontally. A broken line connects the determinations of palynomorphs of the same slide. The actual values of the spectral ratio Q are represented by dots; the absorption values are indicated with small crosses in order to show the spread which is found with either method. Each value represents the fluorescence or absorption value of one pollen grain or spore measured.

[1]) Light absorption measurements are kindly communicated by Shell Exploration and Production Laboratory, Rijswijk (Z.-H.), The Netherlands, determined according to the method by Gutjahr (1966).

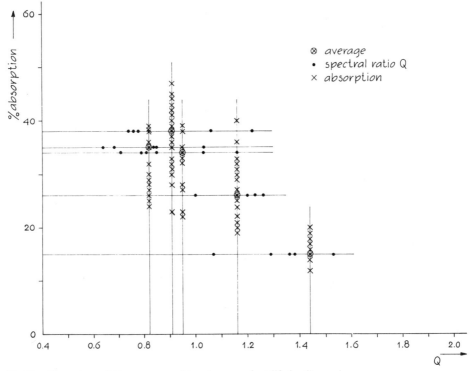

Fig. 34. Comparison of fluorescence and translucency of coalified pollen grains.

The interpretation of fig. 34 meets difficulties due to the small number of fluorescence measurements and samples. More extensive investigations are needed to establish exactly the relationship between fluorescence and light absorption. Nevertheless, the figure suggests that light absorption and fluorescence may be correlated.

Relationship between fluorescence and rank of coal

The correlation between fluorescence and reflectivity, and between fluorescence and light absorption, can be summarized in a general correlation between fluorescence and rank of coal of fossil palynomorphs. In comparison with the large number of reflectivity data, which have been published by various authors, much more determinations of the fluorescence are needed for the continued investigation of this subject. Besides, geochemical coalification forms a complex problem, of which various aspects may be important for the study of the chemical character and the fluorescence of palynomorphs.

Some remarks have to be made with regard to figs. 33 and 34.

The comparative measurements of both figures suggest that the fluorescence determinations show only significant differences in the average Q of various samples in a range of the spectral ratio of approx. 0.90 or more. From the large spread in those samples with an average Q value of less than 0.90, the decrease in spectral ratio seems to be slight. On the other hand, the spread appears to be different in various samples. When fluorescence analyses will be used as a standard to the rank of coal of palynomorphs, it must be known whether these differences are significant or not. Besides, the question must be studied, why the spread is different for various rock samples of similar age and rank.

Fossil palynomorphs from coalified deposits seem to be "older" in fluorescence (fig. 33) than those from non-coalified rocks of the same age (Plate III). A slight increase in rank of younger (Tertiary) deposits may cause a greater change in fluorescence than that caused by a considerable increase in rank of older coalified rocks. This phenomenon may be related with the hypothesis of the chemical composition of sporopollenine (see next Chapter).

CONCLUSIONS

Some preliminary conclusions may be drawn from the fluorescence analysis.

In each group of a certain age in fig. 33 the fluorescence changes into orange or brown colours with increasing reflectivity (i.e. rank). The influence of geochemical coalification may be expressed in this figure by the broken line, which connects the average Q values. The factor time is represented for the ages mentioned by the standard average Q values, derived from Plate III.

The importance of time and rank for the fluorescence phenomena may be illustrated as follows:

(1) The largest shifting in fluorescence colour occurs in the low-rank part of the coalification series (peat and lignite), where geological time plays the prominent role at the chemical alteration of palynomorphs. A much smaller change in fluorescence took place at higher ranks between 48 and 75 % F.C.

(2) The Miocene lignites, mentioned in fig. 33, are nearly equal in age, but they differ in rank and fluorescence. The samples nrs. XII/B-XIV/B are taken from the upper part of the main browncoal seam ("Hauptflöz") in the Lower Rhine area in W. Germany, which belongs to the Lower Helvetian or Upper Burdigalian (von der Brelie, 1967). The lignite of Meiszner (sample nr. XV/B) is somewhat older in age and can be correlated with the lower part of the "Hauptflöz", which has been deposited earlier in the Burdigalian (von der Brelie, 1967). Hence it appears that the decrease in spectral ratio of *Pinus sylvestris* pollen grains with increasing reflectivity of the vitrinite of these browncoals must have been caused by temperature rather than by difference in age.

(3) The fluorescence of exosporia of *Punctatisporites* from the Lower Carboniferous browncoal of the Moscow Basin (sample nr. V/B) appears to be only slightly different from that of similar spores from the more strongly coalified Carboniferous coals (samples nrs. I/C-IV/C). Apparently the fluorescence has been shifted here so much, that even a large increase in rank resulted only in a small decrease of the average Q.

(4) A comparison of the Moscow material with Miocene lignites from the Rhine area (samples nrs. XII/B-XV/B), on the other hand, shows a considerable difference in fluorescence of the palynomorphs. As in Carboniferous coals the fluorescence of vesiculate forms appears to be equal to that of *Punctatisporites* exosporia, the difference between Carboniferous and Miocene fluorescence cannot be explained by the fact that various forms from both ages were measured. The rank of coal differs not much in both cases, as they show only a small difference in reflectivity, but the difference in age of these browncoals amounts to approx. 300 million years. Therefore it may be assumed that the difference in fluorescence between the Miocene and Carboniferous palynomorphs is determined by time rather than by rank.

(5) No important difference exists between the fluorescence colour of vesiculate types and microspores of *Punctatisporites* from the same Carboniferous coals. Likewise the fluorescence colour of *Pinus sylvestris* pollen grains appears to be similar to that of *Pinus haploxylon*-types from Eocene and other Tertiary deposits. This supports the above-mentioned opinion (Chapter V) that the differences in fluorescence colour between various forms become gradually smaller with increasing age and ultimately disappear. Increase of rank has a similar result.

Resuming it may be established that the influence of time alone during biochemical coalification and of time and temperature during geochemical coalification result in a similar change in fluorescence colour of palynomorphs towards the red part of the spectrum. It may be assumed that time forms the most important factor at the colour change in low-rank deposits and that both time and temperature are determining the fluorescence colour of palynomorphs in stronger coalified rocks, whereas time may have had a greater influence on the chemical nature of exines and exosporia in higher rank browncoals and coals than has been expected. It must be noticed, however, that a short exposure time with higher temperature may result in a more rapid disappearance of translucency and fluorescence of palynomorphs than a long exposure time at low temperature. Much more comparative investigations like those described in this Chapter, are needed to get a clear insight into this subject.

Regarding the application of fluorescence palynology to coalification studies, the limitations of this method must be taken into account. Like the light absorption method of Gutjahr, fluorescence photometry covers a range of the coalification series, corresponding to an F.C. content of less than approx. 75 %, contrary to the reflectivity method, that can be used for the range between approx. 48 and 85 %. The disappearance of translucency and fluorescence coincides with the point at which the palynomorphs become unrecognizable.

Nevertheless, fluorescence palynology can be used in the study of low-rank coalified deposits. Besides, it may be useful to investigate the fluorescence of various types of coalified palynomorphs in relation to the question of the chemical and physical properties of sporopollenine (see next Chapter).

VIII. FLUORESCENCE AND CHEMICAL CHARACTER OF SPOROPOLLENINE

INTRODUCTION

The afore-described phenomena of fluorescence can be explained and discussed only when attention is paid to the chemical properties of the sporopollenine. In living plants many biochemical compounds and tissues show characteristic fluorescence colours, depending on their chemical nature and on the conditions during the observation (Goodwin, 1953 and others). The fluorescence of pollen and spore walls is also related to the chemical composition of exine and exosporium respectively.

This composition is difficult to study for palynomorphs, because they are very resistant and difficult to examine chemically. The reason why this fluorescence shows differences during geological history, must be due to changes in their chemical nature, caused by processes in the earth's crust. Such processes are very complicated and still poorly understood, though this subject was studied extensively by many investigators. In order to explain the processes and factors, by which the fluorescence phenomena are governed, the chemical nature of fresh and fossil sporopollenine must be known exactly. It must be established whether one or more fluorescent components in the exine and exosporium are present, how these are composed and how they change or disappear. It appears, however, from the data in literature that the exact chemical nature of fresh and fossil palynomorphs is in fact unknown. The fresh ones, showing bright fluorescence, have been investigated by numerous workers. Though many properties of these objects have become known and some investigators even gave bruto formulae for their composition, these studies deliver not any information on the origin of fluorescence. On the other hand more extensive analyses were made of fossil megaspores from the Carboniferous in which strongly fluorescent substances were found, like resin, paraffine and mainly naphthene,

but these spores are not fluorescent at all! Unfortunately data of exines from intermediate ages are not available. Nevertheless it is useful to pay some attention here to the facts and suppositions of the chemical data, the fossilization processes and their connection with the fluorescence phenomena. So much has been published, however, about coalification, fossilization and chemical analysis that the present knowledge of these subjects must be discussed here in a short review, in which only the most important studies are mentioned. For a complete information, reference should be made to a special paper (van Gijzel, 1968), dealing also with various aspects of the fluorescence phenomena.

PREVIOUS CHEMICAL ANALYSES

Fresh pollen and spores

Erdtman (1954) remarked in his "Introduction to Pollenanalysis" that fresh pollen and spores possess a high resistance to most chemical reactions, except to oxidation and strong heating, and are insoluble in nearly all solvents. Thanks to these properties fossil pollen and spores may be well preserved in deposits under anaerobic conditions, but on the other hand their resistance makes the study of the chemical composition far from easy.

Zetzsche and his co-operators succeeded in isolating an insoluble substance of fresh pollen and spores from the easily soluble constituents as fat, proteins, acidic and colouring matter. It was termed sporonine or sporopollenine and should be characteristic for exosporia and exines, respectively (Zetzsche and Huggler, 1928; Zetzsche and Vicari, 1931; Zetzsche and Kälin, 1931). Its presence has been established in fossil megaspores from the kaustobiolithes of Moscow and Tasmania (Zetzsche, Vicari and Schärer 1931). Zetzsche and Vicari even gave bruto formulae for this substance, which is different in composition for various species of pollen and spores. After Zetzsche it is highly unsaturated and reacts easily with halogens like bromine. The difference in chemical nature between the fossil megaspores and fresh *Lycopodium* spores should be caused by microbiological action. Recently, more data have become available by various studies. Rowley (1962) noticed that the resistance to acetolysis [1]) is regarded at present as a positive test for sporopollenine. Shaw and Yeadon (1966) found that membranes of *Lycopodium clavatum* spores and *Pinus sylvestris* pollen are similar. They consist approximately of cellulose (10—15 %), an ill-defined "xylan" fraction (10 %), a lipid fraction (55—65 %) and a lignin-like fraction (10—15 %). Heinen (1960, 1963) and Linskens (1962) supposed that sporopollenine should be related with cutine, which is the main constituent of cuticles.

Fossil megaspores

Furthermore fossil megaspores only have been subject of chemical study. The fact that certain coal-layers often show an abundance of fossil spores and possess a divergent content of volatile matter drew the attention of coal chemists. However, literature on coal chemistry contains only a few data on the composition of these megaspores.

Zetzsche was the first one who succeeded in cooperation with Vicari and Schärer (1931) in separating and analysing fossil megaspores, occurring in an oil-shale of Tasmania and a browncoal of the Basin of Moscow. He pointed out that biochemical reactions during the coalification left the (undetermined) oxygen of the sporopollenine undisturbed, but changed on the other hand the amount of hydrogen and hydroxyl groups. The fossil exosporia from both deposits should show a decrease in these components. Later on, these suppositions were confirmed by the experiments of Zetzsche and Liechti (1937), at which fresh exines of *Lycopodium* and *Corylus* could be changed in a similar way by chemical processes.

The analyses of Macrae and Sprunk and co-workers are far more extensive.

Sprunk, Selvig and Ode (1938) investigated megaspores from a Michigan spore coal in the U.S.A. They noticed that former workers have separated spores from coal by chemical treatments, but the results of analyses of such spores are subject to criticism, because the spores are affected by the chemicals used. Therefore they cleaned the spores mechanically. The gas, distilled from megaspores, contained a high content of hydrogen, methane and ethane. The amount of carbonized residue proved to be low.

The most extensive analyses were made by Macrae (1943), who investigated spore exines from a bituminous coal from the Upper Beeston seam in West Yorkshire, Gr. Britain, by means of thermal de-

[1]) The term acetolysis is not used here in the chemical sense, but means a pollen preparation method after Erdtman (1954, 1960).

composition experiments. First the spores were extracted with hot aceton in an atmosphere of nitrogen. The extract, containing resins, appeared to be strongly fluorescent. The thermal decomposition of the exines yielded a high content of heavy oils, consisting of carboxylic acids, phenols, bases, and neutral oils. The latter contained many components as unsaturated hydrocarbons, paraffins and napthhenes.

Van Krevelen and Schuyer brought these analyses in connection with other investigations. In their handbooks of coal chemistry (van Krevelen and Schuyer, 1957; van Krevelen, 1961) they noticed that the reaction mechanism of the coalification is extremely complicated from a chemical point of view. Its final product, coal, possesses a complex, chemically heterogeneous structure, which often cannot be determined exactly. The chemical composition of cuticles, spore membranes and corky tissue, which are closely associated with each other, show an intermediate position between those of plant waxes and of lignin. Van Krevelen and Schuyer used the method of statistical constitution analysis by which the average structure of a substance can be determined. They supposed that the spore membranes, etc. are combinations of lignin units with polymerization products of waxy alcohols and waxy acids, built up according to a pattern, similar to wood, which consists of complexes of cellulose with lignin as binding agent. The cellulose should have been replaced in the spores by the strongly water-repellent waxy alcohols.

Conclusions

Some critical remarks must be added to the cited analyses. Although the work of Zetzsche and his co-workers commands much respect, their conclusions must be considered as rather insufficiently founded. The insoluble residue of fossil spores begins to decompose at 200° C or slightly lower, as appears from the work of Sprunk and others. But Zetzsche analysed the samples at a temperature of 300° C, while the analyses of others were made at much higher temperatures (up to 500° C).

Zetzsche treated the pollen and spores with phosphoric acid to extract the cellulose and with hydrofluoric acid to remove the sand particles, while other investigators avoided these treatments. The question arises if these methods did not influence the results of Zetzsche's analyses. It seems not improbable that sporopollenine consists of various compounds, which are not mentioned by Zetzsche. His purpose of analysis, moreover, was quite different from that of Sprunk and Macrae. Hence it appears that the conclusions of Zetzsche are not simply comparable with those of the other authors and that the various analyses cannot be reconciled with each other.

Furthermore the question arises, how much the results of the analyses are influenced by the fact that the distillation was carried out at rather high temperatures, by which it becomes uncertain if all the components found were really present in the examined exines.

Hence, it appears that fossil exines possess a very complex character. Probably the fresh ones are more complicated in nature. Recapitulating, we may say that the composition of exines is still unknown and therefore it is impossible to define Zetzsche's term sporopollenine exactly. The present author uses this term in the sense that it includes all chemical components of palynomorphs, which are different from other plant constituents such as cutine, cellulose, etc.

Taking the fluorescence phenomena into account, it cannot be proved that the fluorescent resins, extracted from the Carbonifereous megaspore walls can be responsible for the fluorescence of the whole exines, because as far as known megaspores are not fluorescent at all. This also holds for other substances of the exosporium, found by Sprunk and Macrae, some of which are strongly fluorescent. On the other hand it may be possible that similar compounds, being absent here but present in younger exines, bring about the fluorescence. Another cause may be formed by the presence of lignin in the exines, except in the most strongly coalified ones (van Krevelen and Schuyer, 1957; van Krevelen, 1961, 1963). It is wellknown that this substance is fluorescent (see Brauns, 1952).

Fluorescence of fossil algae in boghead coal was found by Wolf (1966). She supposed that the fluorescence should be related to the situation of the C-bonds of the organic compounds.

Other chemical data

The photochemical processs of autoxidation, by which fresh pollenine takes up oxygen and fades if it is subjected to prolonged action of air, was discovered by Zetzsche and Kälin (1931). In the author's opinion it plays only a minor role at the fossilization of pollen, but may be responsible for the large variations in fluorescence colour of subfossil pollen.

Recently, interesting details of the properties of sporopollenine have become known by electron-microscopical studies. Afzelius (1956) sectioned fresh pollen for this purpose and observed that the exine can be affected by oxidation with benzoylperoxide at the preparation of the slides. Erdtman (1952) described a similar affection of spores by chlorination. In this reaction the sporopollenine should be transformed to chlorosporopollenine, changing its fine structure.

Chromic acid rapidly demolishes sporopollenine. The lamination and the amorphous granulation are destroyed and an almost hyaline appearance of the exine occurs (Afzelius, 1956).

Williams, Backus and Russel (1951) supposed that a single macromolecule of sporopollenine should exist of approximately fifty units of sporonine, as proposed by Zetzsche. Its molecular weight amounts 60,000 to 100,000.

Larson et al. (1962) published an excellent review of the results of electronmicroscopy of fresh exines. It is certain now that in exines at least two layers can be distinguished, which are different in chemical nature and react differently on acetolysis treatment. This is in accordance with many fluorescence observations.

The ectexine and endexine show in many cases various fluorescence colours.

Other important properties of the exine have become known by the work of Havinga (1962, 1964) and Sangster and Dale (1961, 1964) on corrosion phenomena. Havinga stated a corrosion susceptibility series, based on results of experiments in order to explain the corrosion phenomena of fossil pollen and spores in the soil. Oxidation treatment resulted in a very marked corrosion of the grains in an acidic environment, but proved to be very slight in an alkaline solution. This series shows much resemblance with the fluorescence colour series of subfossil pollen and spores (see Chapter IV). This selective corrosion is also in accordance with the results of palynological comparison of sands and peats.

RELATIONSHIP OF FLUORESCENCE TO CHEMICAL CHARACTER

As was noticed above, both properties are closely related. Too little is known, however, about the chemical nature of palynomorphs to establish this relation exactly. Nevertheless, some suppositions about their chemical composition may be derived from the fluorescence phenomena.

The first hypothesis concerns a division of sporopollenine into two or more components, that may be derived from the fluorescence spectra, obtained by means of the Berek photometer and the UV-microspectrograph.

The fluorescence colour of a spore or pollen grain may be considered to be a mixture of two or more different fluorescence spectra: a blue to green spectrum and an orange to red one, which represent two or more different substances in the sporopollenine. The mutual proportion of these components should determine the fluorescence colour observed. The fluorescence of a pollen type at a certain geological age is determined then by the ratio of these substances, and may vary from type to type with differences, which should depend on the chemical composition and on the resistance against decomposition by geological processes. At increasing age this should result in a relative decrease of the bluish green substances in favour of the orangisch red fluorescent compounds. This hypothesis is not supported, however, by the fluorescence spectra of fresh pollen and spores, measured in suspension (Chapter II, fig. 19b). Such spectra show one distinct maximum only in their curves, that may be situated at various wave-lengths. It is unknown whether these compounds are various modifications of one substance, sporopollenine, or not. It may be possible that they represent lignin and other compounds, mentioned by van Krevelen and Schuyer (1957), which should be present in the maceral exinite in coal. On the other hand, very small concentrations of contaminating elements in the sporopollenine may be responsible for the fluorescence phenomena. This important subject is still under investigation now.

Till more chemical data have become available, the term sporopollenine can be used for all palynomorphs. In the present study no distinction has been made into pollenine and sporine, as done by Zetzsche. Probably the exines and exosporia are not fundamentally different in chemical character.

A second assumption is the constancy of the original fluorescence colour of all fresh spore and pollen walls during geological history, according to the geological principle of actualism. For instance Miocene pollen grains of *Pinus sylvestris*, when they have belonged to the same recent species, must have been bluish white in fluorescence before they became fossilized in the Miocene, just like at present. This supposition is in accordance with the conclusion of White (1933), who found no evidence that the principal peat forming plant compounds in the Paleozoic differed, even in any detail of their chemical composition, from those of the coal-forming vegetation from similar climates in later times.

The third hypothesis about the nature of fossil sporopollenine is the assumption that the change in chemical composition during geological time, resulting in a change in fluorescence colour, should be continuous, not discontinuous, as appears from the standard diagrams of Plate III and fig. 24 (see Chapter V).

Experiments to test preparation techniques for fluorescence palynology (Chapter III) deliver some information on these assumptions. It appeared that the use of acids and oxidation treatment causes a remarkable change in fluorescence colour towards the red part of the spectrum. Continued action of such reagents results finally in complete annihilation of palynomorphs. Oxidation experiments, made by Havinga and others, pointed out that sporopollenine is destroyed by strong oxidation or at least corroded. In UV-light such corroded grains show a divergent colour, similar to those from coarse sands. Similar changes occur in test preparations of the oxidation treatment (Chapter III).

An exact description of the term sporopollenine cannot be given now, because it contains a rather large number of organic compounds. Therefore the relation between chemical character and fluorescence still cannot be explained. Various causes of the fluorescence have to be taken into account.

RELATIONSHIP OF FLUORESCENCE TO FOSSILIZATION

Corrosion

After the sedimentation the exines can be affected by corrosion, being a factor which can hamper fossilization of pollen and spores. This mainly occurs in coarse deposits and takes place rapidly above the groundwater level and very slowly or not underneath this level.

Havinga (1962) and others paid special attention to this subject. Owing to oxidation and biological activity in the soil, a selective disappearance takes place. Some pollen types are exceptional susceptible to corrosion. Pollen of *Quercus* and other types (Table III) is largely destroyed in sands under dry conditions, but is well preserved in a wet sandy soil. In pollen diagrams of sands the older pollen zonings often are not represented because their pollen has been destroyed entirely. Havinga (1962) stated that only a small number of the local tree pollen is preserved when a relative dry soil had a cover of oak forest during the Holocene.

Humification

When after sedimentation the conditions in a deposit were more favourable, the pollen and spores were more or less subjected to the sequence of the processes of humification, coalification and diagenesis.

In general the plant fragments firstly underwent humification or were transformed into peat. The genesis of most coal deposits starts in a living peat bog. The dead plant material settles down on the swamp bottom and is transformed into peat by microbiological activity. During the peat formation, decomposition products are formed, as for instance methane, the humates and humic acids, an important group of degradation products, the structure of which has not yet been clarified. An example of humification is the conversion of a *Sphagnum* bog into the slightly humified Young *Sphagnum* Peat.

When the humification begins, some instable and easily soluble constituents of pollen and spores as intine, fats, proteins, acidic and colouring substances begin to disappear. Under unfavourable conditions pollen grains with a rather weak exine are even completely destroyed.

The pollen surviving humification, is involved furthermore in the coalification process. In fact humification forms the first step in the coalification series.

Biochemical coalification

This study of fluorescence phenomena of fossil palynomorphs bears mainly on the coalification processes, in which they were involved. A clear distinction must be made between the biochemical and geochemical coalification, which differs fundamentally; the result of both is similar and their boundaries cannot always be drawn sharply.

Various authors stated that microorganisms are present in nearly every rock type, even under unfavourable conditions. These organisms cause the processes of biochemical coalification, during which the remaining weaker plant tissues and substances are further decomposed in so far as they survived peat formation and humification. This occurs for instance when a well-structured peat is broken down into a structureless mass with remains of very resistant fragments (wood, seeds, resin) only.

Many microorganisms are adapted to live under anaerobic circumstances as in peat and gyttja, and break down the plant fragments during the complex system of organic reactions, resulting in the formation of humates, humic acids and other compounds. At increasing age this coalification results in a relative decrease of the H and O content of the deposit in favour of the number of C atoms, while the structure of the peat often disappears and volatile matter evades. The action of microorganisms causes decomposition of the carbon combinations as well, and may even bring about affection of the minerals in a deposit.

White (1933) established the gradual decomposition of plant substances at various water levels during the peat formation and summarized it in a coalification diagram (fig. 35), which shows various stages of biochemical decomposition. Plant material shows a divergent resistance to microbiological attack. At a

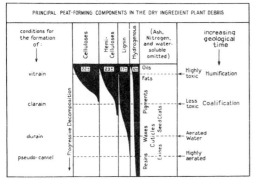

Fig. 35. Relative resistance to microbic decomposition and order of disappearance of principal peat-forming plant substances (modified after White, 1933).

decrease in acidity and toxicity of the environment during peat deposition an increase in the activity of microorganisms takes place, during which the material will be progressively decomposed. Successively the following substances will disappear: (1) oils, fats, celluloses, and part of the hemicelluloses, (2) the remaining hemicelluloses, pigments, part of the cuticles and seed coats, (3) the lignin, waxes and all seed coats and (4) part of the resins and exines and remaining cuticles. The most resistant fragments, as the resins and the exines, are partly preserved morphologically.

Although White's diagram concerns the relative resistance of principal plant components to subaqueous decay, determining in coals the type of coal macerals as vitrain, clarain, etc., it gives also an impression of the relative resistance of various substances to geochemical decomposition. It appears from this diagram that exines belong to the most resistant plant remains, during both biochemical and geochemical coalification.

The important role of biochemical coalification has become known by the geomicrobiological investigations of many authors. For a summary, reference should be made to the papers by Schwarz and Müller (1958), Abelson (1963) and Colombo and Hobson (1964) and to various handbooks of organic geochemistry (Kuznetsov et al., 1962; Breger, 1963). It appeared that microorganisms are present not only in humic deposits, but in other sediments as well. They show an important activity in the processes of oil formation, coalification, oxidation and desulfurication in the lithosphere.

Van Krevelen and Schuyer (1957) and other investigators, however, believe that microbiological decomposition can only continue as long as Fungi and bacteria are capable of participating in the attack on the material. As Fungi do not occur below a depth of about 40 cm, browncoal formation cannot have been influenced by these organisms. Besides, in their opinion, the effect of bacterial action also rapidly decreases with increasing depth. At greater depths bacterial conversion should be completely impossible. Van Krevelen and Schuyer supposed that, after the humification stage and certainly after the browncoal formation, only metamorphosis can have played a role. This view, however, is not shared by all investigators. The processes of geochemical coalification are fundamentally different. The result of their action has been described above (Chapter VII).

CONCLUSIONS

Summarizing, it may be stated that, although a close relationship between fluorescence and chemical character of fossil palynomorphs must exist, it cannot be exactly established, due to our lack in knowledge of this character. Chemical analysis of fossil megaspores only delivers some information on the occurrence of fluorescent substances in exosporia. The fluorescence of pollen and spores may be caused by various organic compounds, in which possibly the situation of their C-bonds may play a role.

It appears from the interpretation of the fluorescence spectra that, in case more compounds are responsible for the fluorescence, they may possess various fluorescence colours. Their mutual proportion may determine the colour of pollen and spores in UV-light.

The question arises if and how the fluorescence phenomena of fossil palynomorphs are related to the aforementioned fossilization processes.

The standard diagrams of fluorescence and geological age show that the change in fluorescence colour of all types of pollen and spores reveals a gradual trend for those deposits, which are not coalified geochemically (Chapter V). Hence, it appears that geological time must be considered as the most important factor at the process of colour change in these deposits.

Time only, however, does not explain these phenomena. The chemical transformation of fossil exines and exosporia during geological time, which appears from the fluorescence spectra, forms a very complex question. Not only their complicated chemical character, but also the lack in knowledge about the role of microorganisms in their fossilization and transformation, makes it difficult to explain exactly what happened.

Biochemical coalification (including humification) may be responsible for the fluorescence phenomena at increasing geological age for the following reasons: (1) geological time forms for both one of the most important factors, (2) both concern decomposition of plant substances, in which microorganisms may have participated, as Zetzsche presumed already for the alteration of sporopollenine in browncoal, (3) biochemical coalification and fluorescence colour change are wide-spread processes, occurring simultaneously in the same deposits.

On the other hand it may not always be expected that the biochemical activity has been constant during geological history. It may vary in different deposits and depends on temperature, pressure and ecological factors. This is not in full accordance with the gradual change in fluorescence colour. But accessory chemical reactions, which came into action by microorganisms may form the real cause as well.

The small ups and downs in the average Q connection line in the standard diagrams may be caused by other factors as well, for instance by corrosion or other sediment properties.

The explanation of the fluorescence phenomena still presents a problem which calls for further investigation in order to obtain essential information about chemical nature and microbiological susceptibility of fossil sporopollenine. Therefore new data must become available.

IX. GENERAL CONCLUSIONS

Review of the results

It appears from this study that fluorescence microscopy, being a well-known technique, applied on a large scale in other sciences for quite a long time, now can be applied succesfully to palynology as well, on the condition however, that equipments for microscopical fluorescence colour observation and determination are used, being sufficient accurate and sensitive. For the investigations, dealt with in this paper, an adapted Berek photometer and an UV-microspectrograph have been used, the results of which can be summarized in the following principles, which determine the primary fluorescence phenomena of fossil pollen grains and spores:

I. Each type shows at a certain geological age a specific fluorescence colour. The differences in colour are clearly observable mainly in material from Tertiary and younger deposits, as for instance is demonstrated by a number of fluorescence spectra of Young Holocene pollen grains and spores (Plate I). Differences in chemical nature between various forms must be responsible for it. The more or less variation in colour may be connected with the state of preservation of the pollen grains. At increasing geological age these variations decrease gradually.

II. Each type shows fluorescence colour changes with increasing geological age from the bluish green to the orange-red part of the spectrum. This relationship has been obtained from spectral fluorescence determinations of vesiculate pollen types, which were studied in about one hundred samples of sediments of various facies, ranging in age from subrecent to Upper Paleozoic. An explanation of this phenomenon may be found in the processes of humification and biochemical coalification. Complex chemical reactions and biochemical action may have changed the chemical character of the exines from the rocks involved in these processes, the final results of which appear from the fluorescence colours (Plate II and III). A relatively small number (ca 17 %) of the deposits measured show a fluorescence, being slightly too high or too low in comparison with the general tendency. Here again the preservation of the exines may be important.

III. The third principle of fluorescence palynology has been derived from a number of fluorescence determinations of vesiculate pollen grains from lignites and coals of various stages in the coalification series. It concerns the change in fluorescence to red and brown colours with increasing rank of coal. The final stage at increasing age and coalification is the complete extinction of fluorescence, as the translucency of the exines has disappeared. This is in accordance with other coalification studies of fossil pollen grains and spores under normal illumination, carried out by Wilson (1961, 1964) and by Gutjahr (1966). The relationships between fluorescence colour and translucency of exines to reflectivity of coal have been investigated. Here again the relation between fluorescence colour and chemical nature of the pollen and spore walls is obvious.

IV. Furthermore it appeared that the preparation techniques of pollen slides for fluorescence microscopy may cause difficulties when strong oxidation or acids, as for instance fluoric acid, Schulze's treatment and acetolysis are used. Then a colour change appears to red or brown colours, similar to that with increasing age. A competitive examination of samples of various sediment types revealed some interesting facts about the following question.

V. The chemical character of fresh and fossil pollen and spore walls appears to be a very complicated problem. The fluorescence spectra suggest that sporopollenine may exist of at least two compounds, being not equal in resistance. Those, being orange or red in fluorescence colour, should be more resistant against oxidation, biochemical and geochemical coalification than the blue and green components. The chemical composition forms still an important question, however, as well as the changes in their nature by the above-mentioned geological processes.

Applications
In reference to the relations of fluorescence palynology the most important applications of these principles are as follows:
1. The difference in fluorescence colour of various forms of pollen grains and spores can be used in pollen morphological studies of fresh and fossil material. It may be valuable to observe the various layers of the exine of different types under the fluorescence microscope.
Continued study of fluorescence spectra of fresh and subrecent material may provide a useful tool in pollen morphology.
2. Age determination of deposits of unknown age, viz. from the Cenozoic, when conventional pollen analysis fails due to the absence of significant pollen associations or specific zonings in the pollen diagram. By comparison of fluorescence measurements with a standard diagram of fluorescence and geological age (Plate III) it has become possible to estimate the age of such deposits with an accuracy of more than 83 %. Fluorescence palynological dating can be applied in particular in the range between 70,000 years B.P. and the Middle Tertiary.
3. Deposits, contaminated with rebedded pollen, can be dated in a similar way. The greater the difference in age of autochthonous and reworked pollen, the more easily both groups can be distinguished (Photoplate, fig. 7). Besides, more groups of secondary material can be recognized, which is very important in case the sources of contamination are unknown. It has become an important question, occurring at any age. A number of examples of contaminated Pleistocene sediments could be dated succesfully, as for instance the problems of the "potklei" and "Lauenburger Ton" in the N. Netherlands and N.W. Germany (Plate V), of which the age of the contamination sources could also be stated.
4. Similar fluorescence determinations can be used to make long-distance correlations, as has been attempted for Pleistocene sections from various continents (van Gijzel et al., 1967). It appeared that fluorescence changes gradually with increasing depth corresponding with increasing geological age. When comparing the spectral ratio values of vesiculate grains from various stratigraphical stages with the average fluorescence values of the standard diagram, dating and correlation are possible, on the condition, however, that a large number of measurements are made.
5. Determination of the rank of coal of coalified pollen bearing deposits may be carried out, unless they are so strongly coalified that exines can no longer be distinguished, (at an F. C. content of more than 75 %), because they have become black and indeterminable. This application may form — together with absorption measurements of exines according to the method of Gutjahr (1966), colour determinations by Wilson (1964) and reflection determinations of other coal substances — an expedient in coalification studies.
In general, the state of pollen preservation causes great difficulties, as Wilson (1961) and Gutjahr (1966) pointed out, when pollen grains are coalified so intensely that they have become opaque and have lost any structure. Exines which are only less coalified, can be expected in deposits with a rank of coal in the range from high-volatile A bituminous coal or at a lower rank. Therefore, all these methods, including fluorescence palynology, are restricted in use up to this stage of the coalification series.
6. Fluorescence investigations of other palynological objects are possible, as, for instance, of acritarchs, hystrichosphaerids and dinoflagellates. The first mentioned are even more resistant than exines and show, due to their different chemical nature, no colour change with increasing geological age. They react differently upon the fossilization and coalification processes in comparison with fossil exines.
The main applications mentioned are now under continued investigation by the present author.

Discussion of the results
The question may arise, if all phenomena of fluorescence colour change with increasing geological age have been caused by biochemical or by geochemical coalification, in other terms, whether temperature and pressure have played an important role. There is no doubt, however, that their influence must be neglected there, for which the following arguments can be advanced:
Firstly, not any indication exists that factors as tectonical pressure or temperature have had an influence on the fluorescence colour change of the pollen grains from Upper Tertiary and Quaternary deposits which have been chosen for the standard diagrams. They are provided by sediments, being situated just below the surface, and, if they are obtained from borings, their depth amounts to less than 300 m. Above this level the influence of the temperature gradient in this part of the North Sea Basin (3° C/100 m) may be neglected. It may be excluded that a thick mass of overburdening rocks has covered these beds. In the collection areas magma intrusions or volcanic activity have been absent during the Cenozoic, and no strong tectonical movements occurred. Hence, it appears that regional metamorphism can be left out of consideration. Therefore geological time only must be considered to be the main factor in these fluorescence colour change phenomena.
Secondly, not any relation has been found between the fluorescence colour or intensity and the temperature of the atmosphere, which has fluctuated considerably and many times during the Cenozoic in N.W. Europe.
The results of this study accent the distinction between the biochemical and geochemical coalification. Although both have taken part in the formation of lignite and coal, they are very different in origin, nature and action. At the transformation of dead plant fragments into peat the first one only played a role. In biochemical coalification geological time forms a very important factor. In the geochemical coalification,

time, temperature and/or pressure are the main factors. The boundaries between both processes cannot be drawn sharply, because they pass often into each other. A browncoal or coal forms the final result of their combined action; but the part of each of the factors time, temperature and pressure cannot always be stated exactly.

Experiments, carried out in the laboratory to establish the separate influence of these factors will always be insufficient, because geological time can never be imitated. Nevertheless, such attempts may deliver new informations about coalification problems. In the authors' opinion, much more may be expected from extensive and comparable studies of the natural conditions, under which peat, lignite and coal have been formed. By preference those deposits must be chosen, of which may be supposed that they have been altered by one of the mentioned factors only. One of the most important conclusions of the present paper is, that fluorescence palynology forms a useful expedient for these studies.

[*Editors' Note:* The acknowledgments and summaries in French and German have been omitted.]

REFERENCES

Aario, L., 1940. Waldgrenzen und subrezente Pollenspektren in Petsamo, Lappland. An. Acad. Sci. Fennicae, A/54/8, 1—120.

Abelson, P. H., 1963. Paleobiochemistry. Proc. 5th Internat. Cong. Biochem., 3, 52—68.

Afzelius, B. M., 1956. Electron-microscope investigations into exine stratification. Grana Palynol., 1/2, 22—37.

Arkhipov, S. A. & Matveyeva, O. V., 1960. Spore-pollen spectra of the Pre-Samarova anthropogene deposits of the glaciation zone in the Yenisei region of the West Siberian Lowland. Dokl. Acad. Sci. U.S.S.R., Earth Sci. Sec., 135/6, Engl. Translat. (1961), 1204—1206.

Asbeck, F., 1955. Fluoreszierender Blütenstaub. Naturw. Deutschl., 42/23, 632.

Bennett, A. J. R., 1963. Origin and formation of coal seams, a literature survey. Commonw. Austral., Commonw. Sci. & Industr. Res. Organ., Misc. Rep., 239, 1—14.

Berger, F., 1934. Das Verhalten der Heufieber-erregenden Pollen im filtrierten ultravioletten Licht. Beitr. Biol. Pflanzen, 22/1, 1—12.

Brauns, F. E., 1952. The chemistry of lignin. New York, Acad. Press Inc., 1—808.

Bräutigam, F. & Grabner, A., 1949. Beiträge zur Fluoreszenzmikroskopie. Wien, Fromme Verl., 1—131.

Breger, I. A., 1963. Organic Geochemistry. Sympos. Publ. Div., New York, Pergamon Press, 1—658.

Brelie, G. von der, 1967. Quantitative Sporenuntersuchungen zur stratigraphischen Gliederung des Neogens in Mittel-Europa. Rev. Palaeobot. Palynol. 2, 147—162.

Brosius, M., 1958. Tegelen Braunkohle von Fürstenhage bei Hessisch-Lichtenau. Notizbl. hess. Landesamt Bodenforsch., 86, 316—319.

Brouwer, A., 1948. Pollenanalytisch en geologisch onderzoek van het Onder- en Midden-Pleistoceen van Noord Nederland. Leidse Geol. Med., 15, 260—346.

Brown, C. A., 1960. Palynological techniques. Library Congress, Catalogue Card 60-14297, 1—188.

Bukatsch, F., 1941. Einige Anwendungsgebiete der Fluoreszenzmikroskopie. Zeitschr. Gesamte Naturw., 7, 288—296.

Cepek, A. G., 1960. Eisrandlagen, Interglaziale und "Präglazial" in Brandenburg.7. Jahrestag. Geol. Ges. D.D.R., Exkursionsf. Brandenb., 115—134.

Chiguryayeva, A. A. & Voronina, K. V., 1960. Data on Upper Pleistocene vegetation in the North Caspian area. Dokl. Acad. Sci. U.S.S.R., Earth Sci. Sec., 131/6, Engl. Translat. (1961), 362—364.

Clark, G. L., 1961. The encyclopedia of microscopy. New York, Reinhold Publ. Co., 1—693.

Colombo, U. & Hobson, G. D., 1964. Advances in organic geochemistry. Proc. Internat. Meeting Milan (1962). New York, Pergamon Press Sympos. Publ. Div., 1—488.

Cushing, E. J., 1964. Redeposited pollen in Late-Wisconsin pollen spectra from East-Central Minnesota. Am. Jour. Sci., 262, 1075—1088.

Davis, M. B., 1961. The problem of rebedded pollen in Late-Glacial sediments at Taunton, Massachusetts. Am. Jour. Sci., 259/3, 211—222.

Diener, S., 1960. Pleistozän südlich und südwestlich von Berlin. 7. Jahrestag. Geol. Ges. D.D.R., Exkursionsf. Brandenb., 135—150.

Dumait, P., 1962a. L'action des ultrasons sur les pollens. Note préliminaire. Pollen et Spores, 4/1, 175—180.

—. 1962b. Le vibroséparateur. Pollen et Spores, 4/2, 311—316.

Eberl, B., 1928. Zur Gliederung und Zeitrechnung des alpinen Glazials. Zeitschr. deutsche Geol. Ges., 80, 107—117.

Edelman, C. H., 1933. Petrologische provincies in het Nederlandsche Kwartair. Thesis, Amsterdam, 1—104.

Eder, H. & Fritsche, H., 1963. Zur Farbphotographie fluoreszenzmikroskopischer Objekte. Leitz Mitt. Wiss. Techn., 2/5, 143—145.

Emiliani, C., 1966. Paleotemperature analyses of Caribbean cores P6304-8 and P6304-9 and a generalized temperature curve for the past 425,000 years. Jour. Geol., 74/2, 109—126.

Erdtman, G., 1937. Pollen grains recovered from the atmosphere over the Atlantic. Meddel. Göteborgs Bot. Trädg., 12, 185—196.

—, 1952. Pollen Morphology and Plant Taxonomy. Angiosperms. Waltham, Mass., Chron. Bot. Co., 1—539.

—, 1954. An Introduction to Pollen Analysis. Waltham, Mass., Chron. Bot. Co., 12, 1—239.

—, 1960. The acetolysis method. A revised description. Svensk Bot. Tidskr., 54, 561—564.

Faegri, K. & Iversen, J., 1950. Textbook of Modern Pollen Analysis. Copenhagen, Munksgaard, 1—168.

Felix, C. J., 1963. Mechanical sample disaggregation in palynology. Micropaleontol., 9/3, 337—339.

Florschütz, F. & Jonker, F. P., 1939. A botanical analysis of a late-pleistocene and holocene profile in the Rhine delta. Rec. Trav. Bot. Néerl., 36, 686—696.

Gibson, M., 1963. Ultrasonic disaggregation of shale. Jour. Sedim. Petrol., 33/4, 955—958.

Gijzel, P. van, 1961. Autofluorescence and age of some fossil pollen and spores. Kon. Ned. Akad. Wet., Proc., B, 64/1, 56—63.

—, 1963. Notes on autofluorescence of some Cenozoic pollen and spores from the Netherlands. Med. Geol. Sticht., Nw. Ser., 16, 25—32.

—, 1966. Die Fluoreszenz-Photometrie von Mikrofossilien mit dem Zweistrahl-Mikroskopphotometer nach Berek. Leitz Mitt. Wiss. Techn., 3/7, 206—214.

—, 1967a. Palynology and fluorescence microscopy. Rev. Palaeobot. Palynol., 2, 49—79.

—, 1967b. Glacial morphology of the Middle and Upper Pleistocene in E. Groningen (The Netherlands). Leidse Geol. Med., in press.

—, 1967c. Palynology of Holocene deposits in the Oldambt area, E. Groningen (The Netherlands). Leidse Geol. Med., in press.

—, 1968. A review of the chemistry of fresh and fossil sporopollenine. Chem. Geol., in press.

Gijzel, P. van, Hammen, Th. van der & Wijmstra, T. A., 1967. Long-distance correlation of Pleistocene deposits in South America, Spain and The Netherlands. Palaeogeogr. Palaeoclimatol. Palaeoecol., 3, 65—71.

Gischler, C. E., 1967. A semi qualitative study of the hydrogeology of the North Netherlands. Transact. Royal Geol. Min. Soc. Neth., 24, 1—75.

Gleason, S., 1960. Ultraviolet guide to minerals. Princeton, N.J., van Nostrand Co., 1—244.

Goodwin, R. H., 1953. Fluorescent substances in plants. Ann. Rev. Plant Physiol., 4, 283—304.

Gottschewski, G. H. M., 1954. Die Methoden der Fluoreszenz- und Ultraviolett-Mikroskopie und Spektroskopie in ihrer Bedeutung für die Zellforschung. Mikroskopie, Zentralbl. mikrosk. Forsch. Meth., 9/5/6, 147—167.

Gripp, K., 1933. Geologie von Hamburg und seiner näheren und weiteren Umgebung. Hamburg, Gesellsch. Hartung, 1—154.

—, 1952. Inlandeis und Salzaufstieg. Geol. Rundschau, 40, 74—80.

Gripp, K., Dewers, F. & Overbeck, F., 1941. Das Känozoikum in Niedersachsen. Geol. Lagerst. Nieders., 3, 1—503.

Groot, J. J., 1966. Some observations on pollen grains in suspension in the estuary of the Delaware river. Marine Geol., 4, 409—416.

Groot, C. R. & Groot, J. J., 1964. The pollen flora of Quaternary sediments beneath Nantucket Shoals. Am. Jour. Sci., 262/4, 488—493.

Groot, J. J. & Groot, C. R., 1966. Marine Palynology: possibilities, limitations, problems. Marine Geol., 4/6, 387—395.

Gutjahr, C. C. M., 1966. Carbonization measurements of pollen-grains and spores and their application. Leidse Geol. Med., 38, 1—29.

Haberlandt, H., 1942. Luminiszenzanalyse und Lagerstättenforschung. Zeitschr. prakt. Geol., 8, 101—104.

Hacquaert, N., 1960. Palynologisch onderzoek van de cenozoische zanden (Scaldisien en Merxemien) van het Hansadok te Antwerpen. Natuurwet. Tijdschr., 42, 65—112.

Haitinger, M., 1938. Fluoreszenzmikroskopie. 2nd ed. Leipzig, Akad. Verl. Ges. Geest & Portig (1959), 1—168.

Hammen, Th. van der, & Gonzalez, E., 1964. A pollen diagram from the Quaternary of the Sabana de Bogotá (Colombia) and its significance for the geology of the northern Andes. Geol. & Mijnb., 43, 113—117.

Havinga, A. J., 1962. Een palynologisch onderzoek van in dekzand ontwikkelde bodemprofielen. Thesis Wageningen, Uitg. Veenman & Zn., 1—165.

—, 1964. Investigation into the differential corrosion susceptibility of pollen and spores. Pollen et Spores, 6/2, 621—635.

Hedemann, H. A. & Teichmüller, R., 1966. Stratigraphie und Diagenese des Oberkarbons in der Bohrung Münsterland 1. Zeitschr. deutsche Geol. Ges. (1963), 115/2—3, 787—825.

Heinen, W., 1960. Ueber den enzymatischen Cutin-Abbau; I. Mitt.: Nachweis eines "Cutinase"-Systems. Acta Bot. Neerl., 9, 167—190.

—, 1963. Ueber den enzymatischen Cutin-Abbau; V. Mitt.: Die Lyse von peroxyd-Brücken im Cutin durch eine Peroxydase aus *Pennicillum spinulosum* Thom. Acta Bot. Neerl., 12/1, 51—57.

Heuvelen, B. van, 1959. Potklei en gumbotill. Boor & Spade, 10, 105—116.

Holmes, A., 1959. A revised geological time-scale. Trans. Edinburgh Geol. Soc., 17/3, 183—216.

Hultzsch, A., 1960. Steine und Erden im Raum Hoyerswerda-Bernsdorf. 7. Jahrestag. Geol. Ges. D.D.R., Exkursionsf. Brandenb., 185—195.

International Committee for Coal Petrology, 1963. International handbook of coal petrography. Centre Nat. Rech. Sci., Paris, 2nd ed.

Iversen, J., 1936. Sekundäres Pollen als Fehlerquelle. Eine Korrektionsmethode zur Pollenanalyse minerogener Sedimente. Danm. Geol. Unders., IV. R, 2/15, 1—24.

Jacob, H., 1961a. Neuere Ergebnisse der Braunkohlen- und Torf-Petrographie. Gedanken zur Mazeral-Klassifikation. Geol. Rundschau, 51/2, 530—546.

—, 1961b. Ueber bituminöse Schiefer, humose Tone, Brandschiefer und ähnliche Gesteine — Ein Beitrag zur Frage der Erdölgenesis aus kohlenpetrologischer Sicht. Erdöl u. Kohle, Erdg. Petrochem., 14/1, 2—11.

Jux, U., 1961. The palynologic age of diapiric and bedded salt in the Gulf Coastal Province. Louis. Geol. Bull., 38, 1—46.

Karweil, J., 1956. Die Metamorphose der Kohlen vom Standpunkt der Physikalischen Chemie. Zeitschr. deutsche Geol. Ges., 107, 132—139.

Kirchheimer, F., 1933a. Die Erhaltung der Sporen und Pollenkörner in den Kohlen sowie ihre Veränderungen durch die Aufbereitung. Bot. Archiv., 35, 134—187.

—, 1933b. Der Erhaltungszustand des Pollens in den Glanzbraunkohlen. Ber. schweiz. Bot. Ges., 42, 246—251.

—, 1934. Die Beschaffenheit unter erhöhtem Druck thermisch beanspruchter *Lycopodium*-Sporen. Ber. schweiz. Bot. Ges., 43, 19—29.

—, 1935. Die Korrosion des Pollens. Beih. Bot. Zentralbl., A/53, 389—416.

Kisch, H. J., 1966a. Chlorite-illite tonstein in high-rank coals from Queensland, Australia: Notes on regional epigenetic grade and coal rank. Am. Jour. Sci., 264, 386—397.

—, 1966b. Zeolite facies and regional rank of bituminous coals. Geol. Mag., 103/5, 414—422.

Kötter, K., 1960. Die mikroskopische Reflexionsmessung mit dem Photomultiplier und ihre Anwendung auf die Kohlenuntersuchung. Brennst. Chem., 41/9, 263—272.

Krevelen, D. W. van, 1961. Coal: Typology-Chemistry-Physics-Constitution. Amsterdam, Elsevier Publ. Co., 1—514.

—, 1963. Geochemistry of Coal. In: Breger, I.A.: Organic Geochemistry. New York, Symp. Publ. Div., Pergamon Press, 183—247.

Krevelen, D. W. van & Schuyer, J., 1957. Coal Science — Aspects of Coal Constitution. Amsterdam, Elsevier Publ. Co., 1—352.

Krieg, A., 1955. Eine neue Einrichtung für Fluoreszenzmikroskopie im durch- und auffallenden Licht. Zeitschr. wiss. Mikrosk. & mikrosk. Techn., 62, 256—264.

Kröger, C., 1966. Betrachtungen zum Inkohlungsablauf. Erdöl u. Kohle, Erdg. Petrochem., 19/9, 638—645.

Kuznetsov, S. I., Ivanov, M. V. & Lyalikova, N. N., 1962. Introduction to geological microbiology. New York, Mc Graw Hill Co. (Engl. Translat., 1963), 1—252.

Larson, D. A., Skvarla, J. J. & Lewis Jr., C. W., 1962. An electron microscope study of exine stratification and fine structure. Pollen et Spores, 4/2, 233—246.

Linskens, H. F., 1962. Cutinase Nachweis in Pollen. Zeitschr. Bot., 50/4, 338—347.

Macrae, J. C., 1943. The thermal decomposition of spore exines from bituminous coal. Fuel Sci. & Pract., 22/5, 117—129.

Maier, D., 1959. Planktonuntersuchungen in tertiären und quartären marinen Sedimenten. Neues Jahrb. Geol. Paläont., Abh., 107/3, 278—340.

Maier, D. & Wetzel, W., 1958. Fluoreszenzmikroskopie geologischer und paläontologischer Objekte. Zeiss Mitt., 1/4, 127—131.

Mehner, W., 1960. Lagerstättenkundliche und bodengeologische Probleme im Braunkohlentagebaugebiet von Senftenberg. 7. Jahrestag. Geol. Ges. D.D.R., Exkursionsf. Brandenb., 163—183.

Menéndez Amor, J. & Florschütz, F., 1962. Un aspect de la végétation en Espagne Méridionale durant la dernière glaciation et l'holocène. Geol. & Mijnb., 41/3, 131—134.

—, 1964. Results of the preliminary palynological investigation of samples from a 50m boring in southern Spain. Bol. R. Soc. Esp. Hist. Nat. (Geol.), 62, 251—255.

Meuter, F. de, 1965. Étude paléo-écologique des Foraminifères des Sables d'Edegem (Miocène Moyen) à Terhagen (prov. d'Anvers). Bull. Soc. Belge Géol. Paléont. Hydrol., 74/1, 53—59.

Overbeck, R., 1964. Ueber das Fluoreszenzverhalten einiger Pflanzenfossilien. Mikrosk. Österr., 19/9, 255—262.

Overbeck, F. & Schmitz, H., 1931. Zur Geschichte der Moore, Marschen und Wälder Nordwestdeutschlands. I. Das Gebiet von der Niederweser bis zur unteren Ems. Mitt. Provinzialst. Naturdenkmalpfl. Hannover, 3, 1—179.

Patteisky, K. & Teichmüller, M., 1960. Inkohlungs-Verlauf, Inkohlungs-Maszstäbe und Klassifikation der Kohlen auf Grund von Vitrit-Analysen. Brennst. Chem., 41/3, 79—84, 97—104, 133—137.

Pflug, H. D., 1963. The palynologic age of diapiric and bedded salt in the Gulf Coastal Province by Ulrich Jux. Bull. Am. Assoc. Petrol. Geol., 47/1, 180—181.

Polak, B., 1936. Pollen- und torfanalytische Untersuchungen im künftigen nordöstlichen Polder der Zuidersee. Rec. Trav. Bot. Néerl., 33, 313—332.

—, 1963. The sub-soil of the "Bleeke Meer" compared to the fluvio-glacial deposits of Speulde. Med. Geol. Sticht., Nw. Ser., 16, 39—47.

Polak, B., Maarleveld, G. C. & Nota, D. J. G., 1962. Palynological and sedimentary petrological data of a section in ice-pushed deposits (southern Veluwe, Netherlands). Geol. & Mijnb., 41/8, 333—350.

Przibram, K., 1962. Über die Fluoreszenz organischer Spuren in anorganischen Stoffen und ihre Verbreitung in der Natur. Forsch. & Fortschr., 36, 1—2.

Ralska-Jasiewiczowa, M., 1960. Pleistocene flora from Zablocie on the river Bug. Folia Quat., Polska Akad. Nauk. Oddz. WKrak., 2, 1—9.

Rowley, J. R., 1962. Stranded arrangement of sporopollenin in the exine of microspores of *Poa annua*. Science, 137/3529, 526—528.

Ruch, F., 1960. Ein Mikrospektrograph für Absorptionsmessungen im ultravioletten Licht. Zeitschr. wiss. Mikrosk. & mikrosk. Tech., 64, 453—468.

—, 1961. Ultraviolett-Mikrospektrographie. Leitz Mitt. Wiss. Techn., 1/8, 250—255.

Ruch, F. & Bosshard, U., 1963. Photometrische Bestimmung von Stoffmengen im Fluoreszenzmikroskop. Zeitschr. wiss. Mikrosk. & mikrosk. Techn., 65, 335—341.

Sacchi Vialli, G., 1962. Ricerche sulla fluorescenza dei fossili. I. Osservazioni sullo smalte dei denti di alcuni Vertebrati. Atti Instit. Geol. Univ. Pavia, 13, 23—54.

—, 1964. Ricerche sulla fluorescenza dei fossili. III. Osservazioni comparative chimiche e di fluorescenza sulla constituzione dei denti di *Carcharodon megalodon* Ag., in condizioni naturali e sperimentali. Atti Instit. Geol. Pavia, 15, 89—145.

Sangster, A. G. & Dale, H. M., 1961. A preliminary study of differential pollen grain preservation. Canad. Jour. Bot., 39, 35—43.

—, 1964. Pollen grain preservation of underrepresented species in fossil spectra. Canad. Jour. Bot., 42, 437—449.

Schneider, W., 1965. Zur faziellen Entwicklung im "Oberbegleiter des Lausitzer Unterflözes" im Tagebau Spreetal. Freib. Forschungsh., 16. Berg- & Hüttenm. Tag. 1964, Freib., 203—224.

Schucht, F., 1908. Der Lauenburger Ton als leitender Horizont für die Gliederung und Altersbestimmung des nordwestdeutschen Diluviums. Jahrb. Preusz. Geol. Landesanst., 39/2/1, 130—150.

Schwarz, W. & Müller, A., 1958. Methoden der Geomikrobiologie. Freib. Forschungsh., C 48, 1—40.

Shaw, G. & Yeadon, A., 1966. Chemical studies on the constitution of some pollen and spore membranes. Jour. Chem. Soc., (C), 16—22.

Shellhorn, S. J., Hull, H. M. & Martin, P. S., 1964. Detection of fresh and fossil pollen with fluorochromes. Nature, 202/4929, 315—316.

Sitte, P., 1960. Die optische Anisotropie von Sporodermen. Grana Palynol., 2/2, 16—37.

Sprunk, G. C., Selvig, W. A. & Ode, W. H., 1938. Chemical and physical properties of spores from coal. Fuel Sci. & Pract., 17/7, 196—199.

Stanley, E. A., 1965. Use of reworked pollen and spores for determining the Pleistocene-Recent and the intra-Pleistocene boundaries. Nature, 206/4981, 289—291.

—, 1966a. The problem of reworked pollen and spores in marine sediments. Marine Geol., 4, 397—408.

—, 1966b. The application of palynology to oceanology with reference to the northwestern Atlantic. Deep-Sea Res., 13, 921—939.

Staplin, F. L., Pocock, S. J., Jansonius, J. & Oliphant, E. M., 1960. Palynological techniques for sediments. Micropaleont., 6/3, 329—331.

Stevens, C. H., Jones, D. H. & Todd, R. G., 1960. Ultrasonic vibrations as a cleaning agent for fossils. Jour. Paleont., 34/4, 727—730.

Streel, M., 1964. Utilisation des ultrasons à divers stades de la technique d'extraction des spores d'une roche detritique. C.R. 5. Congr. Strat. Géol. Carbon. Paris (1963), 3, 1239—1248.

Teichmüller, M. & Thomson, P. W., 1958. Vergleichende mikroskopische und chemische Untersuchungen der wichtigsten Fazies-Typen im Hauptflöz der Niederrheinischen Braunkohle. Fortschr. Geol. Rheinld. Westf., 2, 573—598.

Teichmüller, M. & Teichmüller, R., 1954. Die stoffliche und strukturelle Metamorfose der Kohlen. Geol. Rundschau, 42, 265—296.

—, 1958. Inkohlungsuntersuchungen und ihre Nutzanwendung. Geol. & Mijnb., 20/2, 41—66.

—, 1965. Die Inkohlung im Saar-Lothringer Karbon, verglichen mit der im Ruhrkarbon. Zeitschr. deutsche Geol. Ges., 117, 243—279.

—, 1966. Inkohlungsuntersuchungen im Dienst der angewandten Geologie. Freib. Forschungsh., C 210, 155—195.

Teunissen, D., 1961. Het middennederlandse heuvelgebied. Publ. Geog. Instit. R.U. Utrecht, B 23, 1—153.

Teunissen, D. & Florschütz, F., 1957. Over een pollenhoudende kleilaag op de "Duno" bij Oosterbeek. Tijdschr. Kon. Ned. Aardr. Genootsch., 74/3, 413—421.

Thomson, P. W., 1935. Vorläufige Mitteilung über die spätglaziale Waldgeschichte Estlands. Geol. Fören. Stockh.-Fören., 57, 84—92.

—, 1952. Sekundäre Umlagerung pflanzlicher Mikrofossilien in klastischen Sedimenten. Geol. Rundschau, 40, 286—287.

Urban, J. B., 1961. Concentration of palynological fossils by heavy-liquid flotation. Okla. Geol. Notes, 21, 191—193.

Veenstra, H. J., 1963. Microscopic studies of boulder clays. Thesis Groningen, Uitg. Stabo, 1—211.

Viete, G., 1960. Zur Entstehung der glazigenen Lagerungsstörungen unter besonderer Berücksichtigung der Flözdeformationen im mitteldeutschen Raum. Freib. Forschungsh., C 78, 1—257.

Wagenbreth, O., 1960. Neue quartärgeologische Beobachtungen im Tagebau Profen bei Zeitz. Freib. Forschungsh., C 80, 25—57.

Wee, W. M. ter, 1962. The Saalian glaciation in the Netherlands. Med. Geol. Sticht., Nw. Ser., 15, 53—76.

Wetzel, W., 1939. Lumineszenzanalyse und Sedimentpetrographie. Zentralbl. Min. Geol. Paläont., A (1939), 225—247.

—, 1959. Das lumineszenzmikroskopische Verhalten von Sedimenten. Neues Jahrb. Geol. Paläont., Abh., 107/3, 261—277.

—, 1962. Lumineszenzmikroskopische Studien an chilenischen Sedimenten. Neues Jahrb. Geol. Paläont., Mon. h., 6, 303—308.

White, D., 1933. Role of water conditions in the formation and differentiation of common (banded) coals. Econ. Geol., 27, 556—570.

Williams, R. C., Backus, R. C. & Russell, L. S., 1951. Macromolecular weight, determined by direct particle counting. Jour. Am. Chem. Soc., 73, 2062—2069.

Wilson, L. R., 1961. Palynological fossil response to low-grade metamorphism in the Arkoma Basin. Tulsa Geol. Soc. Digest, 29, 131—140.

—, 1964. Recycling, stratigraphic leakage, and faulty techniques in palynology. Grana Palynol., 5/3, 425—436.

Woldstedt, P., 1950. Norddeutschland und angrenzende Gebiete im Eiszeitalter. Stuttgart, Koehler Verlag, 1—434.

Wolf, M., 1966. Observations pétrographiques sur les schistes boghead d'Autun. Sci. de la Terre, 11, 7—18.

Young, M. R., 1961. Principles and technique of fluorescence microscopy. Quart. Jour. Microsc. Sci., 102, 419—449.

Zagwijn, W. H., 1957. Vegetation, climate and time-correlations in the Early Pleistocene of Europe. Geol. & Mijnb., Nw. Ser., 19, 233—244.

—, 1960. Aspects of the Pliocene and Early Pleistocene vegetation in the Netherlands. Med. Geol. Sticht., C/3/1/5, 7—78.

—, 1961. Vegetation, climate and radiocarbon datings in the Late Pleistocene of the Netherlands. Part I: Eemian and Early Weichselian. Med. Geol. Sticht., Nw. Ser., 14, 15—45.

—, 1963. Pollen-analytic investigations in the Tiglian of the Netherlands. Med. Geol. Sticht., Nw. Ser., 16, 49—71.

Zagwijn, W. H. & Veenstra, H. J., 1966. A pollen-analytical study of cores from the outer silver pit, North Sea. Marine Geol., 4, 539—551.

Zetzsche, F. & Huggler, K., 1928. Untersuchungen über die Membran der Sporen und Pollen, 1. *Lycopodium clavatum* L. Ann. Chem., 461, 89—108.

Zetzsche, F. & Kälin, O., 1931. Zur Autoxydation der Sporopollenine. Helv. Chim. Acta, 14.

Zetzsche, F. & Vicari, H., 1931. Untersuchungen über die Membran der Sporen und Pollen, 2. *Lycopodium clavatum* L., 3. *Picea orientalis, Pinus silvestris, Corylus avellana*. Helv. Chim. Acta, 14, 58—67.

Zetzsche, F., Vicari, H. & Schärer, G., 1931. Untersuchungen über die Membran der Sporen und Pollen. IV.3. Fossiles Sporopollenin aus dem Tasmanit und der Moskauer Braunkohle. Helv. Chim. Acta, 14, 67—68.

Zetzsche, F. & Liechti, J., 1937. Biochemisch veränderte Sporopollenine. XII. Mitteilung über die Membran der Sporen und Pollen. Brennst. Chem., 18/14, 280—281.

Zonneveld, J. I. S., 1958. Litho-stratigraphische eenheden in het Nederlandse Pleistoceen. Med. Geol. Sticht., Nw. Ser., 12, 31—64.

Part II

SPORES AND POLLEN GRAINS – CLASSIFICATION AND STRATIGRAPHIC APPLICATION

Editors' Comments on Papers 6 Through 10

6   KUYL, MULLER, and WATERBOLK
    The Application of Palynology to Oil Geology with Special Reference to Western Venezuela

7   SCHOPF, WILSON, and BENTALL
    Excerpts from *An Annotated Synopsis of Paleozoic Fossil Spores and the Definition of Generic Groups*

8   POTONIÉ and KREMP
    Excerpt from *Die Sporae dispersae des Ruhrkarbons. Teil I*

9   TRAVERSE
    Excerpts from *Pollen Analysis of the Brandon Lignite of Vermont*

10 GERMERAAD, HOPPING, and MULLER
    Excerpts from *Palynology of Tertiary Sediments from Tropical Areas*

The papers in this section were selected either because they have all contributed to the development of modern ideas on the identification and classification of fossil miospores, or because they have greatly advanced our understanding of the use of spores and pollen in stratigraphy. The literature on both classification and stratigraphic use of miospores is already extensive and the output of papers continues to increase each year (Manten, 1969). It is probably true now that the majority of palynological publications are in English, but a very considerable body of literature exists in French, German, and Russian. In the period up to 1945, most papers were written in national languages, and most palynologists were employed either as academics or industrially in the coal and brown coal industries. Three geological time intervals attracted the greatest amount of attention, i.e., the Northern Hemisphere Carboniferous Coal Measures, the Tertiary, and the Quaternary. Unfortunately, the Quaternary is outside the scope of this volume, the reader is referred to papers by von Post (1918), Iversen (1936), and Godwin (1934) for a general introduction to the principles of Quaternary palynology.

German workers predominated in the field of Tertiary palynology in the 1930s, largely because of their country's large reserves of brown coal. The most distinguished of these was undoubtedly Robert Potonié, whose contribution to our understanding of the composition and stratigraphy of Central European Tertiary spore and pollen assemblages was quite outstanding (Potonié, 1931, 1934, 1935, etc.). Potonié's work on the Tertiary was continued after the end of the Second World War (1950, 1951) and his colleagues and students have also made major contributions. In other parts of the world, interest in Tertiary spores and their stratigraphy was a great deal more academic. However, several great palynological traditions were founded in various countries, as, for example, in the USSR (Naumova, 1937), and in Australia (Cookson, 1947).

In Great Britain, Raistrick and Simpson (1933) began very early to use spores from coal seams for Coal Measures (Upper Carboniferous) correlation. Raistrick used a simple letter and number system for "naming" his spores, thus bypassing many of the still unresolved problems of classification and nomenclature. The first attempt at making a systematic classification of fossil spores was published in 1944 by Schopf et al. (Paper 7), who selected certain guide-lines for Palaeozoic spores. They suggested that the International Rules (now "Code") of Botanical Nomenclature should serve as a systematic and procedural basis for fossil spore classification. Relationships between spores had to be based on the possession of "significant characteristics in common." Schopf et al. did not, however, favor a solely morphographic approach; they believed that the palynologist should apply his judgment to the relative importanc of various characteristics, so that taxonomy of spores should not be reduced to a mere cataloging procedure. In contrast, the later morphographic classification by Potonié and Kremp (Paper 8) takes the view that subjectivity should not enter into the *use* of a classification system (although the *construction* of a classification system will inevitably introduce subjective elements).

Schopf et al. were concerned that classification should not obscure biological relationships. These are particularly difficult to determine because plant fossils are usually found as separate organs, so that it is not normally easy to reconstruct the original, entire plant; miospores are an extreme case of plant disaggregation. While changes in spore types can be observed with time, the significance of these changes with respect to the evolution of the parent plants is difficult to evaluate. Furthermore, evolutionary convergence of spore morphology can add to the difficulties of classification. Despite these problems, Schopf et al. correctly predicted that spores and pollen would eventually be of greater practical value in geology than larger plant fossils.

Potonié and Kremp's morphographic system is taxonomically more complete than that of Schopf et al.; moreover a purely morphographic

Editors' Comments on Papers 6 Through 10

suprageneric classification—now used in one form or another by most palynologists—is laid out in their paper. They considered miospores to be organ or form genera, not biological genera whose natural affinities are understood; the very act of identifying genera of the latter type presupposes natural relationships, which may not exist. Placing these genera in Families and Orders compounds this forcible fitting of fossil genera into a grouping which may in reality be far from natural. Potonié and Kremp therefore proposed a completely new set of suprageneric form (or organ) taxa—based on the nongenetic word "Abteilung (= Division), with Higher and Lower Divisions (Ober- and Unterabteilungen). In order to omit all genetic considerations, the terms Turma (Formabteilung) and Subturma (Formunterabteilung) were proposed (the German verb *türmen* means to pile up or form a superstructure). Despite the stated intention of using a morphographic system of classification, all three of the major divisions created by Potonié and Kremp—Oberabteilung (= Anteturma) SPORONITES (Fungi), Oberabteilung SPORITES, and Oberabteilung POLLENITES—surely have biological implications. Potonié himself later came to the conclusion that SPORITES would be sufficient at Anteturma level, because all pollen grains are biologically microspores. The Turmae are based on apertural characters and presence or absence of equatorial features. Although there has been considerable subsequent debate about the validity of using more than one characteristic feature at each level of taxonomy, the overall scheme is used by most palynologists today.

Both the Potonié and Kremp and the Schopf et al. schemes were designed to classify Palaeozoic spores. About the mid-1950s, as a result of the application of palynology to stratigraphy in oil exploration, it became necessary to extend our knowledge of miospores to Tertiary and Mesozoic strata. In the 1930s, Tertiary miospores had been identified as far as possible with spores and pollen from living plants, and given generic names which indicated these similarities, with the ending of *-sporites* or *-pollenites* e.f. *Quercus-pollenites*, etc. Classification of these genera was under the Orders and Families of the plant kingdom. This approach proved to be highly unsatisfactory; several morphographic systems were proposed as alternatives (van der Hammen, 1956: Thomson and Pflug, 1953). However, each had considerable shortcomings like those described by Schopf et al. for Palaeozoic spores.

Traverse's paper (Paper 9) on the Oligocene Brandon Lignite of Vermont represented a major step forward, with his proposal of a combined morphographic plus natural system. Traverse was concerned with using spores and pollen stratigraphically; he had the specific purpose of trying to determine an accurate age for the Brandon Lignite. It is important to make use of all botanical information on Tertiary miospores Traverse believed, since a great deal can be learned by studying living spores and pollen for easier identification of unknown Tertiary specimens, and also because many living plants in fact have a geological history reaching back to the early Tertiary. The employment of *index fossils*, as suggested by Potonié (1935, 1951), would be less useful than in the Palaeozoic, where evolutionary changes seemed to take place much faster. Climatically induced migration of plant groups and other ecological factors are more likely causes of the changes seen in Tertiary assemblages. In consequence, the relative abundances of different pollen types can be used as a means of stratigraphical correlation.

However, naming the pollen presents great difficulties. Traverse pointed out that organ genera might easily be created for pollen which had not been recognized as belonging to an extant genus, and equally, pollen from extinct genera which closely resemble some other modern pollen might be misidentified. Traverse made four proposals for taxonomic procedures, substantially those that are followed for Tertiary miospores today. Pollen indistinguishable from a modern genus is referred to that genus, but modern species names are never used. New generic names are made for pollen that is believed to be certainly a representative of a living family. Abundant and distinctive pollen is put into organ (form) genera, and pollen which cannot be identified with existing and fossil taxa is placed in an "unknown" category.

Miospores from Mesozoic assemblages were systematically and stratigraphically studied by Couper (1958). His approach was somewhat similar to that of Traverse, but instead of using reference collections of modern pollen, Couper looked at preparations made from spores and pollen *in situ* in sporangia and cones. Although it is perfectly feasible to apply a morphographic system to Mesozoic miospores (see Dettman, 1963), Couper applied a system of a type similar to Traverse's for the Tertiary. His approach has not proved popular; at the time of writing, most palynologists use instead a version of Potonié and Kremp's morphographic system for Mesozoic assemblages.

Most of the systems of classification in use in Pre-Quaternary palynology have been designed to give as much stratigraphic information as possible. Spores and pollen grains are now used routinely to date samples either from outcrop or from drill core (cuttings, core, or sidewall core) in particular in the oil and gas industry (see also Part I). They have also been used stratigraphically in the mining industry, especially in coal mining, but also in studies of manganese ore (Kedves and Simonicsics, 1964) and other materials such as salt etc.

The paper written by Kuyl et al. (Paper 6), which covers many aspects as well as stratigraphy, foreshadowed most later developments in palynology. As well as suggesting a careful morphographic approach for classifying Tertiary pollen, the authors describe preparation techniques, distribution of spores and pollen, and the effects of weathering and burial on pollen grains (see Part I). Kuyl et al. distinguished three types of stratigraphic-palynological zones, respectively characterised by (a) types which are confined to one zone or whose first and last appearances coincide with the upper or lower zone boundary, (b) types which are most numerous in one zone, or whose main distribution begins or ends at zone boundaries, (c) constant assemblage zones. The main controls on changes in assemblage are believed to be evolution and floral migration caused by alterations in climatic and edaphic conditions.

The ideas on the use of palynological assemblages in stratigraphy were refined and extended in Germeraad et al.'s paper (Paper 10) on tropical Tertiary assemblages. Here the controls responsible for different types of zone have been identified. Germeraad et al., using marker species which are more or less equivalent to Potonié's (1935, 1951) "Leitfossilien," were able to define three kinds of markers: pantropical, transatlantic (S. America, Africa), and those which had an intracontinental distribution. The zonations delineated by the markers can be used independently of zoopalaeontological time-stratigraphic correlation. Evolution of new plant groups is thought to account for the changes in the pantropical zonation; climatic boundaries produce a more refined subzonation, but one which is more limited in its geographic extent of applicability. Immigration of plants is considered to be of regional value only; though it may well produce clear and sharp boundaries which are extremely useful locally it is unlikely to be useful on a wide scale. All the factors which control or affect assemblages are fully discussed and, because of the large amount of information, statistical treatments had to be applied. This paper shows how, even with tropical Tertiary assemblages (where several hundred species might occur in a single sample), good repeatable zonations can still be erected for stratigraphic purposes.

In the temperate Tertiary, great difficulties arise because climatic fluctuations appear to play a dominant role in controlling assemblages, and unfortunately, no major work of the scope of Germeraad et al.'s paper is yet available to clarify the situation.

Despite the fact that the majority of writings on pre-Quaternary palynology are principally concerned with stratigraphy, a number of authors have attempted to glean some information on major events in the history of the land plants. Chaloner (1970), and Chaloner and Allen (1970) have examined the morphological evidence from early miospore assemblages bearing on the colonization of the land. Pettitt (1965) has looked for evidence of megaspores and for the earliest seeds. The origin of the angiosperms remains an unsolved mystery, but pollen evidence sheds some light on the problems (Muller, 1970: Doyle, 1969).

Thus, although many problems of classification remain, aspects of stratigraphy and overall plant evolution can be illuminated by studies of spores and pollen grains.

REFERENCES

Chaloner, W. G. (1970). The rise of the first land plants. *Biol. Revue* **45**:353-377.

Chaloner, W. G., and Allen, K. C. (1970). Palaeobotany and phytochemical phylogeny. In: J. B. Harbourne (ed.), *Phytochemical Phylogeny*. London and New York: Academic Press, pp. 21-30.

Cookson, I. C. (1947). Plant microfossils from the Lignites of Kerguelen Archipelago. *B.A.N.Z. A.R.E.* Rept. Series A, 2 (8):127-142.

Couper, R. A. (1958). British Mesozoic Microspores and Pollen Grains. A Systematic and Stratigraphic Study. *Palaeontographica* B **103**:75-179.

Dettmann, M. E. (1963). Upper Mesozoic microfloras from south-eastern Australia. *Proc. Roy. Soc. Victoria* **77**:1-148.

Doyle, J. A. (1969). Cretaceous angiosperm pollen of the Atlantic coastal plain and its evolutionary significance. *J. Arnold Arboretum* **50**:1-35.

Godwin, H. (1934 a and b). Pollen analysis: an outline of the problems and potentialities of the method. 1 and 2, *New Phytologist* **33**:278-305, and 325-358.

Iversen, J. (1936). Sekundäres Pollen als Fehlerquelle, eine Korrelations methode zure Pollenanalyse minerogener Sedimente. *Danmarks Geol. Underøgelse* IV **2**(15):1-24.

Kedves, M., and Simonicsics, P. (1964). Microstratigraphy of the carbonate manganese ore layers of the shaft III of Urkut on the basis of palynological investigations. *Acta Miner. Petrograph. Univ. Szegedensis*, **16**(2):3-48.

Manten, A. A. (1969). Bibliography of Palaeopalynology 1836-1966. *Rev. Palaeobotan. Palynol.* **8**:1-570.

Muller, J. (1970). Palynological evidence on early differentiation of angiosperms. *Biol. Revue* **45**:417-456.

Naumova, S. N. (1937). The Spores and Pollen of the coals of the U.S.S.R. *Internat. Geol. Congr., 17th, Moscow, Absts. Papers*, pp. 60-61.

Pettitt, J. M. (1965). Two heterosporous plants from the Upper Devonian of North America. *Bull. Brit. Mus. Nat. Hist. Geol.* **10**:83-92.

Potonié, R. (1931). Pollenformen der miozänen Braunkohle. *Sitz. Ber. naturforsch. Freunde* **1930**:24-28.

Potonié, R. (1934). Zur Mikrobotanik des eozänen Humodils des Geiseltals. *Arb. Inst. Paläobot.* **4**:25-125.

Potonié, R. (1935). Pollen und Sporen als "Leitfossilien" der Braunkohlenflöze. *Braunkohle* **34**:681-685.

Potonié, R. (1950). Stand der mikropaläobotanischen Tertiärstratigraphie. *Z. dt. geol. Ges.* **100**:366-378.

Potonié, R. (1951). Pollen und Sporenformen als Leitfossilien des Tertiärs. *Mikroskopie* **6**:271-283.

Raistrick, A., and Simpson, J. (1933). The microspores of some Northumberland coals and their use in the correlation of coal seams. *Trans. Inst. Min. Eng. (London)* **85**:225-235.

Thomson, P. W., and Pflug, H. D. (1953). Pollen und Sporen des Mitteleuropäischen Tertiärs. *Palaeontographica*, B., **94**:1-138.

van der Hammen, Th. (1956). A palynological systematic nomenclature. *Bol. geol. (Bogota)* **4**:63-101.

von Post, L. (1918). Skogsträdspollen i sydsvenska torvmosselagerföljder. *Forh. Skan. Naturforskeres* **16**, møte 1916, 432-465.

ADDITIONAL REFERENCES

Potonié, R. (1956, 1958, 1960, 1966, 1970). Synopsis der Gattungen der *Sporae dispersae*. I-V. *Beih. Geol. Jb.* **23**:1-103; **31**:1-114; **39**:1-189; **72**:1-244; **87**:1-222.

Tschudy, R. H., and Scott, R. A. (1969). *Aspects of Palynology*. New York, London, Sydney, Toronto: Wiley Interscience.

6

Copyright © 1955 by The Royal Netherlands Geological and Mining Society

Reprinted from *Geologie en Mijnbouw* 17(3), New Ser.:49-75 (1955)

THE APPLICATION OF PALYNOLOGY TO OIL GEOLOGY, WITH SPECIAL REFERENCE TO WESTERN VENEZUELA

O. S. KUYL [1], J. MULLER [1] and H. T. WATERBOLK [2]

[*Editors' Note:* Figure 8 is not reproduced here.]

CONTENTS

		p.
I.	Introduction	49
II.	Morphology and taxonomy	50
	a. Introductory	50
	b. Pollen of Angiosperms	52
III.	The distribution of pollen and spores	55
IV.	Preservation in sediments	55
V.	Technique	56
VI.	Other mocrofossils	57
VII.	Palynological zones	57
	a. General statement	57
	b. Evolution	58
	c. Migration	62
VIII.	Stratigraphic results in western Venezuela	65
IX.	Bibliography	70

SUMMARY

After a short introduction on the history of palynological studies in relation to economic geology, the morphology of pollen grains and spores is discussed in Chapter II and illustrated by Figures 1—4 and Plates 1—8. It is demonstrated that the morphological conceptions of Iversen and Troels-Smith, based on European Quaternary pollen, are also valid for Tertiary and Upper Cretaceous pollen of angiosperms.

The distribution of pollen and spores by wind and water, and the relation of the pollen rain to the actual vegetation, are briefly discussed in Chapter III. In Chapter IV some remarks are made regarding the preservation of pollen and spores in the sediments, and the effects of weathering and burial at great depth. Some technical questions (sample treatment, arrangement of type collections), and the occurrence of other microfossils, are dealt with in Chapters V and VI.

In Chapter VII the establishment of palynological zones is described, as well as the background of the floral changes which are found. The main factors determining floral differentiation are evolution and migration. The effect of both factors is illustrated by examples taken from practice. The supposed dicotyledonous pollen grains from beds older than the Upper Cretaceous (*Tricolpites troedssonii*) are discussed in some detail, and it is shown that they are just as likely to belong to the gymnosperms.

Evolution forms the backbone of the palynological zonation, but climatic and edaphic conditions seem to have provided additional variations in the flora which are of more importance for local correlation work. Some practical results attained in W. Venezuela are given in Chapter VIII. Diagrammatic sections (Figs. 6 and 8) show the divergence between palynological correlation, which is thought to represent time-stratigraphic correlation, and facies correlation based on benthonic foraminiferal assemblages or on lithology.

Finally, a palynological correlation (Fig. 7) is given between different areas round Lake Maracaibo.

I. INTRODUCTION

Since the beginning of the twentieth century the study of fossil pollen grains and spores has constantly increased in importance.

The examination of pollen ("pollen analysis") in Quaternary bogs, which was initiated by Von Post, has reached an advanced stage; archaeology, climatology, and Quaternary geology have all profited by it.

Pollen investigation has also been applied to other studies. Some investigators specialized in the morphology and classification of pollen and spores. The study of pollen present in the air attracted the attention of the medical profession, in connection with the combating of hay fever; the origin of honey proved to be capable of determination by means of the pollen present in it. The name palynology is now used for all these different branches of research.

At first, interest in pre-Quaternary pollen grains and spores was purely palaeobotanical, but the possibilities of stratigraphic application, principally for the correlation of coal

---

[1] Palynologists, Companía Shell de Venezuela Ltd., Maracaibo (Royal Dutch/Shell Group).

[2] Palynologist, N.V. De Bataafsche Petroleum Maatschappij, The Hague (Royal Dutch/Shell Group).

measures, were soon realized. Mention may be made of Raistrick's work on the British Carboniferous (1934). At the present time, palynological investigation is a matter of routine, for instance, in lignite mining in Western Germany (Rein, 1950).

In studying Quaternary sediments, emphasis has recently come to lie (mainly due to the researches of Iversen) on investigation of limnic and brackish water sediments. Admittedly, in comparison with terrestrial bogs, these have the disadvantage of the possible presence of reworked pollen originating from older deposits, but their great advantage is that their pollen content is free from the influence of local vegetation, and is therefore better adapted for regional correlation.

It has long been known that older clastic sediments can also contain pollen. Wodehouse (1933), for instance, mentions the great richness in pollen of the Eocene Green River oil shales of Colorado. Reissinger (1938) investigated a series of clay samples varying in age from Cambrian to Recent, and found a rich microflora in some of them. The possibility of correlating sedimentary rocks by palynology was pointed out by Wilson (1946). He says that fossil spores and pollen grains occur in "most coals, many shales and some sandstones".

It is obvious that petroleum geology might also be able to benefit by the new method, particularly in areas of predominantly terrestrial character or with brackish water deposits, where correlation by means of other microfossils is generally difficult. The Standard Oil Company of California (1952) and Carter Oil Company (Armstrong, 1953) have stated that they have taken up palynological research.

In 1938, on the instructions of N.V. de Bataafsche Petroleum Maatschappij, R. Potonié examined some samples from Mexico. In 1939 and subsequent years, Florschütz (Leyden) analyzed, at the request of this Company, material from the Malay Archipelago and the Caribbean area, which contained coaly clays as well as coals.

After the war, systematic research was also undertaken by the Royal Dutch/Shell Group itself, at first only in Maracaibo (1947). At present, palynologists are also working in British Borneo and Nigeria, while preparatory investigations of other operational areas, such as Colombia, Trinidad and New Guinea, are being carried out in The Hague.

In the present article some general aspects of palynological research are discussed.

The writers are aware that their paper covers an extensive field, ranging from botanical to applied geological observations, which will not have the same interest for all readers. Some fundamental palynological aspects are dealt with in Chapters II (Morphology and taxonomy), III (Distribution of pollen and spores), IV (Preservation in sediments), V (Technique) and VI (Other microfossils). The palynological zones, based on evolutionary trends or changes in climatic and edaphic conditions, are discussed in Chapter VII. Finally, Chapter VIII gives a review of the stratigraphical results achieved in western Venezuela.

II. MORPHOLOGY AND TAXONOMY

a. *Introductory*

In the initial stages of pollen research, investigators were satisfied with a rather superficial examination of the grains. In addition to the size and shape, the principal distinguishing characteristics were considered to be the number and arrangement of pores and furrows, and some easily recognizable features of the wall structure. Nor was any need felt for more details, since the pollen from the important trees of the late and post-Glacial ages could be readily recognized.

23—29: Monocotyledonous pollen classes
23. *Monocolpatae*
24. *Monoporatae*
25—27. *Operculatae*
28—29. *Trichotomocolpatae*

30—57: Dicotyledonous pollen classes
30. *Diporatae*
31. *Triporatae*
32—33. *Stephanoporatae*
34—36. *Periporatae*
37. *Dicolpatae*
38. *Tricolpatae*
39—40. *Stephanocolpatae*
41—43. *Pericolpatae*
44. *Dicolporatae*
45. *Tricolporatae*
46—47. *Stephanocolporatae*
48—50. *Percolporatae*
51. *Syncolpatae*
52. *Heterocolpatae*
53—54. *Zonoratae*
55. *Dyadeae*
56. *Tetradeae* (also to be found in groups 1—12, 13—16, 17—22, 23—29)
57. *Polyadeae*

Generic names of groups 1—12 and 13—16 after Schopf, Wilson and Bentall (1944) and Kosanke (1950), pollen classes mainly after Faegri and Iversen (1950).

Application of Palynology to Oil Geology

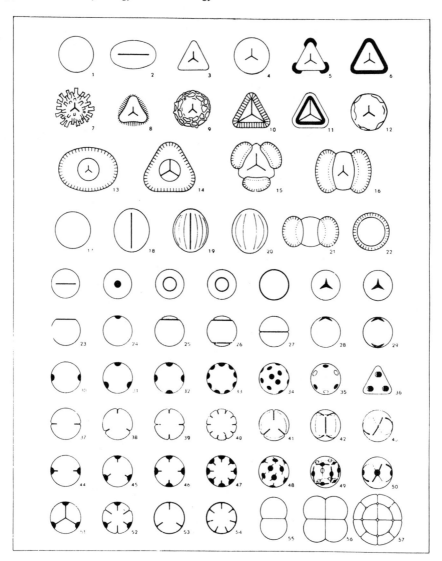

Fig. 1 — Schematic drawings of some important spore and pollen classes.

1—12: Microspores.

1. *Aletes*, etc.
2. *Laevigatosporites*, etc
3. *Granulatisporites*
4. *Punctatisporites*
5. *Triquitrites*
6. *Densosporites*
7. *Raistrickia*
8. *Rheinschospora*
9. *Reticulatisporites*
10. *Cirratriradites*
11. *Lycospora*
12. *Calamospora*

13—16: Prepollens

13. *Florinites*
14. *Endosporites*
15. *Alatisporites*
16. *Illinites*

17—22: Gymnospermious pollen classes (polar and equatorial view)

17. *Inaperturatae* (also to be found in groups 23—29, 30—57)
18. *Monocolpatae* (also to be found in group 23—29)
19—20. *Polyplicatae*
21—22. *Vesiculatae*

105

In this way it was possible to work quickly, and the main events in forest history after the Ice Age, throughout large areas of Europe, were soon known.

Little attention was paid to the pollen of herbs and less common trees. *Ericaceae, Gramineae, Compositae, Chenopodiaceae,* and a few other groups were known, but all the rest were classified under *varia*.

Various investigators have brought about a change in this situation. Fundamental research into the morphology of Recent pollen was carried out by Wodehouse (1935), Erdtman (1943, 1953, etc.), Iversen and Troels-Smith (1950) (cf. Faegri and Iversen, 1950), etc. The first-named worker paid much attention to structural plan and symmetry; the others also analyzed the structure of the exine, which proved to be of the greatest importance.

Many new pollen types could soon be recognized; and a number of old types proved to be heterogeneous, as they comprised pollen from several different plants.

It is now possible, thanks to the work of the writers mentioned, to give accurate descriptions of pollen types. Furthermore, study and comparison of Recent types have made it possible to get an idea of the characteristics important in classification, and of the degree of variability to which these are subject. It is

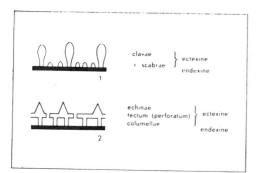

Fig. 2 — Schematic sections through an intectate (1) and a tectate (2) pollen wall.

obvious that these results are of the greatest importance for the study of fossil pollen.

The first attempt at systematic classification of fossil pollen grains naturally encountered great difficulties at first. Following Potonié (1934), relationships with Recent types were sought for, but the observations had to be based on inadequate morphological knowledge; nevertheless an effort was made to express the assumed relationship in the nomenclature. Numerous publications, especially those by German investigators, bear witness to the great difficulties which arose as a result of the differing views expressed by various authorities.

Thomson and Pflug (1953) now propose to classify fossil pollen types according to the pollen classes of Iversen and Troels-Smith (see below). Such an artificial or empirical classification is the only way of reducing to order the multitude of forms encountered in regions where the Recent pollen flora is little known, and especially when studying Lower Tertiary and older deposits.

b. *Pollen of Angiosperms*

The pollen classes of Iversen and Troels-Smith are mainly based on the number and position of furrows and pores, and on pore-furrow combination. The classes sometimes coincide with taxonomic units (such as the polyads in the *Mimosaceae*); but most classes occur in many different families, although individual genera usually belong to one single class. The system is therefore artificial to a high degree, and is to be considered as a practical tool only. Some of the important angiospermous pollen classes, together with a few suggested additions, are shown in Fig. 1: 23—57.

Three further groups of characteristics are important: the sculpture and structure of the outer cell wall (exine), the form and structure of the apertures, and the shape of the grains.

Iversen and Troels-Smith reduce all exine structures to one basic plan, consisting of an endexine and an ectexine. The former is generally structureless, the latter is formed by isolated elements which may fuse at their top wholly or partly into a so-called tectum. Any sculpturing present is to be found either directly on the endexine (intectate), or on the tectum (tectate) (Fig. 2). This basic plan, according to our experience, is equally valid for the fossil pollen of angiosperms, and the morphological terms of Iversen and Troels-Smith are therefore suitable for describing pre-Quaternary pollen. Moreover, these terms are more sharply defined and fewer in number than those of Erdtman (1953).

Pflug (1953), in his study of fossil pollen from the European Tertiary, holds a very different view. His conception of a "lamellarer Aufbau" appears to be unnecessarily complicated. Moreover, some of the layers described by this author, are, in our opinion, non-existent, being only due to optical effects[*].

---

[*] Compare, for instance, Pflug's descriptions of the following species or species groups with our schematic interpretations in Fig. 4, and microphotographs on Pl. 3, of identical types from samples kindly given to us by Dr. Pflug.
Extratriporopollenites thiergartii: Fig. 4: 16; Pl. 3: 9, 10.
Extratriporopollenites sect. *Basaloidea:* Fig. 4: 22.
Extratriporopollenites sect. *Conjunctoidea:* Fig. 4: 23; Pl. 3: 2, 3.
Extratriporopollenites pompeckji: Fig. 4: 24; Pl. 3: 13.

Application of Palynology to Oil Geology

The shape of the grains has been somewhat neglected by many investigators. It is admittedly very variable, but, with due reservations for flattening effects, it is certainly an important feature of a pollen type. In Fig. 3 a number of polar and equatorial views are given, together with the terms used in the Royal Dutch/Shell laboratories.

Aperture structures are very important (Fig. 4). In the majority of dicotyledons these are composite. A common type is formed by a pore in the endexine and a meridional furrow in the ectexine (Fig. 4: 1, 2, 6, 7, 8). The furrows may fuse at the poles (Fig. 4: 5), or be so short as to become a small ectexinous pore (Fig. 4: 9, 10, 11) (exopore, acc. to Pflug). In the latter case two round pores of different diameter may be superposed (Fig. 4: 10; example: *Myrica*). The endexinous pores are

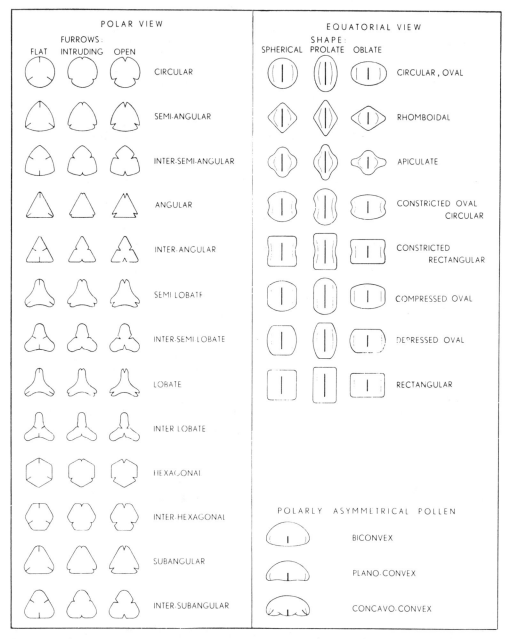

Fig. 3 — Polar and equatorial views of radially symmetrical pollen grains.

often equatorially elongated, thus forming transversal furrows (Fig. 4: 3), which may fuse into one equatorial zone (Fig. 4: 4, zonorate, e.g. *Rhizophoraceae*, Pl. 2: 4—7). In a few cases the pores are meridionally elongated, giving rise to an endexinous furrow which might be called endocolpus (Fig. 4: 13; Pl. 3: 2, 3, 4, 5). In one case (*Extratriporopollenites pompeckji*) these endocolpi fuse at the poles (Fig. 4: 14; Pl. 3: 13).

c. Pollen of Gymnosperms

The pollen of gymnosperms, although — with a few exceptions — easily recognizable as such, presents difficulties in many ways. A number of highly characteristic classes of taxonomical value can be distinguished (Fig. 1: 13—22), but the exine is either structureless, as in the *Ephedra* type, or is difficult to describe, as in the complicated internal structures of the bladders of the winged conifer types. Furthermore, distinct pores and furrows are absent. Many Palaeozoic types ("prepollens") still have a tetrad scar (Fig. 1: 13—16).

d. Spores of Pteridophytes

The spores of pteridophytes are either alete, monolete or trilete (Fig. 1: 1, 2, 3—12). When investigating Tertiary deposits, there is no need for a more detailed classification, but pre-Tertiary (especially Carboniferous) sediments contain such a diversity of forms that subdivision is necessary. Schopf et al. have dealt exhaustively with (mainly Palaeozoic) microspores. A number of genera are distinguished, based on both basic plan and exine features.

The exine is generally structureless, but it shows a great variety of sculpturing, which can easily be described with the aid of Iversen and Troels-Smith's terminology (see, for instance, the gemmate spore of Pl. 4: 7). Apart from the scars, apertures are absent. The shape may be difficult to interpret when the grains are laterally flattened [1]

---

[1] Pflug's genera *Triplanosporites*, *Poroplanites* and *Duplospores* most probably represent folded specimens of ordinary trilete spores.

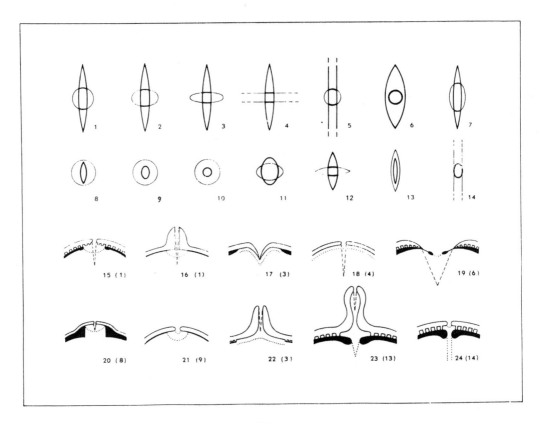

Fig. 4

1—14. Composite aperture types of colporate pollen grains. Thick lines: apertures in ectexine or in exposed part of endexine.

15—24. Some examples of sections through composite apertures. The aperture type is indicated between brackets.

III. THE DISTRIBUTION OF POLLEN AND SPORES

Pollen grains produced by anthophytes during flowering are transported mostly by the wind or by insects. Microspores, produced by pteridophytes, are mainly distributed by the wind. The small size of both pollen and microspores (generally between 10 and 100 μ), and the enormous quantities released, make possible wide and effective dissemination. The production of one hectare of wind-fertilized alder trees has been estimated at more than 2×10^{13} grains per annum (Erdtman, 1943). Most plants flower on sunny days, and, especially in the tropics, a great part of the pollen is then taken up by the rising currents of air which occur, for instance, under cumulus clouds. Such clouds may travel undispersed for 50—100 km., carrying the pollen with them, and accordingly, the potential range of distribution of the pollen is that distance, at any rate. The maximum potential dispersal limit is of course much greater. Proof exists of transportation of pollen over distances of more than 1000 km. (Erdtman, 1943), but this "long-distance" transport is not important quantitatively.

In calm weather, or during showers of rain, the pollen collects as a sediment. Several hundred grains per square cm. per year have been measured in wooded inland areas. Offshore, the quantities decrease, of course, and only a relatively restricted area in coastal, marine, deltaic, lagoonal and swamp environments will receive a pollen rain dense enough to make their sediments a profitable source of pollen. Naturally, the quantity of inorganic sediment deposited within a certain time also plays a part.

The "pollen rain" does not always give a true picture of the vegetation from which it was derived, for the following reasons:

a. the quantitative proportions of the pollen rain may be entirely different from those of the producing vegetation, owing to great differences in production, floating capacity and exposure to the wind;

b. the pollen production of various types of vegetation is mixed;

c. pollen-producing plants in the immediate neighbourhood may play a disproportionately large part;

d. pollen may be transported by water;

e. pollen may be reworked from older sediments and re-deposited.

Regarding b and c, it has been empirically established that, under certain conditions, atmospheric mixing is surprisingly good; this applies particularly to the pollen and spore content of sediments from basins enclosed by land and without autochthonous vegetation. The great success of pollen correlation in the European Quaternary is largely due to this fact. If, however, the basin of sedimentation is covered or surrounded by a vegetation cover which is different from that further inland, atmospheric mixing will be insufficient to produce a homogeneous pollen rain, and local over-representation occurs. Examples of this are well known, e.g. *Ericales* in a *Sphagnum* bog, and *Rhizophora* in a mangrove swamp.

As regards d, Florschütz and Jonker (1939) found a good 10 % *Abies* in sub-boreal sediments in the Rhine delta, while the nearest natural sites in which these trees could have grown were in the Black Forest, about 400 km. upstream. Evidently it is possible for a considerable quantity of pollen to be conveyed by rivers from an area with a totally different flora.

As regards e, pollen from older strata may be liberated during the erosion of rivers; and this must be taken into account in studying all clastic sediments. The quantity concerned may be large, as was found in an investigation of Recent sediments in the Orinoco delta.

IV. PRESERVATION IN SEDIMENTS

The wall of pollen grains and spores consists of two layers. The inner layer (intine) soon decays, at the same time as the living cell content. The outer layer (exine) is highly resistant to all kinds of chemical influences, but not to oxidation. Mottled sediments and red beds, which may have been aerated during deposition or later, are therefore usually barren. Sediments which accumulate in a reducing environment are most suitable for pollen analysis; these sediments may range in type from fully marine shales to coals. General rules for pollen content are hard to lay down, as there is much local variation. A high lime content and coarse grains are unfavourable. Weathering or excessive overburden may destroy pollen and spores in a rock.

Weathering is most active in tropical areas. In certain tropical arid countries, suitable samples have only been found at a depth of more than 40 feet below the surface. However, slight weathering need not necessarily be harmful; it may even facilitate treatment of the sample. Kosanke (1950) describes weathered Carboniferous coals which were capable of being treated immediately with caustic potash solution without the usual previous oxidation in Schultze's mixture.

Overburden effect was found to an increasing extent with increasing depth in sediments from certain geosynclinal areas. The plant tissue fragments, and gradually also the pollen grains and spores, first lose their typical structure, and may even be transformed into amorphous black fragments. The cause of this metamorphic process is perhaps to be found in the pressure of the sedimentary column and increased temperature at depth. Areas where extensive folding or magma intrusion have occurred may also be similarly affected. An indication of the correctness of the hypothesis is seen in the fact that no well-preserved flora is ever found again anywhere below a "black zone". The phenomenon also occurs in surface samples; the obvious assumption is that the sediments concerned were subjected to high temperatures and/or pressure at an earlier period in geological history. In some areas the disintegration of the pollen grains and spores occurs in strata as young as Miocene, elsewhere not until the Cretaceous or the Carboniferous. Comparable metamorphic changes in pure coals have been studied in great detail for some time past. M. and R. Teichmüller (1954), in a recent survey, mainly stress the influence of high temperature and time.

V. TECHNIQUE

Both well and surface samples may be used, and a quantity of 50 g. is amply sufficient. It is useless to start the investigation with ditch cuttings, but side wall samples and cores are highly satisfactory. The former have the advantage that the place from which the samples are to be taken can be selected on the Schlumberger log at lithologically favourable levels. The samples should be carefully cleaned from adhering drilling mud, and only used when it is clear that an autochthonous piece of the wall has been caught. They can also usually be retaken, or, when necessary, a certain important interval can be more extensively resampled. Conventional core samples are ideal for pollen investigation, as they generally yield a wider variety of lithological types than side wall cores.

When sampling a formation from outcrops, all types of sediment present should be collected, until the particular type which carries the best flora has been found; weathered samples should not be taken.

Chemical treatment is carried out at the Royal Dutch/Shell laboratories more or less according to the method which has been repeatedly described in the literature of the subject (cf. Erdtman, 1942; Wilson, 1949; Kosanke, 1950). The most important processes are:

1. dissolving any carbonate present (with HCl);
2. dissolving the minerals (with HF);
3. oxidation of the organic material (with Schultze's mixture);
4. dissolving humic acids (with KOH);
5. removal of the mineral components still remaining (by means of an alcohol-bromoform mixture of s.g. 2.2).

The treatment varies according to the type of rock to be dealt with. Use of hot reagents enables a preparation to be obtained in about half an hour, but slower preparation with cold reagents may be preferable. This takes several days; one laboratory assistant can prepare about 200 samples per month.

In addition to chemicals, glassware, etc., a large fume cupboard, together with a centrifuge, should be available. During preparation of the samples the greatest care should be exercised to avoid contamination through dirty glassware, while contamination with Recent atmospheric pollen may also be an annoying source of trouble.

The final residues are stained and stored in glycerine, in small phials. Examination is made under a binocular microscope allowing magnifications from ×125 to ×1600.

Owing to their small size, pollen grains and spores are difficult to trace for re-examination. The type collections are therefore built up of single-grain preparations (cf. Faegri, 1939; Erdtman, 1952). In Maracaibo it has been found possible to preserve these for more than six years to date. In order to study the grains in different positions or states of preservation at least 5 specimens of each type should be collected, if possible. Documentation is aided by descriptions and microphotographs. For the latter, an oil immersion objective should be used, and care taken to obtain the best lighting conditions and the correct combination of film type, filter and stain.

Type collections in tropical areas may, in the case of the Tertiary, easily contain 500 different types.

The results of the examination of the samples are plotted in different ways. In the first research stage the ordinary type of foraminifera distribution chart is used; later on, for more detailed quantitative work, percentage graphs ("sawblade" diagrams) are to be preferred.

VI. OTHER MICROFOSSILS

After the samples have been chemically treated, certain other plant remains are left, in addition to pollen and spores. Fragments of epidermis and tissue are generally best represented. The quantity varies considerably in different sediments. In some coals it is so substantial that the slides become relatively poor in pollen content; in other cases, such other remains are lacking, and the residue may then contain practically nothing but pollen grains or fragments of them. The state of preservation is usually better in younger sediments than in older sediments. A striking phenomenon is presented by the occurrence of numerous fragments of tracheids in pre-Eocene strata of Venezuela. Accurate cuticular analysis might perhaps yield stratigraphically valuable results, but it should be borne in mind that transport of these types takes place almost exclusively by water, so that their sensitivity to facies conditions would be great.

In certain sediments an important part is played by the *Hystrichospherideae*, a group of marine, now extinct (?) unicellular organisms, the exact taxonomical position of which is uncertain. They are of the same order of magnitude as pollen grains and spores and possess a highly resistant wall, provided with branched projections. It is doubtful whether they are sufficiently differentiated to be of use for stratigraphic purposes, and they, too, are sensitive to facies conditions. Experience with Quaternary sediments has shown that they are capable of bearing reworking remarkably well.

In addition remains of *Peridineae* and other *Algae*, and of *Fungi*, may be mentioned. The latter may be found in very large numbers, especially in certain coals.

The chitinous inner walls of microforaminifera are not highly differentiated, but may be of value for facies studies.

VII. PALYNOLOGICAL ZONES

a. *General Statement*

In a region in which an attempt is to be made to approach certain correlation problems by palynological methods, the investigation should start with the examination of standard sections through formations the relative stratigraphic position of which is well known. In order to be able to recognize possible reworking, and for a general understanding of the floral succession, it is important to extend this orientative investigation to strata older and younger than those which are considered to be of immediate interest from the point of view of petroleum geology.

Careful collection of samples from the standard section, either from wells or from surface sections, the preparation and production of single-grain slides, formation of a type collection, and accurate analysis of the residues, takes a great deal of time. Much attention must, however, be paid to this, because the future success of the investigation is largely dependent on it.

A possible differentiation in the floras found may lead to formulation of stratigraphic-palynological zones. These zones may be characterized by:
a. types which occur only in one particular zone, or whose stratigraphic highest or lowest occurrence is to be found at the boundaries of that zone;
b. types which occur most numerously in one particular zone, or whose main distribution ends or begins at the boundaries of that zone;
c. a fairly constant association of types, the lowest and highest occurrences of which are usually to be found elsewhere in the stratigraphic column.

The number of zones will, of course, be different in different regions. According to our experience, in the Tertiary one may expect at least 8, mainly qualitatively defined, major zones, each of which may prove to be subdivisible on quantitative grounds. The absolute age of the zones can only be determined by reference to other fossils, but the boundaries need not necessarily coincide with those of the current geological time scale.

The main question is whether a zone has a regional value or can be followed only locally, and whether it has a time-stratigraphic value. Before a zonation is applied in practice, its reliability in these respects should be carefully considered. This can be done either empirically or by investigating as thoroughly as possible the floral changes reflected in the diagrams. The latter method obviously depends on the possibility of determining the pollen and spore types; and the older the formation the more one has to resort to empirical methods.

Two main factors determine the differentiation in the pollen floras, i.e. evolution and migration. Evolution manifests itself in the

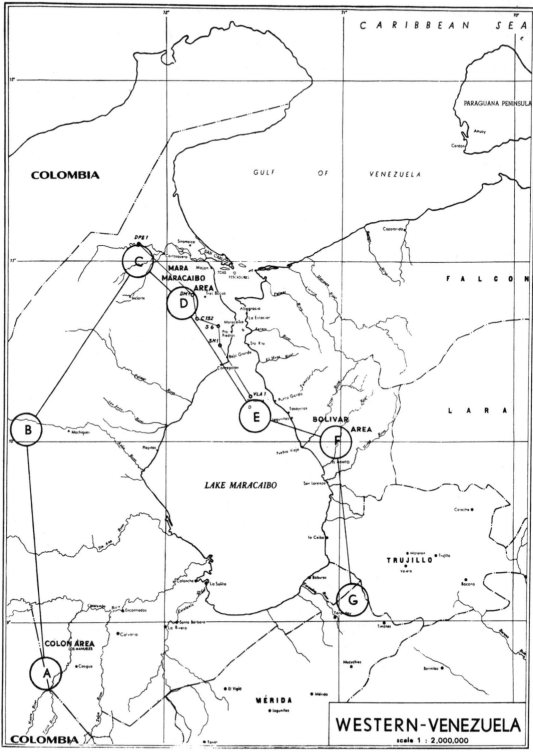

Fig. 5 — Map of western Venezuela, showing locations of the wells and sections in Figs. 6, 7 and 8.

sudden appearance of new types, the gradual modification of existing types, and the extinction of types. Migration is caused by changes in climatic and edaphic conditions and may be responsible for temporary or lasting changes in the composition of the floras.

b. *Evolution*

The development of pollen and spores runs parallel to that of the major units in the plant kingdom. Carboniferous sediments are thus characterized by a great diversity of spores of pteridophytes and primitive phanerogams, whilst genuine angiosperm pollen is lacking. Some spore genera occur which are absent from younger sediments and probably belong to extinct groups; *Florinites* and *Endosporites* (Pl. 7), for example, belong to the *Cordaitales* (Schopf et al., 1944). A remarkable feature is the sudden disappearance, at about the same time, in Europe and North America, of *Densosporites* (Upper Westphalian) and *Lycospora* (Stephanian).

From the Permian onwards gymnosperm pollen of more advanced types plays an important part, together with the spores of pteridophytes. It may be mentioned here that, in the younger Mesozoic of such tropical regions as Venezuela, Colombia and Nigeria, the winged coniferous types are absent, whilst grains of the *Ephedra* type are of very common occurrence (Pl. 1: 2—6). The number of ridges varies greatly; sometimes a furrow is present, just as in the case of the Recent *Welwitschia*. It seems, therefore, that in the tropics the *Coniferales* of the temperate regions are replaced by the *Chlamydospermae*.

A point of great importance, both from the scientific and the practical points of view, is the first appearance of pollen from angiosperms.

Erdtman (1948) has described a tricolpate pollen type from the Lower Jurassic *(Tricolpites troedssonii)* which should, according to the author, undoubtedly belong to the dicotyledons. This view is generally accepted (cf. Thomson, 1953), but it is questionable whether it is correct. At any rate, in the case of many almost identical grains from the Upper Jurassic of the Netherlands (Pl. 6: 1—7), one of the colpi is longer than the other two, and is provided with rounded-off ends. The distance between the short furrows may be smaller than that between a long and a short furrow. Finally, the grains are not ellipsoidal, but flattened. Accordingly, radial symmetry which, in our opinion, is a most important feature of dicotyledonous pollen, is absent. The grains can perhaps best be considered as monocolpate (cf. Pl. 6: 8) with two additional furrows on the proximal side. In view of this, a relationship should rather be sought with *Chlamydospermae (Ephedra, Welwitschia)*. As has been pointed out already, some fossil *Ephedra* types still show a furrow. The complete absence of a visible wall structure is another indication in this direction.

Nevertheless, the resemblance of *T. troedssonii* to dicotyledonous pollen types is so striking that it might well be considered as a forerunner. This, then, should be a strong argument in favour of assuming a chlamydospermous ancestry for part of the dicotyledons.

On the other hand, macrofossils which are undoubtedly angiospermous have been reported in the Aptian, and probable cases have been reported in the Middle Jurassic of England and Germany and even in the Rhetic of Greenland. The presence of angiospermous pollen types might, therefore, very well be expected in Lower Cretaceous and Jurassic sediments. Pl. 6 (9—11) shows a few specimens of a type which, in the authors' opinion, could belong to the angiosperms. The grains are often united to tetrads, they have a rudimentary tetrad scar on the proximal pole, a large, scabrate operculum on the distal pole and a broad intrastriate zone around the equator. Such types occur frequently in the Upper Jurassic of the Netherlands and the Middle East and have apparently also been found by Pflug in the Lias of Siegelsum [5]. It is possible that Klaus's "Tetradensporen" (1953, Fig. 1) and *Pollenites reclusus* of Thiergart (1949) also represent this or a related type.

The differentiated exine and the distal aperture would point strongly either to the monocotyledons or to those dicotyledons which possess distalipolar apertures, such as *Nymphaeaceae* and *Magnoliaceae*. However, the type is probably related to that found in the male flowers of *Cheirolepis münsteri* from the Rheto-Liassic (Hörhammer, 1933; cf. Reissinger, 1951). The family *Cheirolepidaceae* belongs to the conifers, but it is considered to occupy an isolated position among them (Moret, 1943).

The great onset of undoubtedly angiospermous pollen does not take place until the Upper Cretaceous. In the Senonian many important pollen classes are already present, together with some classes, modern representatives of which are unknown (Fig. 1: 26, 27, 29). The absence of *pericolporatae, periporatae* and *pericolpatae* may be noted. Although sculpturing shows variations which are already great, highly composite exine structures like those in the *Compositae* are lacking.

In the Tertiary the flora gradually assumes a modern aspect. *Alnus* has been found from the Lower Eocene onwards both in Europe and the U.S.; *Bombacaceae* (Pl. 2: 11) in the Paleocene of Venezuela and Colombia; *Olacaceae (Anacolosa* type, Pl. 2: 8) in the Lower Eocene of Nigeria and Venezuela; *Gramineae*

[5] Pflug did not recognize the operculum; he classified the grain partly among *Classopoll's* (l.c., Taf. 16, inter alia 29—31 and 39—41), partly among *Circumpollis* (l.c., Taf. 17: 28—36).

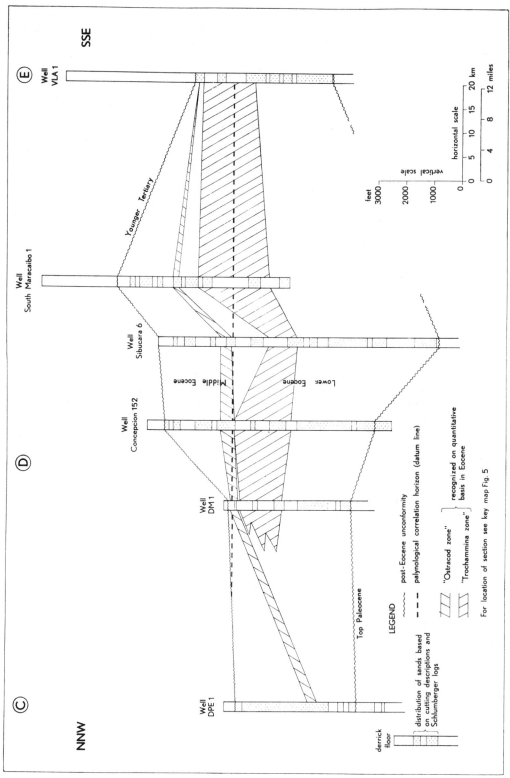

Fig. 6 — Diagrammatic cross section Paez district — Mara-Maracaibo district — N. Lake area, showing relations between lithological, faunal and palynological zonation in the Eocene.

Application of Palynology to Oil Geology

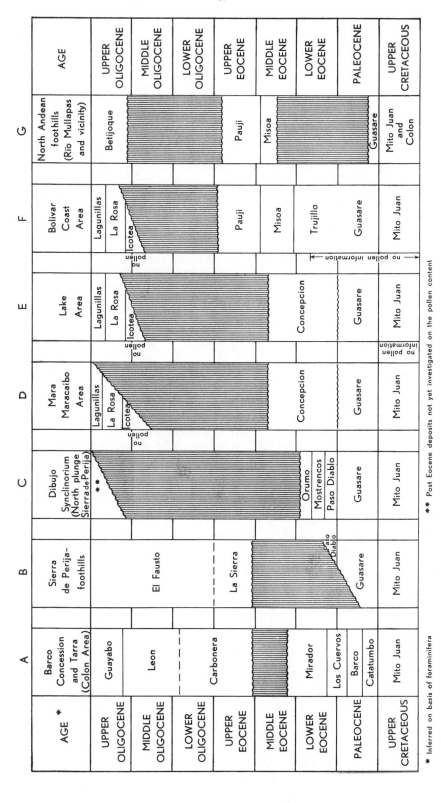

Fig. 7 — Palynological correlation of sections round Lake Maracaibo, western Venezuela. For locations see Fig. 5.

(Pl. 2: 9) from the Middle Eocene onwards (both in Nigeria and Venezuela). Pollen of Compositae — both Liguliflorae and Tubuliflorae — has never been found before the Miocene. Other groups disappear in the course of the Tertiary, such as the peculiar triangular types with very pronounced thickenings around the pores and furrows (Pl. 3: 2—5, 7, 9—11) from the Upper Cretaceous and Lower Tertiary of Europe and the United States, and the types with equatorial furrows (Pl. 5: 1—4) that are so common in the Lower Tertiary of Nigeria and S. America.

The outward appearance of some types (e.g. Alnus) remains remarkably constant, from their first occurrence onwards; others show gradual modifications. As an example the Ctenolophon type may be mentioned. It is characterized by conspicuous endexinous thickenings, which fuse at the poles, and a fine reticulum covering the whole grain. In the case of Senonian specimens (Pl. 2: 10), the endexine is relatively thin; in the case of Tertiary specimens (Pl. 2: 12, 13), the relative thickness of the endexine has increased. The thickenings may form a ring on the poles, which ring, in younger Tertiary specimens, is often irregularly divided, just like the grain shown by Erdtman (1952, Fig. 244).

It is clear from the foregoing that qualitative floral changes based on evolution are eminently suitable for defining the regional framework of the zonation. They may lead to correlations between different sedimentary basins, or even to intercontinental correlations.

c. *Migration*

For more detailed work, however, the quantitative boundaries, revealed when a diagram is constructed on a basis of percentages, are much more attractive from the practical point of view. These boundaries are caused by vegetational shifts which may be the result of changes in climatic or edaphic conditions. Such factors have, of course, ultimately the closest possible connection with geological events in or near the place of sedimentation, such as orogenies, transgressions and regressions, etc. Some of these events may have produced a temporary or local effect on the vegetation cover, others may have caused major and lasting changes in the flora.

The most striking changes in the vegetation cover are those which are caused by changes in climate. In a region such as Venezuela the first climatic changes to be considered are transitions from relatively arid to more humid climates. It will only be possible to reach certainty on this point when climate indicators are found among the fossil pollen and spore types. Additional indications can be found perhaps in the distribution of thick coal layers, which strata are only formed in the tropics under extremely humid conditions. Another phenomenon which points strongly in the direction of alternating climates is the recurrence of certain floral elements at different levels in the stratigraphic column. This alternation of floral elements is on a much larger scale than the small-scale facies change to be discussed below. Such a type of floral boundary is among the most valuable for stratigraphical correlation, because climatic changes usually take place within a relatively short period, as compared with evolutionary changes. They occur over large areas and directly influence the whole vegetation cover on dry land, and to a lesser degree also in swamp areas. Climates, however, change regionally, especially in areas of diversified topography. Some caution is therefore necessary when long-distance correlation of this type of floral change is attempted.

Changes in edaphic conditions, such as small transgressions in swamp areas, peneplanation of a mountainous area, with development of savanna vegetation, leaching of the soil, are bound to have a smaller regional influence than changes caused by climatic factors.

In general, pollen grains from plants belonging to coastal vegetations will exhibit greater quantitative fluctuations in marine

---

PLATE 1

Some identified fossil pollen and spore types (magnification 1000 x).
1. *Parkeriaceae (Ceratopteris)*; U. Tertiary, Borneo.
2, 3. *Ephedra distachya* type; U. Eocene, U.S.A.
4. *Ephedra strobilacea* type; M. Cretaceous, Iraq.
5. *Ephedra strobilacea* type; U. Cretaceous, Venezuela.
10. *Ephedra strobilacea* type; U. Cretaceous, Nigeria.
6. *Mimosaceae*; Miocene, Nigeria.
7. *Cyatheaceae (Hemitelia* type); U. Tertiary, Trinidad cf. Erdtman, 1941, pp. 146—147.
8, 9. (same grain) *Ericales*; U. Cretaceous, Nigeria.

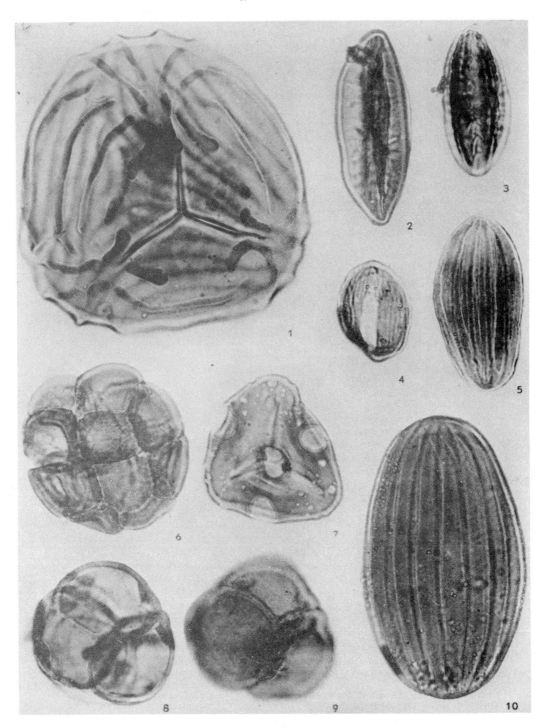

PLATE 1

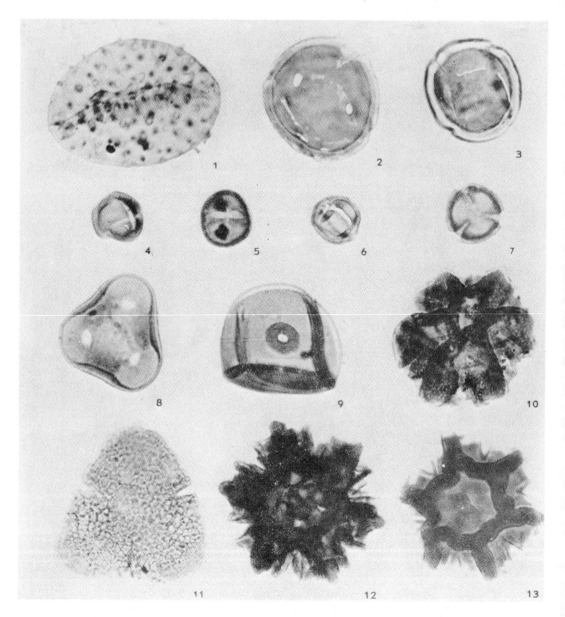

PLATE 2

Some identified fossil pollen types (magnification 1000 x).
1. *Palmae (Mauritia* type); Tertiary, Venezuela.
2, 3. *Malpighiaceae (Aspicarpa* type); U. Tertiary, Trinidad. (12 furrows, arranged in cube pattern, partly provided with pores, which are generally placed near the furrow ends; cf. Erdtman, 1952, pp. 265—266).
4—7. *Rhizophoraceae;* Miocene, Nigeria. (Zonorate, cf. Erdtman, 1952, p. 378.)
8. *Olacaceae (Anacolosa* type); U. Tertiary, N. Guinea. (3 pores on each hemisphere, cf. Erdtman, 1952, p. 295.)
9. *Gramineae;* M. Tertiary, Nigeria.
10. *Ctenolophonaceae (Linaceae* s.l.); U. Cretaceous, Nigeria.
11. Bombacaceae; L. Tertiary, Colombia.
12, 13. (same grain) *Ctenolophonaceae (Linaceae* s.l.); Miocene, Nigeria (cf. Erdtman, 1952, p. 244).

sediments than pollen grains from species which occur farther inland. These are less sensitive to facies and are therefore more suitable for regional correlation.

An extreme example is provided by the pollen grains of the *Podocarpus* type. These are found sporadically in Venezuela from the Eocene onwards, but reveal a striking increase in the Oligo-Miocene sediments. *Podocarpus* trees are restricted nowadays to mountainous districts above about 2,000 m. in height. It would therefore appear obvious to connect this increase with the origin of a mountain range, perhaps the Venezuelan Andes.

Most caution has to be observed when dealing with sediments in which a local over-representation of certain floral elements, e.g. swamp vegetation, is possible. This is, of course, applicable to lignites and coals in the first place, but also to sediments in respect of which such over-representation is not immediately obvious from the lithology. From the Oligocene onwards in Venezuela, Nigeria, New Guinea and Borneo pollen grains of the *Rhizophora* type, which point to a mangrove vegetation, often occur in enormous quantities. In view of the great "facies sensitivity" of this type of vegetation, increases and decreases here will be determined first and foremost by local conditions, and will be valueless for correlation.

It is thus clear that an understanding of the botanical relationship of the fossil types is of the greatest importance, because it can provide the key to an ecological interpretation of the phenomena observed, and to a subsequent separation of regional from local floral change. Without this key empirical work has to be resorted to. Such a case presented itself in Venezuela, in the Middle Eocene of the Lake Maracaibo basin. There, one of the main problems was the correlation of different sandstone bodies in a monotonous sandstone-shale facies, devoid of diagnostic animal fossils. The maximum correlative distance was 90 km., and it was soon evident that the floral succession in the different well sections was not identical. Some pollen types had an erratic distribution, while others showed a similar succession over the whole area. Evidently the first group had a local, probably facies-bound distribution, while the second group represented a regional floral succession, which was perhaps due to climatic causes. Confirmation of this by study of the botanical affinities of the types involved was not possible, owing both to lack of time and to the fact that the Eocene floras are already quite different from the Recent flora, which makes uncertain any detailed interpretation of ecologic conditions in the area in Eocene times. The disturbing influence of the local vegetation was eliminated by calculating the percentage ratio only of those types which showed the regionally similar succession. The floral zonation was then established on the basis of the characteristics shown in the curves of the "restricted diagrams". The strikingly similar floral succession thus obtained over the whole area makes it highly probable that the floral zones based on these changes have time-stratigraphical value. Similar methods have been followed in the practical application of palynological correlation methods in lignite mining operations in Germany (Rein, 1950).

Summarizing the foregoing, it will be seen that some at any rate of the alterations in the microflora observed in a certain region are synchronous, and are therefore recognizable in all sediments in which the pollen has been preserved, irrespective of its facies. This has proved in practice to be one of the most important advantages of the new method, especially in the case of regional basin studies.

VIII. STRATIGRAPHIC RESULTS IN WESTERN VENEZUELA

The preceding survey of palynological working methods will now be further exemplified by a discussion of some practical results obtained with the new methods in the well known oil field area of western Venezuela (Maracaibo Basin, Fig. 5).

Fig. 6 shows the result of an investigation of correlation possibilities in the Eocene, part of which is oil-bearing, in the Paez-Mara-Maracaibo districts and the northern Lake area. The aim was to correlate the major sand bodies in different well sections. It is clear that locally the palynological correlation line crosses the faunal zones, here reflecting local facies changes. Regionally, however, the assumed time-stratigraphic pollen correlation confirms the earlier conclusions arrived at on the basis of the distribution of the *Trochammina* zone, which gave a fair indication of the time-stratigraphy. The sand correlation is now based on the palynological zonation, and it is believed that in this way a better picture is obtained of the geological development, and especially of the facies changes, than was formerly possible, when only faunal zones were known.

The correlation table of Fig. 7 shows the general results achieved up till now, with the aid of palynological methods, in correlation between the four facies provinces generally recognized in the Tertiary of western Venezuela (Schaub, 1946). These provinces are found in the following areas:

1. Mara-Maracaibo area
2. Colon area
3. Bolivar area
4. Falcon area.

Correlation between these areas has always been very difficult owing to the lack of diagnostic fossils, and also because the centre of the Maracaibo basin is covered by Lake Maracaibo, a fact which restricts geological work very much to the rim of the basin. An identical floral succession has now been established in all four areas[6], and it is believed that this provides a sounder basis for regional correlation than that obtained by other methods. Consequently, it appears that some of the older stratigraphical interpretations will have to be revised.

In the correlation table, the stratigraphic names in the Colon area (with the exception of those for the Mirador formation) are based on the nomenclature published by Notestein (1944) in connection with the Barco concession in Colombia. The stratigraphic names of the other sections have been derived from Hedberg and Sass (1937), with the exception of those for the Betijoque and Mirador formations, which have been described by Liddle (1946), and the Concepcion formation, which was first discussed by the staff of the Shell Caribbean Petroleum Co. (A.A.P.G., 1948). Lithological descriptions of the formations mentioned can be found in the literature cited above. The references to the age of the formations as shown in the correlation chart are slightly generalized. In this connection it must be pointed out, in particular, that the position of the boundaries between the Eocene and Oligocene, and also between the Oligocene and Miocene, has not yet been accurately established, in consequence of which the age of the boundaries between the Leon and Carbonera formations, and between the La Sierra and El Fausto formations, is not exactly known. It is also not very clear which part of the Guayabo and Lagunillas formations belongs to the Oligocene and which part to the Miocene.

Comparison of the correlation presented in Figs. 7 and 8 with the chart published by Mencher et al. (1953) shows that, on the basis of palynological evidence, the Catatumbo and Barco formations, and the greater part of the Los Cuervos formation, are correlated with the Paleocene Guasare formation of the Mara-Maracaibo district. The pollen flora found in those largely non-marine and otherwise unfossiliferous formations in the Colon area is the same as the flora found in the marine Guasare in the Mara-Maracaibo area, and is therefore concluded to be of the same age. This correlation is supported by the regional top of pelagic foraminifera, indicating the top of the Cretaceous, running fairly parallel with the correlation indicated by pollen analysis. A striking feature is that this line too, like the pollen line, crosses the facies unit indicated

[6] Falcon area not depicted on Fig. 7.

PLATE 3

Some fossil pollen types of unknown taxonomic position (magnification 1000 x).
1. Monoporate (!) intectate reticulate. Senonian, Nigeria.
2, 3. (same grain) Tricolporate, tectate verrucate. L. Tertiary, Germany.
4, 5. (same grain) Tricolporate, tectate psilate; endocolpus (see p. 6) very clear. L. Tertiary, Germany.
6. Syncolpate, finely striate. Coniacian, Nigeria.
7. Periporate (3 ectexinous pores on each hemisphere, one endexinous pore per set of 2 pores; surrounded by heavy costae). L. Tertiary, Germany.
8. Syncolpate with equatorial bridges (or rather "diplotrichotomosulcate"), intectate reticulate. U. Cretaceous, Colombia.
9. Tricolporate, tectate psilate-scabrate; short furrows in ectexine provided with heavy thickenings, transversal furrows in endexine, provided with coarse, somewhat irregular costae. Eocene, U.S.A.
10, 11. Tricolporate; endexine very thin, round pores faintly visible. L. Tertiary, Germany.
12. Stephanocolporate (zonorate). U. Cretaceous, Nigeria.
13. "Endosyncolpate"; 3 furrows in the endexine which fuse at the poles, 3 small round pores in the ectexine. L. Tertiary, Germany.
14. Tricolpate, tectate foveolate-perforate. Paleocene, Colombia.

Application of Palynology to Oil Geology

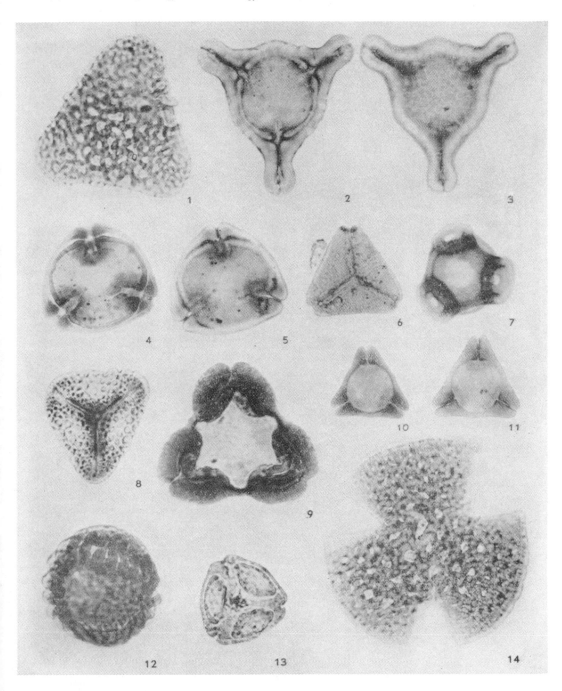

PLATE 3

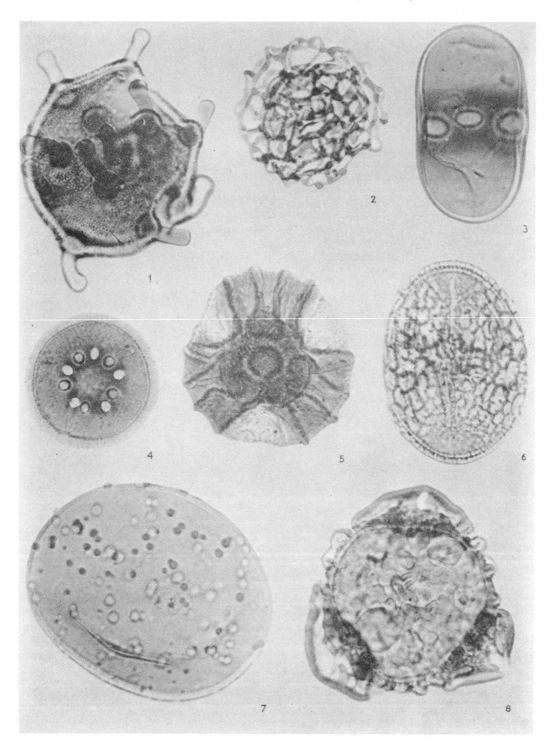

PLATE 4

by the Rotalia and Discorbis bottom-living assemblages; this unit shifts to younger time levels in northerly direction. A further point of interest is the apparent time-stratigraphic correlation obtained from electrical logs, covering the pollen line (see Fig. 8 and also Dufour, 1951). It is concluded that the marine Guasare formation in the north is contemporaneous with the thick coal-bearing sequence of shales and sandstones of the Catatumbo, Barco and Los Cuervos formations in the south, and this appears to give a more logical picture of sedimentation conditions during Paleocene time than we had previously. The facies shift, as mentioned above, into a higher time-stratigraphic level in the direction of off-shore deposition, expressed by the Rio de Oro-Guasare lithology and its benthonic foraminiferal assemblages (*Discorbis* and *Rotalia* assemblages), is noteworthy.

Another point of interest concerns the Mirador formation which, according to palynological evidence, is mainly of Lower Eocene age and can be correlated with the Orumo, Mostrencos and Paso Diablo formations, with the lower part of the Concepcion formation and with the Trujillo formation of Sutton (1946). The Trujillo formation underlies the "second orbitoid level", the age of which is Lower Middle Eocene according to determination of its larger Foraminifera by de Cizancourt (1948) and van Raadshooven (1951). Sutton (1946) correlated the Mirador formation with the Misoa formation and stated that "the Mirador formation has recently been traced around the southern edge of the (Maracaibo) basin into the type Misoa". According to palynological evidence such a correlation is not correct.

Slightly higher in the sections of Fig. 7 it will be seen that the lower part of the Carbonera formation is correlated with the La Sierra formation of the Perija range and with the Pauji formation on the east side of Lake Maracaibo. The disconformity indicated at the base of the Carbonera and La Sierra formation does not extend eastward to the Bolivar coast area. The age of the Pauji formation is partly Upper Middle Eocene (van Raadshooven, 1951); and that of the Carbonera formation is Upper Eocene — Lower Oligocene, and not only Upper Eocene, as stated by Mencher et al. (1953).

As regards the *Hannatoma*-bearing beds, occurring in the Carbonera formation in the southwestern part of the Maracaibo basin, the pollen flora of these beds has not as yet been studied. It has therefore not yet been established on pollen evidence whether this *Hannatoma* molluscan fauna in Venezuela is confined to the Upper Eocene or not.

Continuous deposition, during Paleocene-Eocene time, only took place in the Bolivar coast area, while during Oligocene time continuous deposition took place only in the SW part of the Maracaibo basin. With the exception of a few marine beds which are related to the marine La Rosa incursion of the northern Lake area, all these deposits are continental in origin. It is probable that no sedimentation took place during a large part of the Oligocene in the Mara-Maracaibo Lake and Bolivar area. There is no palynological evidence for the exact stratigraphic position of the Icotea formation, as no pollen has been found in it.

A last important problem for the solution of which palynological correlation provides new evidence is the controversy over the age of the La Rosa and Lagunillas formations. The cause of this controversy lies in the fact that the molluscan faunas in the La Rosa formation were assigned by Hoffmeister (1938) to the Lower Miocene and by some Shell palaeontologists (unpublished reports) to the Middle Miocene, on the assumption that the *Miogypsinae* included were reworked from older deposits. Drooger, however, recently (1952) commented as follows: "The *Miogypsinae* of the samples from the Middle Miocene La Rosa

---

PLATE 4

Some fossil pollen and spore types of unknown taxonomic position (magnification 1000 x).
1. Inaperturate, intectate scabrate + clavate. Miocene, Nigeria.
2. Triporate, reticulum never fixed to the columellae. Miocene, Nigeria.
3. Tricolporate, tectate psilate. Miocene, Nigeria.
4. Periporate (5 pores on each pole), tectate psilate. Miocene, Nigeria.
5. Tricolpate. U. Cretaceous, Nigeria.
6. Tricolporate, tectate reticulate. Miocene, Nigeria.
7. Monolete, gemmate. U. Tertiary, Borneo.
8. Tricolporate, tectate psilate; exine cavate. Miocene, Nigeria.

formation are stated to be reworked from older, Oligocene, deposits. However, each of these assemblages appeared to be homogeneous, no trace of reworking being visible. Therefore, some doubt may be expressed as to the age determinations of these samples, an Uppermost Oligocene age being more suitable".

On the basis of the newly obtained palynological evidence the La Rosa formation should be correlated with the Agua Clara formation in Falcon, which, according to Sem (1940), is of Upper Oligocene age; this was confirmed by Shell palaeontologists (unpublished). On this basis and accepting Drooger's views as correct, the Lagunillas formation, which probably represents the retreating phase of the La Rosa marine ingression, correlates with the Carro Pelado formation in Falcon. The palynological zones recognized in the La Rosa and Lagunillas formations can also be traced in the lower part of the Betijoque formation in the North Andean foothills, and in the Guayabo formation of the Colon area.

IX. BIBLIOGRAPHY

Anon. (1952) — Bulletin of the Standard Oil Company of California, October, pp. 2—7.

Armstrong, T. A. (1953) — New exploration tool. The Oil and Gas Journal 51, No. 44, pp. 64—65.

Caribbean Peeroleum Company (1948) — Oilfields of Royal Dutch/Shell Group in Western Venezuela. Bull A.A.P.G., Vol. 32, pp. 517—628.

Cizancourt, M. de (1951) — Grands foraminifères du Paléocène, de l'Eocène inférieur et de l'Eocène moyen de Venezuela. Mém. Soc. Géol. France, N.S., T. 30, Fasc. 1—2, Feuilles 1 à 9, Mém. No. 64, pp. 1—68.

Drooger, C. W. (1952) — Study of American Miogypsinidae. Thesis, Utrecht, The Netherlands.

Dufour, J. (1951) — Facies shift and isochronous correlation. Proc. 3rd World Petroleum Congress, Section I, pp. 428—437, The Hague.

Erdtman, G. (1943) — Introduction to pollen analysis. Waltham, Mass., U.S.A., 239 pp.

——, (1948) — Did Dicotyledonous plants exist in early Jurassic times? Geol. Fören Förh., Bd. 70, pp. 265—271.

——, (1953) — Pollen morphology and plant taxonomy.

Faegri, K. (1939) — Single-grain pollen preparations. Geol. Fören Förh., Bd. 61, pp. 513-514.

Faegri, K. and Iversen, J. (1950) — Textbook of modern pollen analysis. Copenhagen, 168 pp.

Florschütz, F. and Jonker, F. P. (1939) — A. botanical analysis of a late-Pleistocene and Holocene profile in the Rhine delta. Extr. Rec. Trav. Bot. Néerl., vol. XXXVI, pp. 686—696.

Hedberg, H. D. and Sass, L. C. (1937) — Synopsis of the geologic formations of the western part of the Maracaibo Basin, Venezuela. Min. Fomento de Venezuela, Bol. Geol. y Min., Tomo 1, Nos. 2, 3, 4, pp. 73—112.

Hoffmeister, W. S. (1938) — Aspect and zonation of the molluscan fauna in the La Rosa and Lagunillas formations, Bolivar Coastal Fields, Venezuela. Min. Fomento de Venezuela, Bol. Geol. y Min., Tomo 2, Nos. 2, 3, 4, pp. 103—122.

Hörhammer, L. (1933) — Über die Coniferengattungen Cheirolepis Schimper und Hirmeriella nov. gen. aus dem Rhät-Lias von Franken. Bibl. Bot., Bd. XXVII, Heft 107.

Iversen, J. and Troels-Smith, J. (1950) — Pollenmorphologiske definitioner og typer. Danmarks Geol. Unders., IV. Raekke, Bd. 3, No. 8, pp. 1—54.

Klaus, W. (1953) — Alpine Salzmikropaläontologie (Sporendiagnose). Pal. Zeitschr., Bd. 27, pp. 52—56.

Kosanke, R. M. (1950) — Pennsylvanian Spores of Illinois and their use in correlation. State Geol. Survey, Bull. No. 74, Urbana, Illinois.

Liddle, R. A. (1946) — The geology of Venezuela and Trinidad. 2nd Ed., Paleontological Res. Inst.

Mencher, E. et al. (1953) — Geology of Venezuela and its oil fields. Bull. A.A.P.G., Vol. 37, pp. 690—777.

Moret, L. (1943) — Manuel de paléontologie végétale. Paris.

Notestein, F. B., Hubman, C. W. and Bowler, J. W. (1944) — Geology of the Barco Concession, Republic of Colombia, S. America. Bull. Geol. Soc. America, Vol. 55, pp. 1165—1216.

Pflug, H. D. (1953) — Zur Entstehung und Entwicklung des angiospermiden Pollens in der Erdgeschichte. Palaeontographica, Abt. B, Band 95, pp. 60—171.

Potonié, R. (1943) — Zur Mikrobotanik der eocänen Humodils des Geiseltals. Arbeiten aus dem Inst. f. Paläobot. und Petrogr. der Brennsteine, Bd. 4, pp. 25—125.

Raadshooven, B. van (1951) — On some Paleocene and Eocene larger foraminifera. Proc. 3rd World Petr. Cong., Section 1, pp. 475—482.

Raistrick, A. (1934) — The correlation of coal seams by microspore content. Part. I. The seams of Northumberland. Trans. Inst. Min. Eng. Vol. LXXXVIII, 3, pp. 142—153.

Rein, U. (1950) — Die Bedeutung der Flözengliederung für den rheinischen Braunkohlenbergbau und ihre Anwendung in der Praxis. Braunkohle, Wärme und Energie, pp. 72-78.

Reissinger, A. (1938—1950) — Die "Pollenanalyse" ausgedehnt auf alle Sedimentgesteine der geologischen Vergangenheit. I. Palaeontographica Abt. B, Bd. 84, pp. 1—20. II. Ibidem, Bd. 90, pp. 99—126.

Schaub, H. P. (1948) — Outline of sedimentation in Maracaibo Basin, Venezuela. Bull. A.A.P.G., Vol. 32, pp. 215—227.

Schopf, J., Wilson, L. R. and Bentall, R. (1944) — An annotated synopsis of Paleozoic spores and the definitions of generic groups. Illinois Geol. Surv. Rep. of Inv., 90, 74 pp.

Senn, A. (1940) — Paleogene of Barbados and its bearing on history and structure of Antillean-Caribbean region. Bull. A.A.P.G., Vol. 24, pp. 1548—1610. *(Continued on p. 75)*

Application of Palynology to Oil Geology

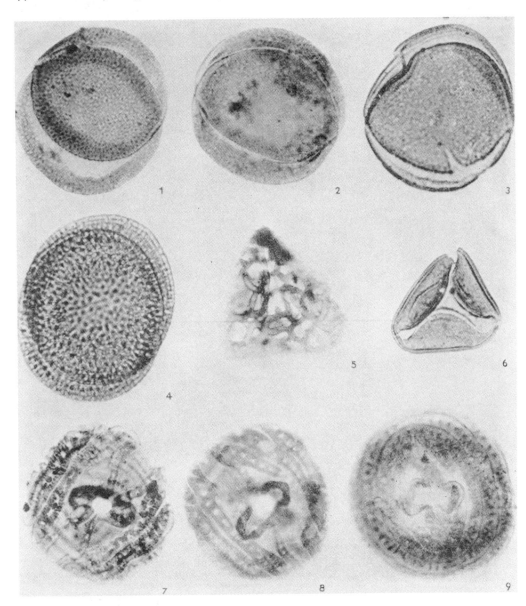

PLATE 5

Some fossil pollen types, belonging to interesting pollen classes.
1—4. Pollen types with one equatorial furrow, surrounding the whole grain and dividing the grain into 2 identical parts.
1. Tectate perforate/intectate reticulate. Paleocene, Colombia.
2, 3. Idem Eocene, Nigeria.
4. Intectate reticulate (M. Tertiary, Nigeria).
5—6. Pollen types with one "trichotomocolpate" aperture *(Palmae?)*.
5. Intectate reticulate. U. Cretaceous, Nigeria. (cf. *Acanthoriza mocinni*, Erdtman, 1952, p. 305.)
6. Tectate psilate. Tertiary, Colombia.
7—9. Highly developed pollen types with, on each hemisphere, one colporate aperture and a complicated system of parallel secondary furrows running at right angles both to the central furrow and to the corresponding furrows on the other hemisphere.
7, 8. (same grain, only upper and lower wall focused). Tectate perforate (one row of perforations between the secondary furrows). U. Tertiary, Trinidad.
9. (only upper part focused). Almost identical type with 3 or 4 furrows running parallel to the equator. U. Tertiary, Trinidad.

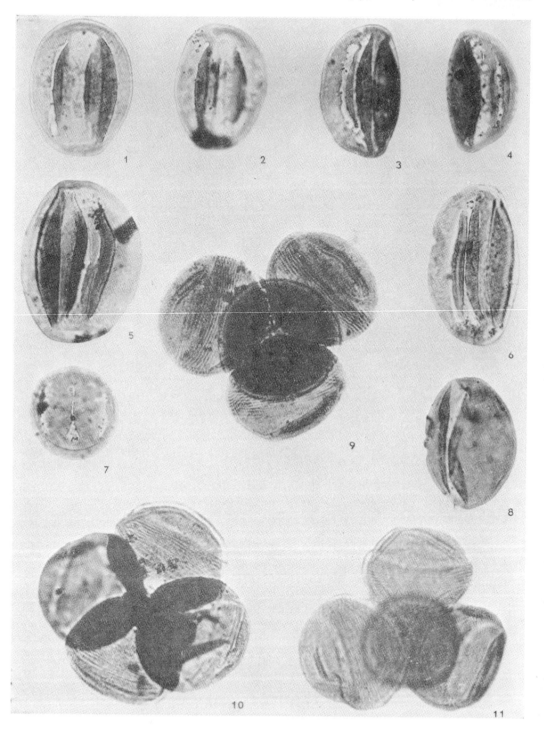

PLATE 6

Application of Palynology to Oil Geology

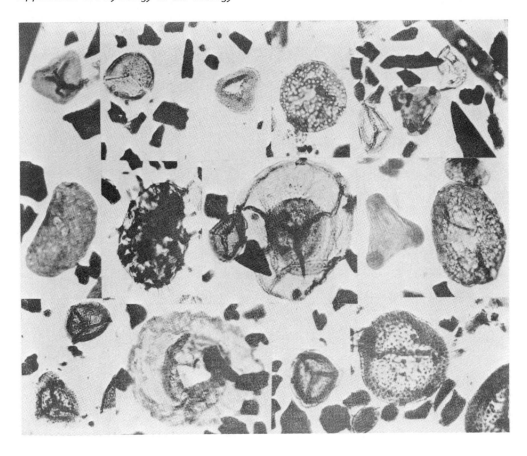

PLATE 7

Composite microphotographs of a slide prepared from a shale in the Dutch well Haaksbergen 1 (947 m), belonging to the Carboniferous (Westphalian D, according to Dijkstra (in the press)) (magnification 500 x).
Top row: from left to right: *Granulatisporites*, *Lycospora*, *Granulatisporites*, *Punctatisporites*, *Triquitrites*;
Middle row: *Laevigatosporites*, *Raistrickia*, *Lycospora*, *Endosporites*, *Triquitrites*, *Florinites*;
Bottom row: *Lycospora*, *Cirratriradites*, *Florinites Lycospora*, *Punctatisporites*.

PLATE 6

Some fossil pollen types from an Upper Jurassic sample from the Netherlands (well Schoonebeek 232, depth 862—863 m) (magnification 1000 x).
1—7. Pollen grains with three furrows. No. 7 probably belongs to a different type. Exine undifferentiated.
8. Monocolpate pollen grain. Exine undifferentiated.
9—11. Tetrads of operculate grains with a large scabrate operculum and a broad intrastriate zone round the equator.

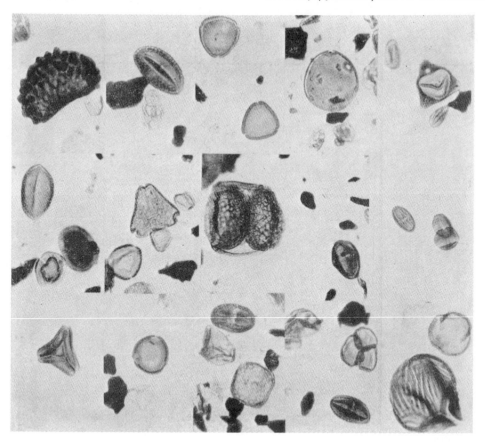

PLATE 8

Composite microphotographs of a slide prepared from a carbonaceous clay sample, collected at Riverside loc., Jackson, Miss. (Upper Eocene, lower part) (magnification 500 x).
Top row: from left to right: monolete spore *(Polypodiaceae)*; tricolporate reticulate pollen; triporate pollen (type 9 of Fig. 4); periporate pollen (cf. *Juglans);* tricolporate pollen.
Middle row: monocolpate pollen; tricolporate, scabrate pollen; vesiculate pollen *(Coniferae);* tricolporate pollen.
Bottom row: syncolpate pollen; triporate pollen (type 10 of Fig. 4); tetraporate pollen *(Ulmaceae);* tetrad: trilete, striate spore *(dorogensis* type); tricolpate, reticulate pollen.

Sutton, F. A. (1946) — Geology of Maracaibo Basin. Bull. A.A.P.G., Vol. 30, pp. 1621—1741.
Teichmüller, M. and R. (1954) — Die stoffliche und structurelle Metamorphose der Kohle Rundschau, Band 47, pp. 265—296.
Thiergart, F. (1949) — Der stratigraphische Wert mesozoischer Pollen und Sporen. Palaeontographica, Bd. LXXXIX, Abt. B, pp. 1—34.
Thomson, P. W. (1953) — Zur Entstehung und Ausbreitung der Angiospermen im Mesophytikum. Paläont. Zeitschr., 27, pp. 47—51.
Thomson, P. W. und Pflug, H. (1953) — Pollen und Sporen des Mitteleuropäischen Tertiärs. Palaeontographica, Abt. B. Band 94, pp. 1—138.
Wilson, L. R. (1946) — **The correlation of sedimentary rocks by fossil spores and pollen.** Journ. Sed. Petrol., Vol 16, 3, pp 110—120.
Wodehouse, R. P. (1933) — **The oil shales of the Eocene Green River Formation. Bull. Torrey** Bot. Club, 60.
——, (1935) — Pollen grains. **New York.**
Wyatt Durham, J. et al. (1949) — **The age of the** Hannatoma Mollusk fauna of **South America,** a symposium. Jour. Pal., Vol. 23, pp 145—160.

Reprinted from pp. 7-10 and 61 of *Illinois Geol. Survey Rep. Inv.* 91:1-73 (1944)

AN ANNOTATED SYNOPSIS OF PALEOZOIC FOSSIL SPORES

AND

THE DEFINITION OF GENERIC GROUPS

BY

J. M. Schopf, L. R. Wilson, and Ray Bentall

INTRODUCTION

The literature on fossil spores is scattered, and in addition there has been much diversity in methods of classification. The objects of the present article are to bring this material together in summary form for convenient taxonomic reference and to evaluate the genera which have been previously proposed. After a study of these microfossils in America and a thorough study of the literature it seems essential also to describe some new genera. It is hoped that a fundamentally sound basis has been provided for studies now in progress in the three laboratories represented by the authors and for future studies.

Guiding principles which relate to the treatment of this material are given below.

1. Only adherence to the systematic principles embodied in the International Rules will give satisfactory results in the study of these microfossils (as in other fields of paleontologic study). Therefore, the International Rules of Botanical Nomenclature, 3rd edition (Briquet), revised by the International Botanical Congress of Cambridge, 1930, and published in 1935,[1] have been used as a basis for taxonomic treatment.

2. Species classified within the same genus or under the same generic name must possess significant characteristics in common. There can be little positive proof of generic identity unless substantial homologous comparisons can be drawn between respective biocharacters of each congeneric species and the type species of the genus. For example many isolated lycopsid megaspores must be classified separately from the microspores found in association with them because there is no adequate basis for close comparison of biocharacters exhibited by these diversified organs.

When affinities become better known so that the difficulty due to uncomparable features may be partially overcome, it then may be possible to relate the forms more naturally in classification within suprageneric groups such as tribe, family, and subfamily depending on how closely information may be correlated. Indeed, assignment to the same genus, or even species, is theoretically possible and may eventually be achieved in some few instances, but in general this implies a much more precise and detailed historical knowledge of interrelationships than is likely to be fully established.

For satisfactory nomenclature it is most convenient that those fossils which possess numerous demonstrably comparable features shall be placed under the same generic name. However, it is unsatisfactory to classify fossils under the same generic name simply because a few arbitrary features are held in common. The essential validity of any classification depends on the relative significance that attaches to the various biocharacters. In fact it is by virtue of such interpretive discrimination that taxonomy is to be distinguished from a cataloging procedure. Interpretations nevertheless must be based on evidence and not on supposition; difficulties are multiplied whenever interpretation exceeds factual bounds.

3. In paleobotanical practice there has been much divergence with regard to the significance and proper interpretation of unusually complete specimens in which parts usually found separated, and therefore generally classified separately, are

---
[1] Gustav Fischer, Jena.

found in organic union. Whatever attitude may be adopted, it is of fundamental importance for later consistent application of taxonomic principles. For this reason it seems desirable to state views applicable to the present work.

The few unusually complete specimens are highly significant in showing beyond question that certain isolated organs at a particular locality and stratigraphic level possess identity of relationship.* This identity must become more dubious, in the case of isolated parts, the further they are removed geographically and stratigraphically from the site of proved union. Stratigraphic discrepancy is most likely to cast doubt on specific identity; geographic discrepancy may be cause for qualification of the presumed relationship in varying degrees of subspecific magnitude (there is no basis for assuming that geographic races were less in evidence in the past than they are at present). If evolution is a more or less continuous attribute of life processes, over a period of time significant alterations in some of the many heritable characteristics are bound to occur. An individual alteration, although of specific significance, will not necessarily involve phenotypic expression in the majority of the other characteristics of the race. For this reason it is inaccurate to postulate the continued specific coordination of an extensive garniture of biocharacters over a period of time, without correspondingly detailed knowledge which proves that the characteristics dealt with were not subject to mutative or other evolutionary processes during that time interval. This detailed information is most readily determined for a restricted group of biocharacters which can be observed on common specimens. Practical reasons dictate that emphasis in systematic treatment be placed on the more common types of fossils rather than on those which show unusual preservation.

Demonstrated connections b e t w e e n spore forms and types of fructifications prove readily that some of the adjacent isolated specimens of either category are also conspecific. The diverse taxonomic groups, diagnosed on the basis of typical isolated specimens are proved, therefore,

---

* No plant, and certainly no single specimen, can have more than one valid name. However, instances arise in which it is difficult to determine which of two or more names should actually be applied. In spite of this, systematic procedure demands that an author follow some consistent usage.

to have coordinate significance through a part of their geologic life as species. During this particular period, identity of relationship must exist. But the most pertinent fact encountered in systematic treatment of the common isolated forms, is that this period of actual synonymy can scarcely ever be defined. As a general thing, due to discontinuity in the fossil record, it is impossible to establish the points at which old characters became coordinated in a new fashion or with other characteristics new to the phyletic succession.

Consequently these overlapping relationships do not lend themselves to taxonomic expression within specific and generic nomenclatural categories. Groups of suprageneric rank, which are less dependent on type specimens for their proper definition, may be better used in expressing such relationships of generalized validity. The present paper, however, does not deal with groups of suprageneric taxonomic status.

4. Various criteria for determination of synonymy have been used by paleobotanists. The practice we prefer is conservative to the extent that unless conspecific relationship is proved for two nomenclatural types, both names are valid. As such, they are available for any purpose of useful nomenclature. There are a large number of "partial synonyms" among the named species of fossil plants to which these remarks apply. The word "synonym" of course is an absolute term—"partial synonyms" are not synonyms; in most instances the names serve useful purposes in indicating groups of different circumscription. None of these names should be lightly considered; neither should a name be used in unqualified application unless its pertinence is evident. Evidences of "partial synonymy" undoubtedly record close natural relationship, but the names should not be regarded as invalid unless shown to be mutually inclusive. Lacking this, they remain applicable to fossils conspecific with their basic nomenclatural types. For the research worker there are advantages in this conservative view of synonymy, as it tends to promote a more precise differentiation of fossils. This basically, is a fundamental reason for the continued use of technical nomenclature, necessary to progress in the study of fossil plants.

5. Specific identification, because of its great significance, must be made critically. Emphasis should be placed on the positive evidence of similarity rather than on negative evidence of lack, real or fancied, of features which cannot be examined critically.

6. Neither spores, leaves, nor any other morphological part of a plant in any stage of its life is here considered to *be* a species. A species of plant, for the purposes of this synopsis, and regardless of how it may be otherwise defined, is regarded as a group of organisms. Any individual organism of such a group, however, may be identified if any diagnostic structure representing any portion of its life cycle is available. Spores in many instances are probably just as definitive of species as any other organ belonging in a coordinated manner to the life cycle of a particular plant individual. They merit study with other types of ancient plant materials, and all data should be recorded on a common basis of taxonomic equality. Because of their unique adaptation for dispersal and their numerical abundance in many sedimentary deposits, it is expected that spores will eventually become of greater practical significance than many of the larger types of plant fossils that are more commonly noted by the geologist.

7. Since the purpose of giving a name to a taxonomic group is not to indicate the characters or the history of the group, but is simply a convenience in reference, all names that are taxonomically valid and pertinent to recognized groups should be continued. The authors agree that the hyphenated generic nomenclature, for example, *Granulati-sporites, Laevigato-sporites, Denso-sporites,* instituted first by Ibrahim (1933), is in poor form, lacks euphony, and tends to be generally misleading. Nevertheless, the status of such names seems reasonably secure in several instances and their continuance is likely to cause less misapprehension than attempts to institute more appropriate nomenclature.

The preceding explanatory paragraphs may serve to clarify some of the policies adopted in this and subsequent reports and explain their consistency with systematic treatment of other types of plant material.

The earliest proposed name requiring consideration is Dawson's *Sporangites*. This name was first proposed in 1863 and has been used ambiguously and in different senses by many authors. It is here regarded as a *nomen ambiguum* and the name *Tasmanites* is adopted for the most frequently encountered forms that have been commonly assigned to *Sporangites*. *Tasmanites* nevertheless is a problematic form, spore-like in many of its characteristics, but actually unassignable to any group of plants now known. All the other genera treated here belong with little doubt to the Cormophyta, and all whose affinity is approximately known belong to the Tracheophyte groups, as mentioned in the respective generic discussions.

Knox (1939, 1941) has suggested that spores of Bryophytes may be present in Carboniferous coals, and some forms that have been described may well belong to this group of plants which at present are very scantly recognized in the Carboniferous. The fact that no spores of Paleozoic Bryophytes are definitely known (*Sporogonites* Halle may be an exception) makes any suggested correlation hazardous at this time.

It is important to realize that many specific features of spores of different geologic age can be matched because of evolutionary convergence as well as because of community of derivation. Consequently unless there exists some corroborative evidence based upon spore forms actually present in contemporaneous fossils whose relationship can be established, all that seems warranted is to direct attention to similarity of the spores of different geologic age, recognizing that the similarities may or may not have phyletic implications.

There is some lack of strict agreement between the morphological nomenclature applied to modern pollen and the morphologic features of fossil forms. For one thing the microspore in modern forms is determined, in a strict sense, solely by the presence of the unicellular male gametophyte.[2] Such a distinction is of course inapplicable to fossil forms. There is also

---

[2] The common indiscriminate reference to all spores of relatively small dimensions as "microspores" is to be lamented. Although true microspores frequently are small, by no means all small spores are microspores. The long established botanical usage of the term "microspore" has reference to fundamentally functional distinctions that are entirely aside from relative or specific size. R. B. Thompson (1927) has in fact demonstrated that in some plants the microspore is *larger* than the actual megaspore.

the further difficulty that no simple correct morphological designation can be applied similarly to the dispersal forms of the male gametophyte bodies in both modern and fossil forms. The dispersal stage rather than the degree of gametophytic development is of greatest practical significance when fossils are considered. The terms perispore, exospore, and endospore might seem equivalent to the pollen structures perine, exine, and intine, respectively. However, in the case of the bladder-equipped fossil gymnospermic pollen, the bladder membrane which seems very similar in morphologic character to the perisporal membrane of certain cryptogamic forms, is regarded by many as equivalent to the exine of angiospermic pollen.

Revision of the morphologic terminology for fossil spores and similar microfossils also seems required. Not only has there been a dearth of descriptive terms available for description of the varied structural features, but a good many deeply rooted terms have become outmoded. Several of these date from the period in which *Selaginella* was presented as a primitive forerunner of modern seed plants, and their homologous implications now oftentimes seem unwarranted. Non-commital descriptive terms, which seem reasonably free of theoretical connotations, have been preferred for use in the present paper. It is hoped that it will be possible later to treat the descriptive morphology of spore forms specifically in greater detail.

In the synoptic lists the species assigned to each genus are given in alphabetical order. The order of genera adopted is chronologic according to the time each was first proposed. Although some intergeneric alliances can be recognized now, and a more logical sequence should be adopted in later works, it is felt that the chronologic order of priority will be of some convenience for taxonomic reference.

Nearly all the references cited (p. 62) are those of particular taxonomic or morphologic pertinence for the fossils discussed. Although we believe this represents a reasonably complete survey of this literature, it does not include all articles that deal incidentally with Paleozoic spores. Various non-taxonomic systems have been used by some workers in designating types of spores in coal thin-sections, notably by Slater and his colleagues in the South Yorkshire laboratory of the British Fuel Research Division, by Th. Lang in Silesia and, most importantly, by Reinhardt Thiessen. These apply very slightly to the present work because species cannot be as reliably distinguished from spores in thin-sections, the significant biological characters are generally more difficult to ascertain, and the informal nomenclature used does not require systematic consideration. Consequently articles of this nature are not considered except in connection with the morphology of certain forms.

On the other hand, there is a considerable literature on isolated spores in the Russian language which needs to be given careful study before nomenclatorial stability can justifiably be hoped for. Thus far only a small portion of this literature has been available to us, chiefly in abstract form, and this is insufficient to provide a satisfactory basis for understanding the quite different nomenclature Russian writers have used. Naumova (1937) mentions that about 400 species have been distinguished based on material ranging in age from the Lower Carboniferous to Tertiary. He mentions genera designated as *Zonotriletes* and *Azonotriletes*. Nikitin (1934) has distinguished the genus *Kryshtofovichia* based on very singular appearing, large spores of Devonian age. Luber (1938) mentions subgroups (genera?) *Azonotriletes*, *Zonotriletes*, *Azonomonoletes*, *Azonaletes* and *Zonaletes* and distinguishes several apparently valid species. In a more recent publication Luber (1939) uses the generic (?) names *Plagulites*, *Turriella*, *Circelliella*, *Saccrimalia*, *Circella*, *Libumella*, *Spinosella*, *Subsacculifer* and *Plicatella*. It is to be hoped that under more favorable circumstances it will be possible to give full consideration to the Russian studies which have already been put to good use in the age determination and correlation of the widely separated coal deposits of the USSR and to integrate the nomenclature they are using with our own.

[*Editors' Note:* Material has been omitted at this point.]

SUMMARY

This report presents the revised classification of plants identified from their isolated fossil spore coats, and deals primarily with those of Paleozoic age. Orthodox taxonomic procedure has been followed as closely as possible for reasons which are presented in the introduction. A cautious policy has been followed with regard to synonymy, and, even though forms are very closely allied, names based on different holotypes have been allowed to stand. In many instances, however, the apparent close relationship, which may later prove conspecific, has been noted. Thus names noted as synonyms are nearly all objective synonyms because they are based on a common holotype. Many difficult problems concerning nomenclature of fossil plants are solvable if nomenclatural types are strictly interpreted.

About 400 named species have been included in the present paper and most of these have been allocated among 23 genera which seem to serve a useful and significant purpose in classification. Additional genera no doubt will require recognition later and new information will modify the views that have been expressed for the genera described here. A number of species described previously apparently do not conform sufficiently to merit inclusion in the same genus with the type species of the group, and at the same time do not show convincing evidence of affinity with other recognized groups. Such forms have been listed as *species excludende*. Attention is directed to the several species excluded from *Reticulati-sporites* which here is interpreted in a considerably restricted sense. Many of these forms require much more careful study in order to arrive at a satisfactory expression of their affinity and classification. Other forms listed under *incertae sedis* also are lacking in sufficiently understood biological characteristics to support a definite systematic allocation. In dealing with plant microfossils it seems unavoidable that many forms worthy of description will nevertheless have such problematic relationship that their assignment under *incertae sedis* is obligatory. The usefulness of fossils nevertheless bears a considerable relationship to the basic and fundamental information available about them and for this reason greatest significance must be attached to species whose relationship has been reliably established.

The authors of this paper and their associates have observed many new types of plant microfossils in preparations from coal and carbonaceous sediments in America. The present synopsis is the outgrowth of a need for a more comprehensive survey of previous work considered from the standpoint of a consistent systematic policy. The essential features of an appropriate policy have been embodied here to serve as a working basis for the great amount of descriptive work yet to be done. We believe that this working basis will require further revision as new information is accumulated and presented. We further believe that such revision can be carried out with greatest efficiency and benefit to all concerned if the orthodox usages characteristic of mature systematic science are adopted and critically applied.

ACKNOWLEDGMENTS

The authors are under particular obligation to Dr. W. H. Camp of the New York Botanical Garden who, at a period when other duties were particularly urgent and pressing, nevertheless found time to read the manuscript and suggest numerous desirable revisions in the manner of presentation.

Mr. R. M. Kosanke of the Illinois Survey, Dr. Gilbert H. Cady, and members of the Survey editorial staff all have contributed in final preparation of this paper for publication. To all of these the authors owe a special debt of thanks.

REFERENCES

Ibrahim, Ahmet Can, 1933, Sporenformen des Aegirhorizonts des Ruhr-Reviers: Dissertation, Berlin; privately pub. 1933, by Konrad Triltsch, Wurzburg, 47 pp.

Knox, E. M., 1939, The spores of Brophyta compared with those of Carboniferous age: Bot. Soc. Edinburgh Trans. and Proc., vol. 32, no. 4, pp. 477-487.

——, 1942, The microspores in some coals of the productive Coal Measures in Fife: Inst. Min. Eng. Trans., London, vol. 10, no. 4, pp. 98-112.

Luber, A. A., 1938, Spores and pollen from coals of the Permian of the U.S.S.R. (in Russian, English summary): Problems of Soviet Geology, vol. 8, no. 2, pp. 152-160.

——, 1939, The correlation by means of spores of coal bearing Upper Paleozoic deposits of the Kuznetsk and Minussinsk basins (in Russian, English summary): Bull. Acad. Sci. U.S.S.R., (Akademia Nauk Otdelenie Matematicheskikh I Estesvennykh Nauk. Izvestiia) Serie 7-A, part 2, 1939, no. 6, pp. 88-104.

Naumova, S. N., 1937, The spores and pollen of the coals of the U.S.S.R.: XVII Internat. Geol. Cong., Absts. Papers, U.S.S.R., 1937 (Chief Editorial Office of the mining-fuel and Geological-Prospecting literature, Moscow and Leningrad), pp. 60-61.

Nikitin, P., 1934, Fossil plants of the Petino hroizon of the Devonian of the Voronzeh region. I. *Kryshtofovichia afracani* nov. gen. et. sp. (in Russuan with English summary and diagnoses): Acad. Sci. U.S.S.R. Bull. (Akademia Nauk Otodelenie Matematicheskiki I Estesvennykh Nauk. Izvestiia) Serie 7-A, part 2, 1934, no. 2, pp. 1079-1091.

DIE *Sporae dispersae* DES RUHRKARBONS

IHRE MORPHOGRAPHIE UND STRATIGRAPHIE MIT AUSBLICKEN AUF ARTEN
ANDERER GEBIETE UND ZEITABSCHNITTE

TEIL I*)

VON

ROBERT POTONIÉ UND GERHARD KREMP

AMT FÜR BODENFORSCHUNG, LANDESSTELLE NORDRHEIN-WESTFALEN, GEOLOGISCHES LANDESAMT KREFELD

MIT TAFEL 1–16 UND ABBILDUNG 1–37 SOWIE 1 TABELLE IM TEXT

Inhalt

	Seite
Zur Art der paläontologischen Bearbeitung der Sporae dispersae	2
Terminologie zur Beschreibung der paläozoischen Sporen und Pollenkörner (Deskriptive Terminologie der Sporographie)	9
Substanz und Schichtung der Sporenwand	15
Zur Ökologie der karbonischen Sporen	20
Morphographische Systematik der paläozoischen Sporae dispersae	23
Systematische Bearbeitung der Sporae dispersae	28
1. Oberabteilung *Sporonites* (R. Pot.) Ibr. 1933 (Fungi: Sporen und Sklerotien)	28
2. Oberabteilung *Sporites* H. Pot. 1893	30
1. Abt. *Triletes* (Reinsch 1881) sowie Historisches zum Begriff *Triletes* und zu den Gattungen Ibrahim's	31
1. Unterabteilung *Azonotriletes* Luber 1935	36
1. Reihe Laevigati (Bennie & Kidston 1886) Pot. & Kr.	36
2. Reihe Apiculati (Bennie & Kidston 1886) Pot. & Kr.	63
3. Reihe Murornati Pot. & Kr.	96
2. Unterabteilung *Lagenotriletes* Pot. & Kr.	117

*) Die Teile II und III (Schluß) der Arbeit erscheinen ebenfalls in Palaeontographica, Abt. B.

Zur Art der palaeontologischen Bearbeitung der Sporae dispersae

Als Sporae dispersae bezeichnen wir Sporen und Pollen, die nicht in situ, d. h. in Fruktifikationen angetroffen werden, sondern verstreut im Sediment. Sie bedürfen besonderer Arbeitsmethoden.

Die paläontologische Bearbeitung der Sporae dispersae strebt nicht nur nach weiteren stratigraphischen Erfolgen, sie verspricht auch viel für die reine Palaeobotanik, insbesondere zur Klärung pflanzensoziologischer, -geographischer, -ökologischer, ja entwicklungsgeschichtlicher und systematischer Fragen. Auch zur Genese der Kohlen läßt sich einiges folgern.

Die bloße Veröffentlichung von Sporenfotos oder Zeichnungen mit knappen morphographischen Angaben und entsprechenden stratigraphischen Bemerkungen genügt auf die Dauer nicht. Die Sporae dispersae müssen, wie das u. a. in der ehemaligen Berliner Geologischen Landesanstalt begonnen worden ist, genau so behandelt werden wie die wertvollsten Megafossilien. Sie müssen eingehend beschrieben und abgebildet sowie in jeder Hinsicht (auch theoretisch) ausgewertet werden, wobei die zu den Abbildungen gehörenden Originale an zugänglicher Stelle für alle späteren Bearbeiter aufzubewahren sind (Int. Code 1954).

Fossile Sporen sind seit langem bekannt. Manche sind auch schon recht bald beschrieben und mit besonderen Formspezies-Namen belegt worden. Ein umfassenderes System zur Ordnung der fossilen Sporae bestand jedoch bis Anfang der dreißiger Jahre nicht. Die verhältnismäßig wenigen Formen, die bis dahin ermittelt worden waren, standen mehr oder weniger unübersichtlich nebeneinander.

Wer an eine Systematik der dispergierten Sporen herangeht, wird sich aus altem Brauch zunächst vornehmen, dabei in erster Linie den Germinalapparat zu berücksichtigen. Dieser zeigt jedoch bei den Sporites viel weniger Unterschiedlichkeit als bei den Pollenites. Auffälligere Sonderformen erscheinen da eigentlich nur in der Unterabteilung Lagenotriletes und in der Abteilung Cystites. Es bleibt also bei der systematischen Ordnung der Formen nichts übrig, als jeweils die Gesamtheit der Eigenschaften sprechen zu lassen, was übrigens auch in der ganzen sonstigen Systematik immer mehr anerkannt wird.

Wollte man ordnend vorgehen, so konnte das nur mit Hilfe eines sog. „künstlichen Systems", besser eines morphographischen Systems geschehen; ähnlich dem von Brongniart für die Pteridophyllen geschaffenen, das sich noch heute bewährt, obgleich daneben die sonstige Systematik besteht.

Seit Beginn der planmäßigen Erforschung der älteren fossilen Sporen hat sich denn auch bei der paläontologischen Bearbeitung die bewußt morphographische Richtung als notwendige Grundlage aller weiteren Forschungsrichtungen, wie besonders der Sporenstratigraphie, erwiesen. Heute, nach mehr als fünfundzwanzig Jahren, von oft durch äußere Umstände unterbrochenen Bestrebungen, stehen wir vor einem Material aus vielen geologischen Zeitabschnitten inner- und außereuropäischer Ablagerungen, so daß es immer wichtiger wird, die Formen übersichtlicher als bisher in ein System einzuordnen. Dieses System muß bewußt ein unbedingt morphographisches sein, wobei Morphographie im Sinne von Henry Potonié (1912, S. 3, 4) verstanden wird, der sich darüber in seiner „Pflanzenmorphologie" äußert:

> Die Gestalt wird studiert, ohne sich zunächst um theoretische Gesichtspunkte zu kümmern, lediglich zu „ihrer praktischen Rubrizierung".

Das Wort Morphologie im ursprünglichen Sinne (Goethe) betraf nicht die bloße Beschreibung der Gestalt, sondern den Vergleich der Gestalten zu theoretischen Zwecken. Erst später wurde der Begriff auf das rein Descriptive ausgedehnt. Manche Forscher sind sich dessen bewußt und bezeichnen deshalb jetzt innerhalb der Morphologie das nur Beschreibende als die „rein morphologische Behandlungsweise" (Niggli). Das

Wort Morphologie mag nunmehr im allgemeinen Sprachgebrauch im umfassenderen Sinn in Anwendung bleiben, also sowohl für das, was als Morphologie im alten Sinne zu bezeichnen ist, als auch für die dann „rein morphologisch" genannte Arbeitsmethode. Wir bevorzugen indessen im folgenden für das „rein Morphologische" den Ausdruck M o r p h o g r a p h i e und sprechen von Morphologie im alten theoretischen Sinne.

Es hat sich bei den fossilen Sporae dispersae gezeigt, daß es zu keiner Übersicht führt, wenn man von vornherein theoretisierend versucht, ein die Verwandtschaftsverhältnisse berücksichtigendes System zu begründen, im Sinne des Systems der natürlichen Pflanzenfamilien. E r s t d u r c h d i e g e w i s s e n h a f t e m o r p h o g r a p h i s c h e A r b e i t s o l l j a h e r a u s k o m m e n , w e l c h e n W e r t m o r p h o g r a p h i s c h e E i n z e l h e i t e n f ü r d i e m o r p h o l o g i s c h e S y s t e m a t i k h a b e n. Selbstverständlich soll, nachdem das System morphographisch steht, möglichst bei jeder seiner Einheiten gesagt werden, wie sie sich zum natürlichen (d. h. morphologischen) System verhalten könnte (R .POTONIÉ 1954).

Nach Anfängen von REINSCH und BENNIE & KIDSTON ist der erste umfassendere Versuch dieser Art, der vor allem karbonische Iso- und Mikrosporen (= Miosporen) aber auch Megasporen zu ordnen trachtete, in der bei R. POTONIÉ durchgeführten Dissertation IBRAHIM's (1933) veröffentlicht worden. Den früheren Bearbeitern lag der Gedanke der Notwendigkeit eines b e w u ß t von der sonstigen Systematik absehenden Systems jedenfalls fern. Nach IBRAHIM haben dann erst wieder NAUMOVA und ERDTMAN den Gedanken ganz klar ausgesprochen.

Inzwischen ist das bewußt m o r p h o g r a p h i s c h e Sy s t e m von einigen Autoren verworfen worden; so z. B. von SCHOPF. Dagegen sind LUBER & WALTZ 1938, NAUMOVA, ERDTMAN, IVERSON & TROELS'SMITH und neuerdings auch P. W. THOMSON & PFLUG sowie KLAUS seinem Sinn entsprechend vorgegangen.

SCHOPF (1938) macht diesem „künstlichen" System den Vorwurf, „that it places together in hopeless confusion the spores of totally unrelated groups of plants" (S. 21). Dasselbe aber könnte man von dem BRONGNIARTschen System der Pteridophyllen sagen. Dennoch benutzen wir es mit Vorteil weiter; es schafft nach nichts als morphographischen Gesichtspunkten eine klare Übersicht, und dabei hat sich im Verlauf der weiteren Ausgestaltung des BRONGNIART'schen Systems schließlich doch eine Reihe von systematischen Einheiten ergeben, die als ± „natürlich" betrachtet werden können. Gleiches ist beim morphographischen System der Sporae dispersae zu sagen.

Wie man das BRONGNIART'sche System in der Praxis weiterbestehen läßt, auch wenn daneben fertile Reste anders eingeordnet werden, so sind die Sporae dispersae — aus später ersichtlich werdenden Gründen — auch dann weiter nach dem morpho g r a p h ischen System zu ordnen, wenn daneben Sporen, die ihnen gleichen, in Fruktifikationen gefunden worden sind. Nur den in situ gefundenen Sporen gebührt der gleiche Name, wie ihn der fertile Pflanzenrest trägt. Diese Sporen bedürfen keines besonderen Namens.

Es war überflüssig, als WICHER Sporen, die BODE in einem *Porostrobus* gefunden hat, den Namen *Poro-(strobo)sporites* gab. Es war falsch, als dann zu *Poro(strobo)sporites* auch noch dispergierte Sporen gestellt wurden, die denen von *Porostrobus* zwar morphographisch entsprachen, jedoch aus jüngeren Schichten stammten, aus Horizonten, wo *Porostrobus* noch nicht gefunden worden ist; solche Sporen sind als *Cystosporites* zu registrieren. Es war nicht zweckmäßig, daß KUBART dispergierte Sporen als *Spencerites* bezeichnete, als er sie denen aus dem Lepidophytenzapfen *Spencerites* SCOTT ähnlich fand. Der Name *Spencerites* ist nur für den Zapfen mit den evtl. noch in ihm vorhandenen Sporen zu verwenden. Dispergierte Sporen gleicher Morphographie sind anders zu nennen; nach dem augenblicklichen Stande der Nomenklatur *Microsporites* DIJKSTRA. Erst neben dieser Bezeichnung kann dann auf die morphographische und vielleicht noch weitergehende Beziehung zu *Spencerites* hingewiesen werden.

Es wäre abwegig, mit SCHOPF, WILSON & BENTALL 1944 eine Gattung dispergierter Sporen als *Calamospora* zu bezeichnen, wenn man dabei an mehr als eine rein morphographische Beziehung dächte und meinte, daß die mit heutigen Mitteln feststellbare morphographische Übereinstimmung der unter *Calamospora* zusammengefaßten dispergierten Sporen mit von HARTUNG und anderen in Fruktifikationen von Calamiten *(Macrostachya, Paleostachya)* gefundenen schon genüge, um auch dispergierte Sporen als restlos zu den Articulatae, geschweige denn zu den Equisetales oder gar zu den Calamitaceae gehörend zu betrachten. Nur wenn wir *Calamospora* rein morphographisch auffassen und lediglich ergänzend hinzufügen, daß darin u. a. Sporen von Calamiten und deren Verwandten enthalten sein dürften, entsprechen wir einer scharfen Methode. Wie auch schon SCHOPF wußte, gehört nicht alles, was den aus Calamiten-Fruktifikationen herausmazerierten Sporen morphographisch entspricht, zu *Calamites*. Da man der Formgattung *Calamospora* S. W. & B. nicht die aus Calamiten-Fruktifikationen gewonnenen Sporen einverleibt, sondern nur, dem Formsinn trauend, dispergierte, d. h. frei im Sediment befindliche, kommen n i c h t oder nur sehr fern miteinander verwandte Individuen zusammen.

Es ist auch nicht richtig, in die Synonymlisten der Sporae dispersae, so wie z. B. DIJKSTRA es tut, die Namen der fertilen Reste einzureihen. Mögen diese fertilen Reste auch Sporen enthalten, die sich mit den dispergierten Sporen morphographisch decken, die Namen der Fruktifikationen usw. gehören bei scharfer Scheidung der Dinge nicht hierher. Sie sind vielmehr erst am Schluß bei der Diskussion der Verwandtschaftsverhältnisse der Sporae dispersae zu erwähnen. Noch lange hinaus müssen die Namen der Sporae dispersae neben denen der fertilen Reste bestehen bleiben. Wir verlieren uns sonst auf Grund bloßer Auffassung und Theorie in verfrühte Vermischungen.

Auf ausgezeichnete allgemeine Untersuchungen gestützt sagt NIGGLI (1949, S. 146): „Heute wird selten eine saubere Trennung rein morphologischer Behandlungsweise gegenüber entwicklungsgeschichtlicher durchgeführt." Diesen Vorwurf wollten wir bei der Bearbeitung der Sporae dispersae von vornherein vermeiden. Auch wir meinten, die Verwandtschaft betreffende Überlegungen hätten neben und nicht in einem morphographischen (= sogenannt künstlichen) System zu stehen. (NIGGLI sagt „r e i n morphologisch" und nicht morphographisch; er meint offensichtlich morphographisch, sonst hätte er das Wort „rein" nicht eingefügt.)

Man hat sich bei der Ordnung darüber klar zu sein, daß selbst morphographisch sehr ähnliche, ja fast übereinstimmende Formen ganz verschiedenen Bereichen des natürlichen Systems angehören können. Um diesem Sachverhalt zu entsprechen, bleibt nur ein ganz konsequent morphographisches System übrig. Nur dann wird sich das dort Geordnete objektiv auswerten lassen. Nur so wird sich zeigen, wo die Grenzen der Erfassung der Sporae dispersae liegen, und bei welchen morphographischen Einheiten man engere Beziehungen zum natürlichen (= morpho*log*ischen) System feststellen kann. Aber auch dann noch muß das morpho*gra*phische System neben dem entwicklungsgeschichtlichen ungestört verbleiben. Überlegungen zur Verwandtschaft haben nicht darin, sondern daneben zu stehen, da sich anders nicht ergibt, welchen Wert die morphographischen Charaktere besitzen. Vor allem aber wird der Stratigraphie so am besten gedient.

Es handelt sich bei der Klassifikation der Sporae dispersae im wesentlichen um ihre Einordnung in Formgenera.

<small>Die Nomenklaturregeln unterscheiden Organgenera und Formgenera. Erstere gestatten eine nähere Angabe betreffend die natürliche Verwandtschaft, so z. B. zur Familienzugehörigkeit, Formgenera dagegen können mehrere Familien umfassen.

Wir benutzen meist den Ausdruck Formgenera, sind uns aber bewußt, daß manche, wie schon ein Teil von ihnen, sich bei weiterer Untersuchung ± als Organgenera herausstellen werden. Deshalb finden wir es auch unzweckmäßig, die Formgenera nicht, wenn möglich, einzuschränken und evtl. zu Organgenera zu machen.

Nach den Nomenklaturregeln soll man „in der Regel" den Namen eines Formgenus in seinem ursprünglichen Sinn gebrauchen. Das soll eben nur „in der Regel" befolgt werden, nicht jedoch dann, wenn die Gestalt des Genotypus es gestattet, die Gattung um den Genotypus so eng zusammenzufassen, daß sie ± ein Organgenus wird.</small>

Formal ist bei den Form- und Organgenera ebenso zu verfahren wie bei den sonstigen Gattungen, nur muß klar bleiben, daß die Form- und Organgenera der Sporae dispersae Formeinheiten sind, die sich mit den anderen Gattungen wohl fast nie absolut decken (vgl. POTONIÉ 1951). Oft enthalten sie mehrere Gattungen oder gar Familien. Manchmal auch verteilen sich Bestandteile einer natürlichen Gattung ja Art auf mehrere Formeinheiten (z. B. betreffend die Mega- und Mikrosporen).

Es entspricht deshalb nur einem erweiterten Brauch, wenn hinter solche Formeinheiten einfach novum genus (n. gen.) gesetzt wird. In Gedanken muß man festhalten, daß es sich nicht um die üblichen Gattungen, sondern zumeist um Formgattungen (gen. form.) handelt. Die Formgattungen stehen dennoch dem Range nach an der Stelle der Gattungen des sonstigen Systems. Sie werden von uns, wie das schon bei IBRAHIM (1933) geschehen, in Einheiten höherer Ordnung zusammengefaßt.

Es ist nötig, diese Dinge ganz klar herauszustellen, da man sich bei der Bearbeitung der Sporae dispersae bis in die jüngste Zeit noch nicht überall zu einem einheitlichen Vorgehen entschlossen hat. Nach wie vor versuchte man bei den dispergierten Sporen, die doch allzu oft nur unsichere Schlüsse auf die Mutterpflanze zulassen:

1. die sofortige Einreihung in das natürliche (= morphologische) System der Pflanzenfamilien; und zwar nicht nur, was verständlich ist, bei Sporen aus dem Pliozän, Diluvium und Postdiluvium, sondern auch bei älteren Formen. Nur bei den jüngeren geschah es z. T. mit Recht;
2. den Ausbau eines rein morphographischen („künstlichen") Systems unter Aufstellung von Formgenera; so vor allem für Formen des Paläozoikums, aber auch des Mesozoikums und Tertiärs;
3. die Verwendung einer sogenannt angenähert natürlichen, halbnatürlichen oder halbkünstlichen Methode, die in Wahrheit ebenfalls eine „künstliche" oder besser morphographische ist und die man vorübergehend für Formen des Tertiär verwandt hat.

Von diesen Methoden kann für unsere Zwecke nur die zweite, nämlich der Ausbau des rein morphographischen Systems, in Frage kommen.

Es ist notwendig, als Grundlage ein System zu besitzen, welches die Formen, völlig unbeeinflußt von der natürlichen Systematik, nach Gestaltsgruppen einteilt. Nicht einmal bei der Namengebung für Gestaltsgruppen und -gattungen sollte man in Zukunft eine Anspielung auf die sonstige Systematik wagen, wenn auch bisherige Namen soweit wie möglich erhalten bleiben müssen.

Auf dem Heerlener Kongreß 1951 ist denn auch empfohlen worden, in den dem System der Sporae dispersae zugeordneten Namen keine Andeutungen der Verwandtschaft oder auch nur der Ähnlichkeit mit den Einheiten des sonstigen Systems zu machen.

Wenn trotzdem noch kurz vordem durch R. POTONIÉ, P. W. THOMSON und FR. THIERGART 1950 bei der Ordnung neogener Sporen das sog. „angenähert natürliche System" ERDTMAN's in Anwendung kam, so geschah dies unter dem Druck von in Liblar stattgehabten Besprechungen, an denen außer den genannten Autoren auch noch W. AHRENS und U. REIN teilnahmen. Die 10 Vorschläge der Sitzung von Liblar sind in der Arbeit von R. POTONIÉ, P. W. THOMSON und FR. THIERGART 1950, S. 37 veröffentlicht worden. Der 1. der Gesichtspunkte lautet dort entgegen dem rein morphographischen Verfahren: „Die monographische Bearbeitung der fossilen Pollen und Sporen ist unter möglichst genauer Einordnung in das natürliche System der Pflanzenfamilien durchzuführen."

Unter die Veröffentlichung der 10 Gesichtspunkte (1950, S. 37) hat aber R. POTONIÉ doch noch den Satz setzen können: „Die vorstehenden 10 Punkte enthalten, wie man sieht, noch viele Unklarheiten." Weiter gelang es ihm, durchzusetzen, daß für die Arbeit POTONIÉ, THOMSON, THIERGART 1950 möglichst viele der dort verwandten Gattungs- usw.-namen wenigstens durch Einschaltung eines „oi" eine Gestalt erhielten, welche deutlich auf den nur morphographischen Charakter ihres Inhalts hinweist, so bei den Namen Cedr*oi*dites, Laric*oi*dites, Pseudotsug*oi*dites usw.

Eine spätere Arbeit, nämlich die von THOMSON & PFLUG (1953), hat die 10 kompromißerfüllten „Gesichtspunkte" von 1950 infolge der Einwirkung von R. POTONIÉ aufgegeben. Sie bedient sich des rein morphographischen Systems, in das nun freilich die Kompromißnamen von 1950 hätten eingebaut werden müssen. Zu viel radikale Änderung ist auch nicht am Platze. Es wäre dabei nur (deutlicher als geschehen) zu betonen gewesen, daß auch diese Namen nichts als eine rein morphographische Bedeutung haben, was ja schon in Liblar (leider noch nicht allen) klar wurde; man denke an den Vorschlag der Einschaltung des „*oi*" hinter den dem natürlichen System entnommenen Gattungsnamen (vgl. oben 1. Zeile).

ERDTMAN sagt 1947 bei seinem Vorschlag einer derartigen Verwendung der Gattungsnamen des natürlichen Systems, nichts sei hiermit gesagt oder angenommen betreffend irgendeiner möglichen Verwandtschaft mit der oder den Ursprungspflanzen der in Frage stehenden Sporomorphen. Mißverständnisse würden vermieden durch die klare Definition des rein morphographischen Charakters der nomina typica concreta.

Erst n a c h der rein morphographischen Einordnung würden diesen Namen theoretische Nebenbemerkungen folgen dürfen, je nach der Besonderheit der mit dem Fundpunkt der Spore verknüpften Tatsachen. Man hätte z. B. zu schreiben, in dieser oder jener Sporengesellschaft handele es sich in der Formgattung *Cedroidites* infolge Beziehungen zu sonstigen Pflanzenresten wahrscheinlich um *Cedrus*. — In verschiedenen Horizonten oder pflanzengeographischen Bereichen kann ja die Beziehung ein und derselben morphographischen Gattung zum natürlichen System verschieden sein. Die Stellung im morphographischen System aber bleibt immer dieselbe.

Zu dem, was PFLUG (in THOMSON & PFLUG 1953) über die Anwendung der Nomenklaturregeln sagt, ist festzustellen, daß es in der gesamten Literatur über t e r t i ä r e Sporae dispersae bis vor kurzem n i c h t üblich war, reguläre Gattungsdiagnosen zu verfassen und Genotypen zu benennen. Man folgte hier bis in die jüngste Zeit noch n i c h t diesem für die Sporae dispersae des K a r b o n s von R. POTONIÉ 1933 durch Vermittlung der Doktorarbeit IBRAHIM's eingeführten Brauch. Trotz vieler Verhandlungen mit Fachgenossen war es R. POTONIÉ bis vor kurzem nicht gelungen, endgültig davon zu überzeugen, daß auch die tertiären Sporae dispersae möglichst durch Formgenera erfaßt werden müßten und daß es nicht gangbar sei, den Sporae dispersae ohne weiteres den Namen einer rezenten oder fossilen Pflanze zu geben. Einer der Autoren, die schließlich diese Gesichtspunkte R. POTONIÉ's annahmen, war P. W. THOMSON. Wie aus seinen Arbeiten hervorgeht, hat er noch bis vor kurzem selbst alttertiäre Sporen unmittelbar mit den Gattungsnamen oder Familiennamen rezenter Pflanzen bezeichnet, ähnlich wie dies auch THIERGART trotz vieler Einwände immer wieder getan hat. Nachdem P. W. THOMSON überzeugt war, hat er auch seinen Schüler PFLUG veranlaßt, entsprechend der Anregung R. POTONIÉ's zu arbeiten. PFLUG benutzt nun merkwürdigerweise in der Schrift von THOMSON & PFLUG 1953 die Gelegenheit, zu beanstanden, daß sich auch R. POTONIÉ früher noch nicht konsequent verhalten habe.

Wie die Sporae dispersae i m b e s o n d e r e n zu behandeln seien, ist erklärlicherweise in den Nomenklaturregeln noch kaum gesagt. PFLUG liest sich dennoch aus den Regeln das ihm Passende heraus. Er beachtet zu wenig, daß in einer neuen Wissenschaft zunächst die historische Entwicklung da ist und daß dieser dann erst die Spielregeln folgen. So haben wir bei der Bearbeitung der t e r t i ä r e n Sporae dispersae selbstverständlich zunächst einmal allerlei Experimente gemacht, die sich zumeist aus Konzessionen ergaben; hieß es doch immer wieder, rein morphographische Einheiten seien da n i c h t am Platze.

Wir waren auch 1950 für die tertiären Sporae dispersae noch nicht zu einer klaren Entscheidung der Frage gekommen, wie der Begriff des Formgenus oder des Organgenus für sie anzuwenden sei. Nur für das Paläozoikum usw. war das ± geschehen.

Deshalb war es begreiflich, wenn für das Tertiär nach wie vor allerlei Zwischenlösungen gesucht wurden. Eine davon war das sog. halbnatürliche System, eine Konzession für diejenigen, die auch damals noch für die tertiären Formen durchaus nicht die im Paläozoikum bewährte Methode benutzen wollten.

Bei den diesbezüglichen in Liblar 1950 erfolgenden Besprechungen legte man R. POTONIÉ den bereits zitierten Satz vor, auf den sich alle anderen Beteiligten schon vorher geeinigt hatten: „Die monographische Bearbeitung der fossilen Pollen und Sporen ist unter möglichst genauer Einordnung in das natürliche System der Pflanzenfamilien durchzuführen." Um dieser Forderung zu entsprechen, wurden dann in der Arbeit von POTONIÉ, THOMSON & THIERGART 1950 Namen des natürlichen Systems der Pflanzenfamilien in der Weise eingeführt, daß man diese mit der Endung -oidites verband (z. B. *Taxodioidites*). Es handelte sich damit um das von G. ERDTMAN für rezente Sporen vorgeschlagene Ordnungsprinzip. Die einzige Änderung an ERDTMAN's Vorschlag war, daß nicht wie bei ihm -idites, sondern -oidites als Endung gewählt wurde.

Ähnlich war es manchmal auch schon vorher gemacht worden. Aber wenn z. B. gesagt wurde Piceae ? -pollenites alatus R. POT. (in POT. & VEN. 1934), *Coryli* ? -pollenites coryphaeus R. POT., dann war das keine Einreihung in ein echtes Form- oder Organgenus, sondern eine lockere Inbeziehungsetzung zum natürlichen System der Pflanzenfamilien. Daher das ?, welches zeigen sollte, daß der Hinweis auf die Verwandtschaft hier nur vorläufiger Natur sei und evtl. durch andere Angaben ersetzt werden könne. Dies ist denn auch in späteren Arbeiten geschehen.

Klarer wurde die Lage erst, als man sich dazu entschloß, dem bei den paläozoischen Sporen schon seit längerem eingeführten Verfahren auch bei den tertiären Sporae dispersae zu folgen und die die Stelle der Gattungen vertretenden Einheiten zu echten Form- und Organgenera mit Genotyp zu machen.

Auch im Paläozoikum war dies noch nicht überall durchgeführt worden. Wir sind jetzt im Karbon so vorgegangen, daß wir jeweils die älteste, die Stelle des Gattungsnamens vertretende Bezeichnung gewählt und dafür als Genotyp bzw. Lectotyp, wo irgend angängig, diejenige Spezies eingesetzt haben, welche die Autoren bei diesem Namen an erster Stelle eingereiht haben. Weiter haben wir eine Gattungsdiagnose geschaffen. Wir haben oft auch d a alte bewährte Namen legitimiert, wo die Nomenklaturregeln uns deren Streichung gestattet hätten. Wir haben sie nicht gestrichen, wenn es erlaubt war, die Lücken entsprechend den Regeln und Empfehlungen des Internationalen Codes (SCHULZE, Berlin 1954) zu schließen. So glaubten wir, der bisherigen Literatur am wenigsten zu schaden.

Somit haben erst wir viele Form- und Organgenera endgültig aufgestellt, aber eben möglichst mit den alten Namen.

Wir glaubten, bei einer erst im historischen Werden begriffenen Wissenschaft nicht so wie THOMSON & PFLUG vorgehen zu dürfen, die einen Namen fast immer verwerfen, wenn sie meinen, in den Nomenklaturregeln die geringste Berechtigung dafür zu finden.

Da wir im Karbon bereits ein echtes System der Form- und Organgenera hatten, im Tertiär aber das sog. halbnatürliche System als angeblich bloß provisorisches Ordnungsprinzip, wollte R. POTONIÉ keine Verwechselung dieser beiden Systeme.

PFLUG schreibt nun 1953, S. 13, ohne die Dinge zu durchschauen, R. POTONIÉ schlage 1952 für die Sporomorphae zwei verschiedene Systeme vor. Indessen wird bei POTONIÉ 1952 nichts v o r g e s c h l a g e n, sondern nur folgendes festgestellt:

„Wir bedienen uns z. Z. bei der systematischen Bearbeitung der fossilen Sporen dreier verschiedener Verfahren."

Es wird dann, um für weitere Verhandlungen eine Grundlage zu haben, der damals gegebene Zustand geschildert und überlegt, wie man sich den nun einmal bestehenden verschiedenen Methoden gegenüber verhalten könnte.

Diese erst 1952 erschienene Zusammenstellung war 1951 (nach den palynologischen Besprechungen in Trier) verfaßt worden, also zu einem Zeitpunkt, als R. POTONIÉ noch glauben mußte, P. W. THOMSON werde bei seinen weiteren Arbeiten bei der von ihm selbst seinerzeit so freudig begrüßten sog. halbnatürlichen Ordnung bleiben.

Es mußte also nach wie vor versucht werden, die Autoren an ein System der Formgenera zu gewöhnen. Deshalb auch heißt es in der von PFLUG 1953 mißverstandenen Arbeit von R. POTONIÉ 1952, S. 170: „Dabei muß das rein künstliche System als das einzige, wirklich alle fossilen Sporomorphen umfassende, stets mit seinem ganzen Inhalt als fester Ausgangspunkt bestehen bleiben. Alle Formen bleiben dort als an dem sichersten Platze."

Nicht zum wenigsten durch R. POTONIÉ's Einfluß haben sich THOMSON & PFLUG zu dem System der Formgenera durchgerungen, so daß es nun unangebracht ist, R. POTONIÉ für die diesem Verfahren vorangehenden Halbheiten, insbesondere für das halbnatürliche System allein verantwortlich zu machen.

Das morphographische System der Sporae dispersae umfaßt die nach ihrer Gestalt zu ordnenden fossilen dispergierten Sporen und Pollenkörner. Sie werden nach Gestaltsgruppen rein morphographisch erfaßt. Da sich die Gestaltsgruppen in Größe und Inhalt nicht mit den Spezies des natürlichen Systems decken, ist von ERDTMAN für die Einheiten des morphographischen Systems der Begriff S p o r o m o r p h a e vorgeschlagen worden. Die Sporites und Pollenites lassen sich eben nicht in Arten aufteilen, wohl aber in Sporomorphae. Die Sporomorphae aber werden formal wie Arten behandelt. Dabei braucht man nicht bei jeder Gelegenheit ausdrücklich zu betonen, daß es sich um Sporomorphen, d. h. um Gestaltsgruppen handelt (ähnlich ist es ja auch bei den Pteridophyllen usw.).

Ein morphographisches System muß so gestaltet sein, daß sich möglichst viele Unterschiede (zunächst einmal ohne alle theoretischen Hemmungen betreffend Bedeutung, Wert und Unwert) festhalten lassen, in einer Weise, welche die Übersichtlichkeit nicht in Frage stellt. Ob dabei ± natürliche Gruppen herauskommen, ist zunächst nicht zu überlegen. Selbst verschiedene Erhaltungszustände sowie heteromorphe Sporen ein und derselben Mutterpflanze wären auseinanderzuhalten. Man würde sonst nichts mit ihnen anzufangen wissen, wenn man ihnen bei stratigraphischer Arbeit isoliert im Sediment begegnet. Theoretische Überlegungen zur Verwandtschaft usw. mögen unmittelbar folgen, den Vorrang aber hat stets die rein morphographische Arbeit, die Ordnung nach nichts als der Gestalt.

Das Wissen um die morphographischen Einzelheiten der Sporae dispersae muß so festgelegt werden, daß es wirklich benutzbar wird. Man findet zwar bei manchen Autoren viele absolute Größenangaben zu einzelnen Formelementen, man vermag aber weniger damit anzufangen, als der Autor hofft. Meist ist nämlich versäumt worden, anzugeben, wie sich die Zahlen zu anderen Körpermaßen desselben Individuums verhalten. Es fördert sehr die Auseinanderhaltung der Formspezies, wenn man u. a. Verhältnisse wie die folgenden ermittelt:

1. Zahl der Zierelemente zu Äquatorumriß
2. Zahl der Zierelemente zu Tectumlänge
3. Zahl der Netzlumina zu Sporenoberfläche
4. Abstand der Zierelemente zu Größe der Zierelemente
5. Durchmesser oder Länge der Zierelemente zu Radiuslänge

6. Kleine Zierelemente zu großen
7. Basisbreite der Zierelemente zu Länge bzw. Höhe der Zierelemente
8. Kontaktarearadius zu Gesamtradius
9. Durchmesser zu Zona- usw.-breite.

Absolute Zahlen haben weniger Wert. Es ist uns meist nicht gelungen, solche Angaben anderer Autoren umzudenken und zur Artdiagnose zu verwerten.

Unsere Angaben für Megasporen beziehen sich zumeist auf Beobachtungen bei auffallendem Schräglicht und bei geringeren Vergrößerungen, die der Iso- und Mikrosporen sind bei durchfallendem Licht ermittelt worden, wie z. T. auch die der Megasporen.

Es sei noch einmal betont, daß bei der Beschreibung neuer Arten fossiler Sporae dispersae unbedingt die Originale zu den Abbildungen in Präparaten aufbewahrt werden müssen. Dies ist selbst in letzter Zeit nicht immer geschehen. Zur wirklich einwandfreien Bestimmung einer Sporenform muß der beschriebene Holotypus im Präparat vorliegen (Int. Code d. Bot. Nom. Berlin 1954, PB. 6 E).

Von großem Nutzen bei der vorliegenden Arbeit sind uns die Mikrofotos gewesen, welche ZEIDLER und andere nach den Holotypen von IBRAHIM und LOOSE hergestellt haben. Daß diese Mikrofotographien genau den Originalen und somit den Abbildungen der genannten Autoren entsprechen, erkennt man u. a. daran, daß Sekundärfalten und Lädierungen, welche von IBRAHIM und LOOSE mitgezeichnet worden sind, auch auf den Mikrofotografien zu sehen sind. Wir veröffentlichen die Mikrofotos der alten Originale neben unseren eigenen Aufnahmen.

Viele Autoren und Interessenten haben uns in liebenswürdiger Weise geholfen. Leihweise sind uns Originalpräparate überlassen worden, sonstige Präparate wurden uns zum Verbleib zugestellt, ebenso Fotos und Fotokopien sowie Kohlenproben. Wir erhielten Auskünfte über schwer zugängliche Literatur sowie manche anderen wertvollen Hinweise. Zu besonderem Dank sind wir folgenden Personen verpflichtet: CH. ARNOLD, Ann Arbor, B. E. BALME, Chatswood, N.S.W., E. BEDERKE, Göttingen, D. C. BHARDWAJ, Lucknow, H. BÖTTCHER, Hervost-Dorsten, S. J. DIJKSTRA, Heerlen, H. DÖRDELMANN, Gladbeck, I. A. DULHUNTY, Sydney, J. A. DURAND, Toulouse, A. EBERT, Krefeld, H. FALKE, Mainz, P. N. GANJU, Aligarh, H. GREBE, Krefeld, P. GUTHÖRL, Saarbrücken, P. A. HACQUÉBBARD, Sydney (Nova Scotia). C. HAHNE, Bochum, F. HEINE, Essen, W. JESSEN, Krefeld, K. H. JOSTEN, Mainz, E. KNOX, Edinburgh, R. M. KOSANKE, Urbana, R. KRÄUSEL, Frankfurt a. M., J. P. MESSERVEY, Halifax, W. NENNO, Marl, K. OBERSTE'BRINK, Essen, H. PUTZER, Criciuma, J. H. ROGER, Paris, H. SCHMIDT, Göttingen, J. M. SCHOPF, Orton Hall, A. H. V. SMITH, Sheffield, H. SPIEKERNAGEL, Gelsenkirchen-Buer, E. STACH, Krefeld, M. u. R. TEICHMÜLLER, Krefeld, P. W. THOMSON, Krefeld, P. VETTER, Decazeville, E. VOIGT, Hamburg, W. WESTERMANN, Duisburg-Hamborn, H. WEYLAND, Wuppertal-Elberfeld.

Terminologie zur Beschreibung der palaeozoischen Sporen und Pollenkörner (Deskriptive Terminologie der Sporographie)

Eine glückliche Systematik der Sporites und Pollenites ist u. a. abhängig von der Pflege ihrer morphographischen Terminologie. Mißlich dabei ist es, wenn eindeutige, bereits seit längerem vorhandene Termini durch neue ersetzt werden, nur weil die alten sprachlich nicht gefallen. Die ältere Literatur, die für die paläontologische Arbeit stets der Ausgangspunkt bleiben muß, wird damit unnützerweise schwerer zugänglich gemacht. Schon DE CANDOLLE 1880 verlangt daher a u c h für die Phytographie Berücksichtigung des erstmaligen Auftretens ihrer Termini „et auquel on remonte lorsqu'un terme a été dénaturé ou employé dans plusieurs sens".

G. M. SCHULZE 1953 erneuert die Forderung einer möglichst einheitlichen Terminologie bei deskriptiven Arbeiten.

Wir folgen bei der Beschreibung der Sporenformen im wesentlichen der von R. POTONIÉ (1934, **4**) benutzten, historisch begründeten Benennung ihrer morphographischen Einzelheiten. Daraus ergibt sich die Methode der von diesem angeregten, zuerst von IBRAHIM 1933 versuchten Klassifikation der paläozoischen Sporae dispersae. Ergänzungen zur morphographischen Terminologie sind J. IVERSEN & J. TROELS-SMITH (1950), ERDTMAN (1952), KOSANKE (1950), DIJKSTRA usw. entnommen worden.

Die F o r m einer Spore ist im einfachsten Fall die einer Kugel. Daneben stehen Gestalten, die sich ± dem Ellipsoid, der Linse, dem Tetraeder, der Bohne usw. nähern.

Im Sediment sind die Sporen meist flachgedrückt (Taf.1, Fig. 1, 7). Ihre Beschreibung kann daher oft nur hiervon ausgehen. Die Gestalt aber, welche sie bei der Abplattung annehmen, ist häufig in kennzeichnender Weise von ihrer ursprünglichen Form abhängig, s o d a ß d i e A n g a b e n, w e l c h e s i c h a u f d e n E r h a l t u n g s z u s t a n d b e z i e h e n, f ü r d i e S p e z i e s b e z e i c h n e n d b l e i b e n. Dies gilt nicht nur für die Abplattung, sondern auch für die Farbe, die Art der Sekundärfaltung, die ev. Zerreißung und Beschädigung, das Aufklaffen der Dehiszenz usw.

Die Form des Fossils wird u. a. durch die Angabe des U m r i s s e s der abgeplatteten Spore erfaßt. Der Umriß deckt sich oft, jedoch nicht immer mit dem Äquator (Taf. 1, Fig. 6) oder einem Meridian (Taf. 4, Fig. 28). Als Ä q u a t o r einer Spore wird die Berührungslinie zwischen Proximal- und Distalhemisphäre betrachtet. Die Äquatorkontur wird als kreisförmig, dreieckig usw. bezeichnet.

Die Lage der Sporen in der Sporenmutterzelle ist verschieden. Bei vier zum Tetraeder zusammengestellten Kugeln berührt jede die drei anderen an drei Punkten. Diese drei K o n t a k t p u n k t e bilden ein gleichschenkliges Dreieck (das in der proximalen Polarhemisphäre liegt). Die vier fertig ausgebildeten Sporen berühren einander mit je drei K o n t a k t a r e e n oder -flächen (s. d.).

Die P r o x i m a l h e m i s p h ä r e kann ± die Form einer dreiseitigen Pyramide behalten, die D i s t a l h e m i s p h ä r e kann ± halbkugelförmig bleiben oder sich weiter aufblähen.

Die Mittelpunkte der vier Sporen können aber in der Mutterzelle auch in einer Ebene angeordnet sein, so daß die Sporen einem Quader oder auch Rhombus einbeschrieben sind. Bei quaderförmiger Anordnung sind die ausgewachsenen Sporen ± bohnenförmig.

Bezogen auf die Anordnung in der Mutterzelle besitzt jedes Korn einen inneren, proximalen und einen äußeren distalen P o l, sowie eine p r o x i m a l e und eine d i s t a l e Hemisphäre. Der proximale Pol ist der A p e x, er liegt dem Mittelpunkt der Mutterzelle am nächsten.

Die Verbindungslinie der Pole ist die Hauptachse oder P o l a r a c h s e; dazu senkrecht stehen die Ä q u a t o r a c h s e n, die je zwei einander gegenüberliegende Punkte des Ä q u a t o r s miteinander verbinden.

Nach der Lösung aus dem Tetradenverband können die Oberflächen der Sporen sich vollkommen glätten, so daß keine Spur des ehemaligen Kontaktes mehr erkennbar bleibt. Es entstehen a l e t e Sporen, d. h. solche ohne Kontaktareen und ohne Dehiszenzmarke.

Die Dehiszenzmarke (Laesura ERDTMAN 1952) zeigt sich bei Sporen aus der Tetraeder-Tetrade als T e t r a d e n m a r k e, t r i l e t e M a r k e oder Y - M a r k e, bei Sporen aus der Tetragonal-Tetrade als m o n o l e t e M a r k e. In den Winkeln der Y-Marke und beiderseits der monoleten Marke können Kontaktareen verbleiben. Die dreistrahlige, also die trilete oder Y-Marke findet sich auch fossil häufig mit geschlossener Sutur (s. d.). Die drei S t r a h l e n oder T e c t a gehen vom A p e x aus (Abb. 3, S. 37). In senkrechter Aufsicht lassen sie zwischen sich etwa gleich große Winkel. Der Apex ist dann der höchste Punkt. Die Tecta sind ± hohe, manchmal auch sehr niedrige, charakteristisch ausgebildete dach- oder wallartige Falten oder Sättel der Exine. Das zeigen am besten Querschliffe der Sporen (Abb. 1; Taf. 1, Fig. 7e; Taf. 9, Fig. 70e; Taf. 13, Fig. 195). Die Scheitellinie, der Grat oder V e r t e x der Tecta ist, wie die Querschliffe des weiteren zeigen, in vielen Fällen geschlossen. Auch in der Aufsicht mazerierter Sporen ist dies zu ermitteln (Taf. 1, Fig. 7a—e). Man bemerkt oft

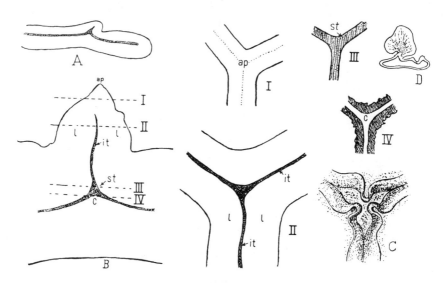

Abb. 1. Bau der Y-Marke. A u. B Skizzen nach Exinen-Vertikalschnitten. I—IV verschiedene Horizontalschnitte durch die Y-Marke, wie sie sich im optischen Schnitt zeigen; ap = Apex bzw. Vertex, l = Labra, it = Intratectum, st = Subtectum, c = Contravertex. C aufgeplatzter Apex einer laevigaten Miospore; die jetzigen Außenränder der Labra waren vor dem Auseinanderweichen zum Vertex verbunden; infolge der Höhe des Apex haben sich löffelartige Lappen gebildet. D Spore aus *Lepidostrobus foliaceus* mit Schwimmblase (nach Scott).

nichts von der S u t u r, „Dehiszenzfurche" oder Naht. Erst wenn sich die Sutur öffnet (Abb. 23), tritt sie in der Aufsicht sowie im Querschnitt der Exine in Erscheinung. Was man an der mazerierten Spore in der Aufsicht oft dafür hält, ist das Intratectum (Abb. 1).

Stellt man das Mikroskop bei längs des Äquators flachgedrückten Mikrosporen auf den obersten Teil des Tectums ein, also auf den Vertex, so zeigt sich oft ein Bild wie Abb. 1 I, die Sutur ist nicht oder kaum erkennbar. Dreht man tiefer, so erblickt man ein Bild wie Abb. 1, II; es zeigt sich das I n t r a t e c t u m (Abb. I B), nämlich der ± schmale Innenraum des Tectums. Das Intratectum wird in Abb. I B umrahmt von dem Querschnitt der Wände des Tectums, von den L a b r a, Lippen, „Wülsten" oder „Dehiszensleisten". Der Name Lippen wäre an sich erst da richtig angebracht, wo sich die Sutur geöffnet hat (Abb. I C), wird aber mit gewissem Recht auch auf die noch in Verbindung stehenden Wände des Tectums angewandt.

Die Seitenwände, die sogenannten Lippen, können dicker erscheinen als die übrige Exine der Spore. Sie können besonders dann dicker aussehen als sie tatsächlich sind, wenn das Tectum durch den Erhaltungszustand nicht mehr ± senkrecht steht, wodurch der optische Querschnitt schräg verläuft (Abb. 1 A).

Bei noch tieferer Einstellung kann das Bild Abb. 1 III folgen. Das Intratectum hat sich u. U. allmählich geweitet; es ist am breitesten im S u b t e c t u m (Abb. 1 B). Die Labra, die sich zunächst ebenfalls verbreitern konnten, sind verschwunden.

Endlich (nicht immer) erscheint Abb. 1 VI; das Subtectum hat hier seine größte Ausdehnung. Manchmal hat es einen unregelmäßigen ± deutlich flexuosen Umriß; in der Mitte seiner Strahlen erscheint hell ein neues Y, der C o n t r a v e r t e x (c), d. i. die Einschmiegung der distalen, der Y-Marke gegenüberliegenden Exine in das Subtectum.

Somit können von ein und derselben Y-Marke recht verschiedene Bilder (auch Fotos) erhalten werden.

Das wird noch vermehrt, wenn sich die Sutur öffnet. In diesem Fall kann entweder eine einfache Fissur erscheinen, dann zeigen die Ränder der auseinanderweichenden Wände des Tectums und auch der Apex keine Eigentümlichkeiten. Manchmal aber zeigen die die Fissur umrandenden Teile der Exine beim Auseinanderweichen eine ganz charakteristische Art der Umgestaltung, der Aufrollung usw. Auch dieses Verhalten ist nicht mit dem Wort Erhaltungszustand im gewöhnlichen Sinne abzutun. Vielmehr haben bestimmte Formspezies bestimmte Erhaltungszustände. Schon ob sich die Sutur ± leicht zur Fisur öffnet oder nicht, ist für die Festlegung der Formspezies nicht immer gleichgültig (Taf. 3, Fig. 16a).

Abb. 1 C zeigt den offenen Apex und die offenen Tecta einer sonst wenig eigentümlichen Form. Nachdem sich die Sutur geöffnet hat, sind die ziemlich hohen Wände der Tecta, die Labra, auseinandergeklappt. Die einander proportionalen äußeren Umrißlinien der Zeichnung Abb. 1 C waren ursprünglich zur Scheitellinie (dem Vertex) der Y-Marke miteinander verbunden. Da der Apex die Tecta etwas überragte, stehen nun an seiner Stelle nach dem Umklappen der Labra drei löffelartige Exinenlappen, eine Ausbildung, die sich nur bei bestimmten Formspezies zeigt; man denke auch an *Lagenicula* (Abb. 35).

Wenn die Tecta (namentlich im näheren Bereich des Apex) sehr hoch werden, entsteht ein Dehiszenzkegel, der als G u l a bezeichnet wird. Man vergleiche die Diagnose der 2. Unterabteilung Lagenotriletes, zu denen die Gattung *Lagenicula* gehört.

Als M a s s a bezeichnen wir einen granulösen ± unförmigen Auswuchs, der etwa kugel- bis kegelförmig, vielleicht auch leicht dreilappig sein kann und der eine Umformung der Tecta der Y-Marke darstellt. Wir verzichten auf den hier bisher benutzten Terminus M a s s u l a, weil dieser bereits Pollenmassen bezeichnet, die wie bei den Mimosae miteinander verbunden ausgestreut werden.

Nicht zu verwechseln mit den Tecta der Monoletes sind die ebenfalls langgestreckten C o l p i der Monocolpates (Abb. 88). Während es sich in den Tecta um wallartige langgestreckte Ausbuchtungen ± ungeschwächter Exine handelt, sind die Colpi der Monocolpates entweder ± langgestreckte primäre oder sekundäre Schwächungen der Exinendicke (T e n u i t a s) oder aber auch nur Furchen oder „Nähte", jedenfalls nicht Wälle oder Tecta. Falls es sich ± um eine Tenuitas handelt, pflegt diese beim Fossil eingebuchtet oder eingefaltet zu sein (Abb 88).

Im Zusammenhang damit, daß die Sporen in der Mutterzelle nach dem Tetraeder, Rhombus oder Quader angeordnet waren, haben sich manchmal beträchtliche Kontaktflächen ergeben. Jede der drei K o n t a k t a r e e n (Area contagionis) der Tetraedersporen ist proximal von je zwei Tecta der Dehiszensmarke begrenzt, distal evtl. von einer der drei C u r v a t u r a e, B o g e n l i n i e n oder B o g e n l e i s t e n (Abb. 7). Die C u r v a t u r a e verlaufen mit einem zum proximalen Pol hin offenen Bogen. Es ist dabei zu beachten, daß die Curvaturae manchmal mit einer in ihrer Bogenmitte befindlichen Spitze in äquatorialer Richtung vorstreben, also nicht genau bogenförmig sind.

Tecta und Curvaturae können ± reduziert sein, oft fehlen die Curvaturae ganz. Gelegentlich auch sind von ihnen nur noch die aus den äquatorwärts gerichteten Enden der Dehiszenzleisten hervorgehenden Anfänge (Curvaturae imperfectae) vorhanden, so daß die Dehiszenzleisten wie Gabelstiele mit zwei ± langen Gabelzinken erscheinen. Dies kann sowohl bei trileten als auch bei monoleten Formen der Fall sein (Taf. 19, Fig. 422).

Bemerkenswert an den Kontaktareen ist, daß die Entwicklung der Skulptur der Oberfläche der Exine in ihrem Bereich durch die Aneinanderpressung in der Mutterzelle ± geschwächt oder verhindert worden sein kann. Die früher geäußerte Meinung jedoch, die unter den Kontaktflächen befindliche Exine sei stets weniger dick als in den übrigen Bereichen der Exine, trifft z. B. bei *Laevigatisporites glabratus* nicht zu; vgl. unsere diesbezüglichen Untersuchungen weiter unten S. 54.

Mit einem zum Äquator offenen Bogen verlaufen die gelegentlich entwickelten Kyrtome oder Tori (siehe Abb. 41). Sie schmiegen sich in die von den Y-Strahlen gebildeten Winkel, den Strahlen etwa parallel verlaufend. Es handelt sich bei den Kyrtomen um konstant angeordnete Exinenfalten, die nach THOMSON & PFLUG 1952, dort Tori genannt, nur gewisse Lamellen (?) ergreifen sollen. Je zwei Kyrtome können sich am Äquator miteinander verbinden. Der Name Tori ist zu vermeiden, weil er bereits vergeben ist. (Was bei den Pollenites als Arcus bezeichnet wird, nämlich Bogenfalten, welche von Apertur zu Apertur verlaufen, ist mit den Kyrtomen nicht gleichzustellen.)

Durch die Abplattung im Sediment entstandene Falten sind als Sekundärfalten zu bezeichnen (Abb. 6, Taf. 11, Fig. 132). Sie treten um so häufiger auf, je dünner die Exine ist, besonders gern bei ursprünglich ± kugelförmigen Exinen.

Bei der Beschreibung der Sporenexine hat man zu unterscheiden zwischen deren Struktur, Infraskulptur (= Interskulptur), und Skulptur.

Von Infraskulptur (auch Intraskulptur) ist zu sprechen, wenn sich z. B. bei der Saccus-Bildung Exinenlamellen voneinander lösen und dabei auf der inneren Saccuswand ein Relief zurückbleibt, welches sich aus Elementen zusammensetzt, die bis dahin als „Struktur" der Isolierschicht zwischen der Exolamelle und der Intexine standen, sie zusammenhaltend (s. Abb. 76 und Tabelle S. 18).

Zur Bezeichnung von Bauelementen, die sich zwischen noch nicht auseinandergelösten Lamellen der Exine befinden, also keine der Exine als Skulptur aufgesetzten Verzierungen sind, doch oft fälschlich als Granulierung, Reticulierung usw. gedeutet werden, haben sich Bezeichnungen wie intra- oder infrapunctat, -granulat, -reticulat (letzterer Ausdruck bei IVERSEN & TROELS-SMITH 1950) ergeben.

Es muß gesagt werden, daß Struktur und Skulptur ineinander übergehen können. Das rührt u. a. daher, daß sich Strukturelemente (nämlich Elemente der Isolierschicht), obschon sie von glatten Lamellen überspannt sind, durchzudrücken vermögen.

Deutsche Ausdrücke für Struktur und Skulptur sind Verzierung und Muster. ERDTMAN (1952, S. 22) unterscheidet LO-Muster von OL-Mustern (L = Lux, O = Obscuritas). LO-Muster zeigen sich bei hoher Einstellung des Objektivs als helle Inseln, getrennt durch dunklere Zonen, und ergeben bei tieferer Einstellung das umgekehrte Bild. OL-Muster zeigen bei hoher Einstellung dunkle, hell umwobene Inseln, und bei tiefer Einstellung die Umkehrung (Abb. 24).

Eine mit Struktur versehene Exine kann also eine ± glatte Oberfläche, d. h. eine glatte Umrißlinie haben (Taf. 11, Fig. 107), sie braucht keine Skulptur aufzuweisen. Dennoch kann sie in der Aufsicht ein deutliches Muster zeigen, eine Punktierung, Fleckung, netzartige Zeichnung (Taf. 12, Fig. 140), flaserige, wolkige, geströmte, chagrenate usw. Beschaffenheit. Das sind Dinge, die sich dann nur auf die interne Struktur, manchmal auch auf die Infraskulptur der Exine beziehen. Daher ist von uns der Ausdruck Punktierung usw. nie für Oberflächenskulpturen verwandt worden, man hat ihn aber, wie gesagt, doch mit Granulierung usw. verwechselt. Zweckmäßig erscheinen uns deshalb die Ausdrücke infrapunctat, infragranulat, infrareticulat (IVERSEN & TROELS-SMITH).

Nur in Fällen, wo in der Aufsicht der Exine punktierte, gefleckte usw. Muster bemerkbar werden, ohne daß eine Außenskulptur vorhanden wäre, hätte man von Infrapunctierung, -fleckung (-maculierung) usw. zu sprechen. Diese Bildungen werden demnach als Strukturen der Exine bezeichnet. Erst als äußerste „Schicht" folgen die der Exolamelle der Exine aufsitzenden Elemente.

Nur im Relief der Oberfläche der Exine hervortretende Formelemente sind Skulpturen zu nennen.

Die Skulpturelemente bestehen aus oft ± regelmäßig angeordneten, ± weit vorspringenden Unebenheiten der Exoexine. So gibt es ± dicht nebeneinanderstehende oder in bestimmter Weise gruppierte Körnchen (Grana, Granula), Wärzchen (Verrucae), Kegel (Coni), Stacheln (Spinae), Stäbchen (Bacula), Keulen (Pila), Haare (Capilli) recht verschiedener Gestalt und Größe. Es kommen auch Kombinationen verschiedener Struktur- und Skulpturelemente vor.

Die Granula oder Grana sind in der Aufsicht ± kreisförmig, von der Seite ± kreisförmig, halbkreisförmig oder halboval. Wir sprechen nur dann von Grana, wenn sie auf dem Sporenindividuum, das sie verzieren, nicht zu sehr in der Größe wechseln, oder doch nur so abändern, daß sich ihre Größe aus einem Bereich der Sporenoberfläche zum anderen bloß allmählich wandelt (Abb. 8, 9).

Die Verrucae sind von unregelmäßigerem Umriß, vor allem aber auf ein und demselben Sporenindividuum von etwas verschiedener Gestalt und Größe; man sieht in der Aufsicht kleinere neben etwas größeren, rundliche neben ungenau rundlichen, gelappte neben bohnenförmigen usw., so daß die Elemente, da wo sie dicht stehen, ein unregelmäßigeres Mosaik bilden als die Grana (Abb. 10). Weiter sind die Verrucae meist nicht so hoch oder nicht sehr viel höher als breit; gewöhnlich auch ist ihre größte Breite nicht viel bedeutender als die geringste; immerhin kann etwa das Verhältnis 2:1 auftreten. Die Oberfläche der Verrucae kann rundlich, flach oder unregelmäßig ausgebildet sein.

Stehen die Verrucae sehr dicht, so werden sie in der Aufsicht polygonal. Ihr Mosaik setzt sich dann aus Elementen zusammen, welche durch ± gerade Klüfte voneinander getrennt sind (Abb. 11). Das Kluftsystem bildet ein negatives Reticulum oder Reticuloid. (Wir unterscheiden also das Reticulum = Extrareticulum [s. unten] vom Infrareticulum und vom Reticuloid oder negativen Reticulum.)

Die Pila zeigen eine terminale Verdickung auf schmächtigerem Hals.

Die Kegel oder Coni (Abb. 14, Taf. 8, Fig. 61) sind von der Seite gesehen spitz-, stumpf- oder abgerundet kegelförmig. Ihre Höhe überschreitet nicht das Maß des doppelten Basisdurchmessers; letzterer kann verbreitert sein.

Die Stacheln oder Spinae (Abb. 18) verjüngen sich langsamer als die Coni, auch sie zeigen oft eine verdickte Basis. Falls dies zutrifft, so ist schon nur die Länge des über der Verdickung befindlichen Teils der Spinae mehr als doppelt so groß denn der über der Verdickung gemessene Durchmesser der Spinae. Manchmal sind die Spinae deutlicher spitz, gelegentlich sind die Spitzen abgebrochen. Die Spinae sind ± geradegestreckt.

Die Stäbchen oder Bacula (Abb. 19, 20) sind cylindrisch; die Verjüngung ist, wenn vorhanden, nur sehr schwach, indessen ist die Basis manchmal etwas kräftiger. Auch die Bacula sind gerade und oft abrupt abgestuzt.

Die Haare, Capilli oder Fimbriae (Abb. 37, 58) sind cylindrisch oder angenähert bandförmig, vor allem aber sind sie nicht geradlinig, sondern ± geschlängelt oder gewunden, knorrig, verästelt, verzweigt oder anastomosierend (Abb. 59, 60), auch wohl partiell verbreitert. Selbst die einfachen ± cylindrischen flexuosen Haare sind wohl nie ganz cylindrisch; ihre Basis kann verdickt sein, evtl. Verjüngung im Hauptteil des Haares ist sehr gering. Manchmal sind sie terminal ziemlich rasch zugespitzt.

Eine besondere Art der Anordnung der Skulpturelemente ist die der C r i s t a e (Abb. 27, Taf. 16 Fig. 295). Das sind aus manchmal seitlich miteinander verbundenen Kegeln, Stäbchen usw. aufgebaute Zäune oder Kämme, die sich auch zu Netzwerken usw. zusammenschließen können.

Die „Reticula" (Abb. 28, 29) bestehen meist aus glatten, niedrigen, sich zu einem Netz zusammenschließenden S t r i e m e n oder höheren L e i s t e n (Muri). In beiden Fällen wird die Exine als e x t r a r e t i c u l a t bezeichnet. Die Umrißlinie des Korns erscheint im zweiten Falle bastionisch (Abb. 30). Sofern besonders hohe Muri vorhanden sind, erscheinen die Bastionzinken durch „Häute", d. h. durch die von der Seite gesehenen Muri miteinander verbunden (Abb. 30). Die Maschen der Netzwerke sind als L u m e n zu bezeichnen. C u r v i m u r a t zu nennen sind ± gewellt verlaufende Muri, f r a g m e n t i m u r a t ± unterbrochene.

Sind die Striemen oder Leisten nicht oder kaum netzig angeordnet und laufen sie z. B. auf langer Strecke etwa einander parallel, so ist die Exine c i c a t r i c o s, dies aber nur dann, wenn die erhabenen Streifen nicht allzu breit werden und die sie trennenden Rinnen nicht relativ ± allzu schmal; in diesem Fall ist die Exine c a n a l i c u l a t zu nennen.

Werden die Lumen der Reticula sehr klein und die Räume zwischen den Lumen breiter als die Lumen, so werden die Lumen zu S c r o b i c u l a e ; größere Scrobiculae heißen F o v e o l a e.

Oft erfordert es einige Aufmerksamkeit festzustellen, ob es sich um ein Reticulum = E x t r a r e t i c u l u m handelt, um ein Infrareticulum oder um ein Reticuloid (= negatives Reticulum). Die Räume zwischen Punkten, Flecken, Körnchen und Warzen machen zunächst vielfach den Eindruck von Striemen oder Leisten (Muri), durch die mehr oder minder große Netzlumen voneinander getrennt zu sein scheinen. Die wahren Verhältnisse kann man z. T. klären, wenn man auf die Umrißlinie (Extrema lineamenta) des Mikrofossils achtet.

Eine besondere Ausbildung erfährt in manchen Fällen (Zonales) die Äquatorregion der Sporen. Im einfachsten Fall kann hier die Exine etwas dicker sein als in den übrigen Bereichen. Es handelt sich dann um eine C r a s s i t u d e. Als c r a s s e x i n i s c h werden durchweg dicke Exinen oder Exinenteile bezeichnet. Die Anschwellung kann an den „Ecken" des Äquatorschnittes der Spore ausgeprägter sein als am übrigen Äquator. Wir sprechen dann von V a l v a e (Abb. 38). Schwellen die Valvae weiter an zu sich vom Umriß absetzenden Kissen oder dergleichen, so wird von Ö h r c h e n, Ohren oder A u r i c u l a e gesprochen (Abb. 39, 40). Setzt sich dem ganzen Äquator eine im Querschnitt oft keilförmige, den äquatorialen Durchmesser deutlich vergrößernde massive Leiste auf, so haben wir ein C i n g u l u m vor uns, auch Reifen oder Arista i. e. S. genannt (Abb. 42, 43). Gelegentlich ist das Cingulum nach außen nicht so wie bei *Lycospora* und *Densospora* ± zugeschärft, sondern es ist im Querschnitt ± kreisförmig mit schmalem Hals; so definitionsgemäß bei *Rotaspora* (Abb. 45). Trägt der Äquator einen breiten, häutigen (im Querschnitt nicht keilförmigen) H a u t s a u m, so handelt es sich um eine Z o n a bzw. F r a s s a, die man auch als Äquatorkragen oder Kragen schlechthin bezeichnet (so bei *Cirratriradites*, Abb. 55). Ist der Äquator mit einem Kranz von Haaren, verzweigten, anastomosierenden Haaren, Fimbriae oder Fransen besetzt, so liegt eine C o r o n a vor (*Reinschia, Superbisporites*, Abb. 56, 59, 60).

Substanz und Schichtung der Sporenwand

Auch das Folgende bezieht sich besonders auf das am fossilen Material noch Erkennbare.

1920 hat R. POTONIÉ nachgewiesen, daß die Exinen von Sporen aus dem oberen produktiven Karbon (z. B. den unteren Ottweiler Schichten, Schwalbacher Flöz, Ensdorfer Schacht bei Griesborn), ebenso wie die Exinen

16 R. Potonié and G. Kremp

von Sporen aus dem Unterkarbon der Moskauer Braunkohle (Tschulkowsky-Schächte bei Skopin) chemisch noch viele Eigenschaften der Exinen von rezenten Pflanzen besitzen. Dies gilt, wie sich dann gezeigt hat, für Sporen aus allen Kohlen bis ± ausschließlich zu einem Inkohlungsstadium wie das der Fettkohle (genauer des Anthrakodits) des Ruhrgebietes, d. h. bis etwa Flöz Katharina (Abb. 2). Es gelingt die HÖHNELsche Cerin-

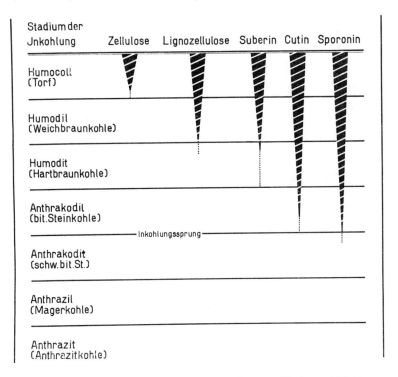

Abb. 2. Abbau der Zellwandsubstanzen im Verlauf der Humifikation und Inkohlung.

säurereaktion (= Auftreten tropfiger Bildungen bei der Einwirkung von $HNO_3 + KClO_3$). Die Widerstandsfähigkeit gegen konzentrierte Chromsäure ist groß, ebenso gegen konzentrierte H_2SO_4. Chlorzinkjod- und Jodjodkalium-Lösung färben die Exinen dunkler gelb bis braun, als sie nach der Mazeration in $HNO_3 + KClO_3$ waren. Weiter gelingt die Färbung mit Gentiana-Violett, Sudan III, alkoholischer Chlorophyll-Lösung und anderen rezente Exinen färbenden Farbstoffen. Die Färbung z. B. mit Gentiana-Violett verbleibt auch nach Auswaschen mit stark verdünnter HCl. Bei der Anfärbung der Exinen mit basischen Farbstoffen (vorzugsweise in alkalischer Lösung) spielen freie Fettsäurereste eine Rolle. Die Färbung z. B. mit Sudan III beruht auf der besseren Löslichkeit des Farbstoffes im Sporonin bzw. Pollenin und Kutin als im alkoholischen Lösungsmittel. Der Eintritt und die Wanderung des Farbstoffs erfolgt in der fettartigen Phase der Exine, jedenfalls auf anderem Wege als der Wasserdurchtritt (O. Härtel 1952). Kochen in Glycerin mit folgender Behandlung in Chromsäure bedingt nur geringe Veränderungen. Kochen in Kalilauge ebenfalls; langes Liegen in Kalilauge verursacht ganz wie bei rezenten Sporen oft intensive Gelbfärbung. Die durch schonende Mazeration in Schulze's Gemisch mit nachfolgender kürzerer Behandlung in Lauge gewonnenen Sporen hellen sich zunächst gegenüber ihrem fossilen Zustand auf, werden ± bräunlich und noch nicht so intensiv gelb wie nach längerem Liegen in Kalilauge.

Die durch die Mazeration entstehende Farbe ist nicht n u r abhängig von der Inkohlung usw., sondern bis zu einem gewissen Grade auch von der Sporenart.

Die natürliche Farbe der Sporenexinen im D ü n n s c h l i f f ist bei jüngeren Kohlen des Ruhrkarbons im durchfallenden Licht bernsteingelb: Sie wird in Richtung auf die Fettkohle dunkler und mehr bräunlich. Im Anschliff bei auffallendem Licht erscheinen die Exinen jüngerer Kohlen dunkel. Sie werden im auffallenden Licht zur Fettkohle hin und weiter hinunter, also mit zunehmendem Inkohlungsgrad, hell, d. h. stärker reflektierend (E. Stach).

Schon die Bernsteinfarbe der Exinen im Dünnschliff z. B. der Gasflammkohle (des Anthrakodils) zeugt von gewissen, wenn auch nicht grundlegenden chemischen Wandlungen des Sporonins gegenüber rezentem Sporonin. Erst durch die Mazeration nehmen die Exinen, soweit sie oberhalb der Fettkohle entnommen wurden, die häutige Geschmeidigkeit rezenten Sporonins ± wieder an. Vor der Mazeration sind die Exinen ± spröde. Das zeigen vertikale Dünnschliffe durch Kohlen. In ihnen lagern die Exinen in der bekannten Weise in Form linsenförmiger Schleifen (Abb. 1A, Taf. 1 Fig. 1, Taf. 13 Fig. 195). Noch während längerer Zeit des Verlaufs der Inkohlung muß die Exine hinreichend geschmeidig gewesen sein, um ohne strukturelle Veränderungen flach zusammengedrückt werden zu können. In einer späteren Phase jedoch macht sich die Sprödigkeit bemerkbar. Es erscheinen hier und dort zerbrochene Exinen, solche mit scharfen, die ganze Exine durchschneidenden Rissen. Bei den Megasporen-Exinen der Steinkohle erkennt man manchmal an den Exinendurchschnitten, gerade an den beiden Stellen, wo die stärkste Krümmung der Exine erzwungen worden ist, eine Brüchigkeit der Haut in Form von (im durchfallenden Licht) sich entsprechend den Beanspruchungslinien aufreihenden dunkelbraunen vitrinitisch ausgefüllten kurzen Rissen. Diese fehlen in den Bereichen der geringeren Biegung (also Beanspruchung).

Die verschiedenen S c h i c h t e n d e r S p o r e n w a n d (S p o r o d e r m i s) sind beim fossilen Material nicht immer scharf auseinanderzuhalten. Fest steht nur, daß manchmal verschiedene Schichten beobachtbar sind, und daß die Autoren bei der Anwendung der für verschiedene Schichten der Sporenwand geprägten Termini nicht immer dasselbe meinen.

In Kohlendünnschliffen erkennt man nur manchmal mehrere Schichten. Zuinnerst liegt gelegentlich eine sehr dünne Schicht, die in Vitrinit umgewandelt ist. Sie wird als die früher aus Zellulose bestehende I n t i n e aufgefaßt. Dann folgt die E x i n e, welche als ± bernsteingelber Exinit in Erscheinung tritt. Die Exine kann wiederum verschiedene Schichten zeigen, die E x o e x i n e und die I n t e x i n e. Die eine kann im Dünnschliff deutlicher punktiert, gefleckt, flaserig, wolkig oder gar prismatisch sein als die andere (Taf. 1 Fig. 1). Nach außen bildet zumeist die Exoexine die Skulpturen der Sporenoberfläche.

Die sogenannte „p r i s m a t i s c h e S t r u k t u r" zeigt sich im Querschnitt der Exine in Form von dicht nebeneinanderstehenden, radial gerichteten helleren „Stäbchen" (s. auch u. unter Columella). Man erkennt dann namentlich im Bereich der Zusammenbiegung der Exine (etwa am Äquator der Spore) und vornehmlich in der Exoexine heller-gelbe, feine, etwa radial gestellte Elemente. Diese Struktur fand sich niemals (auch nicht bei ein und demselben Sporen-Querschliff) überall klar sichtbar. So ist z. B. bei manchen Durchlichtfotos von Mega-

18 R. Potonie and G. Kremp

sporen-Exinen im Querschnitt nur die „Mörtelstruktur" erkennbar. (Diese Struktur zeigt sich jedoch nicht in den der Wand aufsitzenden Zierelementen. Taf. 8 Fig. 61.)

Eine deutliche Grenze zwischen Intexine und Exoexine zeigt sich nicht oft, jedoch sind die Strukturelemente im intexinösen Teil der Exine mehr tangential, im exoexinösen Teil mehr radial oder unregelmäßiger gestellt.

Sporodermis cf. Bischoff 1842, S. 595	Exine Fritzsche 1837 (= Exospor)	Perispor cf. Bischoff 1842, S. 594 (= Perine Strasburger 1882, Erdtman 1950)	
		Exoexine Potonié 1934 (= Ektexine Erdtman 1943 = Sexine Erdtman 1948)	Exolamella Potonié 1934 (= Ectosexine Erdtman 1950)
			Isolierschicht Potonié 1934 (= Endosexine Erdtman 1950)
		Intexine Fritzsche 1832 (= Endexine Erdtman 1943 = Nexine Erdtman 1948)	
	Intine Fritzsche 1837 (= Endospor)		

Die Tabelle zeigt die ideale Schichtung des Sporoderms. Es sind möglichst die ältesten unter den eindeutigen Termini gewählt worden, damit die paläontologischen Arbeiten früherer Autoren verständlich bleiben.

Am mazerierten Material sind demgegenüber manchmal Strukturen zu erkennen, welche der Kohlenschliff nicht aufweist. Die Stäbchenstruktur im Inneren der Exoexine gibt sich hier bei Betrachtung der horizontal ausgebreiteten Exine z. B. als Punktierung zu erkennen.

Meist beobachtet man in Dünnschliffen keine Stäbchen, sondern nur die als Mörtelstruktur bezeichnete Ausbildung, und dies oft am ausgeprägtesten in der Intexine. Als Mörtelstruktur bezeichnen wir (mit Haberlandt) ein Bild, bei dem der Grundmasse oder Matrix der Exine Mikulen oder Puncta eingelagert sind. Die Grenze zwischen Int- und Exoexine ist in Kohlendünnschliffen im durchfallenden Licht wie gesagt nur manchmal erkennbar und kann dann durch einen dunklen feinen Streifen markiert sein. Es ist, als ob hier eine feine Kluft entstanden wäre, welche sich mit einer flockigen dunkelbraunen (also vitrinitischen) Substanz gefüllt hätte, ähnlich der, welche auch in den Bruchstellen der Biegungspartien der Exine (s. o.) zu sehen ist, und wie sie auch in dem im Querschnitt strichförmig erscheinenden Innenraum der Spore und im Intratectum erscheint.

In manchen Fällen zeigt der Dünnschliff für Exo- und Intexine verschiedene Farben. Die Intexine ist etwas bräunlicher oder heller.

Die Termini E n d o s p o r , M e s o s p o r und E x o s p o r werden im allgemeinen nicht von uns benutzt; gelegentlich der Ausdruck P e r i s p o r (= Epispor Russow 1872 = Perine Strasburger 1882). Wir bedienen uns der Einfachheit halber der bei gemeinsamer Beschreibung von Sporen u n d Pollen brauchbaren Ausdrücke der Tabelle S. 18.

Die I n t i n e entspricht dem E n d o s p o r (wobei bemerkt werden muß, daß der Ausdruck Endospor auch wegen z. T. ganz anderer Definition vermieden wird).

Die E x i n e umfaßt den ganzen kutinisierten Teil der Sporen- bzw. Pollenhaut, also das E x o s p o r mit seinem M e s o s p o r (und evtl das Perispor).

Das P e r i s p o r (Perine) kann beim fossilen Material von der Exoexine oft nicht unterschieden werden. Es ist eine sekundäre, zusätzliche, nachträgliche Bildung. Bei der paläontologischen Betrachtungsweise werden zum Perispor äußere Teile der Exine gerechnet, die ± leicht abfällig sind und nach der Loslösung keine Bruchstellen, sondern natürliche Ablösungsflächen hinterlassen. Wir gebrauchen das Wort Perispor nur in diesem Fall.

Benutzt man den Ausdruck Exospor, so ist zu beachten, daß dieser den Begriff Perispor nicht enthält, das Perispor liegt dem Exospor auf.

Das M e s o s p o r indessen ist ein innerer Teil des Exospors und dürfte wohl etwa der Intexine gleichgestellt werden können. Bei Zerndt 1934, 10, gehört das Mesospor nicht zum Exospor. Bei anderen Autoren kommt der Ausdruck M e s e x i n e vor, die zwischen Intexine und Exoexine liegen soll. Bei van Campo ist die Mesexine der innere Teil unserer Exoexine. Die Mesexine löst sich bei der Saccusbildung von der Intexine. Die Mesexine ist also die aus den Columellae bestehende Isolierschicht, die Endosexine Erdtmans.

Von der Exine hörten wir, daß sie aus der Intexine und der Exoexine bestehe. Die Exoexine setzt sich aus einer innen liegenden I s o l i e r s c h i c h t und aus der ihr außen auflagernden E x o l a m e l l a zusammen. Als Isolierschicht ist die Summe der die Exolamelle tragenden, ± locker stehenden, radialen Stäbchen oder C o l u m e l l a e nebst deren Zwischenräumen zu bezeichnen. Die Columellae bilden einen Säulenwald, der gaserfüllten Raum zwischen sich läßt.

Bei der Entstehung der S a c c i (Abb. 71, 72, 76) lösen sich die Basen der Columellae von der Intexine und der gaserfüllte Raum zwischen Intexine und Exoexine weitet sich. Die Exolamellen mit den ihnen innen als Infrastruktur anhaftenden Columellae bilden dann die Haut des Saccus.

Die Sacci können sich dabei entweder blasenartig abgerundet aufblähen (Abb. 75) oder aber (wie bei manchen Monosaccites) durch einen äquatorialen S a u m , den L i m b u s , schirmartig ausgebreitet gehalten werden (Abb. 71, 72). Der Limbus ist ein scharfer randlicher Falz des Saccus, bei dem distale und proximale Saccusdecke auf einem schmalen Saum einander berühren und wohl ± miteinander verschmelzen, wodurch ein stabilerer Reif entsteht, welcher den Schirm ausgebreitet hält.

Eine weitere Ausbildung ist das V e l u m , es besteht sozusagen aus einem auf den gesamten Monosaccus ausgedehnten Limbus (man vgl. *Tsugapollenites igniculus* R. Pot. 1951).

Das Infrareticulum der Sacci der Abietineen kann nach van Campo gestaltet sein:

1. verschieden; ± anastomosierend, d. h., z. T. geschlossene Maschen bildend, z. T. diesem eingeschaltet offene Maschen lassend,
2. unregelmäßig; d. h. fast nur offene Maschen zeigend,
3. regelmäßig; d. h. nur geschlossene Maschen aufweisend.

Zur Ökologie der karbonischen Sporen

Viele der karbonischen Sporen besitzen auffällige Skulpturen und Anhänge. Bei der z. T. ausgeprägten Eigenart mancher dieser Oberflächenbildungen ist man geneigt, Überlegungen über deren Wirkung anzustellen.
Sie können
    1. die Schwebbarkeit in der Luft erleichtern,
    2. das Untersinken im Wasser verlangsamen oder verhindern,
    3. die Verschleppung durch Tiere begünstigen,
    4. die Verankerung am Boden gestatten und
    5. die Verkoppelung der Mikro- mit den Megasporen oder Samenanlagen fördern.

Prüft man die Oberflächenformen der karbonischen Megasporen von diesen Gesichtspunkten aus, so möchte man viele von ihnen im Sinne von Punkt 2 und 4 verstehen.

Zu den auffälligsten Anhangsgebilden der Megasporen gehören äquatoriale Kränze von z. T. anastomosierenden Haaren, welche den äquatorialen Durchmesser der Spore manchmal mehr als verdoppeln. Schöne Beispiele sind *Superbisporites, Rotatisporites* und *Radiatisporites* (Abb. 58—60). Da zwischen den ± peripher miteinander verschmelzenden Haaren Luft haften bleibt, werden solche mit Haarkranz (Corona) versehenen Megasporen (abgesehen vom Lufttransport) im Gegensatz zu den leichter untersinkenden großen und glatten Sporen, zunächst im Wasser schwimmen oder schweben, bis sie, entsprechend Beobachtungen an rezenten Samen, ins flachere getrieben den Boden berühren, um dort infolge des Haarkranzes (vielleicht auch infolge hygroskopischer Bewegung) haften zu bleiben. Sie zeigen damit in einer Zeit, wo die Samen noch zurücktreten, Eigenschaften heutiger Wasser- und Sumpf-Pflanzensamen. *Lepidostrobus foliaceus* enthält sogar Megasporen, die am Apex eine große dünnwandige Blase aufweisen, die, wie bei dem rezenten Wasserfarn *Azolla,* als Schwimmvorrichtung gedeutet wird (Abb. 1D, vgl. auch S. 11).

VAN CAMPO 1951, S. 38, weist auf eine besondere Anpassung bei der rezenten *Trapa natans.* Von den Keimporen der Pollenkörner zu den beiden Polen ziehen kräftige Kämme, die sich auf den Polen wie eine Y-Marke vereinigen und dort einen spitzen Vorsprung bilden. Wenn man diese Pollenkörner in Wasser bringt, zeigt sich, daß in den Winkeln, welche die hohen Kämme bilden, Luftblasen festgehalten werden, was die Pollenkörner wie durch einen Rettungsring „ceinture de sauvetage" schwimmfähig macht.

Wir haben am fossilen Material durch Schwimmversuch festgestellt, daß die besonders hohen Kämme, welche bei vielen Karbonsporen von den Y-Strahlen gebildet werden, eine ähnliche physikalische Wirkung haben. Es seien vor allem die Lagenotriletes genannt, unter ihnen die zu den Lepidodendraceae gehörenden Gattungen *Lagenicula* (Abb. 35) und *Lagenoisporites* (Abb. 36). Auch bei sonstigen Sporen sind die Y-Strahlen oft hoch und brettförmig, so bei *Valvisisporites* (Abb. 40). Bei manchen Formen wird den Luftblasen durch Schlängelung der Y-Leisten noch mehr Haftfläche geboten, so bei *Radiatisporites* (Abb. 58).

Solche Einrichtungen sind verständlich, weil die Karbonwälder Sumpfmoore waren und viele offene Wasserflächen enthielten.

Für das Vorhandensein solcher offenen „Schlenken" im Karbonmoor sprechen auch gewisse Einrichtungen an den S a m e n der Cycadofilicales. *Aetheotesta elliptica* besitzt in der äußeren Schale größere Luftkammern, von denen angenommen wird, sie hätten den Samen zum Schwimmen befähigt; gleiches vermutet man von der haarartigen Bekleidung von *Physostoma elegans*.

Die Vielzahl der in der Kohle noch erkennbaren Sporen ist u. a. ein Hinweis darauf, wieviele (auch Megasporen) an einen zur Keimung ungeeigneten Platz gelangt sind. Diese Sporen zeigen fast alle noch keine oder nur geringere Keimungserscheinungen, wie Kohlen-Anschliff und Mazerationsprodukt beweisen. Sie sind vorzeitig, zumeist s u b a q u a t i s c h eingebettet worden, wie wir denn auch subfossil die größere Frequenz sedimentierter Sporen da finden, wo kurz vor und während der Torfbildung offene Wasserflächen waren (H. POTONIÉ, Rezente Kaustobiolithe). Die Wasserflächen fangen Pollen und Sporen ab und entziehen sie oft der Keimung. Somit wären Schwimmeinrichtungen und Vorrichtungen der Verankerung bei den Megasporen wohl zu verstehen.

Bei *Valvisisporites* IBR., *Triquitrites* WILSON & COE und *Tripartites* SCHEMEL (Abb. 38—40) werden die drei äquatorialen, in Verlängerung der drei Y-Strahlen vorgetriebenen Auriculae wohl nicht dem Lufttransport dienende Schwebeorgane gewesen sein. Sie erinnern eher an Verankerungseinrichtungen. Doch sollen nicht alle Oberflächenformen, die ein Festhalten gestatten könnten, in diesem Sinne gedeutet werden. Erwähnt seien aber die löffelartigen, zugespitzten, hakenartig wirkenden Gebilde, die sich bei der Keimung von *Lagenicula* auseinanderspreizen. Man kann sich vorstellen, daß die Löffel im Beginn der Keimung eine Verankerung herbeiführten und so bei gelegentlicher Überflutung eine Störung der Keimung verhinderten (Abb. 35, vgl. auch 1 C). Vielleicht darf auch an geweihartig aufgespaltene Haare gedacht werden, wie sie z. B. bei *Setosisporites hirsutus* (LOOSE) IBRAHIM vorkommen (Abb. 37).

Vorrichtungen, die der V e r k o p p e l u n g d e r M i k r o s p o r e n m i t d e n M e g a s p o r e n o d e r S a m e n a n l a g e n dienen, darf man in den skulpturierten Oberflächen der Exinen von vielen der im folgenden beschriebenen Formen erblicken. Nach SEWARD 1910, II, S. 192, dienten die Haare der Megasporen dem Fang der Mikrosporen und begünstigten so die Befruchtung. Wenn die Haare nur oder vor allem auf der Y-Marke sitzen, wie bei der Megaspore von *Lycostrobus scotti* NATH. aus dem Lias (vgl. WICHER 1951, Taf. 1, Fig. 1, 2), dann ist gegen diese Ansicht nichts einzuwenden[1]). Oft sind aber gerade im Bereich der Keimung keine oder weit kleinere Haare zu finden. Die Löffel jedoch von *Lagenicula, Lagenoisporites* und *Setosisporites* könnten auch in SEWARD's Sinn gedeutet werden.

D i e p a l ä o z o i s c h e n S p o r e n haben mindestens e i n e Eigenschaft, durch die sie sich grundsätzlich von allen späteren Sporenformen unterscheiden: Die Megasporen der Lepidophyten erreichen Größen, welche deutlich und oft beträchtlich über das spätere Maß hinausgehen. Rezente Megasporen werden höchstens 1,5 mm, senone vielleicht 1 mm, karbonische jedoch bis über 3 mm, so *Laevigatisporites* und *Tuberculatisporites*. Die Samenspore *Cystosporites giganteus* ZERNDT aber zeigt 3,5—11mm. C. WICHER 1951 bildet mesozoische Megasporen von 0,3—0,7 mm Größe ab (Ober-Rät. 0,3—0,7; Unt. Lias 0,46 mm). Zu einer Zeit, wo die Megasporen so bedeutende Größen annahmen, muß die Verbreitungsgeschwindigkeit abgenommen haben. Aber aus Urwalddickichten gelangen die Sporen in jedem Fall schwerer heraus (vgl. S. 22). So mag die beträchtliche Größe vieler Karbonsporen einmal auf den dichten Bestand der Karbonpflanzengemeinschaft verweisen, sodann aber wird sie uns auch veranlassen, manche der früher als Flugvorrichtungen gedeuteten Organe nicht mehr so auszulegen. Für die großen Sporen könnte jedenfalls der Vorteil bestehen, in einem nicht oder nur wenig bewegten flachen Wasser nicht allzuleicht fortbewegt und so nicht im Beginn der Keimung durch nachträglichen Transport gestört zu werden.

---

[1]) Weiter zeigen eine behaarte Y-Marke *Triletes cristatus* CHALONER 1954, Taf. 2 (Mississipian) und die Sporengattung *Thomsonia* MÄDLER 1954, Taf. 5 (Wealden).

Bei einigen der großen karbonischen Megasporen ist die Oberfläche ± glatt, bei *Laevigatisporites glabratus* (Taf. 1 Fig. 4, 5) sogar so glatt und glänzend, daß sie schwer benetzbar gewesen sein wird. Die Exine erscheint unter dem Mikroskop bei Schräglicht wie mit Pailletten bedeckt.

Ähnlich gewöhnlichen Samen verhielten sich die glatten großen dünnhäutig-sackförmigen Megasporen (Samensporen) mancher Lycopsida, so *Cystosporites* (Abb. 69). Diese Megasporen waren mit ihrem distalen Ende mit der Wand des Sporangiums verwachsen (vgl. z. B. *Lepidostrobus major* BRONG.). Die Megasporen wurden hier nicht für sich ausgestreut, sondern verblieben im Sporangium, das seinerseits nur in Verbindung mit dem Sporophyll abfällig war.

Bei heutigen Wasserpflanzen, a u c h bei solchen, die nur mit einem Teil ihres Körpers subaquatisch leben und bei denen die Mikrosporen die weiblichen Fortpflanzungsorgane ü b e r der Wasseroberfläche erreichen, findet man neben anderen besonderen Anpassungen D ü n n h ä u t i g k e i t d e r E x i n e. Wenn auch keine eigentlichen Wasserpflanzen, nennen wir hier die Calamiten (Abb. 6), die außerdem, wie manche rezenten, mit einem ± großen Teil ihres Körpers im Wasser stehenden Gewächse, kugelige, glatte Sporen von weniger vollendeter Form haben. Dünne Exine zeigen auch die Sporen der Sphenophyllen.

Der Pollen der Gymnospermen, soll er sein Ziel erreichen, muß im Gegensatz zu den Iso- und Mikrosporen der Pteridophyten durch die Luft emporgetragen werden. Die Iso- und Mikrosporen der Pteridophyten dagegen keimen auf dem Boden, in dem die zukünftige Pflanze wurzeln wird. Ein Lufttransport ist zwar förderlich, aber nicht notwendig. Bei den Gymnospermen wird er zur Bedingung. Wir sehen bei ihnen denn auch schon bald die besten dem Pollen gegebenen Flugorgane entstehen, die Sacci. Es besaßen Luftsäcke (Abbildung 70—83, 90): *Microsporites* (Lycopsida), *Alatisporites* (Cordaitales?), *Endosporites* (Lycopsida?), *Florinites* (Cordaitales und Coniferales), *Alisporites*, *Vesicaspora* (Caytoniales?), *Illinites* und *Pityosporites* (Coniferales).

Es ist nicht ersichtlich, inwiefern der von BOLKHOVITINA (1952) als Fortschritt betrachtete Weg von den Monosaccites zu den Bisaccites eine flugtechnische Vervollkommnung sein sollte.

Dem Lufttransport dienten wohl auch die geschlossenen Zonae, d. h. die den Äquator der Spore umziehenden ± häutigen Reifen, so bei *Cirratriradites* und *Reinschospora* (Abb. 55, 56).

Nur die Zahl solcher Einrichtungen ist gegen später geringer. Sie dürften erst notwendig werden mit zunehmender Differenzierung der Pflanzengemeinschaften (R. POTONIÉ 1952). Im allgemeinen sind Vorrichtungen für den Lufttransport bei Pflanzen, die sehr eng geschlossenen Gemeinschaften angehören, nicht sehr wirkungsvoll. Aus Urwalddickichten werden nach TH. HERZOG, da dort der Wind keinen Zutritt findet und die umgebenden Laubmassen durch ihre meist feuchte Oberfläche die abfliegenden Sporen auffangen und festhalten, die Sporen nicht herausgetragen. „Diese Absperrung hört auf für die... Baumäste, wo der Wind ungehinderten Zutritt hat." Wir finden denn auch im Steinkohlenwald die Luftsäcke gerade bei den hochragenden Pflanzen der oberen Etage, so bei den Cordaiten. Aber es sind unter den mit Flugorganen ausgestatteten Sporen auch manche, die wir bisher gar nicht bei Gewächsen des karbonischen Hygrophorbiums einzureihen vermochten, so *Reinschospora*. Es darf vermutet werden, daß manches, was zu weitem Fluge befähigt war, aus der weitesten Umgebung des Steinkohlenmoores, vielleicht von den Höhen des Festlandes stammt, von dem uns dann außer diesen Sporen keine weiteren Pflanzenreste hinterlassen worden wären. Solche Abtragungs-Räume bieten für die dort ansässige Organismenwelt keine geeigneten Erhaltungsbedingungen. In der Steinkohle und in deren Nebengestein würden uns demnach diese Elemente als gänzlich bodenfremd entgegentreten.

Hier sind vielleicht auch manche der kleineren und nur deshalb leicht transportablen Sporen zu nennen, wie die der umfangreichen Gattung *Densosporites*. Unter den Samen könnte man an *Gnetopsis* RENAULT denken.

Dieser mittelkarbonische bis permische Same zeigt an der Spitze gefiederte, seine Länge weit übertreffende Flughaare.

Weiteres zur Biologie der Sporen vgl. in POTONIÉ, IBRAHIM und LOOSE 1932 sowie in POTONIÉ, Pal. Z. **29**, 27—32, Stuttgart 1955.

Morphographische Systematik der palaeozoischen Sporae dispersae

„L'esprit va, dans son travail, de son désordre à son ordre. Il import qu'il se conserve jusqu'à la fin, des ressources de désordre, et que l'ordre qu'il a commencé de se donner ne le lie pas si complètement, ne lui soit pas un si rigide maître, qu'il ne puisse le changer et user de sa liberté initiale."

(PAUL VALÉRY „Analecta" Paris 1926.)

Zur systematischen Bearbeitung des Materials gliedern wir die Sporae dispersae in Form-Oberabteilungen, Formabteilungen (= Turmae), Form-Unterabteilungen (= Subturmae), Formreihen (= series), Formgenera und Formspezies. (Die Formgenera können zu Organgenera werden, s. S. 4.)

IBRAHIM schrieb 1933 hinter zwei seiner Abteilungen n. g. und nicht „n. gen." Das hätte nicht zu Irrtümern führen dürfen, da aus seiner Disposition deutlich ersichtlich ist, daß es sich nicht um Gattungen handelt, sondern um Einheiten höherer Ordnung. Im Text (S. 35) hebt IBRAHIM dies denn auch deutlich hervor und spricht von „Gruppen".

Die Gattungsnamen werden von den Abteilungs- usw.-namen zweckmäßigerweise durch Endungen wie -sporites, -sporis, -spora, -pollenites, -pollis unterschieden (Int. Code PB. 6A).

1914 (SEWARD) und 1931 (R. POTONIÉ) wurde damit begonnen, neuen Formgattungen dispergierter Sporen die Endungen -sporites oder -pollenites anzuhängen (S. W. & B. 1944 haben in einigen Fällen -spora bevorzugt). Dieses Verfahren entspricht einem häufigen Brauch der Paläobotanik. Wo einzelne Pflanzenteile in „Form- und Organgenera" eingeordnet werden, erhalten diese Genera oft Namen, aus denen hervorgeht, welcher Pflanzenteil gemeint ist. Dieser Brauch besteht u. a. bei Samen und Hölzern und hat sich als zweckmäßig erwiesen. Man kann so z. B. Listen, die einen ganzen Fossilieninhalt umfassen, besser überschauen. In noch höherem Maße als bei manchen anderen Pflanzenresten ist diese Methode bei den Sporen vorteilhaft. Für die Formgattungen wurden also Namen gewählt, wie *Pityosporites* (1914), *Laevigatisporites* (1933), *Granulatisporites* (1933), *Alnipollenites* (1931) usw. Nur die lediglich dem Ordnungsprinzip dienenden und daher seltener erwähnten Turmae usw. erhielten, um sich von den Formgattungen zu unterscheiden, eine andere, aber ebenfalls feststehende Endung; man vergleiche Triletes, Monoletes.

Unklarheit besteht, wann -sporites, wann -pollenites anzuwenden sei. Bei der Bildung von Gattungsnamen für dispergierten Pollen von tertiären Angiospermen und Gymnospermen ist bisher -pollenites benutzt worden. Bei dem Pollen paläozoischer und altmesozoischer Gymnospermen dagegen -sporites. Es braucht nichts dagegen eingewandt zu werden, wenn der Endung -sporites von manchen Autoren auch dann der Vorzug gegeben wird, wenn es sich um einwandfreien Pollen handelt. Wesentlich ist nur, daß die beiden Oberabteilungen S p o r i t e s und P o l l e n i t e s bestehen bleiben (man vergleiche unsere diesbezüglichen Diagnosen S. 30 und weiter hinten[2])). Hierbei ist, wie S. 31 betont, zu bedenken, daß auch die Unterscheidung in Sporites und Pollenites lediglich morphographischen Charakter hat, daß also nicht gesagt ist, daß alles, was der Diagnose von Sporites entspricht, vom Pollen im entwicklungsgeschichtlichen Sinne abrücke (und umgekehrt). Da aber trotzdem auf morphographischer Basis eine große Gliederung möglich ist, wird diese begrüßt und als nützliches Hilfsmittel empfunden.

Die Gattungen, welche der Diagnose der Oberabteilung Sporites entsprechen und daher unter dieser zu erscheinen haben, sollten jedenfalls möglichst die Endungen -sporites oder -spora erhalten. Bei den Gattungen dagegen, welche zur Diagnose der Oberabteilung Pollenites gehören, ist nichts weiter zu verlangen, als daß sie im System unter der Hauptüberschrift Pollenites stehen. Es werden sich ja trotz aller weiteren Empfehlungen infolge des historischen Werdens des Systems sowieso die ver-

---

[2]) Siehe die späteren Fortsetzungen der Arbeit.

schiedensten Endungen mischen. Wir halten es aber für berechtigt, in Zukunft auch bei ältesten Pollengattungen die Endung -pollenites zu verwenden, um schon hiermit auf die morphographische Eigentümlichkeit hinzuweisen (so bei *Schopfipollenites*).

Eine differenziertere Nomenklatur wird stets die Spuren ihres historischen Werdens bewahren. Es ist leicht, späterhin zu sagen, was besser gewesen wäre. Aber das Sichtbarbleiben des Werdens ist auch nicht ohne Reiz.

IBRAHIM (1933) ließ den ersten Wortteil aller Gattungen seiner Triletes mit i enden: *Puncta*ti*sporites;* der Monoletes mit o: *Puncta*to*sporites,* der Aletes mit a: *Puncta*ta*sporites.*

Wir wollen dies bei diesen drei Turmae im Falle von neuen Formgattungen so fortsetzen.

Iso-, Mega- und Mikrosporen müssen in ein und demselben morphographischen System Platz finden. Erst bei den Formgattungen ist es notwendig, die einen von den anderen soweit dies möglich ist zu trennen.

Zusammenfassender Begriff für Iso-, Mikro- und kleine Megasporen ist Kleinsporen, Mio- oder Oligosporae. Dieser Begriff ist bei den kleineren Sporae dispersae nicht immer zu vermeiden, weil wir oft nicht sagen können, um was es sich im einzelnen handelt. Sehr große Sporen aber sind stets Megasporen[3]).

Der Begriff Miosporae stammt von G. K. GUENNEL 1952. Er nennt so alle fossilen Sporen unter 200 µ. Die Festlegung der Größengrenze ist nicht so wichtig, da es sich ja nur um einen häufig vermeidbaren Verlegenheitsausdruck handelt. Einwandfreie paläozoische Mikrosporen wie *Schopfipollenites* erreichen Größen von 500 µ, Megasporen z. B. von *Archaeopteris latifolia* können unter 200 µ bleiben.

Uns kommt es in der vorliegenden Schrift bei der systematischen Bearbeitung der Sporae dispersae des Ruhrkarbons zunächst darauf an, die Formgenera (bzw. Organgenera) deutlich voneinander zu scheiden. Die stratigraphische Arbeit wird sich nämlich in vielen Fällen nicht sogleich feiner Formarten bedienen können. Oft sind die Unterschiede der Formarten zu fließend oder zu gering. Um so überlegter müssen die Formgenera gefaßt werden. Alle wirklich leicht festzustellenden qualitativen und nicht zu variablen Unterschiede sollten zur Aufstellung von Formgattungen veranlassen, alle diffizileren Merkmale dagegen, insbesondere die quantitativen, wären erst bei der Abgrenzung der Formarten heranzuziehen.

Es ist anzustreben, daß schon die Angabe bloßer Formgattungen stratigraphische Bedeutung habe. Das ist bei allzuvielen der bisherigen Formgattungen nicht versucht worden. Zu viele von ihnen setzen im Paläozoikum ein und reichen bis ins Mesozoikum und weiter. Schon diese Tatsache legt nahe, daß die Formgattungen bisher zu weit gefaßt worden sind, abgesehen von den primitiveren Gattungen, die nach wie vor Vertreter der verschiedensten Formationen enthalten werden.

SCHOPF (1938, S. 16) betont mit Recht, es sei wünschenswert, die Organ-Genera bei den Sporae eng zu fassen. Deshalb sei es wahrscheinlich nötig, in Zukunft in manchen Fällen Genera einzuschränken und neue daneben zu stellen. Er sagt dies in einer Schrift, in der er vorwiegend Megasporen behandelt.

Unsere Arbeit hat gezeigt, daß geeignete Anwendung der Morphographie doch dazu befähigt, stratigraphisch beschränktere Formgattungen zu ermitteln.. Eine so große Formgattung, wie z. B. *Triletes,* darf nicht weiterhin als solche bestehen. Als wir unsere Arbeit begannen, bedeutete *Triletes* zumeist nichts anderes als „trilete Megaspore". Von DIJKSTRA wurden sogar die Megasporen des Mesozoikums schlechthin als *Triletes* bezeichnet, obgleich unter ihnen viele sind, die ausgezeichnete, von den paläozoischen Formen abweichende Einzelheiten aufweisen. Die bedeutende Verschiedenheit der Megasporen ist ja doch gerade auch durch die Arbeiten DIJKSTRA's besonders deutlich hervorgetreten. Bei der Handhabung einer derart großen Formgattung wie *Triletes* mußte sich aber schließlich etwas wie ein psychologisches Moment bemerkbar machen. Man meinte, daß man verwandte Formen schon dann nicht hinreichend beieinander ließe, wenn man sie als verschiedene Formspezies betrachtete. Sie wurden eben nicht durch einen engeren Gattungsbegriff zusammengehalten, ge-

---

[3]) MÄDLER 1954 S. 146 meint, das Wort Megasporen sei anzuwenden, wo sich nicht sagen lasse, ob es sich um Mikrosporen oder „Makrosporen" handele. GUENNEL 1952 empfiehlt dasselbe für das Wort Makrosporen.

hörten alle zu *Triletes,* und innerhalb der großen Einheit *Triletes* verhielten sich die Formspezies schließlich eher wie Formgattungen. Es wurde zwar innerhalb der Formspezies auch auf feinere Unterschiede aufmerksam gemacht, man ging aber andererseits so weit, gute, schon früher durch Namengebung erfolgte Trennungen wieder rückgängig zu machen; wohl nur, um eine Übersichtlichkeit nicht zu gefährden, die in der Tat mangels Einteilungsprinzipien in Frage gestellt war. HORST (1943) und andere Autoren bezeichnen sogar auch noch alle trileten Mikro- und Isosporen als *Triletes* im Sinne eines Gattungsnamens.

Geschickte Aufstellung einer größeren Anzahl guter Formgattungen, nicht nur bei den Iso- und Mikrosporen, sondern auch bei den Megasporen, und Klassifizierung der Gattungen in Reihen, Abteilungen usw., kann die Übersichtlichkeit nur erhöhen, und wird sie auch dann nicht so bald schwinden lassen, wenn nun in Zukunft bei Aufstellung der Formspezies auch feinere Unterschiede berücksichtigt werden. Solche sind ja doch vorhanden. Die Behauptung, sie hätten keine hinreichende Bedeutung, kann bei einem so jungen Gebiet wie dem unseren nicht gelten. Ehe die Unterschiede der verschiedenen Formen nicht durch besondere Bezeichnungen herausgeholt worden sind, kennt man sie auch noch nicht genügend. Solange läßt sich aber auch nicht ermitteln, wie groß ihr paläontologischer und stratigraphischer Wert ist.

Hinter jeder der Gattungsdiagnosen sind außer den im Ruhrgebiet vorkommenden, von uns beschriebenen Formen a u c h s o n s t i g e zur Gattung gehörende sowie (in besonderer Rubrik) v i e l l e i c h t zur Gattung gehörende Formen aufgezählt. Unter diesen s o n s t i g e n Formen dürften manche sein, deren Namen sich später einmal als Synonyma anderer Formen herausstellen werden[4]).

In die eigentlichen Synonymlisten haben wir jedenfalls nur diejenigen Namen aufgenommen, die hierfür hinreichende Sicherheit boten. Uns scheint, als ob man bei der Einziehung mancher Artnamen, mögen auch hier und dort mehr vorhanden sein als notwendig, doch etwas zu schnell vorgegangen sei.

Wie das m o r p h o l o g i s c h e S y s t e m der p a l ä o z o i s c h e n dispergierten Sporae (das also für vorliegende Arbeit zu berücksichtigen wäre) nach dem Stande von 1951 etwa aussah, hat R. POTONIÉ (1952) darzustellen versucht. An dieser Darstellung mußte bei POTONIÉ & KREMP 1954 und bei der vorliegenden Bearbeitung der Sporae des Ruhrkarbons manches revidiert werden, so daß das System gegenüber der l. c. gegebenen Fassung (insbesondere durch die Berücksichtigung inzwischen zugänglich gewordener russischer Literatur) das folgende Bild bietet:

S p o r a e d i s p e r s a e

1. Oberabt. Sporonites (R. POT.) JBR. 1933 (Fungi)
2. Oberabt. Sporites H. POT. 1893 (= ± Rimales NAUMOVA 1937)
    1. Abt. Triletes (REINSCH 1881) POT. & KR. 1954
        1. Unterabt. Azonotriletes LUBER 1935
            1. Reihe Laevigati (BENNIE & KIDSTON 1886) POT. & KR. 1954
            2. Reihe Apiculati (BENNIE & KIDSTON 1886) POT. & KR. 1954
            3. Reihe Murornati POT. & KR. 1954
        2. Unterabt. Lagenotriletes POT. & KR. 1954
    2. Abt. Zonales (BENNIE & KIDSTON 1886) POT. & KR. 1954
        1. Unterabt. Auritotriletes POT. & KR. 1954
            1. Reihe Auriculati (SCHOPF 1938) POT. & KR. 1954

---

[4]) Wo zwischen Gattungs- und Speziesnamen in Klammern ein zweiter Gattungsname steht, ist das der frühere Gattungsname der Spezies, nicht der einer Untergattung oder Sektion, wie das der Int. Code wünscht.

            2. Unterabt. Zonotriletes Waltz 1935
                1. Reihe Cingulati Pot. & Klaus 1954
                2. Reihe Zonati Pot. & Kr. 1954
        3. Abt. Monoletes Jbr. 1933
            1. Unterabt. Azonomonoletes Luber 1935
            2. Unterabt. Zonomonoletes Naumova
        4. Abt. Cystites Pot. & Kr. 1954
3. Oberabt. Pollenites R. Pot. 1931
        1. Abt. Saccites Erdtman 1947
            1. Unterabt. Polysaccites Cookson 1947
            2. Unterabt. Monosaccites Chitaley 1951
            3. Unterabt. Disaccites Cookson 1947
        2. Abt. Aletes Jbr. 1933 (= Napites Erdtman 1947 = Inaperturates Iverson & Troels-Smith 1950)
            1. Unterabt. Azonaletes Luber 1935
            2. Unterabt. Zonaletes Luber 1935
        3. Abt. Precolpates Pot. & Kr. 1954
        4. Abt. Monocolpates Iverson & Troels-Smith 1950

Die obige Aufstellung, bei der die Gattungen weggelassen worden sind, entstand unter peinlicher Verwertung möglichst aller bisher für das morphographische System der dispergierten fossilen Sporae geleisteten Arbeit.

Zur Bezeichnung der 1. Abteilung der Sporites wurde der von Reinsch 1881 geprägte Name Triletes gewählt. Entsprechend der sprachlichen Gestalt dieses Ausdruckes wurden die Namen der weiteren Turmae gebildet, wie schon bei Ibrahim 1933. Spätere Bearbeiter (Schopf, Wilson & Bentall 1944) ließen die Turmae wieder fallen. Wir halten das für unübersichtlich. Die Genannten vermehrten dafür die einander gleichwertigen Formgattungen und machten auch aus zweien der Ibrahim'schen Turmae, nämlich aus Triletes und Monoletes, Formgattungen. Der Name Monoletes ist jedoch nur für die höhere Kategorie (Abteilung) geschaffen worden und muß so fortbestehen. Was zu Triletes zu sagen ist, ergibt sich weiter unten (S. 32). Die Namen Triletes und Monoletes sind jedenfalls aus der Reihe der Formgattungen wieder herauszunehmen. An ihre Stelle rücken u. a. bereits vorhandene Namen geringeren Grades.

Hinzugefügt sei noch, daß Naumova (1937) die Triletes, Zonales und Monoletes als Rimales zusammenfaßt. Die Aletes sind bei ihr die Irrimales. Letzterer Begriff bezieht sich jedoch bei Naumova nicht auf Pollenkörner, sondern nur auf Sporen s. str.

Bei den Gattungen der Sporites und Pollenites handelt es sich um Form- und Organgenera, von denen den Paläobotanikern bekannt ist, daß für sie die Nomenklaturregeln nicht minder gelten als für sonstige Genera. Das ist nicht immer beachtet worden. So haben Luber & Waltz 1938 unbedenklich alle alten, mit Gattungsdiagnose und Genotyp versehenen, ihnen sehr wohl bekannten Formgattungs-Namen der fossilen Sporae dispersae fallen lassen und dafür neue Namen geschaffen. Den Artnamen gegenüber aber verfuhren sie sehr korrekt. Auch Pflug (in Thomson & Pflug 1952) hat bei der Bearbeitung tertiärer Sporomorphen ältere Namen durch neue ersetzt. Er glaubt allerdings, im Sinne der Nomenklaturregeln gehandelt zu haben, weil die älteren Namen keine ausdrücklich genannten Genotypen und keine oder ungenügende Diagnosen hätten. (Aber wie viele alte Namen müßte man bei diesem Verfahren streichen?) U. E. ist es unstatthaft, einen Namen, der in vielen anderen Belangen in Ordnung ist, wegen gewisser Unterlassungen zu verwerfen. Man kann ihn oft durch Ermittlung eines Genotyps nachträglich rechtfertigen. Was die alten Gattungen umfassen wollten, geht meist aus den Diagnosen ihrer Arten hervor. Es ist also doch, ganz abgesehen von den gar nicht so strengen Nomenklaturregeln, Verständnis für das historische Werden des morphographischen Systems am Platze. Spätere Autoren werden hoffentlich nicht nach dem Verfahren von PFLUG vorgehen. Man sollte seine Pflichten mehr im Sinne der baldigen Erzielung eines stetigen Zustandes auffassen. PFLUG hat übrigens die nach seiner Ansicht schlecht definierten Gattungsnamen nur dann fallen lassen, wenn es sich um Namen handelte, welche (wie z. B. *Caryapollenites*) Anklänge an das natürliche System der Pflanzenfamilien zeigten. Auch wir stehen auf dem Standpunkt, daß solche Namen unerwünscht sind. Läßt sich aber ein Typus für solche Gattungen ermitteln, so m ü s s e n sie nach dem Int. Code bestehen bleiben.

Zwei der Empfehlungen des Heerlener Kongresses von 1951 (vgl. Compte Rendu 1952, S. XV) betonen, daß

1. der Name jeder neuen Sporengattung einen Hinweis darauf enthalten sollte, daß es sich um eine Spore handelt (z. B. *Laevigatisporites, Reinschospora*); vgl. auch Int. Code;
2. der Name keine Verwandtschaft zwischen der Spore und irgendeiner lebenden oder fossilen Pflanze andeuten sollte.

PFLUG hat sich bei der Neuaufstellung von Gattungsnamen in dankenswerter Weise nach diesen Vorschlägen gerichtet. Wir meinen aber, er habe dabei die schon vorhandenen Namen zu Unrecht beseitigt.

Auch Namen mit Anklängen an Namen lebender Pflanzen sind weiter zu benutzen, wenn sie die Priorität haben. Wir dürfen dann nur ihre morphographische Bedeutung empfinden und nichts in sie hineindenken, was Beziehungen zum natürlichen System betrifft. Sehr bedenklich ist es, solche Namen (wie PFLUG es tut) mit dem Hinweis zu beseitigen, es seien für sie bei ihrer Einführung keine ausdrücklichen Genotypen genannt und keine Gattungsdiagnosen gegeben worden, also keine Notiz, die sich ausdrücklich als Gattungsdiagnose bezeichnet. Wie oft gibt es nicht auch in neuesten Arbeiten kurze Vermerke, die zwar als Diagnose bezeichnet werden, und die doch nichts besagen. Wie gut kann man in anderen Fällen ersehen, was der Autor mit der Gattung gemeint hat, aus ihrem klaren Inhalt. Die in der Gattung angeführten Arten gestatten dann die Fassung der Gattungsdiagnose und die nachträgliche Wahl eines Genotyps. Es handelt sich also darum, ob man feststellen kann, was die Gattung bedeutet, nicht darum, ob irgendwo der Vermerk Diagnose oder Genotypus steht. Man k a n n mit Hilfe der Nomenklaturregeln sehr gut mit vorsichtiger Hand berichten.

LUBER & WALTZ verteilen die Sporen auf einige wenige Einheiten, die so definiert sind, daß jede von ihnen mehrere, bereits vorher vorhandene Sporengattungen enthält. Genotypen werden nicht genannt. Es hieße nicht im Sinn von LUBER & WALTZ handeln, wollte man ihre Einheiten im Hinblick auf einen neu zu fixierenden Genotyp einschränken, um sie zwischen die alten Gattungen zu stellen. Nach der Definition wollen die beiden Autoren ausdrücklich nur einige wenige, vieles umfassende Einheiten schaffen. Somit haben wir einerseits die Einheiten von LUBER & WALTZ, andererseits die die Priorität besitzenden Gattungen, vermehrt um weitere ergänzende Gattungen. Wir müssen also die legitimen Gattungen wählen. Wenn so die von LUBER & WALTZ benutzten Begriffe als Gattungen fortfallen, kann man sie doch weiterhin als Einheiten höherer Ordnung verwenden, die dann jeweils diejenigen der eigentlichen Sporengattungen umfassen, welche der Definition von LUBER & WALTZ entsprechen.

NAUMOVA hat 1937 eine große Anzahl von Gruppen aufgestellt, die sich ebenfalls häufig mit schon vorher vorhandenen, oder zwar erst später geschaffenen, jedoch durch die Nennung von Genotypen legitimen

Gattungen decken. Die Einheiten von Naumova waren bisher mangels klaren Inhalts unsicher. Einige dieser Einheiten konnten wir durch nachträgliche Nennung von Genotypen zu legitimen Gattungen machen. Naumova selbst hat in ihrer Arbeit von 1937 die Einheiten nicht für die binäre Nomenklatur verwandt, sondern dazu die größeren Einheiten von Luber & Waltz benutzt, so wie Luber & Waltz das schon vorher und auch nach der Veröffentlichung der Arbeit Naumova's getan haben.

1954 hat Pant ein morphographisches System empfohlen. Bei näherer Betrachtung zeigt dieses so manches, was auch wir gewollt haben. Unter anderem sagt Pant (S. 39), Ibrahim's System sei „natürlicher" als die vorangehenden. Er übernimmt, wie wir, möglichst viel von den bisherigen Bezeichnungen. Die „Gattungen" nennt er Sub-groups. Wo er neue einführt, gibt er zwar kurze Diagnosen, beschreibt aber keine Arten und sagt auch nicht, welche Arten einzureihen wären. Von einigen seiner systematischen Einheiten sagt er, sie seien geschaffen worden für den Fall, daß sich solche Sporen einmal finden sollten. Damit sind diese Sub-groups vorläufig als leere Fächer zu betrachten. Im folgenden werden wir trotzdem noch öfter auf Pant's oft gute Überlegungen zurückkommen, mag es sich auch um nomina nuda für Gruppen (nicht legitime Genera) handeln.

Wie nach dem Gesagten das morphographische System der paläozoischen Sporae dispersae z. Z. aussieht, ergibt sich aus der folgenden Bearbeitung der Sporae des Ruhrkarbons.

Systematische Bearbeitung der Sporae dispersae
1. Oberabt. Sporonites (R. Potonié) Ibrahim 1933

Dieser Terminus wurde von Ibrahim 1933, S. 17, eingeschränkt auf die Sporae dispersae fossiler Fungi. Wir stellen dazu auch weitere mikroskopisch kleine Pilzreste, wie z. B. Sklerotien.

Sclerotites Geinitz und andere ± die Pilze betreffende Organgenera hätten also hier Platz zu finden.

Für einzellige usw. Pilzsporen schaffen wir innerhalb der Sporonites vorläufig keine besondere Gattung, da sie noch nicht hinreichend bearbeitet worden sind. Wir bezeichnen sie bis auf weiteres mit dem Abteilungsnamen als *Sporonites*.

Sporonites sp.
Taf. 11, Fig. 101

± kugelig, ± 60 µ groß, Farbe dunkel bis schwarzbraun, Wandung dick, struktur- und skulpturlos, nicht die geringste Infrapunktierung usw. ist erkennbar, Umrißlinie ganz glatt, Haut nur selten aufgeplatzt, in der Art wie auf Taf. 11 Fig. 101.

Vorkommen: Ruhrgebiet, Oberes Westfal B.

Sclerotites Geinitz
Sclerotites angulatus (Ibrahim) nov. comb.
Taf. 11, Fig. 102, 103

1933 *Reticulati-sporites angulatus* Ibrahim S. 35, Taf. 7, Fig. 59.
Holotypus: Taf. 11, Fig. 102 nach Orig. Ibrahim 1933, Taf. 7, Fig. 59; Präp. A 70, e 6 (or).

Diagnose: Größe etwa 80—95 µ, Holotypus 88 µ, Lumina des Geflechts ± kreisförmig, Wände breit, Äquatorkontur ± dreieckig mit zona-artiger Haut und wenigen randlich vorstoßenden Wänden.

Beschreibung: Farbe braun, Äquatorumriß dreieckig, mit konvexen Seiten, die Umrißlinie weist nur hier und dort Vorsprünge auf, Äquatorregion jedoch cingulumartig bis häutig ausgebildet. Es zeigt sich ein Geflecht mit mehr kreisförmigen, nur z. T. polygonalen Lumina. Durchmesser der Lumina bis 6 µ; der der trennenden Wände 2—3 µ.

Vorkommen: Ruhrgebiet, Oberes Westfal B.

[*Editors' Note:* Material has been omitted at this point.]

REFERENCES

Bennie, J., and R. Kidston: On the occurrence of spores in the Carboniferous formation of Scotland. Proc. Roy. Phys. Soc. Edinb. 9, 82-117, pls. 3-6. Edinburgh (1886) 1888.

Campo-Duplan, M. van: Remarques sur les grains de pollen de quelques plantes aquatiques. Bull. Soc. Bot. Nord France 4, 36-39. Lille 1951.

Chaloner, W. G.: Mississippian Megaspores. Contrib. Mus. Paleontol. Univ. Michigan 12, 23-35, 2 pls., 1 fig. Ann Arbor 1954.

Chitaley, S. D.: Fossil Microflora from the Mohgaon Kalan Beds of the Madhya Pradesh, India. Proc. Nat. Inst. Sci. India 17, No. 5, 373-381, 2 pls. Delhi 1951.

Cookson, J. C.: Plant Microfossils from the Lignites of Kerguelen Archipelago. B.A.N.Z. Antarctic Research Expedition 1929-1931, Reports-Series A, 2, 127-142, 3 pls. Adelaide 1947.

Erdtman, G.: Suggestions for the classification of fossil and recent pollen grains and spores. Särtryck ur: Svensk Bot. Tidskr. 41, Heft 1, 104-114. Uppsala 1947.

——— On Pollen and Spore Terminology. The Palaeobotanist 1, 169-176, 7 figs. Ranchi, India, 1952.

Guennel, G. K.: Fossil spores of the Alleghenian coals in Indiana. Indiana Dep. of Conservation, Geol. Surv., Rep. of Prog. No. 4, 1-40, 30 pls. Bloomington, Indiana, 1952.

Härtel, O.: Färbungsstudien an der pflanzlichen Kutikula. Protoplasma 41, 1, 1-14. Wien (Springer) 1952.

Horst, U.: Mikrostratigraphischer Beitrag zum Vergleich des Namur von West-Oberschlesien und Mährisch-Ostrau. Die Mega- und Mikrosporen der haupstächlichen Flöze beider Reviere. 9 Tafeln. Dissertation TH Berlin 1943.

Ibrahim, A. C.: Sporenformen des Ägirhorizontes des Ruhrreviers. Dissertation TH Berlin 1932, 46 S., 8 Tafeln. Konrad Triltsch, Würzburg 1933.

Iverson, J., and J. Troels-Smith.: Pollenmorfologiske definitioner og typer. Danmarks Geol. Unders. 4, 3, No. 8. Kopenhagen (Reitzels Verlag) 1950.

Kosanke, R. M.: Pennsylvanian spores of Illinois and their use in Correlation. State Geol. Surv. Bull. 74, 1-128, 16 pls. Urbana, Illinois, 1950.

Luber, A. A., and J. E. Waltz.: Classification and stratigraphic value of some carboniferous coal deposits in the USSR. Trans. Centr. Geol. Prosp. Inst. 105, 1-45, 10 Tafeln. Moskau 1938. (Russisch, kurze engl. Zusammenfassung.)

Mädler, K.: Azolla aus dem Quartär und Tertiär. Geol. Jb. 70, 143-158, 1 Tafel. Hannover 1954.

Naumova, S. N.: Spores and Pollen of the Coals of the USSR. XVII. Internat. Geol. Congr., Absts. Papers, USSR 1937 (Chief Editorial Office of the mining-fuel and Geological-Prospecting literature, Moskau and Leningrad), 60, 61. 1937.

Niggli, P.: Probleme der Naturwissenschaften, erläutert am Begriff der Mineralart. Basel 1949.

Pant, D. D.: Suggestions for the classification and nomenclature of fossil spores and Pollen grains. Bot. Rev. 20, 33-60. 1954.

Pflug, H.: Zur Entstehung und Entwicklung des angiospermiden Pollens in der Erdgeschichte. Palaeontographica 95, Abt. B, 171 S. Stuttgart 1953.

Potonié, H.: Grundlinien der Pflanzenmorphologie im Lichte der Paläontologie. Jena (G. Fischer) 1912.

——— Zur Morphologie der fossilen Pollen und Sporen. Arb. Inst. Paläobot. Petrogr. Brennst. 4, 5-24. Berlin 1934.

—— Zur Mikrobotanik des eocänen Humodils des Geiseltals. Arb. Inst. Paläobot. Petrogr. Brennst., Preuß. Geol. L.-A. Berlin, 4, 25-125, 6 Tafeln. Berlin 1934.

—— Die Moorpflanzengesellschaft des Karbons und der Rhythmus ihrer Wandlungen. Paläont. Z. 24, 166-183. Stuttgart 1951.

—— Revision stratigraphisch wichtiger Sporomorphen des mitteleuropäischen Tertiärs. Palaeontographica 91, Abt. B, 131-151, 2 Tafeln, 1 Tabelle. Stuttgart 1951.

—— Die Mikrobotanik der Kohle auf dem Internationalen Botanikerkongreß zu Stockholm im Juli 1950. Geol. Rundschau 39, 247-249. Stuttgart 1951.

—— Zur Morphologie und morphologischen Nomenklatur der Sporites. Paläont. Z. 25, 143-154, 3 Abbildungen, 1 Tafel. Stuttgart 1952.

—— Zur Systematik isolierter Sporen. Svensk Bot. Tidskr. 46, 158-173. Uppsala 1952.

—— Die Bedeutung der Sporomorphen für die Gesellschaftsgeschichte. Compte Rendu, Heerlen 1951, 501-506. Maestricht 1952.

—— Gesichtspunkte zu einer paläobotanischen Gesellschaftsgeschichte (Soziogenese). Geol. Jb., Beiheft 5, 115 S., 8 Abbildungen. Hannover 1952.

—— Stellung der paläozoischen Sporengattungen im natürlichen System. Paläont. Z. 28, 103-139, Tafeln 9-13. Stuttgart 1954.

—— Les spores des plantes paléozoiques dans le système naturel (morphologique). Lejeunia, Rev. de Bot. 18, 5-20, 5 pls. Liége 1954.

—— Gibt es angiospermide Eigenschaften an paläozoischen Sporen? Svensk Bot. Tidskr. 28, 328-336, 7 Abbildungen. Uppsala 1954.

—— Zur Biologie der karbonischen Sporen. Paläont. Z. 29, 27-32, 1 Tafel. Stuttgart 1955.

Potonié, R., A. Ibrahim, and F. Loose.: Sporenformen aus den Flözen Ägir und Bismarck des Ruhrgebietes. N. Jb. Min. usw., Beil.-Bd. 67 B, 438-454, 7 Tafeln. Stuttgart 1932.

Potonié, R., and W. Klaus.: Einige Sporengattungen des alpinen Salzgebirges. Geol. Jb. 68, 517-544, 1 Tafel, 10 Abbildungen. Hannover 1954.

Potonié R., and G. Kremp.: Dis Gattungen der paläozoischen Sporae dispersae und ihre Stratigraphie. Geol. Jb. 69, 111-193, 17 Tafeln. Hannover 1954.

Potonié, R., P. W. Thomson, and F. Thiergart.: Zur Nomenklatur und Klassifikation der neogenen Sporomorphae (Pollen und Sporen). Geol. Jb. 65, 35-70, 3 Tafeln, 1 Abbildung. Hannover-Celle 1950.

Potonié, R., and H. Venitz.: Zur Mikrobotanik des miozänen Humodils der niederrheinischen Bucht. Arb. Inst. Paläobot. Petrogr. Brennst., Preuß. Geol. L.-A., 5, 5-54. Berlin 1934.

Reinsch, P. F.: Neue Untersuchungen über die Mikrostruktur der Steinkohle des Carbons, der Dyas und Trias. Leipzig (T. O. Weigel) 1881.

Schopf, J. M.: Spores from the Herrin (No. 6) coal bed in Illinois. Report of Invest. No. 50 of the Geol. Surv. of Illinois, 1-55, 8 pls. Urbana 1938.

—— Two new lycopod seeds from the Illinois Pennsylvanian. Trans. Illinois State Acad. Sci. 30, 139-146. 1938.

Schopf, J. M., L. R. Wilson, and R. Bentall.: An annotated synopsis of paleozoic fossil spores and the definition of generic groups. Illinois Geol. Surv. Rept. Inv. 91, 1-66, 3 pls. Urbana, Illinois, 1944.

Schulze, G. M.: Beiträge zur deskriptiven Terminologie. Bot. Jb. 76, 109-133. Stuttgart 1953.

—— Internationaler Code der botanischen Nomenklatur, angenommen vom 7. Internationalen Botaniker-Kongreß Stockholm 1950. Deutsche Fassung im Auftrage des Botanischen Gartens und Museums Berlin-Dahlem. Berlin 1954.

Seward, A. C.: Fossil plants. Cambridge 1910.
—— Antarctic fossil plants. Nat. Hist. Report British Antarctic ("Terra Nova") Exped. 1910. Geol. 1, No. 1, 1–49. 1914.
Thomson, P. W., and H. Pflug.: Pollen und Sporen des mitteleuropäischen Tertiärs. Palaeontographica 94, Abt. B, 1–138, 15 Tafeln. Stuttgart 1953.
Zerndt, J.: Les mégaspores du bassin houiller polonais. I. Bull. de l'Acad. Pol. des Sci. et des Lettres. Trav. Géol., 1–56, 32 Tafeln. Krakau 1934.

9

Reprinted from pp. 7-10 and 81-90 of *U. S. Bur. Mines Rep. Inv.* **5151**:1-107 (1955) through the courtesty of the Bureau of Mines, U.S. Department of the Interior

POLLEN ANALYSIS OF THE BRANDON LIGNITE OF VERMONT

A. Traverse

[*Editors' Note:* In the original, material precedes these excerpts.]

THE USE OF POLLEN ANALYSIS FOR STUDY OF A TERTIARY SEDIMENT

Fossil pollen and spores are botanically determinable, just as are other plant fossils. These microfossils have readily distinguishable and clearly demarcated characteristics. In addition, preparations of pollen and spores can be made easily from very small samples of sediment collected from carefully marked locations. Each preparation gives a rapid survey of the flora of that part of the sediment, and these preparations can be easily compared. It is true that pollen is produced in different quantities by different species and is differentially destroyed by aerobic decay. It is presumably differentially carried from plants to the basin of deposition, depending on proximity, on whether the plants are anemophilous or entomophilous, and on the size, shape, and density of the grains. But a record of a species, genus, or family _is_ a record, if the pollen is accurately determined, and students of Tertiary plants are _primarily_ interested in the total flora of a deposit - the significance of overrepresentation and underrepresentation is, hence, much less than in the study of Pleistocene and Recent sediments. A rapidly expanding literature on Tertiary pollen shows that alterations of climate in the Tertiary _are_ reflected by changes in the pollen floras deposited, and that fluctuations of local conditions in the basin of deposition - changes in water level and so forth - are also reflected in the microfossil spectrum.

Botanical interest in pollen grains dates at least as far back as the work of the great anatomists Nehemiah Grew and Marcello Malpighi (106, pp. 15-23), and many 19th century botanists, in their eclectic manner, studied pollen more or less assiduously. Within the last 50 years there has been a great expansion of interest in the study of pollen of Quaternary deposits for the purposes of dating and to increase knowledge of the migrations of floras (and faunas, including man) with climatic alteration. However, despite the fact that the first studies of Tertiary pollen were published over 100 years ago by Ehrenberg, in 1838, and Goppert, in 1841, there has been little development of Tertiary palynology until the last 2 decades (31, p. 215). Most of the work has been done on the commercially important Braunkohle of Germany. Much has already been accomplished by investigating pollen and spores in unraveling the stratigraphy of European Tertiary deposits. This research has been fruitful also as a paleoecological tool. Potonié (63) has well summarized this unified outlook of German students of Tertiary pollen. He shows that the stratigraphic importance of the technique of Tertiary pollen analysis has been responsible for its development, but he also points out the great contributions the technique has made to our knowledge of the climate of Europe during the Tertiary. Thomson's papers (87, 89), though written by one who has a great practical interest in Tertiary stratigraphy, are strongly paleoecological in bent. Most of the European investigators have attempted as far as possible to make their results contribute to our total knowledge of the Tertiary.

Use of Tertiary Pollen Analysis for Stratigraphic Purposes in Coal and Petroleum Research

The use of fossil pollen and spores for stratigraphic correlation is rather well developed in Europe, especially in Germany. Although stratigraphic use of plant microfossils was first developed in the Carboniferous coal fields, German coal petrographers and paleobotanists also make use of the abundant pollen in the Tertiary brown coal deposits for identifying and correlating beds. Although the German investigators of pollen have been for the most part people with botanical training and interests, stratigraphic expediency has had a role in the development of German Tertiary palynology. Where stratigraphy is the paramount consideration, there is naturally a tendency for botanic relationships and paleoecologic analysis to be neglected. Using Tertiary pollen and spores in this manner is rather like using palynologic techniques for stratigraphic purposes in investigating Paleozoic or Mesozoic sediments, a technique that is highly developed in the stratigraphic correlation of coal beds of the European Carboniferous and also is being used in North America.

With Carboniferous pollen and spores, however, there is much less opportunity to correlate palynological with other botanical information. It is relatively much more difficult to determine what sort of a plant a given Carboniferous spore came from than it is to determine the botanical origins and relationships of Tertiary pollen. Even from a purely stratigraphic viewpoint, a good argument can be made that the Tertiary palynologist should make full use of the available botanical information about his material. There are two principal reasons for this:

1. Study of the pollen of living plants will help the palynologist of Tertiary sediments be much more competent in interpreting forms found in the sediment, inasmuch as the range of form and structure in Tertiary pollen is very nearly the same as that found in the living flora of the world.

2. The concept of index fossils, so significant and useful in studying Paleozoic sediments, is of much less utility in Tertiary work because many extant genera have been in existence since the beginning of the Tertiary, and migration of genera and alteration of plant associations have been more important factors than extinction of genera and families in the change of the world flora during the Cenozoic era. In other words, the units that the Tertiary palynologist is most likely to find useful are something on the order of index floras rather than index species or genera. This means that a knowledge of Tertiary floral assemblages, both extant and fossil, is very useful to the Tertiary pollen analyst, even if his ultimate interests are in purely stratigraphic study.

Applications of Pollen Analysis of Tertiary Coals to Paleoecology and Coal-Paleobotany

The fact that most Tertiary plants belong to modern taxonomic units increases the versatility of the method for Tertiary sediments along lines more familiar to the student of Pleistocene and Recent pollen. Paleoecology depends on obtaining the maximum possible information about the fossil flora, and the microfossil flora is an integral part of the whole. Because of the speed with which preparations of fossil pollen from a Tertiary coal can be made, palynology is a most rewarding technique in terms of yield of information about the flora that contributed to coal. Paleoecological conclusions derived from knowledge of the total flora of a coal can be expected to contribute to the understanding of many features of a sediment such as a low-rank coal that is almost entirely made of plant material, the deposition of which depended on climatic and edaphic factors intimately related to the nature of the contributing flora. For example, abundance of pollen of certain types of swamp plants in a coal of Tertiary age will allow one to draw conclusions about the contributing flora and about conditions in the basin of deposition during formation of the coal. This may help greatly in understanding the nature and origin of the sediment.

Value of Palynology to Applied Petrography of Tertiary Coals

There are several important ways in which pollen analysis of Tertiary lignites can also make a contribution to coal petrography. As has already been mentioned, there is the importance of the microfossils for stratigraphic purposes. Stratigraphic use of Tertiary pollen and spores is already important in the German brown coal fields. The technique will become more important in the Western American Tertiary coal fields when the western coals are mined on a larger scale.

On the other hand, there is the value of knowing the probable botanic origin of the microstructural components of a coal being investigated. The two primary petrographic or microstructural constituents of lignite are anthraxylon and attritus. Anthraxylon is that part of the coal made of plant tissues, such as wood, bark, endocarps, leaves, roots, etc., providing that these tissues retain cellular structure, are translucent in thin sections, and have not been macerated extensively in the course of deposition and alteration of the original peat. In the Paleocene Fort Union lignite of the Dakotas and Montana, which makes up most of our lignite resources, the anthraxylon is predominantly wood. Attritus is that part of the coal consisting of very small organs such as pollen grains and spores and other waxy tissue parts, especially cuticles; also included in attritus are highly macerated or otherwise altered tissues. Some of this attrital material is translucent in thin section, but some of it is more highly carbonized and is opaque. Investigation of the fossil pollen and spores in a coal can contribute very significantly to the overall knowledge of the flora that yielded these petrographic components.

Knowledge of the original flora may be important in understanding the sedimentary history of a Tertiary coal. Also, inasmuch as the various types of plant material in lignite still retain chemical individuality, owing in part to their botanic origin, knowledge of the contributing flora is likely to help directly in understanding important chemical features of the coal. The possible fruitfulness of this approach to lignite microscopy is due to the fact that living equivalents, or more or less close relatives, of the plants represented in the coal are still extant and can be comparatively studied and also to the fact that the coal is very young, relatively, and chemical individualities of different plant tissues are still a factor to a greater extent than is true of higher rank coal. In other words, the anthraxylon and attritus of which a lignitic coal is composed can have features that are best explained and understood in terms of the sort of plants from which these petrographic components originated. Information derived by pollen analysis about the flora represented in a coal can be helpful in ascertaining the botanic origin of the anthraxylon and attritus.

PREVIOUS AMERICAN WORK IN TERTIARY PALYNOLOGY

Very few publications of results of research on Tertiary spores and/or pollen from American deposits have appeared. Thiessen (100, p. 234) in describing the microscopic structure of an Arkansas lignite from thin sections wrote: "The spores are of great variety in kind and size.... Pollen grains are rather numerous, seemingly both of angiosperms and gymnosperms." But Thiessen made no detailed study of the microfossils. Miner (52) published a paper on Cretaceous and Tertiary coals in which he describes a species of spore of <u>Selagenellites</u> from the Fort Union Formation of Montana (Paleocene). Wilson and Webster (102) have described 1 bryophyte spore, 3 pteridophyte spores and 12 pollen grains of angiosperms and gymnosperms from the same formation. These were determined botanically with reference to living taxonomic units, but no effort was made by the authors in their paper to draw paleoecological conslusions, though this would admittedly be most difficult with such a small flora.

Wilson (101) reported to the International Geological Congress of 1948 on his extensive studies of Tertiary pollen in the western interior of the United States and the paleoecologic and geologic results of the work. This research is as yet unpublished. Winchester (103) noted that C. A. Davis had done some work a few years before on microfossils observed in thin sections of the shales of the Green River formation (Eocene of Colorado, Utah, and Wyoming). Winchester published some notes of Davis, which mention probable moss spores and pollen of the conifers Picea and Pinus as well as pollen of angiosperms. Bradley (13) somewhat expanded the work of Davis on the Green River material using the sections Davis had prepared and additional sections. Bradley made an effort to determine the pollen botanically, but few of his determinations are more than suggestions. He illustrated and briefly described 8 forms of "moss and fern" spores, pollen of Pinus and Thuja, and 8 forms of angiospermous pollen. Later Wodehouse studied Bradley's Green River sections. Wodehouse's two papers (104, 105) on the pollen of the Green River formation pioneered in the study of American Tertiary pollen. No similar studies have been published on other American Tertiary deposits during the two decades that have followed publication of Wodehouse's work. Wodehouse made an effort to correlate his findings with what was already known about the Green River flora and to draw whatever paleoecological conclusions he was able to from the pollen flora. He found, for example, that his pollen flora was considerably less tropical in aspect than was the previously described (megafossil) flora of the Green River shales.

Wodehouse did not describe species that he was not able to identify. He commented (105) that all but one of his 43 species could be assigned to living genera and species, and that there were at least twice as many more species awaiting identification, which he did not include in his paper. The writer believes that all of the distinct, well-preserved entities should be described, illustrated, and identified as nearly as possible, for three reasons:

1. For stratigraphic purposes there should be available a complete record of the distinct forms of pollen and spores in the deposit, whether or not identified. (This could be very important, if the age of the deposit is well established or later becomes so established.)

2. Other investigators might be stimulated to aid in identifying the unknowns.

3. For paleoecological purposes it is very useful to know what percentage of the total number of entities the identified entities comprise. (To date the deposit by Barghoorn's method (6), for example, this would be essential.)

AIMS OF THIS INVESTIGATION

Barghoorn and Traverse felt from the beginning of this research that investigation of the pollen and spores of the Brandon deposit ought to add much to our knowledge of the flora and its environment. Investigation was begun along two lines:

1. Identification, to the smallest possible taxonomic units, of the pollen and spores of the deposit. This would add to our knowledge of the total composition of the flora, with a view to determining its probable age and paleoecological characteristics.

2. A survey of the various parts of the deposit to establish whether they are separable by means of the microfossils and if possible, whether this separation permits generalizations about the sequence of events in the basin of deposition.

[*Editors' Note:* Material has been omitted at this point.]

APPENDIX I: PROBLEMS OF NOMENCLATURE

A. Development of Tertiary Pollen and Spore Nomenclature

A very troublesome barrier in research on Tertiary pollen is the "legal" problem of what names to give the fossil pollen. Some sort of nomenclatural system is indisputably necessary to make it possible to refer to microfossils previously described and to permit systematization of knowledge about Tertiary pollen.

Nomenclatural problems are more difficult in Tertiary palynology than in the study of either pre-Tertiary or of Quaternary pollen and spores. Quaternary plant fossils almost always belong to living genera, and an investigator can attach modern generic names to the fossils with confidence, and often, where a limited flora is concerned, he can use modern specific names safely. In Cretaceous and earlier deposits there are few plants that can be placed in modern genera. The paleontologist can therefore put pre-Tertiary pollen and spores in organ-genera, without fear that he has created artificial genera for extant genera that he happens not to have in his reference collection of extant forms.

With Tertiary pollen there is clearly a danger that one will fail to recognize that fossils belong to extant genera, and that organ-genera will be created for pollen beyond question referable to extant genera. This is obviously a taxonomic procedure to be avoided. Certain Tertiary palynologists unaccountably endorse this as a general practice and would place in organ-genera even Tertiary pollen known to belong to extant genera - even though this results in combining unrelated forms of known relationships in one organ-genus (see later discussion).

The Tertiary palynologist who follows good taxonomic practice will also be worried about putting into organ-genera pollen that he suspects may belong to an extinct member of a modern family. But that is admittedly unavoidable until more information is available about extant pollen - and until reference collections are much larger.

Furthermore, there is some danger that Tertiary pollen referred to an extant genus may belong instead to a related, extinct genus with seemingly indistinguishable pollen. Closely similar pollen is sometimes produced by related, or worse, even by unrelated genera. (Careful study, however, almost always reveals that such forms are similar, not identical.) On the other hand, different species of the same genus and even different individuals of the same species may produce pollen that differs to some extent. When one is working with associations of plants that still exist, these are not overwhelmingly difficult problems because one knows what genera to expect. But in Tertiary, especially early Tertiary, deposits, where evolution and extinction have intervened since deposition, it is more difficult to evaluate observed similarities and differences between pollen grains. It is obvious that the nomenclature of Tertiary pollen is a very difficult subject.

There has been much variety of practice in fossil pollen and spore nomenclature and taxonomy - and, inevitably, controversy. But nearly all Tertiary palynologists agree that the fossil pollen forms must have some sort of names to facilitate general palynologic research and for convenience of stratigraphers who use fossil pollen and spores for correlation purposes.

In work with pre-Tertiary pollen and spores, the argument in favor of artificial organ-genera is overwhelming. The only opportunity of learning the botanic relationship of Pennsylvanian spores is to find them in a fructification of known affinity.

When that occurs, the spores must henceforth be known by the name of the fructification. But for the most part it is essential to describe Paleozoic and Mesozoic pollen and spore forms as species of organ-genera independent of the natural phylogenetic system. As Arnold (2, p. 65) has pointed out, stratigraphers make the most use of a system for classification of pre-Tertiary spores and pollen. They find most desirable an artificial system of organ-genera; that is, of genera defined solely by arbitrary morphological features. As the true relationships of the forms are discovered, the species can be transferred to the appropriate natural genera, which have already been described for megafossils. In the foreseeable future, this transfer of species will be on a very modest scale.

Some investigators of Paleozoic pollen and spores have abandoned normal botanic practices altogether and have used systems of letters and numbers for classification (69). Similar symbolic systems are in use by some Tertiary palynologists whose work is strictly stratigraphic. As long as systems of this sort are for the private use of stratigraphers only, there can be no objection. But comparability of results, as well as the aim of eventual piecemeal incorporation of fossil spore and pollen forms into the natural system of classification, demand that monographing of these microfossils follow the normal practices of paleontologic taxonomy.

In Tertiary palynology, the present taxonomic situation is chaotic, principally because some of the foremost authorities on Tertiary pollen have introduced systems that are compromises between a purely artificial system of letters and other symbols, and the normal taxonomic system of paleontology and other branches of biology. But if a system of nomenclature for Tertiary pollen and spores is to be of general biological use, it must conform to the laws of botanical nomenclature. Hybrid systems that use Latin names but do not adhere to the international rules are misleading and would eventually cause chaos.

B. The Principal Systems Proposed and Used in Publications on Tertiary Pollen and Spores

1. Wodehouse (105) in his paper on the Green River Shale, devised a method for altering modern family, genus, and species names. The name of the smallest unit possible of extant plants is changed by adding *pites* to the root; this name is then used as a genus name. For example, *Ericacipites longisulcatus* is an ericaceous pollen form not referable to an extant genus, for which *longisulcatus* is Wodehouse's specific name.

Providing that types are properly designated and the descriptions are adequate, such genera would be valid organ-genera. But the system does not provide for the presumably large number of fossil forms that will eventually be described for some families. Not all ericaceous pollen of unknown affinity to extant genera can logically be put in a single genus *Ericacipites*.

Wodehouse's system, as such, has been little used, but his idea of modifying extant names for use with Tertiary pollen apparently has influenced others.

2. Erdtman (32, 33) favors the use of names based on morphological characters only. He feels that the imponderables are too great for Tertiary pollen to be worked into modern taxonomic groups. He suggests that Tertiary pollen and spores ought to be dealt with in terms of "sporomorphs", that is, "shape groups", because of the variability of form found within one "pollen-species" and, conversely, the possibility that one sort of pollen may represent more than one species or even more than one genus. He writes (32):

> When studying the microfossils of old deposits, such as Tertiary...a
> pollen analyst cannot determine the pollen grains and spores in the
> same way as when dealing with more recent material. Spore and pollen
> determinations in the usual sense of the word must here as a rule be
> replaced by determinations expressed in terms of artificial spore
> and pollen classification....

Speaking of names that attempt to ally Tertiary spores and pollen with living groups, he says that they "...ought to be replaced by names saying more by saying less."

He suggests a "nomen typicum abstractum," such as Hexaporites (six-pored), describing the morphological type of the pollen forms (sporomorphs). This name would correspond to a genus, according to Erdtman. To this would be added "nomen differentiale," such as oblata, corresponding to a specific name. Between these two names a "nomen typicum concretum" can be (but would not always be) intercalated parenthetically. This name would show similarity of the pollen to other pollen. For example, "Fagidites" could be used as a "nomen typicum concretum" to show similarity to Fagus pollen. These names are made by adding the suffix idites to the root of the name of a modern genus or family. The use of these names with the suffix idites, according to Erdtman, implies nothing about the true relationship of the pollen - but merely points out a morphological similarity of the pollen.

The exact relationship of this system to the ordinary biologic system is somewhat mysterious. Erdtman (32) emphasizes that his system is only a preliminary suggestion, and it is often stated that the system is separate from normal taxonomy, but others have employed it for describing Tertiary pollen, and Erdtman himself (33) describes a Jurassic pollen form using the method, namely: Tricolpites (Eucommiidites) Troedssonii nova sporomorpha. Are these "sporomorphae" to be regarded as species or not? In a future revision, is Troedssonii the valid specific name, or not? The method of description is as if a new species were being described. Plenty of illustration is provided, but no type is designated, and the parenthetical "nomen typicum concretum" is taxonomically illegal. Erdtman apparently regards the description of sporomorphs by this method as not part of the regular botanic system. But a difficult problem is created for future paleontologists.

3. Modifications of Erdtman's system.

    a. Dr. Isabel Cookson, the Australian student of Tertiary pollen, has published papers that reflect several different points of view. In one (18), she refers several Tertiary pollen forms to a living genus, Nothofagus, and assigns the several presumed species letters rather than specific names. Hence, this stops just short of normal taxonomic procedure. Presumably, if there is enough reason to separate a form as "a" or "b", it could as well have a specific name.

In another study (19), she adheres strictly to the Erdtman system, using, for example, the name Monoporites (Graminidites) media for an almost certain grass pollen form.

More recently (20, 22, 23, 25, 26, 27), in describing pollen from Australian Tertiary deposits, Dr. Cookson continues to use Erdtman's concept of sporomorphs and his nomina typica abstracta for unidentified forms, but she uses Erdtman's "nomina typica concreta" as if they were ordinary genus names, wherever the relationship is known. Proteacidites tubercalatus, for example, is a proteaceous pollen form, for which tubercalatus is Cookson's specific name. (Erdtman (32) had suggested that the generic name, or "nomen typicum abstractum," Tricolporites, could be omitted, and the "nomina typica concreta" used, out of parentheses, as if it were a genus name.) Cookson uses these names as ordinary genus names in every way and also uses names such as Dacrydiumites mawsonii for pollen known to belong to Dacrydium. Here it is apparent that Dacrydiumites is no "nomen typicum concretum" but is merely a modification of Dacrydium, indicating that the genus is represented by a fossil pollen grain. In still other instances, Cookson (24) describes Tertiary pollen as a species of a modern genus, without modification of the genus name, for example: Acacia octosporites. It is apparent that Cookson is dubious about the practice of putting fossils known to belong to an extant genus in a genus such as Dacrydiumites. In her work with megafossils she unhesitatingly puts Tertiary material in extant genera.

b. Rao and Vimal (71, 72) use a modification of the Erdtman system, in which pollen forms are referred to the appropriate "nomen typicum abstractum" of Erdtman and then given numbers within that group: Triporites sporomorph 1, etc., or Tetracolpites, type No. 29, etc. Since this has no pretension of being valid permanent description, it does no harm. Eventually, the fossils should be properly described and named.

4. Couper (28) recognized the paleontologic morass that would result from widespread use of Erdtman's various "nomina", without clear definition of where the names stand in the botanic system. In his work he has followed conventional paleontologic practice in putting Tertiary pollen and spores in extant genera where possible, in new genera (for example, Palmidites) of extant families where proper, in organ-genera (for example, Perotriletes) where that was called for. Couper also validated names of Cookson's that were of uncertain standing because of imperfect descriptions.

5. R. Potonié's many publications on the Tertiary pollen of Germany have had considerable influence on the nomenclature of Tertiary palynology. His earlier sytem of nomenclature was followed by most of the important German Tertiary palynologists in the 1930's and 1940's. In this system the pollen forms were treated as follows:

a. Where the family or genus relationship was known, the suffix pollenites was added to the family or genus name, and this combined expression was used as a genus. For example:

Ilex-pollenites iliacus R. Pot.
Ericaceae-pollenites roboreus R. Pot.
(The pollenites is often abbreviated poll.)

b. Where the botanic relationship was doubtful, a question mark was used. For example:

Leguminosae?-Poll. quisqualis R. Pot.

c. Where botanic relationship was unknown, species were described as members of the huge form-genus Pollenites. For example:

Pollenites granifer R. Pot.

d. In many other instances where pollen was of known or suspected botanic relationship, no attempt was made to give valid names, but the group to which the pollen was said to belong was stated without additional information. For example:

> Carya
>
> Corylus-Typ
>
> Myriophyllum?
>
> Ericaceae
>
> Oleaceen-Typ
>
> etc.

It is apparent that the pollen forms in group d. have not been botanically described. Furthermore, the compound pollenites names are somewhat irregular, and those with question marks are invalid. Also, the form-genus Pollenites is so all-embracing that its use is not very helpful.

6. R. Potonié, Thomson, and Thiergart (67) proposed a modification of Potonié's earlier system. The publication was stimulated by Erdtman's proposals (32). Potonié wrote the new proposal, and he elaborated some of his ideas in a later paper (66).

Potonié endorses Erdtman's idea of sporomorphs and "shape groups" (Gestalts-Gruppen). He agrees with Erdtman that the difficulty of identifying Tertiary pollen is very great. But Potonié insists that nomenclature of Tertiary pollen should indicate the maximum amount of information about the botanic relationships of the sporomorphs. Potonié would use Erdtman's system of form-genera based on morphological characters for sporomorphs of indeterminable botanic position, but Potonié would assign names indicating natural relationship wherever possible.

This seems reasonable, but Potonié's suggested method of accomplishing this end is very irregular and wholly unacceptable taxonomically. He would proceed as follows:

    a. The purely artificial system of form-genera, slightly modified from Erdtman, would be used for indeterminable forms.

    b. Where botanic relationship is known, pollen forms would be given names in the half-natural system ("angenähert natürliche system" or "halbnatürlichen system.") The names are derived from extant family, genus, and species names. In one paper (67), the suffix oid is used to modify existing family, genus, and species names to show similarity but not identity of the fossil pollen to an extant form. The suffix ites is added to the modified name to make a genus name. In a more recent paper (66), Potonié changes the generic ending oidites to oipollenites, to emphasize that the genus is a sporomorph. Examples:

> Cornaceoidae, a family similar to Cornaceae.
>
> Taxodioidites (Taxodioipollenites), an organ-genus for pollen resembling that of the family Taxodiaceae.
>
> Quercoidites (Quercoipollenites), an organ-genus of pollen showing similarity to that of Quercus.
>
> Fagus silvaticoide Typ, a pollen form of the genus Fagus, similar to the pollen of Fagus sylvatica L.

c. Where possible, pollen forms would be classified in an approximation of the regular taxonomic system. Potonié would use such genera as Nyssa-pollenites, Nyssaceae-pollenites, etc., where it is felt that the pollen clearly belongs to the indicated extant genus or family. The suffix pollenites is important, according to Potonié, to show that a fossil pollen form is described. Although Potonié refers to this as "Das natürliche System," it is evident that conventional taxonomic usage would not recognize the necessity or even permissibility of using Nyssa-pollenites as a substitute for Nyssa. Fossil leaves referred to Nyssa are not, after all, described as species of Nyssa-phyllites: And a genus such as Nyssaceae-pollenites is not a natural genus at all, since it could include forms from a variety of genera.

The most unusual feature of Potonié's newer proposals is that he would in many instances have single pollen forms listed under two names. If a form in a purely artificial form-genus became botanically better known, it would acquire a name in the half-natural system - without being transferred from its artificial organ-genus. In other words, the three systems, natural, half-natural, and purely artificial, would exist side-by-side: "Drei Systeme also stehen nebeneinander, und es ist bei der Bearbeitung der Sporen zu versuchen, stufenweise aus einem in das andere vorzudringen."(66, p. 170.)

Potonié's suggestion is that only the purely artificial names be considered as bound by the rules of nomenclature, priority, etc. (66, p. 170).

> Dabei muss das rein künstliche System als das einzige, wirklich alle fossilen Sporomorphen umfassende, stets mit seinem ganzen Inhalt als fester Ausgangspunkt bestehen bleiben. Alle Formen bleiben dort als an dem sichersten Platze und werden nur behelfsmässig in das halbnatürliche System hinübergenommen und dann wenn möglich je nach der Besonderheit der mit dem Fundpunkt der Spore verknüpften Tatsachen sekundär vom natürlichen System aus betrachtet. Man schreibt dann z. B.: In dieser oder jener Pflanzengesellschaft handelt es sich in Cicatricosisporites spm. bzw. in Mohrioisporites spm., um Mohria.

Cicatricosisporites is an artificial form-genus for Mohria-like spores. Mohrioisporites is Potonié's half-natural name for the same thing.

The confusion that would arise from using different names for the same species is contrary to the purpose of biologic and paleontologic taxonomy, and the use of names not bound by rules of priority would not be helpful and would tend to confuse.

7. Thomson and Pflug (94) have published the most thorough exposition of a system for classification and nomenclature of Tertiary pollen and spores yet to appear. In this system, all Tertiary pollen would be put in artificial genera, regardless of how much knowledge is available about the true relationships of a particular pollen form. (The authors would permit the use of extant generic names where the relationship is certain, but they regard this as exceptional.) Thomson and Pflug call the system the "morphological system." It is essentially the same as Potonié's "purely artificial system." Thomson's and Pflug's system consists of a thoroughgoing artificial nomenclature, in which each unit down to form-species is given an exact morphological definition. The units used are organ-genera and species, but it is intended that these should be equivalent to natural genera and species in size and so forth.

Thomson's and Pflug's main points are:

a. In the mid-Tertiary it is difficult to orient fossil pollen in the natural system; in the older Tertiary and below even family relationship is doubtful.

b. There can be only one nomenclature - this excludes the possibility of using Potonié's dual systems (half-natural and artificial). One form cannot have two names simultaneously.

c. The earlier literature on German Tertiary palynology, principally by R. Potonié, Thomson, and Thiergart, is full of taxonomic errors and invalid procedures. Genus names were changed as ideas of relationship changed, but the changes were not done according to the nomenclatural rules. Potonié's genera have not had sufficient diagnoses and are nomina nuda. Question marks have been used in some names, hence invalidating them. Type specimens have not been designated in most instances.

d. A thoroughgoing, artificial, morphological system is demanded by which all Tertiary pollen and spores can be classified. However, the nomenclature and taxonomy must follow the normal international rules of botanical procedure. The names applied will be strictly morphological, with no reference to taxonomic units of extant plants.

e. Two papers (61, 68) are made the basis for priority.

Specific epithets of authors published before a paper by Thomson and Pflug (94) will be retained if the author provided an illustration of the form. They will be retained even if sufficient descriptions and type specimens are lacking. In the future, adequate description will be required, and descriptions already implied in the name of the pollen form will not be considered valid. Type specimens must be designated in future publications.

f. If a Tertiary pollen form is clearly referable to an extant species, the modern name can be used. If referable to an extant genus but not to a species, the form can be given a specific epithet (organ-species) and put in the modern genus. Thomson and Pflug say that both of these occurrences are rare, and Pflug (94) does not put any of the many forms described in a modern unit, even where the generic identity is obvious.

8. Several systems of spore and pollen classification have been developed for use primarily with pre-Tertiary material and will not be discussed in detail here. Pant (54), for example, has presented a scheme of classification for use with all fossil spores regardless of geologic age. Pant notes, however,

> ....this is essentially a morphological classification meant only for scattered spores and pollen grains about whose affinities, as is true of many isolated spores found in the Paleozoic, it is difficult to say more than what their remaining form and characters...indicate....when we advance upwards into younger...beds we come across an ever increasing number of forms which can be assigned to some fossil or living genera or species.

In instances of such identification to fossil or living genera, Pant recommends using suffixes such as sporites, with the proper genus name - presumably to indicate the nature of the fossil. Pant's subgroup names are partly original and partly taken from other systems. Most of Pant's terms that would apply to Tertiary pollen are taken from the system of Naumova. Pant says,

If the characters are well preserved...and are sufficiently diagnostic, the names of the subgroups in this scheme...should be used as "generic" names, and size and other minor differences of morphology should...demarcate 'species'. If, however, the characters are ill defined and not such as to enable one to go down to a subgroup, the name of the next identifiable higher category may be used as the 'generic' name.

It is apparent that this is not permissible taxonomic procedure. A taxonomic unit cannot be a genus in one publication and something else in another publication.

Seemingly, Pant regards his system as merely a scheme for conveniently classifying microfossils, not a part of paleontological taxonomy as a whole. His statement about pollen of living plants would seem to make that clear: "...One must... realize that even in the case of spores and pollen grains of living plants...a perfect system of classification has not yet been devised...." Yet Pant includes some well established taxonomic units as part of his scheme. It is highly undesirable to confuse and intermingle a mere scheme of classification of fossil objects with valid nomenclature and taxonomy in this manner.

C. Discussion of Thomson-Pflug Nomenclature

Thomson and Pflug have made a considerable contribution to palynological science by their criticism of the chaotic, invalid practices developing in the mushrooming German literature on Tertiary pollen and spores. Their insistence that all new taxonomic units must have sufficient description, including an illustration, and that a holotype must be designated, can only have a beneficial effect. Also most praiseworthy is their stricture against the use of invalid names, such as those containing question marks, and against the careless changing of generic and other names without proper taxonomic procedure. Such practices would have terribly confused the literature if long continued.

Pflug's (94) proposals for morphologically defined organ-genera are quite acceptable and are surely the best contribution to Tertiary palynologic taxonomy yet made for description of forms for which botanical relationships are unknown or very doubtful.

However, the author is totally in disagreement with the idea and practice of putting all Tertiary pollen in artificial organ-genera, even when the botanic relationship is known. Pflug (94, p. 111), for example, describes an obvious Liquidambar pollen grain as Periporopollenites stigmosus (R. Pot.) n. comb. Of this form, Pflug says, "Botanische Zugehörigkeit zu Liquidambar ist sehr wahrscheinlich." A pollen grain obviously that of Tilia is described as Intratriporopollenites instructus (R. Pot. & Ven.) n. comb. (94, p. 89). Of this grain, Pflug says, "...Entspricht morphologisch völlig dem rezenten Tilia-Typus und dürfte der Gattung Tilia angehören."

In another publication (60), Pflug has the following statements:

(p. 122) Die im Jungtertiär bis in die Tegelen-Stufe häufig gefundene und als Subtriporopoll. simplex (R. Pot.) Pflug bezeichnete Pollengruppe gehört mit Sicherheit der Gattung Carya an...

(p. 127) Die Zugehörigkeit...Polyporo-poll. undulosus (Wolff) Pflug bezeichnet, zu den Ulmaceae ist so gut wie sicher.

There are many other examples in the papers similar to the above. Pollen known to belong to Tilia should not be described as belonging to another genus; and in instances where the family is certain but not the genus, the new genus should most certainly be described as a genus of that family and not placed in a genus such as Polyporo-pollenites, into which totally unrelated forms may also be placed.

Pflug reasons that Tertiary pollen is seldom referable to modern units, and that, hence, a thoroughgoing use of the artificial morphological system is safer and preferable to using the morphological system only where botanic relationships are unknown. For example, he writes (60, p. 119, in translation): "⎾although associated megafossils are known⏌ in no case may we speak of Normapolles as Myricales, because it is as a general rule completely inadmissible to place fossils from beds below the middle-Tertiary directly into the system of recent plants." One senses the circular argument here. A pollen form from the lower Oligocene could not be placed in the Myricaceae because lower Oligocene plants cannot be put in modern botanic units. Presumably part of the argument against the occurrence of modern units in the lower Tertiary would be that the Myricaceae do not occur below the mid-Tertiary.

A collaborator of Pflug's, the late Friedrich Mürriger, states:[9]

As the merely descriptive principle of Thomson and Pflug must not be mixed with other principles and must consequently ⎾be⏌ carried through, there can sometimes be the result of putting pollen grains into the same form-genus, although they are known to belong to different families. That is not fine, no doubt, but it is merely a defect of beauty, not a defect in the matter itself and, as I have said above, a necessary inconvenience and at the same time a minor inconvenience.

Pflug somewhat exaggerates the scarcity of lower Tertiary plants referable to modern families and genera, but that is beside the point. Even if only 5 percent of Eocene microfossils were referable to modern units, it would be utterly unsound biologically to put that 5 percent into artificial genera as a gesture to uniformity.

As Potonié has often emphasized, fossil pollen grains are distinguishable plant fossils and should be treated in the same manner as other fossils. He wrote (63, p. 113):

Hierzu muss man vor allem bei der Voraussetzung bleiben, dass ein Autor, welcher eine Pollenform beschreibt, dabei in derselben Weise vorzugehen hat, als beschreibe er ein Makrofossil.

(62, p. 371):...ich habe übrigens bei den Pollenform keine geringeren Unterscheide gesehen als bei meinen paläozoologischen Arbeiten, z.B. bei manchen Ammoniten.

Couper (28) has also emphasized that pollen and spore nomenclature must follow good paleontologic practice, and the nomenclature in his monograph (28) is an advance in that direction. Artificial organ-genera are used where relationship is unknown, but botanic relationship is shown wherever known. For example, if a pollen form is known to belong to an extant family but not to an extant genus, it is described as a species of a fossil genus of that family. These genera are, however, of the half-natural type, as per Wodehouse, Potonié, Cookson, and others. For example: Haloragacidites, a genus of the Haloragaceae.

---

[9] Letter from Friedrich Mürriger, 1953.

The half-natural generic names have two serious disadvantages. As Thomson and Pflug (94, p. 11) pointed out, if a half-natural genus such as Betulaceoipollenites is set up for pollen that subsequently proves to be myricaceous, one is forced to retain the valid but highly inappropriate and confusing name. Secondly, the half-natural names tend to be more inclusive than a genus. For example, Cookson's genus Podocarpidites is obviously larger than a genus, since it includes a wide variety of podocarpaceous pollen.

New genera of fossil pollen belonging to extant families should be described just as a new genus of the extant family would be described from new plant collections. The names of the genera should be distinct from other generic names in the family, both extant and fossil, and the genus should be a genus within the ordinary meaning of the term. So-called genera that are set up to include all fossil pollen believed to represent a certain extant family are not genera at all; they are neither organ-genera, form-genera, nor natural genera. Rather than set up such units the micropaleontologist would do better in doubtful cases to describe fossil pollen without formal reference to a genus.

D. Present Proposals

The most reasonable procedure is to follow botanic taxonomic norms. This is substantially what the Palynologic Conference in Stockholm, 1950 recommended (65).

In this paper the procedures have been:

1. Pollen inseparable from an extant genus is referred to that genus, with an appropriate specific name, the same procedure that would be followed with specifically distinct material collected from modern plants.

   Examples: Vitis forestdalensis sp. nov.
   Alangium barghoornianum sp. nov.

Note: The author has been conservative in not putting any of the Brandon pollen forms into extant species. Although there is no hard and fast rule that Tertiary pollen should not be placed in extant species, the author feels that such identification should not be made unless the corollary evidence is very strong.

2. Pollen felt to be certainly representative of an extant family but not in a known genus, fossil or extant, is referred to a new genus of that family. In coining the name of the genus, conventional botanic practice is followed.

   Example: Horniella gen. nov.
   Horniella clavaticosts sp. nov. (Rutaceae)

3. A well-characterized pollen form occurring abundantly in the deposit but of uncertain botanic relationships, is described as a member of an organ-genus, according to Pflug's (94) system. But note that the particular organ genera proposed by Pflug can only be used where they have been validly described. Kremp[10]/ states that many of Pflug's genera are invalid because the type species have no holotype. Other organ genera, validly described, must replace such invalidated names.

   Example: Tetradopollenites laxus sp. nov.

4. There are many pollen forms in the deposit not identifiable with extant or fossil genera or families and about which there is doubt because of inadequate information as to range or size and structure. These are figured and described but not formally referred to a species. All the fossil pollen forms in this monograph were given catalogue numbers, and these "unknowns" can be referred to by number.

   Example: BT-65.

---
[10]/ Discussion with Dr. Gerhard Kremp, 1955.

[*Editors' Note:* Material has been omitted at this point.]

REFERENCES

2. ARNOLD, C. A. Megaspores From the Michigan Coal Basin. Cont. Museum Paleontol. Univ. Michigan., vol. 8, No. 5, 1950, pp. 59-111.
6. BARGHOORN, E. S. Age and Environment: A Survey of North American Tertiary Floras in Relation to Paleoecology. Jour. Palentol., vol. 25, No. 6, 1951, pp. 736-744.
13. BRADLEY, W. H. Origin and Microfossils of the Oil Shale of the Green River formation of Colorado and Utah. Geol. Survey, U.S. Department of the Interior, Prof. Paper 168, 1931, pp. 45-46.
18. COOKSON, I. C. Pollens of Nothofagus Blume From Tertiary Deposits in Australia. Proc. Linnean Soc. New South Wales, vol. 71, Nos. 1-2, 1946, pp. 49-63.
19. ———. Plant Microfossils From the Lignites of Kerguelen Archipelago. B.A.N.Z.A.R.E. Rept., ser. A., vol. 2, pt. 8, 1947, pp. 127-142.
20. ———. Fossil Pollen Grains of Proteaceous Type From Tertiary Deposits in Australia. Australian Jour. Sci. Research, ser. B., vol. 3, No. 2, 1950, pp. 166-177.
22. ———. Difference in Microspore Composition of Some Samples From a Bore at Comaum, South Australia. Australian Jour. Bot., vol. 1, No. 3, 1953, pp. 462-473.
23. ———. The Identification of the Sporomorph Phyllocladidites With Dacrydium and its Distribution in Southern Tertiary Deposits. Australian Jour. Bot., vol. 1, No. 1, 1953, pp. 64-70.
24. ———. The Cainozoic Occurrence of Acacia in Australia. Australian Jour. Bot., vol. 2, No. 1, 1954, pp. 52-59.
25. COOKSON, I. C., and PIKE, K. M. A Contribution to the Tertiary Occurrence of the Genus Dacrydium in the Australian Region. Australian Jour. Bot., vol. 1, No. 3, 1953, pp. 474-484.
26. ———. The Tertiary Occurrence and Distribution of Podocarpus (Section Dacrycarpus) in Australia and Tasmania. Australian Jour. Bot., vol. 1, No. 1, 1953, pp. 71-82.
27. ———. The Fossil Occurrence of Phyllocoladus and Two Other Podocarpaceous Types in Australia. Australian Jour. Bot., vol. 2, No. 1, 1954, pp. 60-68.
28. COUPER, R. A. Upper Mesozoic and Cainozoic Spores and Pollen Grains From New Zealand. New Zealand Geol. Survey Paleontol. Bull. 22, 1953, 77 pp.
31. ERDIMAN, G. An Introduction to Pollen Analysis. Chronica Botanica, Waltham, Mass., 1943, 239 pp.
32. ———. Suggestions for the Classification of Fossil and Recent Pollen Grains and spores. Svensk Bot. Tidskrift, Bd. 41, H. 1, 1947, pp. 104-114.
33. ———. Did Dicotyledonous Plants Exist in Early Jurassic Times? Geol. Fören. i Stockholm Förh., Bd. 70, H. 2, 1948, pp. 265-271.
52. MINER, E. L. Paleobotanical Examinations of Cretaceous and Tertiary Coals. II. Cretaceous and Tertiary Coals From Montana. Am. Midland Naturalist, vol. 16, No. 4, 1935, pp. 616-625.
54. PANT, D. D. Suggestions for the Classification and Nomenclature of Fossil Spores and Pollen Grains. Bot. Rev., vol. 20, No. 1, 1954, pp. 33-60.
60. PFLUG, H. Zur Entstehung und Entwicklung des Angiospermiden Pollens in der Erdgeschichte. Palaeontographica, Abt. B, Bd. 95, Lfg. 4-6, 1953, S. 60-171.
61. POTONIÉ, R. Zur Mikrobotanik des eozänen Humodils des Geiseltales. Arb. Inst. Paläob. u. Petr. Brennstein 4, Preuss. Geol. L. A., 1934, S. 26-125.

62. ——. Stand der mikropaläobotanischen Tertiarstratigraphie. Ztschr. Deutschen Gecl. Gesellschaft, Bd. 100, 1948, S. 366-378.
63. ——. Zum stand der mikro-paläobotanischen Tertiarstratigraphie. Geol. Randschau, B. 37, 1947, S. 112-113.
64. ——. Pollen und Sporenformen als Leitfossilien des Tertiärs. Mikroskopie., Band 6, Heft. 9/10, 1951, S. 272-283.
65. ——. Die Mikrobotanik der Kohle auf dem internationalen Botanikerkongress zu Stockholm im Juli, 1950. Geol. Rundschau, Bd. 39, Heft. 1, 1951, S. 247-249.
66. ——. Zur Systematik isolierter Sporen fossiler Pflanzen. Bemerkungen zu den Besprechungen auf der Tagung der paläontologischen Gesellschaft in Trier 1951, in: Palynology. Aspects and prospects. III. Svensk Bot. Tidskrift, Bd. 46, H. 2, 1952, pp. 158-173.
67. POTONIÉ, R., THOMSON, P. W., and THIERGART, F. Zur Nomenklatur und Klassifikation der neogenen Sporomorphae (Pollen und Sporen). Geol. Jahrb., Bd. 65, 1950, 5. 35-70.
68. POTONIÉ, R., and VENITZ, H. Zur Mikrobotanik des miozänen Humodils des niederrheinischen Bucht. Arb. Inst. Paläob. u. Petr. Brennsteine 5, Preuss. Geol. L. A., 1934, S. 5-54.
69. RAISTRICK, A., and SIMPSON, J. The Microspores of Some Northumberland Coals, and Their Use in the Correlation of Coal Seams. Trans. Inst. Min. Eng. (London). vol. 85, 1933, pp. 225-235.
71. RAO, A. R., and VIMAL, K. P. Tertiary Pollen From Lignites From Palana (Eocene), Bikaner. Proc. Nat. Inst. Sci. India, vol. 18, No. 6, 1952, pp. 595-601.
72. ——. Preliminary Observations on the Plant Microfossil Contents of Some Lignites From Warkalli in Travencore. Current Sci. (India), vol. 21, 1952, pp. 302-305.
87. THOMSON, P. W. Beitrag zur Mikropaläontologie und Waldgeschichte des Neogens (Jungtertiärs) von Niedersachsen und Schleswig-Holstein. Neues Jahrb. Mineral., Monatsh., Abt. 13, Heft. 9-12, 1948, S. 364-371.
89. ——. Die Entstehung von Kohlenflözen auf Grund von mikropaläontologischen Untersuchungen des Hauptflözes der rheinischen Braunkohle. Braunkohle, Heft. 3/4, 1950, S. 39-43.
94. ——. Pollen und Sporen des mitteleuropäischen Tertiärs. Palaeontographica, Abt. B, Band 94, lfg. 1-4, 1953, S. 1-138.
100. WHITE, D., and THIESSEN, R. The Origin of Coal. Bureau of Mines, U.S. Department of the Interior, Bull. 38, 1913, 390 pp.
101. WILSON, L. R. Plant Microfossil Studies of Early Tertiary Deposits in the Western Interior of the United States (Abstract). International Geol. Cong., Rept. of 18th sess., Great Britain, 1948, Part 10, Proc. Sec. J., Faunal and Floral Facies and Zonal Correlation, London, 1952, p. 10.
102. WILSON, L. R., and WEBSTER, R. M. Plant Microfossils From a Fort Union Coal of Montana. Am. Jour. Bot., vol. 33, No. 4, 1946, pp. 271-278.
103. WINCHESTER, D. E. Oil Shale of the Rocky Mountain Region. Geol. Survey, U.S. Department of the Interior, Bull. 729, 1923, p. 32.
104. WODEHOUSE, R. P. Tertiary Pollen—I. Pollen of the Living Representatives of the Green River Flora. Bull. Torrey Botan. Club, vol. 59, 1932, pp. 313-34.
105. ——. Tertiary Pollen—II. The Oil Shales of the Green River Formation. Bull. Torrey Botan. Club, vol. 60, 1933, pp. 479-524.
106. ——. Pollen Grains. McGraw-Hill Book Co., New York, 1935, 574 pp.

10

Copyright © 1968 by Elsevier Scientific Publishing Co.

Reprinted from pp. 189-192, 203-206, 221-222, 230-248, 263-268, and 346-348 of *Rev. Palaeobotany and Palynology* 6:189-348 (1968)

PALYNOLOGY OF TERTIARY SEDIMENTS FROM TROPICAL AREAS

J. H. GERMERAAD, C. A. HOPPING AND J. MULLER[1]

Bataafse Internationale Petroleum Maatschappij, The Hague (The Netherlands)

(Received February 8, 1968)

[*Editors' Note:* The material omitted up to page 263 consists of figures 1 through 14 and 17, 18, and 19.]

SUMMARY

This article deals with the most important aspects of nearly twenty years of intensive study of the pollen-and-spore content of Tertiary sediments in some parts of tropical South America, Africa and Asia.

For a proper evaluation, the character of the data, including the selection and preparation of the samples, the diversity of previous recording and the statistically achieved uniformity in presentation of these basic data needs a full exposition, given in the introduction. This is directly followed by an explanation of the process of elimination of all stratigraphically unimportant species. The resulting interpretation of climatic and topographical influences on the dispersal of pollen and spores is illustrated with examples from the fossil record. The disturbing effect of redeposition forms a problem, which in some cases can be solved. Now that the main ways of dispersal of pollen and spores are understood, the characteristics of the three major depositional environments can be distinguished by purely statistical analysis, without necessarily having any botanical information from probably related Recent plant species. Additionally botany and palaeontology may bring supporting evidence. This many-sided approach leads to the discrimination between local and regional features of environmental or time-stratigraphical significance which is needed for the evaluation of long-distance correlation.

As a result the marker species can be classified into: (*1*) a restricted number of pantropical marker species; (*2*) a larger number of marker species which occurred in both the South American and west African regions, tropical today (transatlantic distribution); and (*3*) a still greater quantity of species which are of significance only within a single botanical province (intracontinental distribution).

Thus a broad stratigraphical framework on a pantropical scale is established, which may be further subdivided regionally. These three systems of subzonation are compared with independent zoopalaeontological time-stratigraphical correlation and discussed in great detail, with special emphasis on the Carribean data.

---

[1] Present address: Rijksherbarium, State University, Leiden (The Netherlands).

The major palynological changes marking the boundaries of the pantropical subzonation are thought to reflect the evolution of new groups of plants. They are mostly marked by a gradual incoming of pollen types. Extinction of plants is stratigraphically of less value, since they may have survived longer in one area than in another.

Climatic boundaries are next in importance, but in general they are more restricted to specific regions. Similarly the immigration of plants, although producing sharp and useful boundaries, is only of regional value. Of least significance for regional correlation are the locally restricted boundaries which are caused by changes in habitat or dispersal. They may still be valuable for studies within one basin.

An intriguing aspect of the palynological studies is formed by the possible affinity of the fossil type with Recent botanical species. Such affinities are obviously present in many fossil types. Whereas most are restricted to the level of family relationship, some interesting cases of much closer affinity are recorded here. In exceptional cases the morphogenetic development and migration of a restricted group of related pollen types can be traced.

In the final section of this paper the species selected for this study have been formally described and illustrated; they include several new ones. The study is further documented by distribution charts and sections showing the stratigraphical significance of the marker types, as discussed in detail in the stratigraphical section.

INTRODUCTION

This paper presents some of the results obtained in twenty years of palynological investigation of Tertiary sediments from tropical areas by companies of the Royal Dutch/Shell Group. The primary purpose of this investigation was to arrive at a better interpretation of the stratigraphy of those Tertiary sedimentary basins where other means of correlation failed. Of course it is not possible in this paper to do more than present an outline of the large amount of work done and to select for discussion a few topics which appear to be of more general interest. Documentation will also be restricted to selected examples.

MATERIALS AND METHODS

The restriction of the subject of this paper to the Tertiary of tropical areas is due to a combination of factors. Firstly, the tropics represent a natural geobotanical unit. Secondly, palynological research by Shell was initially mainly concerned with the Tertiary sediments of northern South America, Nigeria and Borneo. Thirdly, the amount of detailed information on the Upper Cretaceous palynological succession in these areas is still very scanty and the base of the Tertiary represents, therefore, at the present state of knowledge, a logical lower limit. Pleistocene strata have been studied too, but Holocene sediments have been left out of consideration because of lack of information.

Geographically the main emphasis will be put on the South American data, which are by far the most extensive, covering Colombia, Venezuela, Trinidad and the Guianas. From Africa only the Nigerian Tertiary succession is adequately known. In the Far East, Borneo has supplied valuable information, but here the stratigraphical coverage is incomplete, data on the Eocene being very limited.

The number of samples on which the study is based, is very large. Only samples collected at the surface and cores and sidewall samples taken in wells have been studied. As far as possible sections were chosen which contained, or were situated in close proximity to, samples dated by other means.

Tropical Tertiary pollen floras are very rich in species and the average type collection may easily contain 800–1,000 different species. For stratigraphical purposes generally less than 200 are of importance per area. For a comprehensive review, such as this, a further reduction is desirable and only 49 species are discussed.

These are, firstly, the species used to establish the major zonation, and some which are of importance for elucidating local correlation problems. In addition a few species of intrinsic botanical interest are discussed.

Fossil species have as far as possible been referred to published ones[1]. The introduction of new names has thus been kept to a minimum.

Preparation of samples was mainly by standard techniques, modified according to local needs. For Neogene material treatment with HF, followed by bromoform separation, is generally sufficient, although when much plant debris is still present acetolysis may be helpful. For Palaeogene sediments slight oxidation with Schulze's mixture may be advantageous.

The majority of the sediments examined consisted of clay and shale, silt or siltstone with varying amounts of carbonaceous material. Sandy sediments or coals were generally avoided.

The basic data are presented on distribution charts accompanied by all pertinent lithostratigraphical and additional palaeontological information (Fig.1–14). Again it must be emphasized that only examples are given. A complete documentation would have been impossible because of lack of space. However, the range chart (Fig.15) presents a compilation of all the data assembled for this study.

Unfortunately, when palynological work was started in the different areas, no uniform system of recording data was adopted. Sometimes percentages were

---

[1] In this connection we should like to express our gratitude for the facilities extended to us by Dr. T. van der Hammen (Amsterdam) and his staff, which enabled us to study the type collections not only of already described species but also of types to be described in future publications.

calculated on the basis of a pollen sum; in other cases a rough estimate of the abundance of the species was given.

In order to create uniformity in the presentation of these quantitative data on our distribution charts, the occurrences are expressed as a value for the probability of re-observation, i.e., if a specified number of additional specimens from a new sample of the same rock or stratum were to be investigated. This re-observation of the species may be in any number of specimens (grains), including one only.

This chance of re-observation is computed according to the following formula:

$$P = 1 - (1 - \frac{a}{N})^M$$

where a = the amount, that is the number of grains of the single species observed in the old sample; N = the sum of the amounts of all species, that is the total number of grains or pollen sum observed in the old sample; M = the sum of the amounts of all species in the new sample.

Depending on the accuracy required by the palynological investigation and the time and sample material available, the value of M may be taken larger or smaller than 100. For technical reasons related to the type of computer at present in use for palynological data processing, the value of M is here taken to be 88. This conforms approximately to the current sample size of 70–100 specimens of the selected species. The advantage of such a relatively small sample size lies mainly in the possibility of examining large numbers of samples within a short time, which is essential for routine work, both stratigraphically and statistically, and is far more informative than a few samples with large pollen sums.

In general the counting of the marker species takes place in two phases. First, an analysis is made of the whole flora in which the dominant species like *Rhizophora* are also counted. The second part of the microscopical investigation is concerned only with the more rare but important markers, among which, e.g. Asteraceae (Compositae). To save time dominant species are no longer counted, but statistically it is not permissible to multiply the amount counted to get the corresponding figure. Therefore, the probability of re-observation of, for instance, *Rhizophora* pollen is calculated from the first pollen sum, but the probability of re-observation of the rare markers like *Echitricolporites spinosus* from the second pollen sum.

Example: I = count of the whole flora, including, inter alia, *Rhizophora* = 100 grains.
II = count of the special selection, excluding *Rhizophora* = 84 grains.
Amount of *Rhizophora* in "I" = 80 grains, probability of re-observation = $1 - (1 - 80/100)^{88} = 1.00$.
Amount of *Echitricolporites spinosus* in "I" = 1 grain, probability of re-observation = $1 - (1 - 1/100)^{88} = 0.590$.
Amount of *Echitricolporites spinosus* in "II" = 4 grains, probability of re-observation = $1 - (1 - 4/84)^{88} = 0.986$.

Since *Echitricolporites spinosus* is part of the special marker selection, the probability of re-observation of this species in I is of little interest and is not recorded on the distribution charts presented. This example also shows the importance of further counting of the selected species.

In order to avoid the impression of high accuracy and to facilitate visual evaluation of the charts, the probability values on the distribution charts (Fig.1–14) were grouped into classes and each class was indicated in the following way:

$P\ 0.01–0.75 = .$
$P\ 0.75–0.90 = /$
$P\ 0.90–0.95 = \circ$
$P\ 0.95–0.99 = \odot$
$P\ 0.99–1.00 = \bullet$

An interrupted vertical line on some distribution charts indicates a distribution of a species which was not counted but which appeared common to abundant in most samples of the interval examined.

INTERPRETATION OF DATA

The practical purpose of our investigations can be briefly stated as the establishment of a time-stratigraphical zonation, adapted to the scale of the geological problem under investigation. To achieve this goal a discrimination between stratigraphically reliable and unreliable pollen and spore types is necessary.

The problems may vary from detailed correlation between wells a few kilometres apart in order to solve, for instance, tectonical complications, to broad correlation between sediments of widely differing facies hundreds or thousands of kilometres apart in different basins of deposition. It is of special importance to evaluate, for every part of the zonation, the scale on which it is valid, together with an estimate of the degree of accuracy.

The selection of the 49 species discussed and the zonation established in this paper were preceded by a detailed study of the factors which influenced the regional significance of these markers.

It is of course impossible to present a full account of this elimination process. Instead an attempt will be made in this section of the present paper to outline the principles involved with the aid of a few concrete examples. First the factors influencing the composition of the pollen record in the rocks will be analysed. Then the data from other disciplines, viz. botany, geology and zoopalaeontology, which permit a better understanding of microfloral changes, will be discussed, leading finally to the establishment of a zonation.

The pollen spectrum, obtained by calculating the percentages in which the various species occur in a sample, is the end result of various factors which influence production and dispersal.

Primary factors control the presence of the plants which produced the pollen grains and spores, and secondary factors affect dispersal prior to fossilization.

Evolutionary change and migration

Primary factors are evolutionary change and migration or increases and decreases in response to changes in climate or topography.

Evolutionary change may obviously be expected to yield the most reliable criteria for the establishment of a time-stratigraphical framework of regional significance. Examples are the gradual development of Poaceae (Gramineae) and Asteraceae (Compositae), as reflected in the stratigraphical distribution of the corresponding pollen types. Since these are large and varied groups of plants with at present a virtually world-wide distribution, their first appearance and subsequent development have proved to be of wide time-stratigraphical significance.

An example of evolutionary development on a subregional scale is the evolution of *Sonneratia* pollen types in the Indo-Malesian area. As will be discussed in detail later, it was possible here to trace morphogenetic changes within a single lineage.

However, such cases are exceptional, and pollen and spore species generally appear fairly suddenly in the stratigraphical record without any indication of their phylogenetic origin. Unless coinciding with a hiatus in the stratigraphy, these relatively sudden appearances must have been caused by changes in the environment which enabled the parent plants involved to increase in numbers and extend their geographical range. The main factor of interest here is climatological. The problem of how to recognize the effect of changes in climate in the stratigraphical distribution patterns of the many species studied can be approached in a direct way by botanical identification of fossil pollen and spore types with living taxa, which have well-defined climatological tolerances.

So far not much success has been achieved in this direction, since none of the identified lowland species discussed in this paper are known to be indicative of a specific climate except in so far as they are adapted to the tropical climate in general. However, it is sometimes possible, by reference to climatologically significant lithological changes, to detect the relative climatic requirements of certain fossil markers. Such a case was present in the Oligo–Miocene (*Magnastriatites howardi* Zone) of western Venezuela. Here a local accumulation of anhydrite in shales, contrasting strongly with coal-bearing underlying strata, was associated with a peculiar pollen flora not encountered elsewhere in this composition. Here the inference of a climatological change towards greater aridity was possible.

Another case in which a climatological cause may be detected is the rather sudden disappearance of certain pollen species. The extinction of *Nypa*-type pollen at the Eo–Oligocene boundary in Venezuela is approximately contemporaneous with a widespread increase in the incidence of mottled shales, indicating a change to a climate with more pronounced seasonal rainfall. It is conceivable that such a change could have caused *Nypa* to disappear, since the plant at present is mainly restricted to the ever wet tropics in the Indo-Malesian area.

The often rather sudden immigration of certain species may also have been caused by a change in climate which enabled the plants involved to extend their range. An example which could be explained this way is the rather sudden appearance in the Eocene of western Venezuela of *Retitricolporites guianensis* and *Cicatricosisporites dorogensis*. However, such phenomena may also be due to the development of new pathways of migration. A good example of this is the southward migration along mountain chains of montane elements such as *Alnus*. In Borneo this could be attributed to a mid-Tertiary phase of mountain building. In the Caribbean area the cooling of the climate at the beginning of the Pleistocene must have facilitated the migration of *Alnus* along existing mountain chains from the north.

An analogous case is the dispersal across oceans of plants which are adapted to a coastal environment, such as mangroves.

If such floral changes are considered for the definition of a regional zonation, the question of dispersal speed must be taken into account. Although little is known quantitatively from Recent plants, it must be assumed that differences in this respect exist and that they depend on factors such as adaptation and viability of the seeds.

The conspicuously sudden appearance and subsequent rapid increase of Asteraceae (Compositae) pollen types in the Lower Miocene of northern South America, Nigeria and the Far East could reflect the adaptation to easy dispersal of Asteraceae seeds.

In contrast one distinct case of delayed dispersal has come to light: *Symphonia* pollen appears earlier in the stratigraphical record in Africa than in South America and this obviously is a consequence of the difficulty of crossing the Atlantic Ocean. It will be clear that such cases can only be detected when independent dating is available, and they should of course not be utilized for the definition of a regional time-stratigraphical zonation.

The majority of the floral changes observed on the distribution charts will, however, have been influenced to a varying degree by local facies factors.

For instance, the varying abundance in which mangrove pollen occurs in the sedimentary record, once it has made its first appearance in the area, largely reflects the acreage covered by mangrove, which is determined by the local topography.

It will now be clear that, if climatological and topographical influences are of restricted lateral extent, the interplay of the environmental conditions on the one hand, and the variable tolerances of different species on the other, combine to produce complicated stratigraphical distribution patterns in which time-stratigraphical correlation lines are not easily detected.

Dispersal of pollen grains and spores

In addition, the secondary factors controlling pollen and spore dispersal may further obscure the patterns. The majority of the samples investigated are deltaic–marine silts, clays and shales, which means that the individual grains were generally transported, mixed, sorted according to size and possibly redeposited once or several times.

Before giving a few examples of such complicated patterns, it is necessary to discuss the influence of the secondary factors in greater detail. Relatively soon after starting the investigation of Tertiary clastic sediments in the tropics, it was realized that the conventional picture of a fairly homogeneous pollen rain, falling impartially in various sedimentary environments, was inadequate for an understanding of the complicated stratigraphical distribution patterns encountered. Studies of Recent tropical sediments have since shown that, in a humid environment at least, water transport may be of far greater quantitative importance than wind transport (MULLER, 1959).

In the dispersal picture which has now evolved all species pass initially through a phase of air transport. This phase may be of very short duration as, for instance, in tropical peat swamp forests, where most of the pollen produced by the trees was found to be deposited within a very restricted area. If the peat formed here is preserved in the geological record as coal, no further transport takes place, except when lumps of coal are later eroded and transported as pebbles.

At the other end of the scale, the pollen grains and spores produced by the montane vegetation in the tropics are known to be transported in large quantities over considerable distances by high-altitude atmospheric currents. This is especially true of wind-pollinated trees such as *Alnus*, *Pinus*, and of certain ferns.

Sooner or later, however, mainly depending on wind strength, turbulence and wash-out by raindrops, all pollen grains and spores land on the earth's surface. If they are not soon carried away by water, they are mostly destroyed by corrosion and escape fossilization.

A different fate, of course, awaits that portion of the pollen rain which is caught in river systems or the sea. It makes a great difference whether further transport takes place by river systems or whether pollen grains and spores settle directly on the surface of the sea with further transport exclusively by sea currents.

In the first case, a small proportion of the pollen gathered from the entire drainage area will be deposited in the alluvial and coastal plain sediments in the delta area, but most of it will be concentrated at the river mouth and discharged together with the sedimentary load into the sea. In this way concentrations of pollen in fluviomarine sediments close to river mouths are formed, such as have been found in front of the Orinoco delta (MULLER, 1959) and in the Gulf of California (CROSS, 1966). Size sorting will hardly be noticeable during this phase and tidal currents and turbulence will generally cause efficient mixing of the final pollen suspension entering the sea. Also, the largest number of species will be present in the fluviomarine sediments.

In the second case the composition of the pollen load is quite differently affected. First of all sea currents act as slow transporting agents over wide fronts with relatively less turbulence further offshore; the result is gradual settling and deposition of pollen grains and spores, accompanied by size sorting. Wind-transported pollen may be added and this may considerably modify the composition of the pollen load in areas far offshore. Another factor about which little is known is selective corrosion of pollen and spores during seaborne transport.

One final complication has to be evaluated and this is the probability that not all pollen is travelling in a single move from anther to final resting place, but that temporary entrapment in sediments is followed by subsequent erosion and that renewed transport takes place. This recycling of pollen grains and spores may take place within what is geologically speaking an insignificant period of time, as is the case in the average delta, where meandering streams may erode sediments deposited only a short while before. Such recycling, of course, only increases the degree of mixing of the pollen load eventually supplied to the sea and does not affect stratigraphical distribution patterns to any appreciable degree. However, when the time lag between deposition and recycling becomes greater, serious difficulties may arise, culminating when older pollen-bearing formations are eroded in the source area of a river system.

When a clear difference in preservation between such older reworked pollen grains and spores and autochthonous ones is present, they can be easily excluded from the pollen sum, but unfortunately this is usually the exception rather than the rule. Generally there is no clear-cut difference, and the presence of many slightly different corrosion patterns in one single sample may be the only clue indicating the presence of several generations of reworked pollen grains and spores. In such cases a detailed knowledge of the basic floral succession in the area of investigation may enable one to spot the scattered presence of anomalous pollen associations and to recognize them as reworked assemblages. Here general geological considerations, indicating the probable origin of the sediment particles have also to be taken into account. Recently VAN GIJZEL (1961) has demonstrated that fluorescence microscopy may be able to discriminate between autochthonous and reworked pollen.

In the Palaeogene sediments of western Venezuela, for instance, reworked pollen was scarce because the bulk of the sediments deposited during that time had been derived from the Guiana shield, where no pollen-bearing sediments were eroded. In younger sediments the occurrence of reworked pollen of a Palaeogene age proved to be clearly related to distinct unconformities.

In contrast, in the Neogene sediments of northwestern Borneo no such clear-cut unconformities exist and since the sediments had been recycled many times during the course of the Tertiary, a rather diffuse distribution of reworked pollen was the result. It was not unusual to recognize in a single sample up to three different generations of reworked pollen and spores, which created considerable difficulties in the interpretation.

This situation has practical consequences for the establishment of a zonation, since the oldest occurrence of species obviously is more reliable than the youngest. In fact the regional zonation presented here is exclusively based on such oldest occurrences. However, when cuttings have to be investigated for routine purposes, the youngest occurrences may be of greater practical value.

Examples

The practical application of the above principles will now be illustrated with a few examples.

The first case refers to base *Monoporites annulatus* Zone in northern South America. Within the Eocene of western Venezuela, which was the first area investigated, two floral boundaries were apparent in approximately the same stratigraphical interval: aal well-defined increase in *Echitriporites trianguliformis* and, slightly higher, a less conspicuous base of *Monoporites annulatus*. Later the absence of *Echitriporites trianguliformis* in the Eocene of Colombia was noted. This absence could not be attributed to a stratigraphical hiatus or to a different climate, since the remainder of the floral assemblage proved to be essentially similar. However, the Colombian Eocene sediments are more terrestrial than those in Venezuela, and it became clear that *Echitriporites trianguliformis* was derived from a coastal plant with mainly waterborne dispersal of its pollen in seaward direction. This reduced its value for regional correlation and, since *Monoporites annulatus* appeared to have a wider distribution, because it was derived from a more inland environment, its increase in the Early Eocene both in marine and more terrestrial deposits is preferred as the regionally valid time-stratigraphical correlative horizon. The marked increase in *Echitriporites trianguliformis* retains, however, its practical correlative value within the restricted area of the Lower Eocene in the Maracaibo basin.

In the case of the top in the occurrence of *Proteacidites dehaani* (Upper Cretaceous) it was soon apparent that this event could be recognized both in marine deposits in the northern Maracaibo basin and in alluvial plain deposits in Colombia and the southern Maracaibo basin. Regional time-stratigraphical value could, therefore, be assumed for this floral boundary.

In both these examples botanical identification of the pollen species was impossible or did not provide sufficient evidence to permit an independent check on the environmental requirements of the parent plants. The method outlined above was the only feasible one for the recognition of time-stratigraphical value.

In the examples from the Neogene, described below, on the other hand, this sort of conclusion is indeed supported by direct botanical evidence.

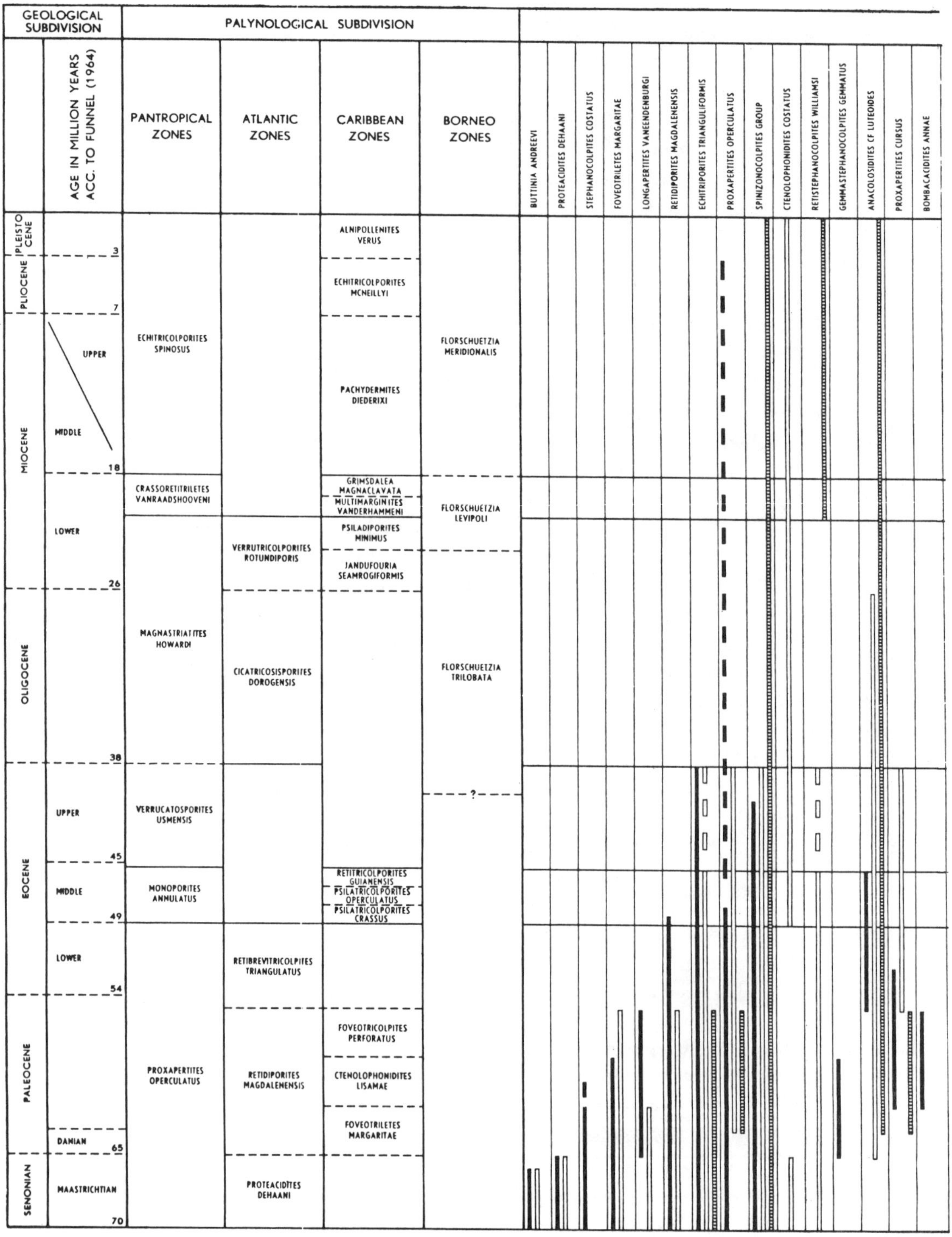

Fig. 15. Range chart.

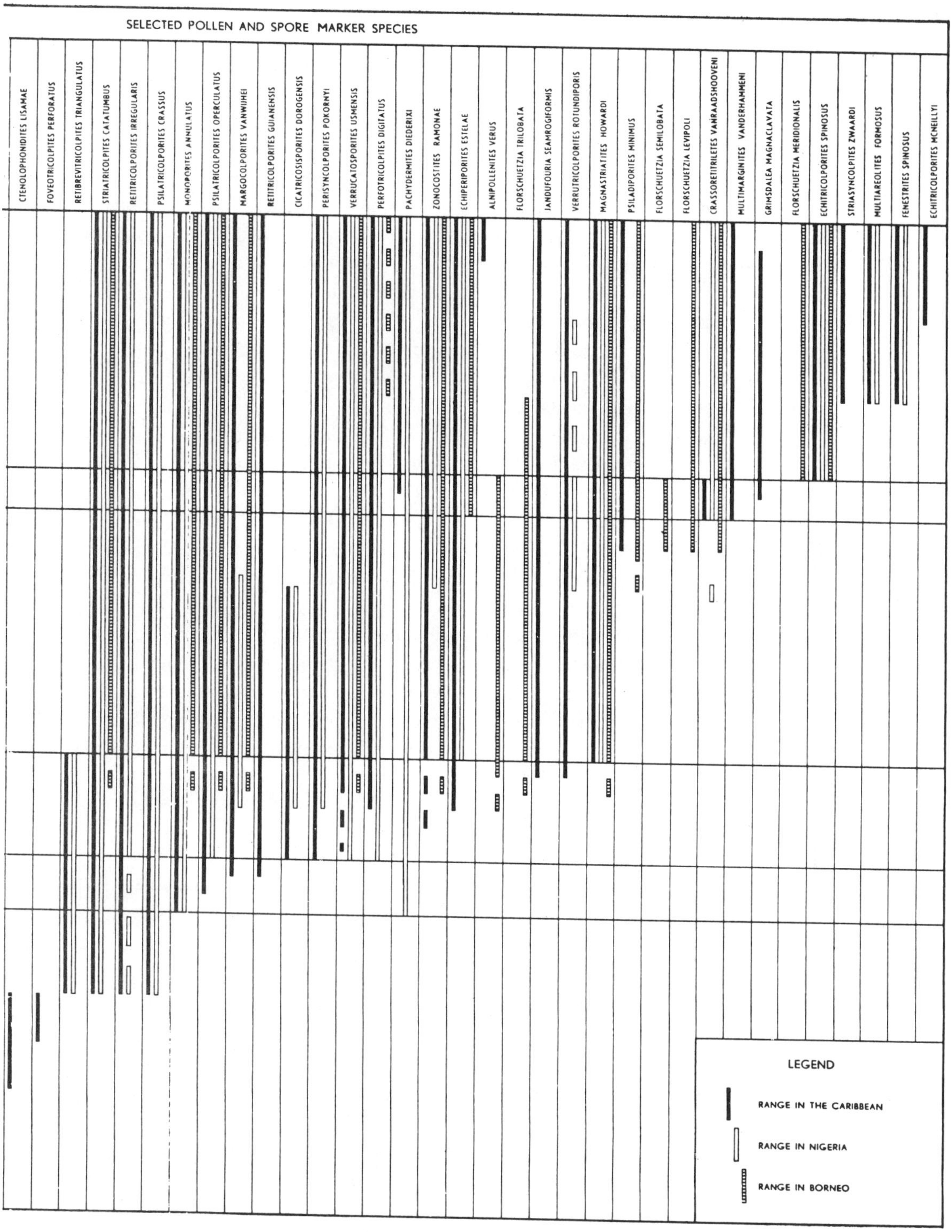

SELECTED POLLEN AND SPORE MARKER SPECIES
CTENOLOPHONIDITES LISAMAE
FOVEOTRICOLPITES PERFORATUS
RETIBREVITRICOLPITES TRIANGULATUS
STRIATRICOLPITES CATATUMBUS
RETITRICOLPORITES IRREGULARIS
PSILATRICOLPORITES CRASSUS
MONOPORITES ANNULATUS
PSILATRICOLPORITES OPERCULATUS
MARGOCOLPORITES VANWIJHEI
RETITRICOLPORITES GUIANENSIS
CICATRICOSISPORITES DOROGENSIS
PERISYNCOLPORITES POKORNYI
VERRUCATOSPORITES USMENSIS
PERFOTRICOLPITES DIGITATUS
PACHYDERMITES DIEDERIXI
ZONOCOSTITES RAMONAE
ECHIPERIPORITES ESTELAE
ALNIPOLLENITES VERUS
FLORSCHUETZIA TRILOBATA
JANDUFOURIA SEAMROGIFORMIS
VERRUTRICOLPORITES ROTUNDIPORIS
MAGNASTRIATITES HOWARDI
PSILADIPORITES MINIMUS
FLORSCHUETZIA SEMILOBATA
FLORSCHUETZIA LEVIPOLI
CRASSORETITRILETES VANRAADSHOOVENI
MULTIMARGINITES VANDERHAMMENI
GRIMSDALEA MAGNACLAVATA
FLORSCHUETZIA MERIDIONALIS
ECHITRICOLPORITES SPINOSUS
STRIASYNCOLPITES ZWAARDI
MULTIAREOLITES FORMOSUS
FENESTRITES SPINOSUS
ECHITRICOLPORITES MCNEILLYI
LEGEND
RANGE IN THE CARIBBEAN
RANGE IN NIGERIA
RANGE IN BORNEO

In the Oligo–Miocene the first increase in the *Rhizophora* pollen type *(Zonocostites ramonae)* can only be reliably observed in coastal and offshore marine sediments, while the approximately contemporaneous first increase in the *Ceratopteris* type *(Magnastriatites howardi)* can be detected in terrestrial environments as well. However, the small fresh-water ferns of the genus *Ceratopteris* can only flourish in open vegetation and will hardly occur under a closed forest canopy, which restricts the dispersal of the spores to water transport mainly. *Ceratopteris*-type spores thus occur more regularly in inland environments, where *Rhizophora* pollen can be extremely rare, and therefore their oldest occurrence has been taken as the more reliable time-stratigraphical horizon. In practice of course both floral changes support each other and are used in conjunction.

Another example from the Neogene of western Venezuela concerns the recognition within the same body of strata of three sets of correlation lines:

(1) A topographically-influenced succession of mangrove dominance alternating with more inland vegetation.

(2) A subregional succession, probably of a climatological nature.

(3) A top occurrence of *Cicatricosisporites dorogensis*.

The correlation lines derived from *(1)* cross at an angle those from *(2)*, while none of these were observed in the adjoining fully marine succession. However, the top occurrence of *Cicatricosisporites dorogensis* proved to be recognizable regardless of environmental influence, and thus was judged to be of major time-stratigraphical significance.

In general the principle may be formulated that any palynological change which is paralleled by a facies change is of questionable time-stratigraphical value. If, on the other hand, the pattern of palynological changes crosses the environmental correlation lines derived from lithology or benthonic faunal distribution, then that pattern is likely to be of time-stratigraphical value at least within the area of study.

Statistical analysis of the palynological data

In the absence of positive information from lithology, faunal or botanical affinities or climatic indications, statistical analysis of the palynological data can be developed into a powerful tool for interpretation.

A first method is to measure the percentage variation, which is high in a terrestrial environment where pollen dispersal is restricted and low in a marine environment in which the pollen load has become thoroughly mixed. This variation is best measured by calculation of the standard deviation, for which a minimum of twenty samples is necessary. This calculation is, however, applicable only to common species.

Secondly, an attempt can be made to separate statistically associations of species which have a common source area. For instance, if alluvial-plain deposits are being investigated, the percentages of the species inhabiting this environment will vary greatly, depending on the local edaphic factors, while species associations from outside this environment and transported into it in a better-mixed composition will always show a constant ratio. If successful, such a statistical analysis will enable the palynologist to separate, for instance, mangrove, river bank, peat swamp, alluvial swamps and marshes, upland and montane associations. The proportion in which these associations occur will then be mainly a function of the general topography of the source area as a whole, while any changes occurring within associations may have climatic or evolutionary significance and will be of greater time-stratigraphical significance.

Association tests can be adapted for this approach.

Thirdly, the variation in occurrence between rarer types is best measured with a run length test, which measures the variation in distance and length of series of single occurrences. To be reliable a minimum of 80 consecutive samples are needed for this test, which can also be applied to more common species when a minimum percentage truncation is chosen.

All these statistical tests can be so devised that the significance of the results is expressed in probability values. One disturbing factor may be the presence of reworked pollen as erratic assemblages.

For large numbers of samples, calculation by means of mechanical data processing is, of course, the only practical method of applying these statistical tests.

Summary of interpretation experience

Summarizing twenty years of experience, it may be stated that for most cases tools are available which allow discrimination between local and regional events of time-stratigraphical value on the one hand and effects caused by time-crossing events such as transgressions and regressions on the other. This information then allows the establishment of a zonation adapted to the particular stratigraphical problem at hand. For short distances and short time intervals the effects of striking topographical events may provide the correlative framework; for correlation over larger distances climatological changes will produce the best means, while on an intercontinental scale correlation on evolutionary change appears to be the only reliable method.

In general, however, the larger the distance, the more difficult it becomes to define and trace sharp floral boundaries. In particular boundaries defined on evolutionary change are notoriously hard to place accurately in thick continuously-deposited sediments. This often poses a practical problem, since the geologist generally requires correlation lines, while the palynological events on which the zonation is based are by nature mostly gradual. In fact any sharp floristic boundary is more probably indicative of a hidden unconformity than of anything else.

It is worth ending this chapter by listing below a few more of the limitations encountered in applying palynological correlation methods.

(*1*) *Determinable pollen and spores absent or scarce.* This may be due to: (*a*) original scarcity in sediments such as sands, conglomerates, limestones; (*b*) destruction by oxidation shortly after deposition, e.g., mottled clays; (*c*) destruction by oxidation after weathering of outcrops, common in the more arid parts of the tropics; (*d*) destruction due to high temperature effects caused by deep burial, volcanic activity, proximity of intrusions, heat generated by tectonic friction.

(*2*) *Statistical errors.* These fall mainly under two headings: (*a*) the use of inadequate counts; (*b*) the use of an unsuitable selection of species which comprises the pollen sum, if percentage ratios are calculated.

As regards (*2a*) the main practical problem is to achieve a maximum of results with a minimum count. In practice a count of 70–100 selected marker species with a sampling distance of 75–100 ft. has proved to be entirely adequate for the solution of most correlation problems. Here, the number of grains counted per sample has been sacrificed to the number of samples investigated.

If more than one investigator is counting the same material, the influence of determination and counting errors should be eliminated by carefully designed counting schemes, which will average out individual errors.

Errors grouped under (*2b*) are bound to influence the results especially when floral changes are indistinct, as in the case of thick sedimentary series deposited within a short interval of time, or when local influences in the pollen spectra obscure the occurrence of scarce regional marker species.

Then it becomes imperative to test various percentage sums in order to arrive finally at the combination of species which, in their changing ratios, will reflect most distinctly the events of time-stratigraphical value on which the correlation will be based.

This is an important point and it should be clearly understood that the percentage calculation system employed in the present study is adapted to the recognition of the zones described here. Other studies, for instance of a localized area in a restricted interval only, will start with a different pollen sum.

In every area one of the principal problems in the early phase of developing a zonation in the extremely rich tropical Tertiary microfloras has been to decide which marker species to include in the pollen sum. Generally the marker species have been divided into two groups. Group A comprises all species which are of importance for the major regional zones, while group B contains those types whose fluctuations are of more local significance. Again one of the advantages of mechanical data processing is that this allows the rapid calculation, plotting and comparison of percentages based on different pollen sums.

ZONATION

The zones recognized in this paper are biostratigraphical units in the sense of the AMERICAN COMMISSION ON STRATIGRAPHIC NOMENCLATURE (1961). They are, therefore, not a priori to be considered as time-stratigraphical units and in fact some of the boundaries delimiting the zones probably are not contemporaneous over large distances. However, within the areas in which the zones are recognized, their succession is identical in all sections investigated.

According to their lateral extent, the zones are grouped as follows: (*1*) pantropical zones; (*2*) transatlantic zones; (*3*) intracontinental zones.

The definition and description of the zones will be documented by a restricted number of type sections, mainly taken from northern South America (Fig.1–14). The range chart (Fig.15) illustrates the ranges of all species discussed, based on evidence from a much larger number of sections. The independent dating of the zones is discussed in the section "Independent dating".

Pantropical zones

The oldest pantropical zone recognized is the *Proxapertites operculatus* Zone characterized by the regular co-occurrence of *Proxapertites operculatus*, *Proxapertites cursus*, the *Spinizonocolpites echinatus* group, and *Echitriporites trianguliformis*. Its base, which as yet could only be studied in Nigeria and Borneo, is provisionally defined by the qualitative base of the occurrences of *Echitriporites trianguliformis* and the *Spinizonocolpites echinatus* group. Further study is necessary to define and evaluate more accurately the base, which is probably of Senonian age.

The zone is well-developed in the Rubio-road section (Venezuela) (Fig.3), well Egoli-1 (Nigeria) (Fig.9), and in the Lundu-Kayan section (Borneo) (MULLER, 1968). The transition to the overlying *Monoporites annulatus* Zone is marked by the first regular occurrence of *Monoporites annulatus*. This is not a sharply defined event, since the increase in *Monoporites annulatus* is rather gradual with alternating periods of higher or lower abundance.

In the Rubio-road section (Venezuela) (Fig.3) the base of *M. annulatus* is rather well defined, but this is a stratigraphically condensed section. In well Icotea-1 (Venezuela) (Fig.4), the increase is much more gradual. Since it is known that this well penetrates a very thick Eocene section without any marked gaps, the record shown here is probably a good reflection of the rather irregular increase in *Monoporites annulatus* in this part of the world.

In Nigeria the *M. annulatus* Zone in the Ovim Bende section (Fig.11) also shows a well-pronounced base, but in the Imo-river section (Fig.10) the base is less distinct. In Borneo the base has not yet been accurately fixed stratigraphically. *M. annulatus* is absent from all Upper Cretaceous and Paleocene sediments examined so far and is known to occur throughout the Neogene. Its base can, therefore, be expected within the Eocene there also.

The boundary with the overlying *Verrucatosporites usmensis* Zone is defined by the first occurrence of *Verrucatosporites usmensis*. However, there is a fairly wide interval in which transitional forms between *Verrucatosporites usmensis* and the presumed ancestral forms with a low verrucation occur. Moreover, there is evidence that the increase in number of *Verrucatosporites usmensis* takes place earlier in Nigeria than in the Caribbean area. Consequently this boundary is not a sharp one. In the Caribbean area the base in the occurrence of *Cicatricosisporites dorogensis* appears consistently to be below the base of *V. usmensis* and serves as a useful additional criterion.

In the Paz del Río section (Colombia) (Fig.1), the *Verrucatosporites usmensis* Zone is thus characterized by regular occurrences of *Cicatricosisporites dorogensis* and scattered occurrences of *Verrucatosporites usmensis*.

In the Ovim Bende and Imo-river sections (Nigeria) (Fig.10, 11), the boundary between the *Monoporites annulatus* Zone and the *Verrucatosporites usmensis* Zone is sharply defined as these sections are incomplete, while in Benin West-1 (Fig.12) the change is slightly more gradual.

In Borneo the boundary has not been sufficiently studied as yet.

The boundary between the *Verrucatosporites usmensis* Zone and the overlying *Magnastriatites howardi* Zone is marked by the first appearance of *Magnastriatites howardi*.

In the Paz del Río section (Colombia) (Fig.1) this boundary is sharply defined, in Chafurray-3 (Colombia) (Fig.2) less so.

In Benin West-1 (Nigeria) (Fig.12) a well-defined boundary is present, but in the Imo-river section (Fig.10) recognition of the *Magnastriatites howardi* Zone depends on a rare occurrence of *M. howardi* in one sample only. In Borneo the stratigraphical position of this boundary has still to be clarified.

The boundary with the overlying *Crassoretitriletes vanraadshooveni* Zone is marked by the base of the regular occurrence of *Crassoretitriletes vanraadshooveni*.

In Chafurray-3 (Colombia) (Fig.2), this boundary is weakly pronounced, but in B-188 (Venezuela) (Fig.6), a sharply defined base is present.

In Okoloma-1 (Nigeria) (Fig.13), the base is rather well defined. In Borneo the base of the occurrence of *Crassoretitriletes vanraadshooveni* more or less coincides with the base of *Florschuetzia levipoli* and is slightly older than in the Caribbean area.

The youngest boundary is the base of the *Echitricolporites spinosus* Zone and is marked by the base of the regular occurrence of *Echitricolporites spinosus*. In B-188 (Venezuela) (Fig.6) and in CO-85 (Trinidad) (Fig.7) the boundary is well defined.

In Lubara Creek-2 (Nigeria) (Fig.14) the increase in *Echitricolporites* is gradual and fairly irregular, and the location of the boundary is, therefore, difficult to determine.

Transatlantic zones

The pantropical *Proxapertites operculatus* Zone can be subdivided in the Caribbean area and in Nigeria into three transatlantic zones. The oldest is the *Proteacidites dehaani* Zone.

This zone is characterized mainly by the co-occurrence of *Proteacidites dehaani* and *Buttinia andreevi*, together with high percentages of *Foveotriletes margaritae*. The boundary with the overlying *Retidiporites magdalenensis* Zone is taken at the top occurrence of *Proteacidites dehaani*. In the Rubio-road section (Venezuela) (Fig.3) this boundary is well defined.

In Egoli-1 (Nigeria) (Fig.9) the boundary is also rather well marked, mainly, however, by the decrease in *Buttinia andreevi*, which was absent in the previous section, although it is known to occur elsewhere in the Caribbean area within the zone.

The *Retidiporites magdalenensis* Zone is mainly negatively characterized by the absence of *Proteacidites dehaani* and *Buttinia andreevi*. *Foveotriletes maragritae* is still fairly frequent, although it may diminish in abundance in the higher part of the zone. The zone is further characterized in both areas by the regular presence of *Retidiporites magdalenensis*, *Echitriporites trianguliformis*, and *Proxapertites operculatus*.

The boundary between the *Retidiporites magdalenensis* Zone and the overlying *Retibrevitricolpites triangulatus* Zone is marked by the first appearance of *Retibrevitricolpites triangulatus*, *Striatricolpites catatumbus*, and *Psilatricolporites crassus*.

In the Rubio-road section (Venezuela) (Fig.3) this boundary is clearly defined.

In the Ovim Bende section (Nigeria) (Fig.11) the boundary is also fairly distinct.

In western Venezuela this boundary is further marked by the first appearance of *Retitricolpites irregularis* and the disappearance of *Retidiporites magdalenensis*, *Proxapertites cursus*, *Bombacacidites annae*, *Ctenolophonidites lisamae*, and *Foveotricolpites irregularis*. It is assumed that this pronounced floral change coincides with an unconformity.

The pantropical *Magnastriatites howardi* Zone can further be subdivided both in northern South America and Nigeria into two transatlantic zones, the *Cicatricosisporites dorogensis* Zone and the overlying *Verrutricolporites rotundiporis* Zone. The boundary between these two zones is marked by the more or less simultaneous decrease in *Cicatricosisporites dorogensis* and increases in *Verrutricolporites rotundiporis* and in *Zonocostites ramonae*.

In Chafurray-3 (Colombia) (Fig.2) this boundary is well marked, but in Benin West-1 (Nigeria) (Fig.12) only one sample was available, which could be assigned to the *Verrutricolporites rotundiporis* Zone, and the evidence here is not as strong.

Preliminary results indicate that the major increase in *Zonocostites ramonae* in Borneo takes place roughly at an equivalent stratigraphical level. The other two marker species do not occur here.

Intracontinental zones

Caribbean area

In the Caribbean area the transatlantic *Retidiporites magdalenensis* Zone can be subdivided into three units. The oldest unit is the *Foveotriletes margaritae* Zone, which is characterized by the co-occurrence of frequent *Stephanocolpites costatus, Foveotriletes margaritae, Longapertites vaneendenburgi,* and *Gemmastephanocolpites gemmatus,* and by the absence of *Bombacacidites annae* and *Ctenolophonidites lisamae.*

The boundary with the overlying *Ctenolophonidites lisamae* Zone is taken at the first occurrence of *Ctenolophonidites lisamae* and *Bombacacidites annae,* and this zone is further characterized by the regular presence of *Gemmastephanocolpites gemmatus,* diminishing quantities of *Foveotriletes margaritae,* and the regular presence of *Proxapertites cursus.*

The base of the youngest *Foveotricolpites perforatus* Zone is defined by the first occurrence of *Foveotricolpites perforatus,* which species is restricted to the zone. *Bombacacidites annae* and *Ctenolophonidites lisamae* are frequent in this zone, while *Stephanocolpites costatus, Foveotriletes margaritae* and *Gemmastephanocolpites gemmatus* have virtually disappeared.

This subdivision is clearly visible in the Rubio-road section (Venezuela) (Fig.3).

The *Monoporites annulatus* Zone can be subdivided in the Caribbean area into four units. The boundary between the *Psilatricolporites crassus* Zone and the overlying *Psilatricolpites operculatus* Zone is taken at the lowest occurrence of *Psilatricolpites operculatus,* as visible in Icotea-1 (Venezuela), (Fig.4) and the Prevención section (Venezuela) (Fig.5).

The next higher boundary between the *Psilatricolpites operculatus* Zone and the *Retitricolporites guianensis* Zone is placed at the base regular occurrence of *Retitricolporites guianensis.* This boundary is present in the Prevención section (Venezuela) (Fig.5).

The *Verrutricolporites rotundiporis* Zone can be subdivided into the *Jandufouria seamrogiformis* Zone and the overlying *Psiladiporites minimus* Zone. The boundary between these two zones is marked by the base of the regular occurrence of *Psiladiporites minimus,* as is visible in B-188 (Venezuela) (Fig.6). The *Crassoretitriletes vanraadshooveni* Zone can be subdivided in northern South America into the *Multimarginites vanderhammeni* Zone and the overlying *Grimsdalea magnaclavata* Zone. The boundary between these two zones is marked by the base of the occurrence of *Grimsdalea magnaclavata.* This is clearly visible in B-188 (Venezuela) (Fig.6).

The *Echitricolporites spinosus* Zone can be subdivided into three zones.

The boundary between the *Pachydermites diederixi* Zone and the overlying *Echitricolporites mcneillyi* Zone is marked by the base of the regular occurrence of *Echitricolporites mcneilly*.

This boundary is distinct in CO-85 (Trinidad) (Fig.7) and Paria-1 (Venezuela) (Fig.8).

The boundary between the *Echtricolporites mcneilly* Zone and the *Alnipollenites verus* Zone is placed at the base of the regular occurrence of *A. verus*. In Paría-1 (Venezuela) (Fig.8) this boundary is well developed.

Borneo

In Borneo a local subdivision can be established, based exclusively on the evolutionary development within the genus *Florschuetzia* (Fig.16).

The lowermost zone, *Florschuetzia trilobata* Zone, is characterized by the presence of *F. trilobata* and the absence of younger forms. The zone probably covers the upper part of the *Verrucatosporites usmensis* Zone and the lower part of the *Magnastriatites howardi* Zone, but the stratigraphical position of its base has still to be determined precisely.

The first development of *Florschuetzia semilobata* and *F. levipoli* marks the base of the next higher *Florschuetzia levipoli* Zone, which coincides approximately with the weakly defined base of the *Crassoretitriletes vanraadshooveni* Zone. In this zone *Florschuetzia trilobata* decreases in numbers, while *F. semilobata* disappears. The average size of *F. levipoli* increases also from 30 μ to 35 μ. The overlying *Florschuetzia meridionalis* Zone is characterized by the regular presence of *F. levipoli* and, in increasing quantities, of *F. meridionalis*, while the average size of both species shows further increases. *Florschuetzia trilobata* is still present in the lower part of this zone, but disappears soon.

The boundary between the *Florschuetzia levipoli* and *Florschuetzia meridionalis* Zones coincides approximately with the base of the *Echitricolporites spinosus* Zone, and the first regular occurrence of *E. spinosus* and *Florschuetzia meridionalis* can be taken as a criterion for its recognition.

Both species have the disadvantage that in the very thick Miocene sediments of northwestern Borneo, their increase in number is gradual, which makes it impossible to recognize a sharp boundary.

INDEPENDENT DATING

In the virtual absence of any age-indicating plant macrofossil evidence with which the palynological data could be correlated, associated animal fossils provide the only check on the age of the proposed zones. For this purpose most reliance has been placed on the age-indicating Foraminifera and, to a lesser extent, on

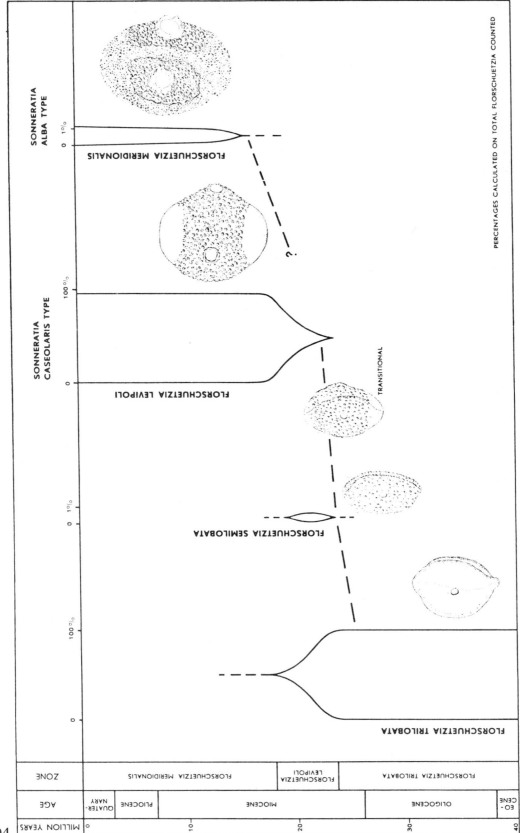

Fig.16. Development of *Florschuetzia* pollen types in Borneo.

the Mollusca. Not all areas investigated have produced a sufficient number of check points and it has also only rarely been possible to tie a flóral change closely to a faunal change. In general, however, the available evidence indicates that the floral zones recognized are not markedly diachronous over the areas in which they are considered valid. The boundaries between the zones, however, can only rarely be sharply defined, and may not be exactly isochronous over larger distances.

Future detailed studies will no doubt provide further evidence to narrow down the gaps in knowledge which still exist.

In the account given below the floral zones will be discussed chronologically, a list being given of the age-indicating animal fossils, the faunal zones of which these are characteristic and, as far as possible, the standard time-stratigraphical units with which the zones are correlated. For the Palaeogene this is reasonably well established, but for the Neogene considerable uncertainty still exists, and no attempt will be made to relate the palynological zonation to the European stages. Instead the planktonic zonation established by BOLLI (1966) will be taken as yardstick and his subdivision in terms of Oligocene, Lower, Middle and Upper Miocene and Pliocene is adopted as a fair approximation. For Borneo reference is made to the well-known letter classification of Van der Vlerk and Umbgrove (ref. LIECHTI, 1960). Nomenclature of Foraminifera is according to LOEBLICH and TAPPAN (1964).

It was not possible to present a full documentation for this part of the present paper, except for those sections shown for the purpose of defining the zonation and in which the occurrence of age-indicating fossils is recorded. However, in all other instances where a palynological zone is mentioned as associated with certain animal fossils, this is based on actual co-occurrence in the same sections.

(a) Proteacidites dehaani Zone

In western Venezuela and Colombia this zone is associated with the smaller Foraminifera *Globotruncana gansseri, G. lapparenti, G. stuarti, Guembelitria cretacea, Siphogeneroides bramletti,* and the ammonite *Sphenodiscus* sp. In Nigeria it is associated with the smaller Foraminifera *Bolivina afra, Rugoglobigerina rugosa,* and striate *Heterohelix* spp., with the ammonites *Didymoceras* sp., *Libyoceras ismaeli* and *Sphenodiscus* sp., and with the lamellibranchiat *Inoceramus* sp.

This faunal list clearly indicates a Maastrichtian age. So far no evidence of an older age has turned up, but since the lower limit of the *Proxapertites operculatus* Zone has not been properly studied, the lower part of the *Proteacidites dehaani* Zone may of course be older than Maastrichtian.

Unfortunately, both in Nigeria and in the Caribbean area, the top of the *Proteacidites dehaani* Zone is present in a predominantly coastal facies without sufficient marine fossils, and, therefore, cannot be closely correlated to the top of the Maastrichtian. However, the presence in Nigeria of Danian fossils in the lower part of the overlying zone suggests that this top closely coincides with the top of the Maastrichtian.

(b) Retidiporites magdalenensis Zone

The presence, in the lower part of the *Retidiporites magdalenensis* Zone of Nigeria, of the smaller Foraminifera *Globigerina compressa* and *G. daubjergensis*, which are markers for Stolk's *Globigerina compressa* range Zone, indicates a Danian age for part of this interval (STOLK, 1963).

A slightly younger age is indicated by the presence in the higher part of the *Retidiporites magdalenensis* Zone in Nigeria of the smaller Foraminifera *Globorotalia pseudomenardii*, *G. velascoensis* and *G. acuta*, markers for Stolk's *Globorotalia acuta/Globorotalia velascoensis* range zone, of Paleocene age.

This and the fact that the lower part of the overlying *Retibrevitricolpites triangulatus* Zone still is of Paleocene age, indicates a Danian/Paleocene age range for the *Retidiporites magdalenensis* Zone.

In Borneo a Paleocene age for part of the *Proxapertites operculatus* Zone is indicated by the presence of the smaller Foraminifer *Globorotalia velascoensis* and the larger Foraminifera *Miscellanea miscella* and *Nummulites nuttalli*.

(c) Retibrevitricolpites triangulatus Zone

In Colombia the lower part of this zone is associated with the larger Foraminifer *Actinosiphon barbadensis*, indicating a Paleocene age. In Nigeria the lower part contains the smaller Foraminifera *Globorotalia velascoensis*, and *G. acuta*, also indicative of Paleocene, while in the higher part the smaller Foraminifera *Globorotalia formosa* and *Globorotalia rex*, which are markers for Stolk's *Globorotalia formosa* range zone, indicate an Early Eocene age.

In view of the fact that in Venezuela the overlying *Psilatricolporites crassus* Zone carries already a Middle Eocene fauna, while in Nigeria the overlying *Monoporites annulatus* Zone is associated with a Lower–Middle Eocene fauna, it is suggested that the *Retibrevitricolpites triangulatus* Zone ranges from Late Paleocene to Early Eocene in age.

(d) Monoporites annulatus Zone (not subdivided)

In Nigeria this zone is associated with the smaller Foraminifer *Cassigerinelloita amekiensis* and associated planktonic Foraminifera indicative of Stolk's *Cassigerinelloita amekiensis* range zone of late Early–Middle Eocene age.

In the Caribbean area a more detailed tie-in of the subdivision has been possible, and this will be discussed separately for each subzone below.

(e) Psilatricolporites crassus Zone

In Venezuela the lower part of this zone is associated with the larger Foraminifera *Linderina floridensis*, *Helicostegina gyralis* and *Lepidocyclina* sp. A, while the upper part was found to contain *Helicolepidina spiralis* form C. According to VAN RAADSHOOVEN (1951), these faunal associations indicate an early Middle and a late Middle Eocene age respectively.

(f) Psilatricolpites operculatus and Retitricolpites guianensis Zones

In Venezuela these zones are associated with *Helicolepidina spiralis* form C and are, therefore, of late Middle Eocene age.

(g) Boundary Monoporites annulatus Zone/Verrucatosporites usmensis Zone

Since the lowermost part of the overlying zone in Venezuela still is of late Middle Eocene age, this boundary must be situated in the Middle Eocene.

(h) Verrucatosporites usmensis Zone

In Venezuela the presence in the lowermost part of *Helicolepidina spiralis* form C is still indicative of a Middle Eocene age, but for the upper part the larger Foraminifera *Lepidocyclina pustulosa*, *Pseudophragmina mirandana*, *Nummulites striatoreticulata* and *Helicostegina soldadensis* and, in Colombia the smaller Foraminifer *Bulimina jacksonensis*, indicate a Late Eocene age.

In Nigeria the presence of the smaller Foraminifera *Chiloguembelina martini* and *Truncorotaloides rohri*, indicative of Stolk's *Chiloguembelina martini–C. cubensis* concurrent range zone, indicates a late Middle–Late Eocene age.

The well-known Late Eocene mollusc fauna, originally described by Bullen Newton (ref. REYMENT, 1965) from the higher part of the Bende–Ameki Group, also falls within this zone.

In the absence of any evidence of an Oligocene age, an age range for the *Verrucatosporites usmensis* Zone from Middle to Late Eocene, possibly including the earliest Oligocene, appears reasonable.

(i) Cicatricosisporites dorogensis Zone

In the Caribbean area the presence of the smaller Foraminifera *Globigerina ampliapertura*, *G. ciperoensis ciperoensis*, *Globorotalia opima opima* and *G. kugleri*, and in Nigeria *Globigerina ciperoensis ciperoensis*, *G. ciperoensis angulisuturalis* and *Globorotalia kugleri* indicates an age range for the *Cicatricosisporites dorogensis* Zone spanning the interval *Cassigerinella chipolensis/Hastigerina micra*

Zone–*Globorotalia kugleri* Zone of BOLLI (1966). Accepting Bolli's age assignment of this interval, it is concluded that the *Cicatricosisporites dorogensis* Zone largely coincides with the Oligocene.

(j) Boundary Florschuetzia trilobata Zone/Florschuetzia levipoli Zone

This well-marked boundary in Borneo was found to be situated at a level slightly above the boundary Tertiary e 1–4/Tertiary e 5 on a combination of evidence from larger and smaller Foraminifera. In the present paper the latter boundary is taken as coinciding with the Oligo–Miocene boundary, but it is realized that this needs further confirmation.

(k) Jandufouria seamrogiformis Zone

This zone is associated in the Caribbean area with the smaller Foraminifera *Catapsydrax stainforthii* and *Globigerinatella insueta*, and approximately covers the *Catapsydrax dissimilis*, *C. stainforthi* and part of the *Globigerinatella insueta* Zones of Bolli (Lower Miocene).

(l) Psiladiporites minimus Zone

In this zone the smaller Foraminifera *Globigerinatella insueta*, *Globorotalia fohsi barisanensis*, *G. fohsi fohsi* and *Orbulina universa* indicate that part of the *Globigerinatella insueta*, the whole of the *Globorotalia fohsi barisanensis* and part of the *Globorotalia fohsi fohsi* Zones of Bolli are covered (Lower Miocene).

(m) Verrutricolporites rotundiporis Zone (unsubdivided)

In Nigeria the presence of *Globorotalia fohsi fohsi* and *Orbulina universa* indicates that Bolli's *Globorotalia fohsi fohsi*, *Globorotalia fohsi lobata* and *Globorotalia fohsi robusta* Zones may possibly be covered, but since in the *Psiladiporites minimus* Zone of the Caribbean area no evidence of the latter two zones is present, it seems likely that in Nigeria only the *Globorotalia fohsi fohsi* Zone is represented in the upper part of the *Verrutricolporites rotundiporis* Zone. Due to unsuitable facies conditions, the *Globigerinatella insueta* and *Globorotalia fohsi barisanensis* Zones have not been found in Nigeria.

Stratigraphic position of the base of Orbulina universa

In view of the considerable importance which is generally attached to this datum, its location in the different areas will be discussed separately.

In the Caribbean area, as well as in Nigeria and Borneo, it has been found to lie in the upper part of the *Magnastriatites howardi* Zone. In the Caribbean area it can be pinpointed more closely and was found to lie in the *Psiladiporites minimus* Zone. In Nigeria it is present in the *Verrutricolporites rotundiporis* Zone, while in Borneo it was found in the middle part of the *Florschuetzia levipoli* Zone. In Bolli's scheme the base of *Orbulina universa* occurs in the upper part of the *Globigerinatella insueta* Zone.

(n) Multimarginites vanderhammeni Zone

The presence of the smaller Foraminifera *Globorotalia fohsi fohsi* and *G. fohsi lobata* is indicative of Bolli's *Globorotalia fohsi fohsi* and *Globorotalia fohsi lobata* Zones (Lower Miocene).

(o) Grimsdalea magnaclavata Zone

This zone was found to contain the smaller Foraminifera *Globorotalia fohsi robusta, G. mayeri* and *G. menardii* and thus covers approximately the *Globorotalia fohsi robusta, Globorotalia mayeri* and *Globorotalia menardii* Zones of Bolli. The boundary between Lower and Middle Miocene would thus be present within the zone.

(p) Crassoretitriletes vanraadshooveni Zone (not subdivided)

In Nigeria the presence of the smaller Foraminifera *Globorotalia fohsi fohsi* and *Orbulina universa* indicates a possible age range covering Bolli's *Globorotalia fohsi fohsi, Globorotalia fohsi lobata* and *Globorotalia fohsi robusta* Zones. This corresponds well with the more detailed evidence discussed under the two preceding Caribbean subdivisions of the *Crassoretitriletes vanraadshooveni* Zone.

(q) Florschuetzia levipoli Zone

The age range of this Bornean zone is determined by the presence, in the lower part, of the smaller Foraminifer *Globigerinatella insueta*, indicative of Bolli's *Catapsydrax stainforthi* and *Globigerinatella insueta* Zones, in the middle part, of the larger Foraminifer *Flosculinella bontangensis*, indicative of Tertiary f 1, and of the smaller Foraminifera *Globorotalia fohsi barisanensis, G. foshi fohsi* and *Orbulina universa*, indicative of Bolli's *Globorotalia fohsi foshi* Zone, and, in the upper part, of *Globorotalia fohsi lobata, G. fohsi robusta* and *G. menardii*, indicative of Bolli's *Globorotalia fohsi robusta* and *Globorotalia menardii* Zones.

(r) Echitricolporites spinosus Zone (not subdivided)

In Borneo the lower part of this zone carries the smaller Foraminifer *Globorotalia menardii*, indicating an age younger than Bolli's *Globorotalia fohsi lobata*

Zone, while in the upper part *Pulleniatina obliquiloculata* occurs, which is generally taken to indicate a Pliocene age.

In the Caribbean area the latter Foraminifer is not found below the *Echitricolporites mcneillyi* Zone, suggesting that the Mio–Pliocene boundary may be close to the base of the last-mentioned zone. The youngest boundary in the Caribbean area, between the *Echitricolporites mcneillyi* and *Alnipollenites verus* Zones is, in view of the climatological interpretations, taken as closely coinciding with the Plio–Pleistocene boundary.

Summarizing the results presented in the foregoing account, it may be stated that the top Maastrichtian, top Paleocene, top Lower Eocene, top Middle Eocene, top Upper Eocene and top Oligocene are well established in correlation with the palynological succession. The subdivisions in the Miocene up to the Mio–Pliocene boundary are less accurately known, mainly due to uncertainties remaining in the correlation of BOLLI's (1966) planktonic zonation and of Van der Vlerk and Umbgrove's letter classification (ref. LIECHTI, 1960) in the Far East with the standard European succession.

On the general range chart (Fig.15) the proposed correlation of the standard time-stratigraphical subdivision with the palynological subdivision is presented. Zones have been drawn approximately to an absolute time scale, derived from FUNNEL (1964).

Two points should be mentioned here in connection with this chart.

In the first place, the bases of the occurrences of *Verrucatosporites usmensis* and *Crassoretitriletes vanraadshooveni* have proven to be slightly diachronous, as discussed earlier. In the former case, the base occurrence in Nigeria has been taken as the lower limit of the *Verrucatosporites usmensis* Zone, while the time-equivalent horizon in the Caribbean area is approximately given by the base of *Cicatricosisporites dorogensis*. In the latter case, the base occurrence in the Caribbean area has been taken as the lower limit of the *Crassoretitriletes vanraadshooveni* Zone, since this is the best defined change in relation to the planktonic subdivision.

Secondly, the detailed subdivision of the Eocene and Lower Miocene reflects local stratigraphical needs more than rapid floral change. The relatively long, unsubdivided Oligocene interval probably indicates lack of information due to a hiatus in the stratigraphical record of both the Caribbean area and Nigeria. In Borneo the Eocene interval has not yet been adequately investigated and no subdivision is possible at present.

STRATIGRAPHICAL APPLICATION

In this chapter the practical application of the palyno-stratigraphical evidence will be discussed and illustrated with three geological cross-sections from the Caribbean area (Fig.17–19).

The sections chosen have been constructed as far as possible on the basis of wells or measured surface sections in which a fairly detailed palynological succession had been established. In a few cases a composite section had to be constructed by piecing together various scattered shorter sections, so that thicknesses of formations are consequently less reliable.

Nomenclature of geological formations follows as far as possible generally accepted terminology. For Venezuela reference is made to the respective volume of the International Stratigraphic Lexicon (SCHWARCK ANGLADE, 1956), for Colombia and Trinidad to more scattered literature.

The environment of deposition has been broadly classified into three units: (*1*) *upper coastal plain (fresh water)*, and more inland environments characterized by the absence of any brackish or marine faunal elements; (*2*) *lower coastal plain (brackish water)*, deposits which may carry typical faunal associations indicating low and variable salinity; (*3*) *marine*.

Section I (Fig.17) covers the Palaeogene from east-central Colombia to western Venezuela, running along the northern slopes of the Andes, circling around Lake Maracaibo to turn southwards along the Perija foothills as far as the Tarra area. A large part of the area covered by this section is underlain by marine Cretaceous deposits.

The Colombian sections have not reached these, but in Venezuela the Colón, Mito Juan and their lateral equivalent the La Paz Shales have been sampled and found to be restricted to the *Proteacidites dehaani* Zone. In the north a slightly younger marine deposit is the Guasare, covering the *Foveotriletes margaritae* and *Ctenolophonidites lisamae* Zones.

The regression shown to have taken place by the change towards a transitional or even terrestrial environment has evidently not been contemporaneous, a phenomenon earlier described in a generalized way by KUYL et al. (1955). South of Lake Maracaibo, the sediments assigned to the *Proteacidites dehaani* Zone were already deposited in part in a transitional environment, as indicated by the Umir in Colombia and the lower part of the Orocué in Southwestern Venezuela. Further north in the Dibujo area (Riecito Maché), the change from marine to terrestrial deposits takes place at the base of the *Foveotricolpites perforatus* Zone.

The effect of tectonic movements during the Paleocene is also clearly visible. The interval *Foveotriletes margaritae* Zone–*Foveotricolpites perforatus* Zone is thickest in the Lebrija section and shows very regular thinning to the south and northeast.

Along the Andean foothills (Fig.17, columns 3–7) and in the northern Lake and Concepción areas (columns 8–10) the Paleocene has been truncated, indicating a minor orogenic phase separating Paleocene from Eocene. This truncation is least noticeable in the Riecito Maché section, where the possibility of continuous sedimentation is not excluded, although the rather sharp and distinct floral change at the base of the *Retibrevitricolpites triangulatus* Zone still suggests the presence of a hiatus. It increases southwards since, in Concepción, the *Foveotricolpites perforatus* Zone is missing, while in the North Central Lake area the *Ctenolophondites ilsamae* Zone is also missing, and it reaches maximum values along the northern Andean foothills. In Rio Mullapas and Rio Perdido the *Proteacidites dehaani* Zone is found underlying the Eocene, and the whole Paleocene appears to be missing.

In a southwesterly direction the magnitude of the hiatus rapidly decreases again and in Quebrada La Mora the *Foveotricolpites perforatus* Zone is found again. Further south and in the Tarra area the top of the Paleocene interval coincides with the base of the Mirador Sandstone, and it is likely on geological grounds that a minor unconformity is present at this level (SCHAUB, 1948, p.225), coinciding with the base of the *Retibrevitricolpites triangulatus* Zone. In Colombia sampling gaps around this level preclude any definite statements on this problem.

The Eocene sedimentation cycle starts with the *Retibrevitricolpites triangulatus* Zone. (As mentioned in the previous section of this paper the lowermost part of this zone may be of Paleocene age.) Sediments assigned to this zone are thickly developed in the northeast and in Colombia, but they are absent along the northern Andean foothills between Rio de Oro and Rio Mullapas. The next higher *Psilatricolporites crassus* Zone is absent between Rio Perdido and Rio de Oro and in the Colombian sections. The overlying *Psilatricolpites operculatus* Zone is generally rather thin and its absence in sections with reduced sedimentation such as Rio Mullapas and Quebrada La Victoria may be due to sampling gaps. In contrast to the rather restricted areal distribution of the older Eocene zones, it is clear that the overlying *Retitricolporites guianensis* and especially the *Verrucatosporites usmensis* Zones have a much wider distribution. The former is thickly developed in the Rio Mullapas and Rio Lebrija sections, while the latter is found in more or less uniform thickness in all the northern Andean foothills sections as well as in Táchira. The absence of one or both zones in Rio Mullapas, North Central Lake and Concepción is due to truncation by the well-known post-Eocene unconformity.

From these observations it may be deduced that during the Eocene progressively younger sediments transgressed over an eroded Paleocene surface, first localized in the deeper parts of the sedimentary basin, but rapidly transgressing over virtually the whole area of investigation in the upper part of the *Retitricolporites guianensis* Zone and especially during the time of deposition of the *Verrucatosporites usmensis* Zone. This last transgressive phase coincides with the widespread deposition of the marine Pauji in the east, which sharply contrasts with the transitional environment prevalent in the older Eocene.

It must be noted that in the southwestern part this transgressive younger Eocene rests unconformably on Lower Eocene sediments of *Retibrevitricolpites triangulatus–Psilatricolporites crassus* Zone age, indicating a minor orogenic phase, probably resulting in a slight subsidence accompanied by a southwestern tilt of the depositional basin. Thus in Quebrada La Mora the *Verrucatosporites usmensis* Zone is resting on the *Ctenolophonidites lisamae* Zone, in Rio de Oro on the *Foveotricolpites perforatus* Zone and in the Rubio road section already on the *Psilatricolporites crassus* Zone.

In general Eocene sedimentation was thickest in the northeast, as shown by the Catanejo–Prevención section. Continuation of the Eocene sedimentary basin towards Falcón is likely, but no palynological data are available from this area due to carbonization of plant material.

At this point it may be mentioned that, although the strong vertical exaggeration of the section suggests the presence of marked angular unconformity at many places, in reality this can hardly ever be observed in the field. Moreover, the similar lithologies of the formations on both sides of the unconformities and the absence of other age-indicating fossils made it virtually impossible to unravel the geological history in detail before the advent of palynological studies (cf. SCHAUB, 1948; MENCHER et al., 1953).

The situation was further complicated by the presence of various sandstone bodies, which had been correlated merely on lithological similarities supported by weak photogeological evidence. The three main sandstone bodies are the La Sierra in the northwest, the Misoa in the southeast and the Mirador in the southwest. Palynological studies revealed that the bulk of the Mirador Sandstone had been deposited in the interval *Retibrevitricolpites triangulatus–Psilatricolporites crassus* Zone and was, therefore, roughly contemporaneous with the C-sand group of the Eocene terminology used in the Lake Maracaibo area, while both the La Sierra and Misoa Sandstones are mainly deposited in the interval upper *Psilatricolporites crassus–Retitricolporites guianensis* Zone, thus being markedly younger and equivalent to the B-group of sands in the Lake area. In addition, as already pointed out by SCHAUB (1948, p.224), the Mirador Sandstone cannot be traced in the field along the northern Andean foothills beyond Táchira, and earlier attempts to do so were based on miscorrelation either with underlying Paleocene sandstone bodies equivalent to the Barco Sandstone member or with basal Upper Eocene sandstones transgressing over the eroded Paleocene surface. As can be seen on section I in the area between Rio de Oro and Rio Perdido, (Fig.17), due to the considerable gap shown to be present by palynology, these sandstones are close together and can easily be confused in the field or by photogeology.

A further complication is present in the Tarra area southwest of Lake Maracaibo, where the onlapping La Sierra Sandstone comes into contact with the underlying Mirador Sandstone. It has been shown by palynological dating that in some areas, Quebrada La Victoria for example, the uppermost part of what had been called Mirador Sandstone is in fact the age-equivalent of the lowermost La Sierra Sandstone and that locally a minor hiatus may thus exist in the upper part of the Mirador.

These observations also have some consequences for the stratigraphical terminology. First of all, the boundary between the Carbonera and the Orocué Formations is distinct only when these formations are separated by the Mirador Sandstone, as is the case in Táchira and the Tarra area. However, the disappearance of the Mirador in a northeasterly direction along the Andean foothills and the similar facies of the Orocué and Carbonera make it virtually impossible to separate these formations in the field. On the section the boundary between the formations has in such cases been taken to coincide with the stratigraphical hiatus revealed by palynology and indicated with a dotted line.

Similarly the lithological separation of the Mirador and La Sierra Sandstones in the area of overlap is difficult, and recognition of these different sandstone bodies may only be possible when shale intercalations carry sufficient pollen for dating.

Recently GONZALEZ (1967) has studied the transition Los Cuervos–Mirador in the Barco area (Colombia). He reports the presence of *Cicatricosisporites dorogensis* rather low in the Mirador Sandstone, which suggests that a relatively large part of this formation can be assigned to our *Verrucatosporites usmensis* Zone. However, the lack of data from the upper 100 m of the Mirador Sandstone, the absence of *Retitricolporites guianensis* and *Perisyncolporites pokornyi*, several other anomalies in the floral succession and the rather restricted number of samples of unknown quality preclude a more definite interpretation of his results.

The final phase of Eocene sedimentation, where preserved, indicates general regression of the sea. In the Catanejo–Prevención area the marine-deltaic Eocene sediments are succeeded by the continental La Victoria Formation. In the area between Rio Lebrija and Rio Perdido the Esmeralda and Carbonera Formations show no sign of transgression and in the Paz del Rio section the transitional San Fernando Formation, which is roughly time-equivalent to the marine Paují transgression phase, is succeeded by the Oligocene continental Margua Formation.

The regression here has taken place during the time of deposition of the *Cicatricosisporites dorogensis* Zone and is, therefore, younger than in the Catanejo–Prevención area where the La Victoria Formation is restricted to the *Verrucatosporites usmensis* Zone. In the Rio de Oro and Rio Perdido sections the León Shale at the transition between the *Verrucatosporites usmensis* and *Cicatricosisporites dorogensis* Zones indicates a localized more pronounced subsidence, without marine influences.

Section II (Fig.18) covers the older Neogene part of the stratigraphical succession and runs from the western Llanos area in Colombia to southwest Venezuela, crosses Lake Maracaibo and traverses Falcón from west to east.

The top of the San Fernando Formation, which in most sections can be found around base *Cicatricosisporites dorogensis* Zone, appears in the Rio Cubugón to lie close to the top of the zone, thus indicating that the León Shales of the La Fria area may be a lateral equivalent of the upper part of the San Fernando. In

Falcón marine deposits characterize the *Cicatricosisporites dorogensis* Zone, while in well B-188 no evidence of the occurrence of this zone could be found. Possibly the barren Icotea Formation represents this interval.

In the south the period of deposition of the San Fernando and of the León Formations is followed by a regressive period, starting earliest in Voragine-1. No trace of this regression can be detected in Falcón. Then, approximately in the middle of the overlying *Verrutricolporites rotundiporis* Zone (*Jandufouria seamrogiformis* Zone, upper part), a rather pronounced transgressive phase can be discerned. This is the well-known La Rosa transgression from the Bolivar coastal area of Lake Maracaibo, which can be followed from the marine Agua Clara in Falcón, via the La Rosa of the Maracaibo basin, towards the Uracá fossil horizon of La Fria, which here occurs intercalated in the terrestrial Guayabo Formation. In Colombia this transgression is visible as a marine intercalation in the lower part of the Cubugón Formation of the Rio Cubugón and in the lower part of the Chafurray in well Chafurray-3. Voragine-1 was situated more inshore and no marine interval occurs here. It has already been pointed out by SCHAUB (1948) and MENCHER et al. (1953) that this La Rosa transgression is nowhere connected with any disconformity. Its contemporaneity shows that during a short time a continuous seaway was present from east Falcón as far south as Chafurray-3 in the Colombian Llanos.

In the upper part of the *Verrutricolporites rotundiporis* Zone (*Psiladiporites minimus* Zone) a regression took place which can be recognized from the south as far as north central Falcón in the transitional-continental deposits of the Chafurray, Caja, Cubugón, Guayabo, Lagunillas and Cerro Pelado Formations, respectively.

In the south, only in the Rio Cubugón, a second transgression is indicated in the *Multimarginites vanderhammeni* Zone, which may be connected with the marine Querales Formation of north central Falcón. In younger zones regressive deposits predominate. Only in Falcón does marine influence persist. While in north central Falcón the marine environment has been interrupted twice during the Neogene by regressive phases, in east Falcón sedimentation was continuously marine, which made it possible to arrive at an accurate tie-in between the palynological zonation and Bolli's subdivision on pelagic Foraminifera. Unfortunately the east Falcón stratigraphical sequence is based on a combination of scattered short sections and stratigraphical thicknesses shown are, therefore, probably not more than rough minimum estimates.

As already mentioned, in contrast to the preceding Palaeogene section, no pronounced unconformities could be detected in the Neogene sediments investigated. Still, rather strong variations in thickness are obvious.

In Colombia the section penetrated by Chafurray-3 is much more reduced than in Voragine-1. In Rio Cubugón it is again slightly thicker than Voragine-1, but in the La Fria and Catatumbo areas maximum thicknesses are observed. Remarkable are the extremely reduced thicknesses observed in B-188, especially since the floral succession appears complete, without any marked gaps.

Thicknesses increase again rapidly, approaching the Falcón basin, where, in north central Falcón especially the *Verrutricolporites rotundiporis* Zone is excessively thick. This second section shows again the great value of palynology as a means of correlating from the marine facies, as present in Falcón, to the mainly continental environment in southwestern Venezuela and the adjoining part of Colombia.

The correlation of the various formations shown on both sections closely corresponds, for the Maracaibo basin, with the correlation chart presented by MILLER et al. (1958).

In section III (Fig.19) the stratigraphy of the younger Neogene deposits is traced from northwestern Colombia to Trinidad. Three separate sedimentary basins are represented. In northwestern Colombia an investigation has been made of the area between Sinu and Bayunca, which is separated from the southern Maracaibo lake and Falcón basin by the Perija mountain range on the border between Colombia and Venezuela. The latter basin is in turn separated from the eastern Venezuela-Trinidad basin by the mountains connecting the Mérida Andes with the coastal range of Venezuela. In these mountainous areas the youngest Neogene deposits have never been present or were eroded during uplift. In view of this geographical separation, the section is best discussed from west to east.

In northwestern Colombia the marine sediments in the Sinu basin and near Bayunca are correlated by the *Pachydermites diederixi* Zone. The total interval covered by these two sections gives a complete succession from the *Multimarginites vanderhammeni* Zone to the *Alnipollenites verus* Zone.

These isolated marine deposits in northwestern Colombia are correlatable with terrestrial sediments in the southern Lake Maracaibo area. In well Catatumbo-1 the *Grimsdalea magnaclavata* and *Alnipollenites verus* Zones can be recognized, but not the boundary between the *Pachydermites diederixi* and *Echitricolporites mcneillyi* Zones, which is due to an unfavourable mottled clay facies and to lack of samples. In general sediments are thicker in this area than in northwestern Colombia, and especially in the Rio Buena Vista section the younger zones reach considerable thickness. In contrast, sediment thickness in the older interval *Jandufouria seamrogiformis* Zone–*Multimarginites vanderhammeni* Zone remains fairly constant, irrespective of facies, as shown by the correlation of Catatumbo-1 with the Rio Onia and Rio Buena Vista sections. This increase in thickness of the youngest sediments in the southern Maracaibo basin is a reflection of tectonic events connected with the approximately contemporaneous rise of the Mérida Andes.

As already shown in section II, the younger Neogene sediments in the southern Maracaibo basin are continuous with those in Falcón, but this sedimentary area was probably separated by a broad land area from the east Venezuelan basin. However, the floral characteristics marking the palynological zones turn up in the same sequence and in a virtually identical expression in both areas, and correlation between the marine Falcón sequence and well OG-2, where a regressive sequence is present, presents no difficulties.

From OG-2 to Trinidad the floristic boundaries again cross facies lines, and in the Catshill-Ortoire area a second tie-in with the zonation on age-indicating pelagic Foraminifera confirms the time-stratigraphical value of the palynological zonation. The younger part of the sedimentary sequence in Trinidad shows rapid changes in thickness and facies over short distances and palynology has been of great value here for unravelling the complex geological history. Of special interest is finally the great thickness attained by the *Grimsdalea magnaclavata* Zone in the Catshill–Ortoire area of Trinidad. This, together with the presence of large quantities of reworked material and the well-known Lengua boulder beds, reflects the final upward movement of the Andean orogeny, resulting in the upthrust of the Trinidad central range.

Of course subdivision of the main floral zones presented here will make it possible to refine considerably the stratigraphical picture presented. This, however, lies beyond the scope of this paper, which was aimed primarily at demonstrating the value of palynological subdivision for regional correlation.

Summarizing the results obtained so far, it may be stated that palynological evidence has:

(*1*) shown the presence and magnitude of previously unsuspected stratigraphical gaps;

(*2*) produced positive evidence for correlation of many Tertiary formations lacking age-indicating animal fossils;

(*3*) shown that pollen boundaries can cross facies boundaries and are then eminently suited for palaeogeographical reconstructions.

[*Editors' Note:* Material has been omitted at this point.]

REFERENCES

AMERICAN COMMISSION ON STRATIGRAPHIC NOMENCLATURE, 1961. Code of stratigraphic nomenclature. *Bull. Am. Assoc. Petrol. Geologists*, 45: 645–665.

ANDERSON, R. Y., 1960. Cretaceous–Tertiary palynology, eastern side of the San Juan Basin, New Mexico. *New Mexico, Bur. Mines Mineral Resources, Mem.*, 6: 58 pp.

BELSKY, C. V., BOLTENHAGEN, E. und POTONIÉ, R., 1965. Sporae dispersae der Oberen Kreide von Gabun, Äquatoriales Afrika. *Paläontol. Z.*, 39: 72–83.

BOLLI, H. M., 1966. Zonation of Cretaceous to Pliocene marine sediments based on planktonic Foraminifera. *Bol. Inform. Asoc. Venezolana Geol. Mineria Petrol.*, 9: 3–32.

BOLTENHAGEN, E., 1967. Spores et pollen du Crétacé supérieur du Gabon. *Pollen Spores*, 9(2): 335–355.

BRATZEVA, G. M., 1965. Pollen and spores in Maastrichtian deposits of the Far East. *Tr. Geol. Inst. Akad. Nauk S.S.S.R.*, 129: 5–42 (in Russian).

CHRIST, H., 1910. *Die Geographie der Farne*. Gustav Fischer, Jena, 357 pp.

COOKSON, I. C., 1950. Fossil pollen grains of proteaceous type from Tertiary deposits in Australia. *Australian J. Sci., Ser. B*, 3: 166–177.

COOKSON, I. C. and PIKE, K. M., 1954. Some dicotyledonous pollen types from Cainozoic deposits in the Australian region. *Australian J. Botany*, 2: 197–219.

COUPER, R. A., 1953. Upper Mesozoic and Cainozoic spores and pollen grains from New Zealand. *New Zealand, Dept. Sci. Ind. Res., Geol. Surv. Palaeontol. Bull.*, 22: 77 pp.

COVAS, G. y SCHNACK, B., 1945. El valor taxonomico de la relacion "longitud del pistilo: Volumen del grano de polen". *Darwiniana*, 7: 88–90.

CROSS, A. T., 1966. Source and distribution of palynomorphs in bottom sediments, southern part of Gulf of California. *Marine Geol.*, 4(6): 467–524.

DRUGG, W. S., 1967. Palynology of the Upper Moreno Formation (Late Cretaceous–Paleocene), Escarpado Canyon, California. *Palaeontographica, B*, 120: 1–71.

ENGLER, A., 1905. Über floristische Verwandtschaft zwischen den tropischen Afrika und Amerika, sowie über die Annahme eines versunkenen brasilianisch–äthiopischen Continents. *Sitz. Ber. Königl. Preuss. Akad. Wiss. Berlin*, 1905: 180–231.

ERDTMAN, G., 1952. *Pollen Morphology and Plant Taxonomy (An Introduction to Palynology, 1). Angiosperms*. Almqvist and Wiksell, Stockholm; Chronica Botanica Co., Waltham, Mass., 239 pp.

ERDTMAN, G., 1954. On the occurrence of olacaceous pollen grains in Eocene deposits in Germany. *Svensk Botan. Tidskr.*, 48: 804–885.

ERDTMAN, G., 1956. Pollen grains of cf. *Ctenolophon* from Tertiary deposits in India. *Grana Palynologica*, 1: 5–7.

FUCHS, H. P., 1967. Pollen morphology and its relation to taxonomy and phytogeography in the family Bombacaceae. *Rev. Palaeobotan. Palynol.*, 3: 119–132.

FUNNEL, B. M., 1964. The Tertiary Period. In: W. B. HARLAND, A. GILBERT SMITH and B. WILCOCK (Editors), *The Phanerozoic Time Scale. A Symposium Dedicated to Professor A. Holmes.—Suppl. Quart. J. Geol. Soc. London*, 120: 179–191.

GONZALEZ GUZMAN, E., 1967. *A Palynological Study on the Upper Los Cuervos and Mirador Formations (Lower and Middle Eocene; Tibú Area, Colombia)*. Brill, Leiden, 129 pp.

GOTHAN, W. und WEYLAND, H., 1964. *Lehrbuch der Paläobotanik*. Akademie Verlag, Berlin, 594 pp.

HARRIS, W. K., 1965. Basal Tertiary microfloras from the Princetown area, Victoria, Australia. *Palaeontographica, B*, 115: 75–106.

HOPPING, C. A., 1967. Palynology and the oil industry. *Rev. Palaeobotan. Palynol.*, 2: 23–48.

HUGHES, N. F. and MOODY-STUART, J., 1966. Descriptions of schizaeaceous spores taken from Early Cretaceous macrofossils. *Palaeontology*, 9: 274–289.
HUTCHINSON, H. and DALZIEL, J. M., 1954. *Flora of West Tropical Africa*. Crown Agents Overseas Governments and Administrations, London, 2nd ed. revised by R. W. J. KEAY, 1(1): 1–294; 1(2): 297–828; 2(1963): 1–544; supplement by A. H. G. ALSTON: *Ferns and Fern-Allies of West Tropical Africa*, 1–89.
JARDINÉ, S. et MAGLOIRE, L., 1965. Palynologie et stratigraphie du Crétacé des bassins du Sénégal et de Côte d'Ivoire. *Mém. Bur. Rech. Géol. Minières*, 32: 187–245.
KEDVES, M., 1960. Études palynologiques dans le bassin de Dorog, 1. *Pollen Spores*, 2: 89–118.
KEDVES, M., 1961. Études palynologiques dans le bassin de Dorog, 2. *Pollen Spores*, 3: 101–153.
KRUTZSCH, W., 1957. Sporen- und Pollengruppen aus der Oberkreide und dem Tertiär Mitteleuropas und ihre stratigraphische Verteilung. *Z. Angew. Geol.*, 3: 509–548.
KRUTZSCH, W., 1959. Mikropaläontologische (sporenpaläontologische) Untersuchungen in der Braunkohle des Geiseltales. *Geologie (Berlin), Beih.*, 3 (21–22): 425 pp.
KUYL, O. S., MULLER, J. and WATERBOLK, H. T., 1955. The application of palynology to oil geology with reference to western Venezuela. *Geol. Mijnbouw*, 17: 49–76.
LANGENHEIM, J. H., HACKNER, B. L. and BARTLETT, A., 1967. Mangrove pollen at the depositional site of Oligo–Miocene amber from Chiapas, Mexico. *Harv. Univ. Bot. Mus. Leaflet*, 21(10): 289–324.
LEIDELMEYER, P., 1966. The Paleocene and Lower Eocene pollen flora of Guyana. *Leidse Geol. Mededel.*, 38: 49–70.
LIECHTI, P., 1960. The geology of Sarawak, Brunei and the western part of North Borneo. *Bull. Geol. Surv. Dept., Kuching*, 3: 1–360.
LINDEMAN, J. C., 1953. *The Vegetation of the Coastal Region of Suriname*. Thesis, State Univ., Utrecht, 135 pp.
LOEBLICH, A. R. and TAPPAN, H., 1964. Treatise on Invertebrate Paleontology, part C, Protista— *Geol. Soc. Am. Spec. Papers*, 1(2): 1–900.
MANTEN, A. A., 1966a. Half a century of modern palynology. Earth-Sci. Rev., 2(4): 277–316.
MANTEN, A. A., 1966b. Some current trends in palynology. Earth-Sci. Rev., 2(4): 317–343.
MAYR, E. (Editor), 1952. The problem of land connections across the South Atlantic with special reference to the Mesozoic. *Bull. Am. Museum Nat. Hist.*, 99, 79–258.
MENCHER, F., FICHTER, H. J., RENZ, H. H. and WALLIS, W. E., 1953. Geology of Venezuela and its oil fields. *Bull. Am. Assoc. Petrol. Geologists*, 37: 690–777.
MILLER, J. B., EDWARDS, K. L., WOLCOTT, P. P., ANISGARD, H. W., MARTIN, R. and ANDEREGG, H., 1958. Habitat of oil in the Maracaibo Basin, Venezuela. In: L. G. WEEKS (Editor), *Habitat of Oil*. Am. Assoc. Petrol. Geologists, Tulsa, Okla., pp.601–640.
MÜLLER, H., 1966. Palynological investigations of Cretaceous sediments in northeastern Brazil. In: J. E. VAN HINTE (Editor), *Proceedings of the Second West African Micropaleontological Colloquium, Ibadan, 1965*. Brill, Leiden, pp. 123–136.
MULLER, J., 1959. Palynology of Recent Orinoco delta and shelf sediments. *Micropaleontology*, 5: 1–32.
MULLER, J., 1964. A palynological contribution to the history of the mangrove vegetation in Borneo. In: L. M. CRANWELL (Editor), *Ancient Pacific Floras*. Univ. Hawaii Press, Honolulu, pp. 33–42.
MULLER, J., 1966. Montane pollen from the Tertiary of northwestern Borneo. *Blumea*, 14: 231–235.
MULLER, J. and HOU-LIU, S. Y., 1966. Hybrids and chromosomes in the genus *Sonneratia* (Sonneratiaceae). *Blumea*, 14: 337–343.
MULLER, J., 1968. Palynology of the Pedawan and Plateau Sandstone Formations (Cretaceous– Eocene) in Sarawak, Malaysia. *Micropaleontology*, 14:1–37.
NAYAR, B. K., LATA, P. and TIWARI, L. P., 1964. Spore morphology of the ferns of west tropical Africa. *Pollen Spores*, 6: 545–582.
POTONIÉ, R., 1934. Zur Morphologie der fossilen Pollen und Sporen. *Arb. Inst. Paläobotan. Petrog. Brennstein.*, 4: 5–24.
POTONIÉ, R., 1956–1960. Synopsis der Gattungen der Sporae dispersae, 1–3. *Geol. Jahrb. Beih.*, 23(1956): 1–103; 31(1958): 1–114; 39(1960): 1–189.

POTONIÉ, R. und GELLETICH, J., 1933. Über Pteridophyten-Sporen einer eocänen Braunkohle aus Dorog in Ungarn. *Sitz. Ber. Ges. Naturforsch. Freunde Berlin*, 1932: 517–528.

RAMANUJAM, C. G. K., 1966. Palynology of the Miocene lignite from south Arcot district, Madras, India. *Pollen Spores*, 8: 149–203.

REYMENT, R. A., 1965. *Aspects of the Geology of Nigeria*. Univ. Press, Ibadan, 144 pp.

SAAD, S. I., 1962. Pollen morphology of *Ctenolophon*. *Botan. Notiser Lunds Botan. Fören.*, 115(1): 49–57.

SCHAUB, H. P., 1948. Outline of sedimentation in Maracaibo Basin, Venezuela. *Bull. Am. Assoc. Petrol. Geologists*, 32: 215–227.

SCHWARCK ANGLADE, A. (Editor), 1956. Lexico Estratigrafico de Venezuela. *Bol. Geol., Publ. Especial., Min. Minas e Hidrocarbon*, 1: 740 pp.

SHORT, K. C. and STÄUBLE, A. J., 1967. Outline. of geology of Niger delta. *Bull. Am. Assoc. Petrol. Geologists*, 51: 761–779.

SIMPSON, G. G., 1952. Probabilities of dispersal in geologic time. In: E. MAYR (Editor), *The Problem of Land Connections across the South Atlantic with Special Reference to the Mesozoic—Bull. Am. Museum Nat. Hist.*, 99: 163–176.

STOLK, J., 1963. Contribution à l'étude des correlations microfauniques du Tertiaire inférieur de la Nigeria méridionale. Dans: *Colloque International de Micropaléontologie, Dakar—Bur. Rech. Géol. Minières Mém.*, 32: 247–275.

TRALAU, H., 1964. The Genus *Nypa* VAN WURMB. *Kgl. Svenska Vetenskap. Akad. Handl.*, 10(1): 1–29.

VAN GIJZEL, P., 1961. Autofluorescence and age of some fossil pollen and spores. *Koninkl. Ned. Akad. Wetenschap., Proc., Ser. B*, 64: 56–63.

VAN DER HAMMEN, T., 1954. El desarrollo de la flora Colombiana en los periodos geologicos, 1. Maastrichtiano hasta Terciario mas inferior. *Bol. Geol. (Bogota)*, 2(1): 49–106.

VAN DER HAMMEN, T., 1956a. Nomenclatura palinologica sistematica. *Bol. Geol. (Bogota)*, 4: 23–62.

VAN DER HAMMEN, T., 1956b. A palynological systematic nomenclature. *Bol. Geol. (Bogota)*, 4(2/3): 63–101.

VAN DER HAMMEN, T., 1956c. Descripcion de algunos generos y especies de polen y esporas fosiles. *Bol. Geol. (Bogota)*, 4: 103–109.

VAN DER HAMMEN, T., 1956d. Description of some genera and species of fossil pollen and spores. *Bol. Geol. (Bogota)*, 4(2/3): 111–117.

VAN DER HAMMEN, T. and GARCIA DE MUTIS, C., 1966. The Paleocene pollen flora of Colombia. *Leidse Geol. Mededel.*, 35: 105–116.

VAN DER HAMMEN, T. and WYMSTRA, T. A., 1964. A palynological study of the Tertiary and Upper Cretaceous of British Guiana. *Leidse Geol. Mededel.*, 30: 183–241.

VAN HOEKEN-KLINKENBERG, P. M. J., 1964. A palynological investigation of some Upper Cretaceous sediments in Nigeria. *Pollen Spores*, 6: 209–231.

VAN HOEKEN-KLINKENBERG, P. M. J., 1966. Maastrichtian, Paleocene and Eocene pollen and spores from Nigeria. *Leidse Geol. Mededel.*, 38: 37–48.

VAN RAADSHOOVEN, B., 1951. On some Paleocene and Eocene larger Foraminifera of western Venezuela. *World Petrol. Congr., Proc., 3rd, The Hague, 1951, Sect. I*, pp.476–489.

VAN STEENIS, C. G. G. J., 1962a. The distribution of the mangrove genera and their palaeogeographical implication. *Koninkl. Ned. Akad. Wetenschap., Proc., Ser. C*, 65: 164–169.

VAN STEENIS, C. G. G. J., 1962b. The land-bridge theory in botany. *Blumea*, 11(2): 235–372.

VARMA, C. P. and RAWAT, M. S., 1963. A note on some diporate grains recovered from Tertiary horizons of India and their potential marker value. *Grana Palynologica*, 4: 130–139.

WILSON, J. T., 1965. Evidence from ocean islands suggesting movements in the earth. In: P. M. S. BLACKETT, E. BULLARD and S. K. RUNCORN (Editors), *A Symposium on Continental Drift— Phil. Trans. Roy. Soc. London, Ser. B*, 1088: 145–167.

WYMSTRA, T. A., 1967. A pollen diagram from the Upper Holocene of the Lower Magdalena valley. Leidse Geol. Mededel., 39: 261–267.

WYMSTRA, T. A. and VAN DER HAMMEN, T., 1967. Palynological data on the history of tropical savannas in northern South America. *Leidse Geol. Mededel.*, 38:71–90.

Part III

REWORKING AND OTHER STRATIGRAPHIC PROBLEMS

Editors' Comments
on Papers 11, 12, 13

11 WILSON
Recycling, Stratigraphic Leakage, and Faulty Techniques in Palynology

12 TURNAU
The Age of Coal Fragments from the Cretaceous Deposits in the Outer Carpathians, Determined on Microspores

13 HAMILTON, HELBY, and TAYLOR
The Occurrence and Significance of Triassic Coal in the Volcanic Necks Near Sydney

Because miospores are usually very small, they are particularly prone to problems of contamination and reworking. Paper 11 by L. R. Wilson gives a clear exposition of many of the hazards of reworking. Reworking does not occur in deposits such as coal and peat, where only authigenic pollen and spores will occur. In clastic sediments derived from preexisting sediments, however, recycling, (or reworking) of pollen and spores can be serious and difficult to recognize. In some cases, reworking has not been recognized with the result that authors have erected new taxa for reworked miospores. However, the presence of reworked material can be used to elucidate the provenance of sediments. In Cretaceous flysch deposits of the Carpathian Mountains, Turnau (Paper 12) determined the age of coal fragments which had been derived from the unroofing of the Silesian coal fields. The coal fragments generally become older as the flysch becomes younger, presumably because of ever deeper erosion in the source area. Turnau was able to recognize the provenance of the coal by a study of the sedimentology of the flysch. In this case, the energy of deposition of the flysch must have been high, because very large boulders of coal occur. A similar situation, although in a different tectonic setting, was illustrated by Traverse (1972), who described palynomorph assemblages from very large boulders in the Franciscan formation of California. The mixture of ages from two fossiliferous boulders supports the contention Hsü and Ohrbohm (1969) that the Franciscan rocks are a melange derived from tectonic mixing of allochthonous blocks.

In cases where reworked spores are freed from their matrix, care is necessary in making identifications, but they can still be used to assist in palaeogeographic reconstructions (Muir, 1967; Bless and Streel, 1976).

An extreme example of reworking was illustrated by Phillips (1974) —an entirely reworked Mesozoic assemblage of spores from rocks which contained only well-preserved lower Tertiary leaves. No contemporaneous pollen was present in the deposit at all!

Wilson's paper also gives examples of stratigraphic leakage, where spores have naturally fallen or been washed down into older sediments. The last paper in this section, by Hamilton et al. (Paper 13), describes Triassic spore assemblages in coal fragments which have dropped down volcanic pipes in the Sydney Basin, and now occur at stratigraphic levels much below the strata from which they are derived. Both Hamilton et al. and Turnau point out that the source rock for the reworked material must have been well coalified before the erosion and/or redeposition took place. In the case of Hamilton et al.'s material, the time interval available for the stratigraphic leakage seems to have been very short, indicating that coalification here has been a relatively rapid process.

Detection of reworking is a difficult problem. In some cases, reworked spores are better preserved than indigenous ones; in other cases, worse. They may stain more or less, be more or less crushed, be older or younger, or be the only spores present. Contamination in collecting, or drilling (either from caving, or from drilling mud), or in the laboratory, can be as difficult as reworking to detect, but, with precautions, can usually be kept to a minimum.

REFERENCES

Bless, M. J. M., and Streel, M. (1976). The occurrence of reworked miospores in a Westphalian C microflora from South Limburg (the Netherlands) and its bearing on palaeogeography. *Meded. Rijks Geol. Dienst*, N.S. **27**(1):1-39.

Hsü K. J., and Ohrbohm, R. (1969). Melanges of San Francisco Peninsula—geologic reinterpretation of type Franciscan. *Amer. Assoc. Petrol. Geol. Bull.* **53**(7): 1348-1367.

Muir, M. D. (1967). Reworking in Jurassic and Cretaceous spore assemblages. *Rev. Palaeobotan. Palynol.* **5**:145-154.

Phillips, L. (1974). Reworked Mesozoic spores in Tertiary leaf-beds on Mull, Scotland. *Rev. Palaeobotan. Palynol.* **17**:221-232.

Traverse, A. (1972). A case of marginal palynology: a study of the Franciscan melanges. *Geoscience and Man* **4**:87-90.

RECYCLING, STRATIGRAPHIC LEAKAGE, AND FAULTY TECHNIQUES IN PALYNOLOGY[1]

L. R. WILSON

Erosion, transport, and redeposition of sediments to form younger sedimentary rocks have been recognized almost since the beginning of geology. Such a process in this paper is termed "recycling". Heavy mineral stratigraphic techniques have for many years utilized the phenomenon of recycling and have made many significant contributions to geology. In the field of paleontology occurrences of plant and animal megafossils in deposits obviously of younger age than those of the fossils are often noted in literature (Jones, 1958). The recognition of palynological fossils as recycled objects has received little notice by most investigators; although Iversen (1936), Grichuk (1950), Ananova (1960), Davis (1961) and others have described specific examples of the phenomenon. This oversight is a serious omission in palynological studies because every clastic rock that is examined is a possible container of recycled fossils, and totally erroneous conclusions can and have resulted from the nonrecognition of such redeposited fossils. When recycled palynological fossils are recognized in an assemblage they are of great value in geomorphic and stratigraphic investigations. The recognition of an understanding of recycled palynomorphs is an important aid in the interpretation of paleogeographic conditions and detection of stratigraphic traps in petroleum geology.

The reason for the lack of earlier general recognition of the occurrence of recycled palynological fossils is probably that the first palynological studies were concerned primarily with peat and coal deposits and not with limestones, shales, siltstones, and sandstones. Peat and coal generally contain only endemic and wind-borne paly-

---

[1] A study resulting from National Science Foundation Grant No. GB-1850.

nological materials; but shales, siltstones, sandstones, conglomerates, and some limestones are derived from preexisting rocks, many of which also contain palynological fossils and therefore can contain fossils of earlier ages in addition to those extant at the time of the sediment's deposition.

The term "palynology" in this investigation is used in its widest sense. Attempts have been made to restrict the use of the word "palynology" to the study of spores and pollen. The name, however, is widely used by the petroleum industry to include all minute fossils observed in sediments or sedimentary rocks, which are studied with the aid of the compound microscope. Although this wider usage may be distasteful, the comprehensive use eliminates the coinage of a multiplicity of discipline names for various groups of fossils and their living counterparts. For industry it is expedient to have a single concise name by which to designate the activities of the laboratory concerned with the "sub-microfossils". Microfossils in geology are generally considered to be the foraminifers, ostracodes, conodonts, and other fossils of comparable sizes. Regardless of the desirability of restricting palynology to spore and pollen investigations, the science cannot remain so cloistered when in economic practice palynology includes both plant and animal fossils found associated on a microscope slide. Here appears to be a case where industry, by popular usage, will dictate the scope of the science.

In the course of the last several years an attempt has been made to collect examples illustrating various problems in palynological fossil recycling and to test criteria by which these problems may be detected and resolved. The results of these efforts are here briefly summarized.

Preservation and duration of palynological fossils in rocks vary with biological groups from which they come. In general those palynomorphs with chitinaceous parts and derived from the animal kingdom appear to preserve better and are identifiable in rocks of greater induration than are those with cellulose or lignaceous composition derived from the plant kingdom. Probably the least durable are thin-walled pollen grains; and the most persistent are the chitinozoans. Within each group of fossil organisms there are morphological structures which are differentially affected by preservation, induration, and recycling. Those Paleozoic spores with fimbriate processes, hyaline equatorial ornament, and spores or pollen with cavate and saccate structures are least persistent in coals changed from low to higher ranks; whereas those with spherical shape, thick walls or heavy ornament are most durable (Wilson, 1961). Subsequent observations have indicated that these same morphological forms will react likewise in the process of recycling, although some may be more resistant because they are in fossil state and when transported are generally protected from abrasion within aggregates of silt or clay rather than being transported unprotected as individuals.

Recycled assemblages are found in both outcrop and bore-hole samples and are natural, but others result from contamination due to faulty collecting or processing techniques. Palynological recycling may be recognized in one or more of the following types of assemblages:

1. Assemblages consisting of fossils of more than one geological age.
2. Assemblages consisting of fossils with different biological stain reactions.
3. Assemblages consisting of fossils showing differential preservation.
4. Assemblages consisting only of fossils recognized to be geologically older than the rocks in which they occur.
5. Assemblages of marine fossils preserved in fresh-water sediments.
6. Assemblages consisting of fossils recognized to be geologically younger than sediments in which they appear to occur.

1. Assemblages of palynological fossils derived from more than one geological age

Rocks in tectonic areas or areas adjacent to them can be expected to contain many examples of recycled sediments, and palynological fossils are commonly found abundantly mixed throughout these rocks. The Ardmore basin in southern Oklahoma contains many excellent examples of this type of palynological recycling. In many cases the "contaminants" are difficult to recognize and to demonstrate the source of their origin; however, the incompatible ages of the fossils and sequences of fossil ages are useful criteria in distinguishing mixed palynological deposits.

The basal part of the Goddard Shale, on the Goddard Ranch in Johnston County, Oklahoma, is an example of recycling that illustrates a geological age inversion of recycled fossils. This Mississippian shale has an abundant spore and pollen flora throughout the section. Near the base it contains approximately a dozen species of *Tasmanites, Quisquilites, Archaeotriletes, Stenotriletes,* and hystrichosphaerids derived from the Woodford Shale, a Devonian formation that was somewhere exposed to erosion in Goddard (Mississippian-Chester) time. Somewhat higher in the Goddard section there occur additional recycled palynological fossils that can be demonstrated as having originated from the Sylvan Shale (Ordovician). These consist of hystrichosphaerids and at least one chitinozoan. The last was described from the Goddard Shale by Wilson and Clarke (1960) but not named because only a single specimen was found. This chitinozoan is similar but probably not congeneric with the genus *Desmochitina*, which is abundant in the Sylvan Shale. However, no chitinozoans are known from either the Devonian or Silurian rocks of the region; whereas they are abundant in several Oklahoma Ordovician formations, and because the hystrichosphaerids are of the same species as those of the Sylvan Shale, it seems reasonable to consider that the chitinozoan is a specimen of a species not reported from that shale.

Another example where two age assemblages are easily distinguished is in the Wills Point Shale (Paleocene) of Saline County, Arkansas (Wilson and Venkatachala, unpublished manuscript). The Wills Point Shale contains an abundant spore, pollen, dinoflagellate, and hystrichosphaerid assemblage of Paleocene age and also a large number of Carboniferous spores. The last were derived from Paleozoic rocks exposed to the north during the deposition of the Wills Point Shale.

An example in Oklahoma where recycling has caused stratigraphic difficulties for some petroleum exploration is in the Anadarko basin where productive stratigraphic traps in the Springer (Mississippian-Chester) are overlain in angular unconformity by Morrowan (Pennsylvanian) rocks. The Springer Formation is commonly identified in the subsurface by commercial palynologists on the presence of *Tripartites, Rotospora,* and several other genera of spores in the rock samples. In the course of Morrowan deposition during Pennsylvanian time a portion of the Springer Formation was removed on the eastern flank of the basin and redeposited toward the west. As a result it is not an uncommon occurrence on the west to find spores of *Tripartites*, etc. in Morrowan rocks one hundred or more feet above the base, and in some cases where no Pennsylvanian spores are preserved. Were it not for the fact that Pennsylvanian type spores are present lower in the bore-hole, the stratigraphic age could be incorrectly determined.

Marine deposits recovered from off-shore localities commonly contain palynological assemblages of more than one age as, for example, those described by Groot (1963) from a core collected on the Biscay Abyssal Plain in the Atlantic Ocean (47°10.8′ N. and 11°25.5′ W.). In this core were found spores and pollen of Paleozoic, Mesozoic, and Cenozoic ages.

2. Assemblages consisting of fossils with different biological stain reactions

When a standard processing and stain schedule is maintained in palynological preparations it is apparent that some fossil assemblages readily absorb color; whereas others are not affected. Factors which appear to determine stain reaction of palynological fossils are one or more of the following: (1) inherent nature of the fossil, i.e. carbonaceous vs. chitinaceous, (2) age, (3) depth of burial, (4) association with tectonic activity, and (5) association with igneous activity.

The Wills Point Shale (Paleocene) noted above also illustrates how recycling of palynomorphs can be detected by biological stain reaction. The older fossils (Carboniferous) are dark brown to black and are difficult or impossible to stain; whereas the younger fossils are yellow to light brown and absorb stain readily. The stain generally used in these investigations is Safranin 0 and it will give a range of colors that, with some experience, can be used to determine the relative age or induration of plant microfossils. This color range is roughly as follows: (1) spores or pollen that have been deeply buried or subjected to moderate induration will not, with general techniques, absorb stain, the wall structure will remain dark brown to black; (2) spores that have not been deeply buried or were not subjected to moderate induration will absorb stain readily and become various intensities of red. Generally, the less the induration or the younger the spores and pollen the brighter will be the color; (3) and finally, modern spores or pollen (except fungus spores most of which are normally brown to black) that occur as contaminants in outcrop samples, or in well-cuttings collected during periods of abundant pollen-rain, will generally absorb only enough Safranin 0 for the preparation to be stained pink. It is not uncommon in parts of Oklahoma, to find in a single palynological preparation of well-cuttings, Pennsylvanian fossils that are unstained, mixed with Cretaceous fossils stained bright red, and modern pollen that are pink. The Pennsylvanian fossils represent the stratigraphic level at which the sample was cut; whereas the Cretaceous fossils were derived by "cave-in" from levels above, and modern pollen from the atmosphere at the well head.

Dark color in palynological fossils appears to result mainly from the depth of burial, or from degree of induration. The latter may be a regional phenomenon, as in areas of the higher-ranked coals and indurated shales, or may be localized to the vicinity of a fault plane (common in the Arbuckle Mountains of Oklahoma), or adjacent to intrusive rocks as dikes and sills (common in Colorado, Wyoming, and Alaska). The shade of color, and persistence of palynological fossils in high- to low-volatile coals of the Arkoma basin in Oklahoma have been described by Wilson (1961). It was shown that in the high-volatile coals (50–60 % fixed carbon) the spores are yellow to light brown; whereas in the medium and somewhat lower volatile coals (60–70 % fixed carbon) the spores are dark brown to black. No spores were recovered from coals with a fixed carbon above 70 %. An attempt has been made to compare the color of the spores at various depths in several deep wells in the Anadarko basin of Oklahoma to determine if a spore-color depth scale could be established. The results are tentative with the following observations: spores yellow to light brown, 2,000 to 6,000 feet; spores light to dark brown, 6,000 to 11,000 feet; spores dark brown to black, 11,000 to 20,005 feet; no spores have been observed below 20,005 feet. The formations penetrated in this area are the Atoka and Morrow (Pennsylvanian), and the Springer (Upper Mississippian). An erosional unconformity separates the Springer and Morrow but it does not seem to be a significant factor in the color shade of the spores. Bore-hole temperature observations at various depths suggest that in the approximate range of 300°F fossil spores and pollen are carbonized to the extent that most are destroyed. At approximately 240°F well hole temperature spores are dark brown but many are still recognizable to species; especially those forms with resistant wall characters. The spores and pollen recovered from samples reported as having 240°F temperature approximate in color and preservation those observed in coals with 60–65 % fixed carbon.

3. Assemblages consisting of fossils showing differential preservation

Assemblages that are known by criteria other than differential preservation to have been recycled, have shown that the perfection character may reveal itself in one of two ways. In some cases the older fossils are less well preserved and in others the younger fossils show greater damage. Generally the closer the two or more assemblages are alike in phylogenetic composition, age, paleoecology and geological physical history the more alike will be the preservation, and adversely the more they will differ.

Probably one of the better illustrations of differential preservation of spores in an assemblage is found in the Hoxbar Group (Pennsylvanian) of southern Oklahoma. The shales contain recycled fossils from the rocks of the Ordovician, Devonian, Mississippian, and parts of the Pennsylvanian periods. In general the Ordovician and Mississippian fossils are better preserved than the Pennsylvanian, probably because they came from rock outcrops slightly more indurated than the others. Many of the Devonian fossils show evidence of abrasion and the Pennsylvanian fossils appear to have been subjected to oxidation or decay before the environment of fossilization was reached. Some of the older fossils appear to have been somewhat brittle and probably were fragmented during recycling. In the Goddard Formation (Mississippian-Chester) of southern Oklahoma, Ordovician and Devonian palynomorphs are generally less well preserved than those of the Goddard, and they are relatively rare fossils in the assemblage.

4. Assemblages consisting only of fossils recognized to be geologically older than the rocks in which they occur

Sediments that are deposited rapidly or under conditions not suitable for plant growth commonly contain few spores and pollen of ages contemporaneous with the deposition. Such sediments are generally the coarser clastics such as sandstone, siltstone, and in some cases, claystones. These sedimentary rocks commonly contain fragments of shale and coal that contain a great abundance of palynological material. Evidence of this nature is present in some of the Pennsylvanian channel sandstone deposits in Nova Scotia and in Iowa. On the north flank of the Uinta Mountains in Wyoming occur Tertiary silty shales that contain only Lower Pennsylvanian or possibly Mississippian spores in the upper part of the rock section and an abundance of Tertiary pollen in the lower part.

5. Assemblages of marine fossils preserved in fresh-water sediments

Marine fossils of palynological status consist of such types as *Tasmanites*, chitinozoans, hystrichosphaerids, some dinoflagellates, the megalospheric stages of foraminifera, marine diatoms, silicoflagellates, discoasterids, etc. Fossils of these organisms normally are deposited in marine sediments and unless the rocks in which they were formed are uplifted above the sea and eroded they do not have an opportunity to be recycled into fresh-water environment. Todays rivers flow across many exposed marine rocks that contain the above fossils and the rocks are eroded and particles transported and deposited, in some instances, far from their place of origin. In the clays of the Mississippi River it is not uncommon to find marine microfossils along with spores and pollen of Pennsylvanian, Cretaceous, Tertiary and Recent plants. One of the remarkable examples of recycling marine palynomorphs occurs in the clays of Lake Michigan in the vicinity of Chicago, Illinois, and Milwaukee, Wisconsin. In certain strata one may find countless millions of *Tasmanites* fossils and virtually nothing else. These fossils appear to have been excavated and transported by glacial ice from exposed Devonian rocks possibly in the Lake Huron region. With the melting of the glacial ice the fossils were deposited in the lake clays.

6. Assemblages consisting of fossils recognized to be geologically younger than the sediments in which they appear to occur

When fossils are found that are younger than the rocks in which they appear to occur the phenomenon is generally referred to as stratigraphic leakage. The emplacement of such fossil assemblages requires special geological conditions. The more common occurrences of this type are associated with the earlier development of cavernous or karst topography in which the subsequent sedimentary fillings contain palynological fossils.

In the United States, the western part of Illinois, eastern and central parts of Iowa contain cave fillings in Devonian limestone that are Pennsylvanian in age. These fillings are of shales and sandstones and commonly contain fragments of cordaitalean wood, leaves of ferns and pteridosperms, seeds and spores. Some of the cave deposits are extensive and presently many feet below the ground, but were closely associated with the surface during Pennsylvanian time. Presently in northwestern Arkansas the Pitkin Limestone (Mississippian) is exposed and in places has karst-like topography. In the widened joint planes of the limestone, and in sinkholes, carbonaceous soils have accumulated which contain in some cases an abundance of modern pollen. This modern example could be visualized as the condition in Pennsylvanian time before the seas advanced upon the Devonian rock topography in Illinois and Iowa.

Another case where younger fossils may be found stratigraphically under older fossils is in association with low-angle thrust faulting. Generally this condition is observed in bore-hole samples and a duplication of fossil assemblages is commonly noted, and in the vicinity of the fault the fossils are destroyed or highly carbonized.

Faulty techniques resulting in real or apparent mixed palynological assemblages

Mixed palynological assemblages arising from faulty techniques generally result in several types that resemble those occurring naturally. When one is aware of the conditions that may produce these mixed assemblages it is generally possible to recognize the true origin of the palynomorphs. The naturally mixed assemblages and those which are the result of contamination have many similarities, as, for example, in both there may be fossils of different ages, spores and pollen with various wall colors, and differentially preserved morphological structures. The stages in palynological work where the contaminants are acquired mostly occur during the collection, storage, and processing of the samples.

Collection of samples

People unfamiliar with palynological techniques, when collecting outcrop samples are likely to overlook the need for collecting only from freshly exposed sections. Surface samples are generally weathered and barren of palynological fossils and almost all contain contaminants such as soil fungi, algae (*Protococcus*), modern spores and pollen. In tropical and warm temperate regions the depth of weathering is at places many feet on the outcrop; but in adjacent drilling abundant fossils occur. The following two examples illustrate the need for careful surface sample collecting. (1) The Ellenberger chert (Ordovician) of Texas is an important rock unit that has been intensely studied for hystrichosphaerids and other palynomorphs, but to date it has proved barren except for a rich and widespread assemblage of algae that has been identified as a modern species of *Protococcus*. All collections that contain the algae were collected at the surface and by one unfamiliar with palynological work. (2) A collection of coal from an abandoned mine tipple in New Brunswick, Canada, proved upon processing to contain not only an abundance of Carboniferous spores, but a large number of *Picea mariana* pollen. The mine is located in a *Picea mariana* forest. Additional coal samples were subjected to charring in an open flame and when processed, the preparation contained only Carboniferous spores.

Much difficulty may be encountered with bore-hole samples if the collector is not reliable. Nearly every subsurface field geologist has experienced having bore-hole cuttings collected for him by a driller who was irresponsible. Samples that are not reliably collected should be ignored. Cores and side-wall cores are best for stratigraphic work in subsurface palynological investigations but generally only cuttings are available. These last must always be suspected of containing fragments of rock which originated higher in the drilling hole and "caved" down the hole. If one studies the samples progressively down the hole he is always aware of what overlies the level he is examining, and in this manner reduces to a minimum the interpretative errors of mixed palynological assemblages.

The popular belief that bore-hole cores are free of contaminating palynological fossils is erroneous. Cores always need thorough cleaning before being sampled because commonly they are covered with clay when removed from the core barrel and those cores consisting of friable rocks are in many cases badly contaminated. Palynological literature contains examples of faulty work done with uncleaned core samples.

A source of mixed assemblages in bore-hole cuttings is some drilling muds currently being manufactured. These commercial muds vary in composition but some contain pulverized brown coal that is rich in spores and pollen. The detection of the contaminants is difficult if one does not know the palynological flora of the coal mixture, and especially if the stratigraphic age of the bore-hole is close to the age of the drilling mud fossils.

Storage

The method of storage of samples by some palynologists is careless and inadequate. The samples should be air dried as rapidly as possible, or placed in a liquid preservative. This procedure prevents wet coals from slaking and inhibits the growth of fungi on shales and clays. The dried samples should not be contained in cloth or paper bags, for the first allow fine particles to sift out and the latter become brittle with age and break open. Wide-mouth quart-size glass jars appear to be the ideal containers because they are rust-proof, prevent contamination, and are easily arranged in a storage room.

Processing of samples

Because palynological fossils occur in a wide variety of sedimentary rocks there are many processing techniques, but generally they all consist of digestion, maceration, and disaggregation steps. This discussion does not intend to describe any of these techniques but rather to emphasize the dangers of overprocessing, and inconsistencies in schedules when close stratigraphic studies are being made. The use of strong bases, excessive heating, too frequent and sustained use of ultrasonic generators are especially destructive to most palynological fossils. Assemblages remaining after the above treatments will be much different from those not as severely processed. Fossils with delicate ornament will be divested, the cavate and saccate spores and pollen will lose their diagnostic features, and where a sample contains more than one age assemblage the one that is more indurated survives the processing, resulting in a possible incorrect age determination.

REFERENCES

ANANOVA, Y. N. 1960. On the redeposited complex of pollen. Moscow Assoc. Naturalists, Bull., Sect. Biology, 65 (3): 132–135.

DAVIES, M. B. 1961. The problem of rebedded pollen in the late-glacial sediments at Taunton, Massachusetts. Amer. J. Sci., 259: 211–222, 1 pl.

GRICHUK, V.P. 1950. The vegetation of the Russian Plain during Lower and Middle-Quaternary. Akad. nauk SSSR, Geogr. In-ta, Trudy, 46.

GROOT, J. J. 1963. Palynological investigation of a core from the Biscay Abyssal Plain. Science, 141, no. 3580: 522–523.

IVERSEN, JOHS. 1936. Sekundäres Pollen als Fehlerquelle: Danmarks Geologiske Undersøgelse, IV. Raekke, Bd. 2, Nr. 15: 3–25. København.

JONES, D. J. 1958, Displacement of microfossils. J. Sedimentary Petrology, 28: 453–467.

WILSON, L. R. 1961, Palynological fossil response to low-grade metamorphism in the Arkoma basin. Tulsa Geol. Soc. Digest, 29: 131–140.

WILSON, L. R. and CLARKE, R. T. 1960. A Mississippian chitinozoan from Oklahoma. Okla. Geol. Survey, Okla. Geol. Notes, 20: 148–149.

The Age of Coal Fragments from the Cretaceous Deposits in the Outer Carpathians, Determined on Microspores

by

E. TURNAU

Presented by M. KSIĄŻKIEWICZ on March 16, 1962

Coal fragments are abundant in some flysch deposits of the Outer Carpathians. Their presence has been known for a long time and described by several authors. The coal fragments, either not rounded or subangular, are of various size, from less than one millimeter to a few meters in diameter. They occur in the Cretaceous and Tertiary flysch deposits of the Silesian, Subsilesian and Skole series, and have not been hitherto found in the Fore-Magura and Magura series (with the exception of the outer part of the latter in the environs of Bystrice near Hostyn [2], p. 115—116, Fig. 46). Tietze and Stur determined the age of some coal fragments as Carboniferous, basing on macroflora. Zerndt [11], basing on megaspores, determined as the Westphalian C or upper Westphalian B the age of coal fragments found at Woźniki near Wadowice and at two localities in the environs of Dynów. Some authors consider that the coal fragments originate from the continuation of the Upper Silesian coal basin in the south and the south-east (e. g. [8], p. 119, Fig. 6).

The coal fragments dealt with in the present paper derive from the Cretaceous flysch deposits of the Silesian and Subsilesian series in the area between the Dunajec and Biała rivers (Fig. 1).

The age of the coal has been determined basing on microspores. The stratigraphical microspore sequence established by Dybova and Jachowicz [3] and by Jachowicz and Żołdani [6] for Upper Silesia has been accepted as standard. The age of coal fragments has been determined on characteristic forms, rather than on characteristic assemblages, and the quantitative composition of assemblages has been considered less important. This was implied by the nature of the material obtained. The results of analysis of small coal fragments (one to six cm. in diameter) are not wholly comparable with the results of analysis of mean samples taken from the coal seams *in situ*. This is due to the vertical variability of spore assemblages in a single coal seam, connected with the natural succession of assemblages of plants.

It has been attempted to obtain at least a few fragments of coal from particular flysch layers. Each fragment has been analyzed separately. Fragments of coal obtained from the same layer, and even from layers a few score meters distant in the profile, never revealed any difference in age.

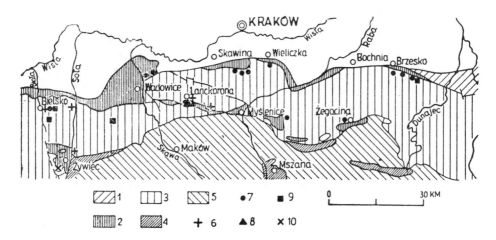

Occurrence of exotic coal ragments in the flysch Carpathians between the Dunajec and Biała Rivers

1 — Skole series, 2 — Subsilesian series, 3 — Silesian series, 4 — Foremagura series, 5 — Magura series (after Książkiewicz), 6 — coals in Upper Cieszyn shales, 7 — coals in Grodischt sandstones, 8 — coals in Gaize Beds, 9 — coals in Lgota Beds, 10 — coals in Szydłowiec sandstones

In Table I, the list of the investigated occurrences of coal fragments is presented, with the results of analyses.

It may be seen from the above list, that the analyzed fragments derived from the deposits of the uppermost Namurian C or lower Westphalian A, of the Westphalian A, and — most commonly — of the Westphalian B.

The coals representing the upper Namurian C or the lower Westphalian A contain microspores which in Upper Silesia appeared in the Westphalian A, i.e. *Florinites antiquus* Schopf, *Alatisporites* sp., *Densosporites decorus* (Loose) Dyb. and Jach., together with microspores which in Upper Silesia disappeared in the Namurian C, i.e. *Anulatisporites coronarius* Dyb. and Jach., *Schulzospora rara* Kosanke. There occur also microspores common to the Namurian and to the lower Westphalian, such as *Microsporites karczewskii* (Zerndt) Dijkstra, *Simozonotriletes intortus* (Waltz) Potonié and Kremp, *Cingulizonates tuberosus* Dyb. and Jach., *C. asteroides* (Kosanke) Dyb. and Jach., *Densosporites spinosus* Dyb. and Jach., and microspores common to the upper Namurian and to the Westphalian, such as *Florinites similis* Kosanke, *Ahrensisporites angulatus* (Kosanke) Dyb. and Jach. Moreover, there occur several species with a large vertical range.

The coals, probably representing the Westphalian A, contain an assemblage of microspores which differs from the above mentioned by the lack of Namurian species, as *Schulzospora rara* and *Anulatisporites coronarius*.

TABLE I

Series	Beds	Locality	Age of coal
Silesian	Upper Cieszyn shales	Stream Ubionka, Sułkowice near Lanckorona	Westphalian B
		Lanckorona	Westphalian A or B
		Bujaków east of Bielsko	Westphalian B
		Stream Lipnik near Bielsko	Westphalian B
		Żywiec	Westphalian B
	Grodischt sandstones	Żegocina	Upper Namurian C or lower Westphalian A
		Porąbka Uszewska near Brzesko	Westphalian B
		Bocheniec near Brzesko	Westphalian B
		Okocim near Brzesko	Westphalian B
		Wiśniowa east of Myślenice	Westphalian B
		Koźmice near Wieliczka	upper Namurian C or lower Westphalian A
		Stream Lipnik near Bielsko	Westphalian A or B
	Lgota Beds	Porąbka Uszewska near Brzesko	upper Namurian C or lower Westphalian A
		Kaczyna south-west of Wadowice	Carboniferous
		Stream Lipnik near Bielsko	uppper Namurian C or lower Westphalian A
		Straconka near Bielsko	Westphalian A
Subsilesian	Grodischt sandstones	Mogilany south of Kraków	Westphalian B
		Stream at Świątniki south of Kraków	Westphalian B
		Zbydniowice south of Kraków	Westphalian B
		Lanckorona	Westphalian B
		Ryczów near Wadowice	Westphalian B
		Woźniki near Wadowice	Westphalian B
	Gaize Beds sandstones with Bryozoans and Lithothamnium from Szydłowiec	Lanckorona	Westphalian B
		Middle of stream at Radziechowy near Żywiec	Westphalian A

The coals representing the Westphalian B contain numerous microspores which appear in the Westphalian B, i.e. *Calamospora breviradiata* Kosanke, *Punctatisporites obliquus* Kosanke, *P. orbicularis* Kosanke, *Reticulatisporites adhaerens* Kosanke, *Cirratriradites punctatus* Dyb. and Jach., *Florinites ovatus* Dyb. and Jach., *Wilsonia punctata* Dyb. and Jach., *Cancellatisporites cancellatus* Dyb. and Jach. (the last species is considered as characteristic for the Orzesze Beds of Upper Silesia — [3]), together with microspores which disappear in the Westphalian B, i.e. *Ahrensisporites angulatus* (Kosanke) Dyb. and Jach., *Cingulizonates asteroides* (Kosanke) Dyb. and Jach., *Densosporites spinosus* Dyb. and Jach. *Reistrickia pallida* Dyb. and Jach. There occur also several microspores limited to the Westphalian, i.e. *Florinites antiquus* Schopf, *Calamospora flexilis* Kosanke. *Verrucosisporites adenotatus* Dyb. and Jach., *Densosporites decorus* (Loose) Dyb. and Jach., *Alatisporites trialatus* Kosanke, *Tuberculatisporites gigantonodatus* Dyb. and Jach., *Triquitrites auritus* (Kosanke) Dyb. and Jach., and microspores of which the vertical range is still larger.

In some cases it has not been possible to decide whether the coal represents the Westphalian A or the Westphalian B. These coals contained Westphalian microspores such as *Florinites antiquus* Schopf, *Alatisporites* sp., *Densosporites decorus* (Loose) Dyb. and Jach., together with microspores common to the Namurian and to the lower Westphalian, such as *Densosporites spinosus* Dyb. and Jach.

The direction measurements of supply of clastics for the flysch deposits in which coal fragments have been found indicate source areas lying to the south and to the south-east of the Upper Silesian coal basin. The directions of supply ascertained are as follows: for Upper Cieszyn shales — from the west and north-west, for Grodischt sandstones — from the north-west, north-east, south-west and south-east, for the coarse clastics in the Lgota Beds — from the north-west and north-east, for Gaize Beds — from the west, north-west and north, for sandstones from Szydłowiec — from the north-east (Książkiewicz [7]). However, it should be taken into account that it has not been possible to measure the directions of supply of clastics for the particular layers which yielded the coal fragments investigated here, with the exception of the Lgota Beds at Straconka, where A. Radomski (personal information) found drag-casts and cross-bedding indicating a supply from the west and west-north-west.

It is striking that no fragments of coal representing the Namurian A, Namurian B and at least partly Namurian C, have been found. This might mean that till the highest Namurian C the area of deposition of the coal measures did not extend far to the south and south-east of the Upper Silesian coal basin.

The lack of coals from the Namurian A agrees well with the results of sedimentological investigations made by Gradziński, Radomski and Unrug [4]. According to these authors, in the Namurian A clastics have been supplied to the productive formation of Upper Silesia from an area lying to the south-east, the uplift of this area being due to the Sudetic phase.

Nevertheless, it is possible that to the south of the area of sedimentation of the outer Carpathian flysch coal measures have been deposited in the Namurian. According to Šilar [9] in the Upohlav Beds near Púchov occur fragments of coal, the age of which was determined by Havlena [5] as Namurian A. The supply of clastics for the Upohlav Beds was, according to Andrusov [1], from the south, or from areas lying inside the Pieniny Klippen Belt. On the other hand, it might be doubted whether the coal fragments from the Upohlav Beds are really of Namurian age. Havlena's determinations of badly preserved specimens (judging from figures in his paper) seem to be not quite correct; some microspores generally considered as characteristic for the Westphalian (*Punctatisportites obliquus* Kosanke, *Cyclogranisporites leopoldi* (Kremp) Pot. and Kremp, *Verrucososporites* cf. *obscurus* (Kosanke) Pot. and Kremp, are stated by Havlena to occur in the assemblage which he considers as Namurian.

All coal fragments found represent the Westphalian A and B, and possibly also the highest Namurian C. This suggests that in these times the coal measures were well developed in the area to the south and to the south-east of Upper Silesia, and either had been deposited in several small basins, or extended uninterruptedly from Upper Silesia to the Bug basin.

In the Westphalian C and D the sedimentation of the coal measures seems to have been again limited to the Upper Silesian and Bug basins; this is indicated by the complete lack of coals younger than the Westphalian B in the investigated fragments..

Therefore, the deposition of the coal measures in the area to the south and to the south-east of Upper Silesia appears to have been restricted to the uppermost Namurian C, Westphalian A and Westphalian B. On the other hand, in the eastern part of the Upper Silesian basin, the Namurian B and C and the Westphalian A and B are not represented, the deposits of the Westphalian C lying immediately on those of Namurian A. According to Siedlecki [9], this is due to an uplift occurring about the limit of the Namurian and Westphalian connected with the Erzgebirge phase, followed by a negative movement in the Westphalian C. It is possible that the same tectonic movements were reponsible for the initiation and close of deposition of the coal measures in the area lying to the south and to the south-east, from which derived the fragments dealt with in the present paper.

The subject of the present study has been pointed out to me by Professor M. Książkiewicz. I should like to express my gratitude also to those who provided, or helped me to collect the material described here, and particularly to Miss Dr. J. Burtan, Dr W. Nowak, Dr R. Unrug, Dr S. Dżułyński, Dr. S. Gasiorowski.

INSTITUTE OF GEOLOGY, POLISH ACADEMY OF SCIENCES
(ZAKŁAD NAUK GEOLOGICZNYCH, PAN)

REFERENCES

[1] D. Andrusov, *Etude géologique de la Zone des Klippes Internes des Karpates Occidentales*, Slov. Akad. Vied. a Um., Geol. Prace, 34, Bratislava, 1953.

[2] — , *Geologie der tschechoslowakischen Karpaten. I*, Bratislava, 1958.

[3] S. Dybowa, A. Jachowicz, *Microspore Zones in Upper Silesian Productive Carboniferous*, Kwart. Geol., **1** (1957), 1.

[4] R. Gradziński, A. Radomski, R. Unrug, *Directions of transport of the clastic material in the Upper Carboniferous in the Silesian Coal Basin*, Kwart. Geol., **5** (1961), 1.

[5] V. Havlena, *Plant spores of the Carboniferous cannel coal pebbles from the Carpathian Cretaceous*, Čas, Pro Min. a Geol., I/4, Praha, 1956.

[6] Jachowicz, Z. Żołdani, *Notes on Carboniferous microflora from bore-hole "Żebrak"* Kwart. Geol., **4** (1960), 3.

[7] M. Książkiewicz, *Pre-Orogenic sedimentation in the Carpathian Geosyncline*, Geol. Rundschau., **50** (1960).

[8] J. Nowak, *Esquisse de la tectonique de la Pologne*, Cracovie, 1927.

[9] S. Siedlecki, *Paleozoic formations of the Kraków region*, Inst. Geol. Bull., **73** (1954).

[10] J. Šilar, *Vorkommen karbonischer Kohle in senonischen Konglomeraten in Thale des Váh bei Pùchov*, Čas. Pro Min. a Geol., **1**, Praha, (1956), 4.

[11] J. Zerndt, *Versuch einer stratigraphischen Bestimmung von Steinkohlen-Geröllen der Karpaten auf Grund von Megasporen*, Bull. Acad. Polon. Sci. et lettres, Cl. Sci. Mat. et Natur., Sér. B., Sci. Natur., Cracovie, 1932.

The Occurrence and Significance of Triassic Coal in the Volcanic Necks near Sydney

L. H. HAMILTON
School of Applied Geology, University of New South Wales

R. HELBY
New South Wales Department of Mines, Sydney

AND

G. H. TAYLOR
C.S.I.R.O. Division of Mineral Chemistry, Chatswood, N.S.W., 2067

ABSTRACT—The breccia pipes near Sydney contain numerous inclusions of coal. Spores have been macerated from some of this material, the microflora obtained being no older than Hawkesbury Sandstone equivalent. The coal both in the breccia pipes and in the peripheral contorted zones is of bituminous rank, which is evidence that it has not been heated above quite modest temperatures.

Introduction

Although David (1896) suggested that the coal in the Euroka Farm breccia pipe was probably derived from the "Hawkesbury Sandstone", coal fragments in other similar breccia pipes near Sydney have generally been assumed to be of Permian age. New evidence indicates that coal in at least several of these breccia pipes is of Triassic age and is now at a lower position than the strata from which it was derived.

The volcanic necks under discussion are situated in an area extending to about 25 miles north, 43 miles east, and 30 miles south-east of Sydney. Localities are given in the table as Army Grid references for the Sydney Four-Mile Topographic Sheet. Further details on the localities are given by Adamson (1966). Wilshire (1961) has described some of the volcanic necks as layered diatremes. They generally occur as vertical pipes with circular to irregular elongated outcrops ranging in area from a few acres to more than 40 acres. The breccias consist predominantly of altered basaltic fragments, commonly amygdaloidal, set in a matrix of clay and carbonate minerals and scattered quartz grains. They also contain a wide variety of other igneous and sedimentary rock fragments, including coal.

Peripheral Zones in the Pipes

At some localities contorted beds of sedimentary rocks have been exposed between the breccia and the non-deformed country rocks. Coal occurs sporadically in such peripheral zones at Minchinbury quarry and Erskine Park quarry. At an exposure in Minchinbury quarry the contact between the contorted sediments and the wall rock is marked by a small fault. Here the contorted sediments are less deformed and are clearly part of the country rock. The sedimentary beds in the peripheral zones are generally centroclinal.

The coal in the peripheral zones attains a maximum thickness of about one foot. It is a moderately bright humic coal and contains a high proportion of exinite, especially in the form of leaf cuticle. This abundance of leaf cuticle is not typical of the Permian coals of the Sydney basin, but is much more characteristic of some Triassic coals—such as the Ipswich coals of south-eastern Queensland (Cook and Taylor, 1963). The maximum reflectance of the vitrinite in the contorted zones lies within the range $0\cdot87$–$0\cdot97\%$. These low reflectance figures suggest maximum temperatures probably of less than 100° C. Deformation in the contorted sediments has resulted in fragmentation of the coal, with rotation of the fragments and their subsequent bonding by "pressure welding". The optical anisotropy of these fragments, however, is related to the original direction of bedding, indicating that the fragments were not plastic during deformation. This also indicates that the coal was subject to a maximum temperature well below 350° C., at which the vitrinite would become appreciably plastic (Brown, Taylor and Waters,

1965). It also suggests that this coal may have been of sub-bituminous or bituminous rank at the time of deformation.

An abundant, well-preserved, microflora has been extracted from the coal in the contorted sediments. It consists of *Alisporites* spp., *Aratrisporites* spp., *Cadargasporites senectus*, *Converrucosisporites cameroni*, *Dictyophyllidites mortoni*, *Cycadopites nitidus*, *Kraeuselisporites differens*, *Lycopodiumsporites* sp., *Microreticulatisporites* sp., *Neoraistrickia taylori*, *Nevesisporites limatulus*, *Polypodiisporites ipsviciensis*, *Punctatisporites* spp., *Punctatosporites walkomi* and *Verrucosisporites* spp. In particular, the presence of *Cadargasporites senectus* in association

The coalified wood fragments are generally elongated and range in length from about one millimetre to some tens of centimetres. They are commonly surrounded by rims up to several millimetres thick of calcite containing euhedral prismatic crystals of quartz. The presence of the rims suggests that the fragments of woody coal have shrunk, usually to between one-half and three-quarters of their former volume. Also, the even thickness of the rims around fragments of banded coal suggests that they have shrunk uniformly to some 80–90% of their former volume.

A well-preserved microflora has been extracted from two coal specimens from the Hornsby

Localities of Pipes and Reflectances of Coal Specimens

	Pipe	Locality		Maximum Reflectance of Vitrinite at λ of 527 nm.
		Army Grid Reference	Adamson's (1966) Reference	
Fragments in Breccias	Hornsby Quarry	409,837	4	0·98–1·26
	Minchinbury Quarry	383,824	5	0·84–0·92
	Erskine Park Quarry	379,822	6	0·85–1·09
	Richardson's Farm	382,821	31	0·80–0·86
	Norton's Basin	361,817	35	0·76–1·04
	Davidson's Quarry	383,835	7	0·81–0·96
	Bulls Hill	389,818	32	0·92–0·97
	Patonga	428,859	—	0·73–1·78*
	Fitzpatrick's First Quarry	391,824	28	0·93–0·98
	Gilligans Road	405,840	21	0·66–0·91
	Campbelltown	377,786	36	1·80–1·90†
	Bloodwood Road	407,852	14	0·78–0·94
Peripheral Zones	Minchinbury Quarry	383,824	5	0·76–0·96
	Erskine Park Quarry	379,822	31	0·82–1·04

* The material of higher reflectance shows textural evidence of heat alteration, presumably by the basaltic intrusion which occurs in the pipe.
† A single specimen from vicinity of intrusion which appears to have caused abnormal reflectance.

with *Kraeuselisporites differens* and *Nevesisporites limatulus* indicates that the microflora is no older than the equivalent of the Minchinbury Sandstone, i.e., it is of M.–U. Triassic age.

Coal Fragments in the Breccias

Every breccia pipe in the Sydney region so far examined by the writers contains sparsely and unevenly distributed coal fragments. Although the coal fragments in the breccias are now of approximately the same rank as the coal in the peripheral zone (i.e., high volatile bituminous), most of these fragments appear to represent former single pieces of wood.

breccia. It consists of *Alisporites* spp., *Aratrisporites* spp., *Cycadopites nitidus*, *Duplexisporites gyratus*, *Granulatisporites minor*, *Kraeuselisporites pallidus*, *Neoraistrickia taylori*, *Osmundacidites* spp., *Pilasporites plurigenus*, *Polypodiisporites ipsviciensis*, *Punctatisporites* spp., *Punctatosporites walkomi*, *Verrucosisporites* sp., and *Vitreisporites pallidus*. The microflora is dominated by *Kraeuselisporites pallidus* and *Pilasporites plurigenus* and represents a particularly specialized assemblage reminiscent of some of the microfloras extracted from Ipswich coals of Triassic age. The presence of *Duplexisporites gyratus* indicates that the microflora is certainly no older than middle Hawkesbury Sandstone equivalent.

A well-preserved microflora also has been extracted from coal in the Patonga breccia. It consists of *Acenthotriletes* sp., *Alisporites* spp., *Cycadopites nitidus*, *Cycadopites* sp., *Dictyophyllidites mortoni*, *Kraeuselisporites differens*, *Monosulcites* sp., *Neoraistrickia taylori*, *Osmundacidites* spp., *Pilasporites plurigenus*, *Polypodiisporites ipsviciensis*, *Punctatisporites* spp., *Vitreisporites* sp., and *Circulisporites parvus*. The microflora is dominated by *Alisporites* spp. and *Neoraistrickia taylori*. The acritarch *Circulisporites parvus* is also particularly common. Overall, the assemblage is similar to the microfloras described previously, the presence of *Cycadopites nitidus* indicating that it is no older than Hawkesbury Sandstone equivalent.

The coal fragments from which spores have been extracted are petrologically very similar to the coal in the peripheral contorted zones at the margins of the Erskine Park and Minchinbury pipes.

A comparison of coaly material from the various breccia pipes and the peripheral zones is given in the table. The reflectances of woody and banded coals fall within the same range, although woody coal is in general isotropic and banded coal is anisotropic.

The table shows that the maximum reflectances of vitrinite from coal in the peripheral zones mostly lie within the same range as those from the breccias. All of the values are significantly lower than those for Permian coal in the northern part of the Southern Coalfield, N.S.W., where the maximum reflectance values so far recorded lie between 1·35% and 1·51%.

Discussion

The rank of the coaly material at the time of its incorporation in the breccia poses something of a problem. The properties of fragments of banded coal in the contorted sediments and, probably, in the breccia also, point to the coal having been at or close to the bituminous coal stage of rank. However, this would appear to conflict with available stratigraphic evidence, which suggests that an insufficient thickness of sediments could have overlain a Wianamatta-age coal at the time the breccia was formed.

All microfloras recovered from the coal to date indicate a specific Triassic age for the breccia formation. This is consistent with the observation that volcanic detritus is the dominant component of Wianamatta Group sediments above the Ashfield Shale, suggesting contemporaneous volcanic activity. There is also some evidence that at least one of the volcanoes (Richardson's Farm) has been buried by Wianamatta sediments.

None of the coal shows evidence of profound thermal alteration, which, had it occurred, would have been indicated under the microscope by changes in the optical and structural properties of the vitrinite and exinite minerals. In the same way spores are excellently preserved after maceration. The banded coal fragments have undergone very little deformation while in the breccia, and the only effects attributable to their incorporation are shrinkage, as indicated by rims, and a possible slight increase in rank. Some of the woody fragments of coal in the breccia have deformed plastically, apparently prior to a fairly uniform shrinkage. These effects of shrinkage and rank are probably the result of slight increases in temperature. The low temperature indicated is consistent with the hypothesis that cooling may have occurred as a result of adiabatic expansion. This is also suggested by the glassy nature of the basaltic fragments (though much of this is now altered) and by the occurrence of tiny vesicles in the same material.

References

ADAMSON, C. L., 1966. The Crushed Stone and Gravel Industry in the County of Cumberland, N.S.W. *Contracting and Construction Equipment*, October.

BROWN, H. R., TAYLOR, G. H., and WATERS, P. L., 1965. Research into the Production of Metallurgical Coke from Australian Bituminous Coals. *Proc. Eighth Commonwealth Min. Metall. Congr.*, **6** (General), 911.

COOK, A. C., and TAYLOR, G. H., 1963. The Petrography of Some Triassic Ipswich Coals. *Proc. Aust. I.M.M.*, No. 205, 35–55.

DAVID, T. W. E., 1896. Anniversary Address. *J. Proc. Roy. Soc. N.S.W.*, **30** (postscript on p. 69).

WILSHIRE, H. G., 1961. Layered Diatremes near Sydney, New South Wales. *J. Geol.*, **69**, 473–483.

Part IV
MEGASPORES

Editors' Comments on Papers 14 and 15

14 DIJKSTRA
Carboniferous Megaspores in Tertiary and Quaternary Deposits of S. E. England

15 HUGHES
Wealden Plant Microfossils

Paper 14 by Dijkstra, on Carboniferous megaspores bridges the gap between Parts III and IV. Dijkstra was the first to recognize that many megaspores in British Tertiary and Quaternary deposits are, in fact, reworked from Carboniferous deposits. Megaspores are generally thick-walled and tough; they are thus resistant to erosion, transport, and redeposition. They are not now generally used very much for stratigraphy, although in a crude way they can be used for dating on the basis of size. In the Devonian, where they first appear, megaspores are rather small in diameter, although by definition they must be over 200 μm. By the Lower Carboniferous, diameters of over 1 mm are know. Carboniferous megaspores can reach 4–5 mm in diameter, but Permian and Triassic megaspores are smaller, around 600 μm. In the Jurassic and Lower Cretaceous, megaspores again increase in size and reach nearly 1 mm, and from the Lower Tertiary onwards they maintain a maximum size of about 800 μm. Most megaspores are Lycopodialean in affinity; the bizarre forms described by Hughes from the Lower Cretaceous Wealden beds (Paper 15) were probably Lycopodialean.

Megaspores possess many morphological features which are not found in microspores. They often have extensions of the triradiate mark in the form of frills or as separate chambers known as "gulas." The wall ultrastructure is also apparently quite different from that of microspores, more layers being present (Kempf, 1973).

The study of megaspores tends nowadays to be somewhat neglected in palynology. The reader is referred to Spinner (1969) for further information.

REFERENCES

Kempf, E. K. (1973). Transmission electron microscopy and fossil spores. *Palaeontology* **16**:787-798.

Spinner, E. (1969). Megaspore assemblages from the Visean deposits at Dunbar, East Lothian, Scotland. *Palaeontology* **12**:441-465.

Carboniferous Megaspores in Tertiary and Quaternary Deposits of S.E. England. By Dr. S. J. DIJKSTRA, Geological Survey, Heerlen, The Netherlands.

[Plates XVI–XXI.]

IN this Magazine Miss Chandler (1946) drew attention to some large spores which had been found at several horizons in Tertiary and Quaternary deposits, in the hope that someone would be able to throw light on their true nature. These spores were attributed by C. Reid (1899, 1916, 1920) to *Isoetes*, and this error had been repeated without investigation by Chandler (1921) and by E. M. Reid and Chandler (1923). Chandler (1946), however, pointed out that their size (1·55–1·95 mm.) is much larger than any known *Isoetes*, or than any living spore hitherto seen. R. Potonié, who examined a number of smaller specimens from Cliff

End, Mudeford, suggested that they must have been derived from older deposits, and actually compared them with various Carboniferous types. According to Chandler, having regard to their mode of preservation, and to their occurrence in beds of varying Tertiary and Quaternary age in widely-separated localities, such a suggestion is untenable.

The spores have been gathered : (1) from the pre-glacial Cromer Forest Bed and early glacial freshwater beds at Beeston, Norfolk, (2) from the late Pleistocene sub-Arctic beds of Temple Mills and Barrowell Green, Lea Valley, Essex, and (3) one specimen from the late glacial plant-beds of Barnwell Station, Cambridge *.

On first seeing Chandler's figures, although the enlargement ($\times 12$–15) is too small, my opinion was that these spores must be of Carboniferous age. Moreover, their size is too large for recent megaspores which is, as far as I know, at most 1·5 mm. (see Hieronymus, 1900); the size of Senonian spores is at most 1 mm. (Dijkstra, 1949). Several types of Carboniferous spores, however, are larger than 2 mm. (Dijkstra, 1946).

I had the opportunity, during a visit to Great Britain, to see and to study the Chandler collection of spores in the British Museum (Natural History). They belong to different types, all of which are of Carboniferous age. The specimens figured by Chandler (1946), pl. xiv. figs. 1–3, 7, 15, belong to *Triletes mamillarius* Bartlett ; figs. 4–6, 8, 9, 11–14, to *Triletes superbus* Bartlett ; and figs. 10, 16, to *Triletes glabratus* Zerndt. I have not seen the spore from the late glacial plant-beds of Barnwell Station, fig. 17, which is in the Sedgwick Museum, Cambridge.

Besides these spores Miss Chandler showed me some big spores gathered at the following localitites in the Hampshire Basin † :—

1. Lake. Possibly Cuisian.
2. Sandbanks. Lutetian.
3. Cliff End, Mudeford. Probably Auversian.

* For details of localities, see above references.

† Particulars of these localities and their horizons will be appearing in a British Museum Monograph by Miss Chandler. All these spores were collected by Miss Chandler, excepting those from Pagham and Lower Swanwick, which were collected by Messrs. E. M. Venables and D. Curry, respectively.

4. Colwell Bay. Upper Headon.
5. Pagham. Cuisian ? or Ypresian (London Clay).
6. Bursledon Brick Company's Yard, Lower Swanwick, Hants, London Clay.

These megaspores, total 34 specimens, also appear to be Carboniferous. Some of them are weathered, but most of them are easily recognizable. They belong to the following species :—

From Lake : *Triletes glabratus, T. mamillarius, T. prætextus* and *T. superbus*.

From Sandbanks : *T. glabratus, T. mamillarius, T. prætextus*, and *T. superbus*.

From Cliff End, Mudeford : *T. glabratus, T. auritus, T. mamillarius, T. brasserti, T. prætextus* and *T. superbus*.

From Colwell Bay : *T. brasserti*.

From Pagham : *T. hirsutus, Triletes prætextus* and probably *Cystosporites verrucosus* forma *abortivus*.

From Lower Swanwick : *T. horridus*.

Triletes glabratus Zerndt. (Pl. XVII, fig. 2 ; XVIII, fig. 2.)

Diameter 0·4–3 mm., usually about 1·7–2·5 mm. Spore smooth. Triradiate ridges, arcuate ridges, and contact faces clearly distinguishable ; other ornamentation lacking. Triradiate ridges about ½–¾ of the spore-radius.

Occurrence :—

Turkey : Westphalian D.
Poland : Westphalian B, C, D.
Bohemia : Westphalian (B ?), C, E.
Germany : Westphalian B, C.
The Netherlands : Westphalian A, B, C.
Belgium : Westphalian A, C.
France : Westphalian C.
Great Britian : Namurian (?) ; Westphalian A, B, C.
U.S.A. : Westphalian B, (C ?) or D.

Triletes mamillarius Bartlett. (Pl. XVII, fig. 1 ; XVIII, fig. 1.)

Diameter 0·4–2·6 mm., usually 1·4–2·1 mm. Spore-coat covered with numerous mamilliform papillæ. Papillæ hemispheric or taller than broad, breadth 30–100 μ, height

40–100 μ, on the contact faces much smaller (about 15 μ) or lacking. Triradiate ridges and contact faces clearly distinguishable.

Triletes mamillarius varies markedly in size and in the shape and density of the papillæ. In large specimens the spore-coat is more streched than in the smaller ones owing to their full expansion in growth. Therefore the coat has become thinner and the papillæ are separated from each other by a greater distance. It is probable that *T. mamillarius* comprises several species, but so far it has not proved possible to separate them.

Occurrence :—

    Turkey : Westphalian A, B, D.
    Poland : Namurian (A ?), B, C ; Westphalian A, B, C.
    Bohemia : Westphalian A, B, C, D.
    Germany : Westphalian B, C, D.
    The Netherlands : Namurian B ; Westphalian A, B, C.
    Belgium : Westphalian A, C.
    France : Westphalian A, B, C.
    Great Britain : Namurian ; Westphalian A, B, C.
    U.S.A. : Wesphalian B, (C ?) or D.

Triletes auritus Zerndt. (Pl. XX, figs. 1, 5.)

Spore trilobate-subtriangular-rounded in outline. Diameter 0·5–0·7 mm. (1·5 mm.). Triradiate ridges as long as the spore-radius. At the points of contact of the triradiate ridges with the equatorial ridge, which is usually clearly distinguishable, ear-shaped lobes, 0·3 mm. broad, 0·2 mm. long or smaller, are formed. Spore-coat smooth and shining. One form of *T. auritus* has an apiculate surface.

Occurrence :—

    Turkey : Westphalian D.
    Poland : Westphalian A (?), B, C, D.
    Bohemia : Westphalian B, C, D, E.
    Germany : Westphalian B, C.
    The Netherlands : Westphalian A, B, C.
    France : Westphalian A, C, D.
    U.S.A. : Westphalian B, (C ?) or D.

Triletes hirsutus (Loose). (Pl. XIX, figs. 3, 4.)

Diameter 0·5–0·8 mm., the mean being 0·65 mm. Triradiate ridges about two-thirds of the spore-radius, meeting in a pronounced beak at the apex of the spore. Arcuate ridges clearly distinguishable. Spore-coat, excepting the contact faces, covered with pointed, hair-like, 0·1 mm. long, appendages, sometimes branching at their tips and anastomosing at their bases.

Occurrence :—

  Turkey : Westphalian A.
  Poland : Namurian C ? ; Westphalian A, B.
  Bohemia : Westphalian B ?, C.
  Germany : Westphalian B, C.
  The Netherlands : Westphalian A, B, C.
  France : Westphalian A, C.

Triletes brasserti Stach & Zerndt. (Pl. XIX, figs. 1, 2, 5.)

Diameter of spore-body, excluding equatorial zone, 0·8–1·6 mm. Equatorial zone 0·1–0·4 mm. broad. Triradiate ridges often undulating, as long as the spore-radius, 0·1–0·2 mm. high. Contact faces and triradiate ridges ornamented with hemispherical granules. Equatorial zone is readily detached from the spore-body.

Occurrence :—

  Turkey : (Namurian A, B, C); Westphalian A, B, D.
  Poland : Dinantian ; Namurian A, B, C ; Westphalian A, B.
  Germany : Westphalian A, B, C.
  The Netherlands : Westphalian A, B, C.
  France : Westphalian A, B, C, D.
  Great Britain : Dinantian ; Namurian ; Westphalian C.

Triletes prætextus Zerndt. (Pl. XX, figs. 2, 6.)

Diameter 0·9–1·8 mm., the mean being 1·3 mm. Triradiate ridges increase rapidly in height towards the apex, where they form a very characteristic protuberance 0·2–0·3 mm. high. Top area, with exception of the contact faces, ornamented in places with ramifying, hair-like appendages, which are most dense where triradiate ridges and arcuate ridges meet.

In the Turkish Carboniferous we find two forms of *T. prætextus*: a small form which occurs in the Namurian and a large form found in the Westphalian only. In the same manner we could subdivide *T. brasserti* from the Turkish Carboniferous into two forms which differ in size; they also have a different vertical distribution. The small form occurs in the Namurian, the large form in the Westphalian.

Occurrence :—

Turkey : Namurian A, B, C); Westphalian A, B.
Poland : Dinantian; Namurian A, B, C; Westphalian A, B.
Bohemia : Westphalian B, C.
Germany : Westphalian B, C.
The Netherlands : Westphalian B, C.
France : Westphalian B.
Scotland : Westphalian B.

Triletes superbus Bartlett. (Pl. XVI, figs. 1, 2.)

Diameter, including equatorial zone, 1·5–3 mm. the mean being about 2·25 mm. Spore-coat covered with appendages, often with club-shaped and flattened ends; appendages most numerous and longest (up to 1 mm.) on the equatorial ridge, where they anastomose to form a wreath; on the basal area and on the contact faces their length is about 0·05 mm. Badly preserved specimens, which have lost their appendages, show conical projections 0·25 mm. high on the spore-coat.

I believed (1946) I had found one specimen of *T. superbus* among many of *T. brasserti* in a sample of Waverley coal from Scotland. Waverley is a trade name which has been applied to the Peacock and the Corbiecraig coals in the Limestone Coal Group of Midlothian. These coals belong to the Dinantian. I wondered why the vertical distribution of *T. superbus* in Scotland was so different from that in other coal basins. I have since studied more samples of Scottish coal and find that the specimen of 1946 belongs to a variety of *T. brasserti* which was then unknown to me.

Occurrence :—

Poland : Westphalian B.
Bohemia : Westphalian (B ?), C.
Germany : Westphalian C.
The Netherlands : Westphalian B, C.

Belgium : Westphalian (A ?).
France : Westphalian A, B, C.
Scotland : Westphalian B.
U.S.A. : fossil in drift coal.

Triletes horridus (Zerndt). (Pl. XXI, figs. 1, 3 (?), 4, 5.)

Spore oval-bottle-shaped in outline, mostly flattened laterally. Length of the spore, excluding the neck-like projection, 0·4–1 mm., the mean being 0·86 mm. ; breadth 0·57–1 mm., the mean being 0·85 mm. Lips of the neck-like projection 0·35–0·60 mm. long. Arcuate ridge visible. Spore-coat, excluding the neck-like projection, covered with about 150 spine-like hairs. Hairs 0·10–0·15 mm. long, on the top pointed and bent, on the base up to 60 μ broad. Spore-coat about 20 μ thick.

Type 27 of Zerndt contains more than one species. Zerndt (1934) distinguished two types, but later he found forms which, according to him, were transitional. Therefore he included all in his one type 27. During the study of the Carboniferous from South Limburg (1946) we observed only two easily distinguishable forms, which we called *Triletes subpilosus* (Ibrahim), pl. vi. figs. 2, 6, and *Triletes horridus* (Zerndt), pl. vi. fig. 4. The main differences between these two are that *T. horridus* is larger and more robust, its hairs are thicker but less numerous. *T. horridus* resembles *Lagenicula* I. var. *major* of Bennie & Kidston (1886), and *T. subpilosus* resembles *Lagenicula* I. of Bennie & Kidston. The Namurian of Turkey contains two forms. One of them much resembles *T. horridus*, pl. vi. fig. 3 ; however, there are some small differences ; the other form resembles *T. subpilosus*, but its dimensions are much larger. It may be distinguished as *T. subpilosus* forma *major*. In some coal samples from Scotland I found a spore which resembles *T. subpilosus* forma *major*. I found *T. horridus*, pl. vi. fig. 5, in shale from the Yorkian Series (Westphalian A or B) at Preesgeweene Colliery, Shropshire. This spore is identical with those from The Netherlands. The spore found at Bursledon Brick Company's Yard, Lower Swanwick, pl. vi. fig. 1, is *T. horridus* too. Its spore-coat is corroded a little and the tips of most of its hairs are broken off.

As Zerndt did not distinguish the two types in his later studies, and because it seems that there are transition forms, it is not possible to give a complete review of the distribution of *T. horridus*.

Occurrence :—

Turkey : Namurian A (?).
Poland : Westphalian A (?).
Germany : Westphalian B, C (?).
The Netherlands : Westphalian A, B, C.
Great Britain : Westphalian A or B.

Cystosporites verrucosus Dijkstra forma *abortivus*.
(Pl. XX, figs. 3, 4 (?).)

Spore wart-shaped, about 300–800 μ long, 250–600 μ broad, the neck-like portion being 150–320 μ long. Tri-radiate ridges usually indistinguishable. Arcuate ridges conspicuous. Spore-wall near the contact faces has longitudinally-running plications ; wall dull, black, 15–20 μ thick, sometimes ornamented with few small scattered spines up to 30 μ long.

Only one spore of the tetrad of the genus *Cystosporites* was fertile ; it had an elongate sack-shaped form, attained a great size (3–11 mm.), and had a very thin coat. The abortive spores remained small, varying in both shape and size (0·35–1 mm.). Small individuals frequently remain attached to the fertile spore. As a result of the variable shape it is sometimes difficult to determine an isolated abortive spore of this genus. The spore from Pagham most nearly resembles *Cystosporites verrucosus*. The abortive form was described in 1946. Until recently only some fragments of the forma *fertilis* were known. During last year (1949), however, I found several complete specimens of the fertile form in shale from the Westphalian B, C, in The Netherlands, and from the Yorkian Series (Westphalian A or B) at Preesgewene Colliery, Shropshire.

Occurrence :—

The Netherlands : Westphalian B, C.
Great Britain : Westphalian A or B.

Discussion.

Miss Chandler's spores must have been derived from the Carboniferous. Carboniferous megaspores have been found from time to time in younger deposits. The Kunradian (Cretaceous) in South Limburg, The Netherlands, sometimes contains small pieces of coal, debris from the

Carboniferous which has been partly washed out by the sea during the Cretaceous. I have macerated this coal and found megaspores of the Westphalian B or C in it. The river Meuse transports shales and coal with fossils of Carboniferous origin from the Belgian Ardennes and from the basin of Liège to our country. Shale, in contrast with spores, disintegrates very quickly in the air, and in this way it will not be impossible to find Carboniferous megaspores in the Quaternary. Bartlett (1928) studied coal pebbles from glacial outwash deposits at Ann Arbor, Michigan, and found Carboniferous megaspores in them. Zerndt (1933) got the same results with coal pebbles from Carpathians. Besides transport by nature, the activity of mankind is an important geological factor. Prof. T. M. Harris, who studies Jurassic megaspores, showed me some Carboniferous spores found in rocks of Jurassic age. He told me that he gathers his samples in old quarries and prefers dark spots in the wall, because such spots often contain humic matter with cuticulæ and spores. Sometimes these spots are, according to him, probably remains of old fires kindled by labourers.

In connection with the derivation of these spores I asked Miss Chandler if she could tell me the kind of strata from which she and Mr. Reid had collected them. She answered that she did not personally know the Lea Valley beds, nor the Cromer Forest localities, but all the other spores (except those from Pagham) were collected by her. According to Miss Chandler there is no evidence whatsoever of the presence of weathered coal in the deposits which yield them so far as she has observed. The spores are always associated with Tertiary plants, usually embedded in a typical unconsolidated Tertiary sand, which includes no material from other sources. The localities which yield them are at various stations along the coast ranging from the north shore of Poole Harbour (Dorset) to the Selsey area in Sussex. Miss Chandler has washed many samples of Tertiary beds in the hope of finding fruits and seeds, but has never found these spores in any except those which yield a genuinely indigenous Tertiary flora. As for the Lea Valley, it is, according to her, possible that the spores of these localities are derived from material from Lower Tertiary beds, as rafts of material from the Woolwich and Reading Beds containing typical Lower Tertiary angiosperm fruits occur at Broxbourne.

If Miss Chandler is right and these spores are not derived from the Carboniferous but really belong to the Tertiary then we must assume that some spore-bearing plants of Carboniferous type persisted until Tertiary times. Mrs. Knox (1939) found that it is possible to draw broad comparisons between existing and Carboniferous microspores. According to her it is, however, nor possible at this stage of investigation to say that any specific or even generic significance can be attached to them. Nevertheless, the assumption that Carboniferous plants have persisted unchanged until the Quaternary is not very probable, because such plants have never been found in the Tertiary, nor even in the Mesozoic. Mesozoic megaspores are quite different from Carboniferous megaspores; moreover, they are smaller (see Dijkstra, 1949). *All* Chandler's spores have the appearance of Carboniferous megaspores. If they belong to the Tertiary we might also expect some spores, showing some other, non-Carboniferous, structure. But all there spores have a Carboniferous appearance.

Another question is, if Chandler's spores are types which had not changed from the Palæozoic till the Tertiary or till the Quaternary, then they may be the megaspores of *Selaginella*. Herbaceous lycopodiaceous plants which demonstrate a close affinity to the heterosporous *Selaginella* and to the isosporous *Lycopodium* existed during the Carboniferous and perhaps already during the Devonian. Zeiller (1906) instituted the designation *Selaginellites* for fossil herbaceous lycopods which are heterosporous, the genus *Lycopodites* being conveniently retained for isosporous specimens or for such sterile specimens as may with reasonable certainty be regarded as generically identified either with *Lycopodium* or *Selaginella*. For this reason I have studied megaspores of the following species :— *Selaginella amsena, S. apoda, S. cuspidata, S. denticulata, S. digitata, S. epirrhinos, S. helvetica, S. Martensii, S. monospora, S. pallescens, S. Poulteri, S. Schechteri, S. selaginoides, S. umbrosa, S. Watsoni, S. Willdenowii* and *S.* spec. The result of this study is that not one of them at all resembles Chandler's spores *.

* It is very interesting to notice that some of my Aachenian megaspores, especially those which belong to the section *Aphanozonati*, have such a great resemblance to some of these recent *Selaginella* megaspores that it is very probable that they are spores of *Selaginella* or of plants which are very closely related to *Selaginella*.

After I had examined these spores Miss Chandler sent me R. Potonié's letter in which, in 1935, he had written to her : " Die Sporen gehören keinesfalls dem Tertiär an. Es handelt sich um Carbonsporen." In this letter he suggested one definite identification : *Sporites brasserti* ; others he compared with *Sporites silvanus*, *S. diffusipilosus*, *S. primus*, *S. clavatipilosus* and *S. fuscus*. These names in the same arrangement are the same as *Triletes brasserti*, *T. auritus*, *T. prætextus*, *T. glabratus*, *T. superbus* and *T. mamillarius*. For an extensive list of synonymy *vide* Dijkstra (1946). Thus Potonié had recognized most of the types.

Our conclusion is that these megaspores are Carboniferous and not Tertiary. They belong, at least partly, to the Middle Coal Measures, from which they must have been transported. The problem has therefore now resolved itself into one of deciding how these megaspores were introduced into the Tertiary. It seems probable that they must have come from coal or shale which disintegrated long ago. On petrological and faunal grounds it is generally conceded that one or more rivers flowing into the Hampshire Basin during Eocene times drained land in the west of England (Stamp, 1921 ; Boswell, 1923, p. 225). We might therefore look to barren coal Measures of Devon and Cornwall and perhaps to the Coal Measures of Bristol and Somerset as possible sources for the Carboniferous megaspores in the Tertiary deposits of Hampshire.

I wish, in concluding this paper, to express my very warm thanks to Miss M. E. J. Chandler and to Mr. W. N. Edwards for the opportunity they gave me to study these megaspores. I wish to thank Mr. W. N. Croft for the correction of this paper.

EXPLANATION OF THE PLATES.

PLATE XVI.

Fig. 1. *Triletes superbus* Bartlett. Base of the Arctic Bed. Beeston, Norfolk. V. 27151. (See Chandler, 1946, pl. xiv. fig. 12.)
Fig. 2. *Triletes superbus* Bartlett. Aegir Seam, Westphalian B. Emma State Mine, The Netherlands. Both spores have lost their appendages.

PLATE XVII.

Fig. 1. *Triletes mamillarius* Bartlett. Cromer Forest Bed, Beeston, Norfolk. V. 27147. (See Chandler, 1946, pl. xiv. fig. 7.) The surface of the spore is corroded.
Fig. 2. *Triletes glabratus* Zerndt. Westphalian D. Kiliç Mahallesi, Turkey.

PLATE XVIII.

Fig. 1. *Triletes mamillarius* Bartlett. Aegir Seam, Westphalian B. Emma State Mine, The Netherlands.
Fig. 2. *Triletes glabratus* Zerndt. Base of the Arctic Bed. Beeston, Norfolk. V. 27150. (See Chandler, 1946, pl. xiv. fig. 10.)

PLATE XIX.

Fig. 1. *Triletes brasserti* Stach and Zerndt. Westphalian D. Seam at Kireçlik, Turkey.
Fig. 2. *Triletes brasserti* Stach and Zerndt. In Eocene beds. Cliff End, near Mudeford, Hants. V. 28279. The point of attachment of the equatorial zone is distinguishable on the right side of the spore.
Fig. 3. *Triletes hirsutus* (Loose). In Eocene beds. Pagham, Sussex. V. 28261.
Fig. 4. *Triletes hirsutus* (Loose). Seam H, Westphalian B. Emma State Mine, The Netherlands.
Fig. 5. *Triletes brasserti* Stach and Zerndt, forma *minor*. Namurian C. Seam at Kokaksu, Turkey.

PLATE XX.

Fig. 1. *Triletes auritus* Zerndt. In Eocene beds. Cliff End, Mudeford, Hants. V. 28266.
Fig. 2. *Triletes prætextus* Zerndt. Kesmeli Seam, Westphalian A. Ihsaniye Colliery, Turkey.
Fig. 3. *Cystosporites verrucosus* Dijkstra, forma *abortivus*. Seam F 3, Westphalian B. Maurits State Mine, The Netherlands.
Fig. 4. Cf. *Cystosporites verrucosus* Dijkstra, forma *abortivus*. In Eocene beds. Pagham, Sussex. V. 28263.
Fig. 5. *Triletes auritus* Zerndt. Seam N, Westphalian C. Emma State Mine, The Netherlands.
Fig. 6. *Triletes prætextus* Zerndt. In Eocene beds. Sandbanks, Hants. V. 28252.

PLATE XXI.

Fig. 1. *Triletes horridus* Zerndt. In London Clay. Bursledon Brick Company's Yard, Lower Swanwick. V. 28285.
Fig. 2. *Triletes subpilosus* (Ibrahim). Seam C, Westphalian B. Maurits State Mine, The Netherlands.
Fig. 3. Cf. *Triletes horridus* (Zerndt). Seam at Kiliç Dere, Namurian A. Turkey.
Fig. 4. *Triletes horridus* (Zerndt). Unnamed seam between seams M and L, Westphalian B. Emma State Mine, The Netherlands.
Fig. 5. *Triletes horridus* (Zerndt). In shale from the Yorkian Series, Westphalian A or B. Preesgeweene Colliery, Shropshire.
Fig. 6. *Triletes subpilosus* (Ibrahim). Seam C, Westphalian B. Maurits State Mine, The Netherlands.

The enlargement of all the photographs is ×50. The registered numbers refer to specimens in the Collections of the Geol. Dept., British Museum (Natural History).

Carboniferous Megaspores.

Carboniferous Megaspores.

Carboniferous Megaspores.

Carboniferous Megaspores.

Carboniferous Megaspores.

Carboniferous Megaspores.

REFERENCES.

BARTLETT, H. H. 1928. "Fossils of the Carboniferous Coal Pebbles of the Glacial Drift at Ann Arbor." Pap. Michigan Acad. Sci. Arts Lett. ix. pp. 11–28.

BENNIE, J. and KIDSTON, R. 1886. "On the occurrence of spores in the Carboniferous formation of Scotland." Proc. Roy. Phys. Soc. Edinb. ix. pp. 82–117, pls. iii.–vi.

BOSWELL, P. G. H. 1923. "The Petrography of the Cretaceous and Tertiary Outliers of the West of England." Quart. J. Geol. Soc. Lond. lxxix. pp. 205–230.

CHANDLER, M. E. J. 1921. "The Arctic Flora of the Cam Valley at Barnwell, Cambridge." Quart. J. Geol. Soc. Lond. lxxvii. pp. 4–22.

——. 1946. "Note on some Abnormally Large Spores formerly attributed to *Isoetes*." Ann. & Mag. Nat. Hist. London (11) xiii. pp. 684–690, pl. xiv.

DIJKSTRA, S. J. 1949. "Megaspores and some other fossils from the Aachenian (Senonian) in South Limburg, Netherlands." Med. Geol. Stichting, Maastricht (n. s.) iii. pp. 19–33.

DIJKSTRA, S. J., and VIERSSEN TRIP, P. H. 1946. "Eine monographische Bearbeitung der karbonischen Megasporen mit besonderer Berüksichtigung von Südlimburg (Niederlande)." Med. Geol. Stichting, Maastricht (C) iii. 1, pp. 1–101, pls. i.–xvi.

HIERONYMUS, C., and SADEBECK, R. 1900. 'Selaginellaceæ.' In Engler, H. G. A., and Prantl, K. E. A. "Die natürlichen Pflanzenfamilien," i. 4, pp. 621–715.

KNOX, E. M. 1939. "The Spores of Bryophyta compared with those of Carboniferous Age." Trans. Bot. Soc. Edinb. xxxii. pp. 477–487, pls. xxxvii.–xli.

REID, C. 1899. "The Origin of the British Flora." vi+191 pp. London.

——. 1916. "The Plants of the Late Glacial Deposits of the Lea Valley." Quart. J. Geol. Soc. Lond. lxxi. pp. 155–163, pl. xv.

——. 1920. In Marr, J. E. "The Pleistocene Deposits around Cambridge." Quart. J. Geol. Soc. Lond. lxxv. pp. 204–229, pl. xi.

REID, E. M., and CHANDLER, M. E. J. 1923. "The Barrowell Green (Lea Valley) Arctic Flora." Quart. J. Geol. Soc. Lond. lxxix. pp. 604, 605.

STAMP, L. D. 1921. "On the beds at the base of the Ypresian (London Clay) in the Anglo-Franco-Belgian basin." Proc. Geol. Assoc. London, xxxii. pp. 57–108, pls. i., ii.

ZEILLER, M. R. 1906. "Étude sur la flore fossile du bassin houiller et permien de Blanzy et du Creusot." Études Gîtes Min. France, ii. (Flore Fossile), pp. 1–261.

ZERNDT, J. 1933. "Versuch einer stratigraphischen Bestimmung von Steinkohlen—Geröllen der Karpaten auf Grund von Megasporen." Bull. Acad. Pol. Sci. Lett. Cracow (B), 1933, pp. 1–7.

——. 1934. "Les Mégaspores du Bassin Houiller Polonais, I." Acad. Pol. Sci. Lett. Cracow (Trav. Géol.), i. pp. 1–55, pls. i.–xxxii.

Wealden Plant Microfossils

By N. F. Hughes

(PLATES X–XII)

Abstract

One new and several re-interpreted species of Lower Cretaceous age are included in *Pyrobolospora*, a new organ-genus of large spores bearing neck structures ; in *Triletes* (section *Aphanozonati*) are placed two other species of megaspores, one emended and one new. These are a small part of a distinctive Wealden assemblage of plant microfossils, which represents a critical period in plant evolution and which should play its part in stratigraphy. Certain difficulties in nomenclature and morphology are discussed.

I. Introduction

A STUDY which is in progress of the plant microfossils of the English Wealden beds, has inspired the hope that they may prove to be of some use in correlating the various scattered outcrop successions, with the Continental " Wealden " and thus eventually with marine equivalents both in England and abroad. It will be easier, however, to make use of this plant evidence when the ostracod zones of Anderson (1939 and 1940) become available. The plant material is plentiful and is mostly collected from beds containing few ostracods or other fossils ; the two lines of evidence should thus be complementary. The assembling of plant evidence on the scale of ostracod information already obtained will take some time and it is not expected to result in the erection of any scheme of purely floral zones. The problem of the age of the various parts of the Wealden series in England has not been solved over a long period, and there is little chance of this being done even now by one research method alone. It is of course necessary to attempt zonation, but this is bound to be artificial and " strained " if based on one fossil group ; pointers must be sought to the natural overall pattern or direction of geological change which it is believed may be determinable from these sequences and areas, when all the evidence is considered together. Working from the basis of the sedimentary evidence of Allen (1954 and earlier), with the addition of ostracod, plant, and all other fossil information, the effect of the general pattern of geological evolution of the area should become apparent.

The best chance of finding a solution in this case is believed to lie in such a close integration of the simultaneous effort of several individuals. The present paper is concerned solely with some preliminary descriptions.

Plant Fossils and Geology

The use of plant-fossils in the stratigraphy of the Carboniferous system has an unfortunate history (Jongmans, 1952), and little has been attempted in later systems. Although there can be no comparison with the general usefulness of graptolites and cephalopods in zoning, transported plant microfossils are no worse material than many other animal groups which have had to be used; in fact such plants which have floated even for a short time may well be more widely distributed than some benthonic organisms.

The rigid higher-group framework in botanical classification of living plants which has of necessity been used for fossil plants, has had the effect of support for the idea of plant evolution being too slow to be of stratigraphical use; this is illustrated by several contemporary botanical textbooks, which infer that a cryptogenetic group such as the Angiosperms must have had a long undetected history, in order to achieve its high state of perfection on appearance. This view of plant evolution is thought to be misleading and it is suggested that judgment on this problem be withheld.

II. THE WEALDEN FLORA

Seward (1894–5, 1913) described the macrofossils and little has since been added, but his specimens came almost entirely from the Fairlight Clays at the base of the series. The larger plant microfossils, which come from nearly all horizons of the Wealden include megaspores, seeds and similar bodies, and cuticles of various plant organs. The complementary study of British Wealden microspores and pollens has only recently begun, although a very brief survey of German Infra-Valanginian Wealden is given by Thiergart (1949).

Twenty of the twenty-six Wealden megaspore species described from Dutch borings and from Kingsclere by Dijkstra (1951) are now recorded from outcrop sections mainly on the South Coast of England. This includes nine which were only previously known from Holland and there are in addition a number of new forms.

III. COLLECTION OF MATERIAL

Megaspores are sometimes very common, occasionally reaching hundreds in a 2 oz. sample; they are however seldom discernable in the field. It has not yet proved possible to pick out with certainty a megaspore-bearing bed, although many sediments can be rejected. The best lithology is a silt, which is frequently commoner than true clay both in the Hastings Beds and the Isle of Wight Wealden Marls; there should be small plant fragments visible, but these should not be merely pieces of charcoal or be heavily pyritized. If plant fragments (pieces of stems and leaves) are large and abundant, there are usually few spores ; this is presumably due to difference in transportability. Dijkstra (1949, p. 21) pointed out that poor sorting in the sediment increased the chance of survival of plant material, but conditions inferred from poor sorting may also have favoured deposition of this kind of fossil.

Laboratory Treatment

The object is separation and cleaning with a minimum of further destruction. The most tedious operation is that of picking out spores, seeds, etc., individually from the mass of largely unrecognizable plant debris, as no flotation separation is of course possible ; this is most easily done under water, with a low-power binocular.

An adequate average sample is about 50 gms., but this should be decreased if plant material is obviously plentiful and vice versa. The sample is dropped into boiling water to which has been added a small amount (about 3 gms.) of anhydrous sodium carbonate which produces a slow release of carbon dioxide on continued boiling. If the rock is not broken down in 15 minutes, the material should be dried and the process repeated at least once. Dijkstra (1951) describing this process recommended retaining the material on a sieve of mesh size $0 \cdot 1$ mm. and examining when dry ; it has been found easier to use three sieves of 40, 80, and 100 meshes to the inch and to examine the results from each sieve separately under water with a stereoscopic binocular of 20–30 × magnification. Most megaspores apart from a few large ones are held by the 80 sieve and practically none appear on the 100 sieve (approximately 150μ mesh) which is used merely as a check for the few very small types known, such as *Triletes adoxus* Dijkstra. Spores with any kind of appendages must usually be cleaned with hydrofluoric acid after isolation, particularly if they are to be drawn, photographed, sectioned, or dissected.

IV. SYSTEMATIC SECTION

A. Discussion of Genera

Dijkstra (1951) classified Wealden megaspores under three sections of the Genus *Triletes* Reinsch emend. Schopf—*Aphanozonati, Zonales,* and *Lagenicula,* only the last of which is fully discussed in this paper. It is intended here to separate from *Triletes* a new group, *Pyrobolospora,* which will include a new species and several species from the *Lagenicula* section. From a morphographic point of view, none of the Lower Cretaceous species can now belong to the sub-division *Lagenotriletes* or to the genus *Lagenicula* within it as redefined by Potonié and Kremp (1954), although one Upper Cretaceous species could possibly be still included. *Lagenicula* as now restricted requires the characters : polar axis usually greater than equatorial, contact faces well developed, emergences from exine short spines but never blunt appendages. Potonié in addition makes a good case for regarding the *Lagenotriletes* group as megaspores only of the *Lepidodendraceae* and *Bothrodendraceae* excluding even the *Sigillariaceae* (also Chaloner 1953 and 1954); this would seem to be a further reason for erecting a new organ-genus to receive these Cretaceous " megaspores ", which have not yet been reported from the Jurassic or the Tertiary.

B. Division (*Oberabteilung*) Sporites H. Potonié.

Organ-genus *Pyrobolospora* gen. nov.

Diagnosis.—Spores in the megaspore size-range (150μ upwards) with a prominent neck composed of (usually) six segments. Spore-body spherical (and not axially elongated), covered uniformly by sculptine detail which may vary in different species from small granules to appendages 300μ long. Inner spore-wall (intexine) is much thinner and darker coloured than the exoexine. Small tri-radiate tetrad scar is present apically beneath the neck and may only be seen by dissection or in microtome sections.

Type Species.—*Pyrobolospora vectis* gen. et sp. nov.

Derivation of name.—Gr. πυρόβολος—giving forth fire; of a grenade or bomb.

Comment.—This name is given in order to separate what is believed to be a homogeneous group from previous confusing attributions. The genus is not placed in the Division (Abteilung)—*Triletes* of Potonié and Kremp, although it may eventually be included there.

Pyrobolospora vectis gen. et sp. nov.

Plate X, figs. 1-3; Plate XI, fig. 6; Plate XII, figs. 3-8, 10-11; Text-fig. 1, 2

Abstract Diagnosis.—Small ($200-300\mu$) dark thick-walled megaspore with a prominent neck and with the body covered by numerous very long ($150-200\mu$) translucent appendages arising from low ridges (Plate X, fig. 1).

Type Preparations.—S.M. K2229-46.

Locality and Horizon.—Cliff opposite remains of Yaverland sea-wall, N.E. of Sandown, Isle of Wight; locality S 18. Approximately 30 feet below top of Wealden Marls (Osborne White, 1921, p. 19).

Description.—Spore-body roughly spherical; equatorial diameter $185-320\mu$ (mean 263μ), average equatorial diameter slightly greater than axial; total length including neck $300-520\mu$ (mean 418μ, 50 specimens). Length of neck $120-200\mu$ (mean 155μ, 50 specimens), width at base in average specimen 170μ. Neck consisting of six leaf-like translucent segments fused together. Six ridges of the same dimensions as the body ridges but not connected to them, rise from the body up the neck, the leaves being folded inwards between them. When the neck opens, splits run from the apex downwards along these ridges, thus separating the leaf-like segments. Average specimen has lower half fused and rigid; in the opened upper half the separate leaves are often not easily resolved. Spore-body smooth, covered with narrow ridges 10μ broad and 20μ high. These are arranged in a coarse reticulum of $40-50\mu$ diam. or in a nearly regular spiral with an interval of the same size (approximately resembling a *Chara* nucule). Appendages arise at ridge junctions or from slight swellings on the ridges, not less than 50μ apart; they are thin-walled and translucent throughout their whole length, up to 250μ long, 20μ wide at the attachment, and up to 70μ wide in the club-shaped distal end; they appear to have been bladders (from the crushed folded walls, Plate XII, fig. 11) and there were 40–50 per spore.

Spore-coat of three layers; outer exoexine layer 15μ thick, light yellow-brown colour with radially arranged darker bands, giving rise to all appendages and continuous with neck; inner exoexine layer, 10μ thick, yellow-brown colour, adhering closely to the (innermost) intexine; this last is 5μ thick, dark red-brown in colour, entire except for small tri-radiate mark at apex (centrally below neck), and has a smooth inner surface. The outer and inner exoexine which are of the same colour are seen in thin sections under high power to be separated, sometimes by a split which probably occurs in sectioning along a surface of weakness. The inner exoexine is seen in section to be interrupted with the intexine at the triradiate laesurae and it disappears from serial sections at the spore apex (just above the base of the neck), Plate XII, fig. 7. In broken specimens under low power, the combined exoexine appears homogeneous and $20-25\mu$ thick. Tri-radiate laesurae are about 50μ long and without adjacent thickening (Plate XII, fig. 10). Within the neck and above the spore proper is a neck-chamber (Plate X, fig. 2) normally about 60μ high, but although its presence is certain the shape of the chamber is not known from many specimens. The six neck ridges appear to arise in three pairs, each pair straddling one of the tri-radiate laesurae as in the diagram, Text-fig. 1. Serial sections (Plate XII, figs. 3–8) illustrate whole structure.

Comparison.—The nearest resemblance is to *P. hexapartita* (Dijkstra) which however lacks the ridges on the spore-body and differs in the appendages; there is also a difference in proportions of neck and body. *P. medusa* (Dijkstra) (mentioned below) differs in its appendages which are cylindrical, thick-walled (opaque) except at the tip, longer and only 15–30 in number; the spore-body is smooth and without ridges.

Discussion.—The neck-chamber is comparable with the "androcamera" of Nikitin's *Kryshtofovichia africana* at least in section, although externally the Russian spore (of Devonian age) appears to be a *Lagenicula*, but no sections of true Carboniferous *Lagenicula* spores have been examined in this connection.

The tri-radiate scar which suggests the megaspore nature of *P. vectis* is not easily seen. In broken spores it is usually irregularly torn and not identifiable; in specimens with the neck leaves open the spore is also open and usually filled with sediment. Preparations may be made from suitable spores which have not been too much laterally flattened, by removing most of the neck and by gently dissecting away the lower

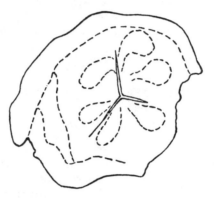

TEXT-FIG. 1.—*Pyrobolospora vectis* gen. et. sp. nov.; diagram × 200 of view from below of spore-apex (dissection), cf. Plate XII, fig. 10. Broken lines show position of body-wall ridges, and outline of bases of six neck ridges seen through the spore-apex.

two-thirds of the spore. This exposes the apex of the intexine from the inside (Plate XII, fig. 10), and the tri-radiate mark is seen when the specimen is mounted as a transparency. As the spore is small and partly brittle, the dissection is best done on a slightly moist surface.

Terata.—In rather more than two hundred specimens of *P. vectis*, three spores were found (from two localities on opposite sides of the Isle of Wight) each bearing two necks. Two specimens from one of the Sandown localities (S18) are large spores of the type shown in Plate X, fig. 3; the sculptine detail is normal with full-length appendages but the two necks have only three leaves each. The other from a Brixton Bay locality (S40B) shown in Plate XI, fig.6, is of normal size but has a second neck at the distal pole; one neck has three leaves and the other probably four. Zerndt (1934, 1937) figures several abnormalities of the tri-radiate scar in Carboniferous *Triletes* which can be attributed to extra mother-cell divisions, but no explanation is offered in the present case. They are mentioned as they are more frequent than might be expected (see also *P. hexapartita*) and they may help in the discovery of the origin of these spores.

Terminology.—The outermost layer, *exo.* in text-fig. 2, of the spore-wall with neck and appendages was at first considered as perispore because it stands clear of the tri-radiate scar, but as a working definition of this term is avoided even by Erdtman (1952b, p. 174), his comprehensive term sculptine is used where necessary. However, this outer layer (*exo.*) has radial dark bands and is therefore interpreted as the outer exoexine (see Potonié, 1952) ; the inner exoexine (*in. exo.*) appears identical except for the dark bands. Possibly the terms " exolamellen "

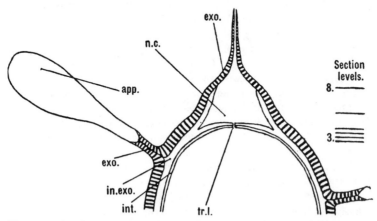

TEXT-FIG. 2.—*P. vectis* sp. nov. ; diagrammatic longitudinal section of proximal half of spore (approx. × 200). *app.*, appendage ; *exo.*, exoexine of neck and ridges ; *in. exo.*, inner exoexine ; *int.*, intexine ; *n.c.*, neck chamber ; *tr. l.*, tri-radiate laesure. Section levels relate to Plate 3, figs. 3–8.

and " isolierschicht " recommended by Potonié could be used. The intexine (*int.*) is quite distinct and is dark brown in colour. This adherence to the priority of Potonié unfortunately excludes the shorter terms used by Erdtman for the same layers ; but Erdtman specifically states that pollen for which his terms are chiefly erected is homologous with *microspores* of heterosporous pteridophytes ; it seems doubtful whether megaspores should still be excluded in view of the subjective implications of function which have become involved in the general use of the name (cf. Doyle, 1953).

Pyrobolospora hexapartita (Dijkstra)

Plate X, figs. 4, 5

1951 *Triletes hexapartitus* Dijkstra. *Meded. geol. Sticht.*, N.S. 5, 14.

Abstract Diagnosis.—Small (150–200μ) dark thick-walled " megaspore" covered with short blunt appendages, and with a prominent long (250μ) flame-like neck (Plate X, fig. 4).

Reference Preparations.—S.M. K2247–58.

Locality and Horizon.—Cliffs of Compton Bay, Isle of Wight—north of the Small Chine fault which repeats the Wealden succession (Osborne White, 1951, p. 12); locality S 47. Wealden marls : Bed 11 of this series, described as " Grey clay with large trunks of trees ".

Emended description.—Nearly spherical spore-body with equatorial diameter 80–252μ (mean 180μ, 50 specimens); average ratio axial to equatorial diameter 10:11. Total length including neck 280–555μ (mean 430μ, 50 specimens). Length of neck 202–353μ (mean 280μ); neck consisting of six leaf-like segments fused by their margins into six ridges running from the spore-coat up the neck; the centre part of each leaf is folded inward the whole forming a long flame-like organ, larger than the spore-body. The leaves usually display about half a rotation of an anti-clockwise twist. The thickness of the leaves decreases from about 10μ at the base. The neck is sometimes seen open (cf. Dijkstra, 1951, plate ii, fig. 7) with the leaves separate to the base; in a few cases the neck has clearly shrivelled without opening. Sporebody smooth, bearing about 50 blunt opaque cylindrical appendages which may vary in different specimens from mere hemispherical swellings to a length of 30μ; they are usually about 15μ broad and are hollow when seen in section. The full appendage, only seen in about half the spores, is longer (up to 50μ) and bears on the end a thin translucent disc (diam. 20–25μ) which appears to be a collapsed spherical bladder. Spore-coat of three layers:—the exoexine is 15μ thick and yellow-brown in colour forming the neck and appendages; it is divisible into two layers as in *P. vectis*—the outer 12μ with slight radial thickening and the inner only 3μ adhering to the intexine which is 3–5μ thick, red-brown in colour and entire. The apex of the spore proper seems to project upwards and fit closely into the base of the neck; the infolded parts of the neck leaves touch, leaving very little central space in the neck (unlike *P. vectis*).

Discussion.—No tri-radiate scar has been seen but the shape of the spore apex and brittleness of the specimens make this spore difficult to investigate by dissection. The general appearance however of the exoexine and intexine in microtome sections closely resembles the thicker-walled *P. vectis*, and suggests affinity. The size-range of the spore body includes three very small specimens of only about 80μ diameter, the minimum otherwise being 130μ. These small spores are fully developed and have necks about 200μ long but it is noticeable that the necks of all the small spores are open (as are a proportion of those of normal spores); since they were found on the 100-mesh sieve, it suggests that perhaps similar small spores with an unopened neck were washed through. These would fall in the range of microspore technique and have not yet been seen; the downward limit of size must be left open therefore and with it the meaning of ' megaspore '.

Note.—Dijkstra's two specimens from Kingsclere have been seen but not studied; his description is adequate for the continuance of his name, which is clearly the valid one. Dijkstra describes the spore-body as hemispherical and in fact a few specimens do exceed the ratio given and thus approach this shape.

Comparison.—This spore differs from *P. vectis* in the shorter thicker appendages arising on the spore-coat independently, and not from a network of ridges. The neck is here always much longer than the spore-body diameter, and the neck-chamber is very small or lacking.

Terata.—In about 50–60 specimens so far extracted there has been one large spore with two (reduced) necks of the same type as the *P. vectis* spore shown in Plate X, fig. 3. Another specimen is apparently two normal spores fused together with slight overlap, and with the necks united into a massive structure with twelve leaves (Plate X, fig. 5).

Pyrobolospora pyriformis (Dijkstra)

Plate XI, figs. 1, 2

1951 *Triletes pyriformis* Dijkstra. *Meded. geol. Sticht.*, N.S. 5, 14.

Abstract Diagnosis.—" Megaspore " 500–700μ diam. with broad neck up to 250μ long; spore body covered with short blunt appendages (Plate XI, fig. 1).

Reference Preparations.—S.M. K2259–68.

Locality.—Half-a-mile N.E. of Fairlight Glen foot; cliffs in Middle Fairlight Clay below Coastguard Station; author's locality 80A. (Osborne White, 1928, p. 33).

Emended Description.—Spore-body originally nearly spherical but usually equatorial diameter exceeds the axial. Equatorial diameter 290–920μ (mean 651μ, 35 specimens or mean 631μ including Dijkstra's specimens). The axial measurement with or without the neck is meaningless, as the majority of specimens are asymmetrically laterally flattened (Plate XI, fig. 1); this suggests that the top of the spore below the neck is the most rigid part. Only average maximum neck dimensions are given as the neck is so frequently incomplete; length up to 350μ, width 350–400μ at extremity, narrowing to the base. The neck arises from the spore-body as three prominent folds and consists basically of three bulging lobes between these folds; the specimen in Plate XI, fig. 2, is a small one with somewhat irregular appendages but it shows at the neck attachment a fold centrally, flanked by two downward bulging lobes, with the two other folds at the margins in this view. In most cases the main part of the neck is opaque and ochreous in colour, and lacking in precise form; broken surfaces reveal the upper part to be composed of a solid mass of sponge-like tissue without any central canal. Certain complete specimens seem to show a six-fold symmetry of leaves of the neck on the top surface, and others in which the neck is reduced to translucency by maceration show the same number. Spore-body covered evenly with blunt simple or compound appendages 30–40μ high, 40–50μ broad, and seldom less than 15μ apart in an average specimen; slight decrease in size of appendages towards the neck.

Spore-coat brown to black, consisting of opaque exoexine layer 25–30μ thick and intexine < 5μ thick and translucent; exoexine layer is likely to be divided as in *P. vectis* but good confirmatory sections have not yet been prepared because of brittleness of specimens. Tri-radiate mark (laesurae 220μ long) seen on inner side of apex of intexine and ? inner part of exoexine in broken specimen. Ideally a neck-chamber is present in the base of the neck but it has not been explored; the swollen lobes mentioned above were seen in some good specimens to be hollow.

Comment.—The tri-radiate scar is believed to be below the neck (as in *P. vectis*) and invisible from outside, but this is not easy to demonstrate clearly. The " tri-radiate ridges " of Dijkstra are probably the wall folds described and his additional plications have been seen in some specimens but they are not regular.

Occurrence.—This spore is never common but has been recorded from several localities in the Fairlight Clays and in the Wealden Marls of the Isle of Wight.

Comparison.—*Triletes argus* Dijkstra is not considered to be a separate species although the specimen has not been studied; *P. pyriformis* is very variable in size and in appearance of the neck. Larger appendages more widely scattered are considered to be a result of greater size; in the very small specimens of *P. pyriformis* the appendages are smaller and more closely packed but probably about the same in total number.

Pyrobolospora medusa (Dijkstra)

1951. *Triletes medusus* Dijkstra. *Meded. geol. Sticht.*, N.S. 5, 14.

Only one further good specimen has been seen from a sample at 777 feet (Wealden) in the Kingsclere No. 1 boring, which extends slightly the range of this rare spore. The neck is not so prominent as the appendages in this species; although Dijkstra describes tri-radiate ridges, they are not clear in his figure or in my specimen. It seems likely that this species belongs to the group in spite of its possibly trilete neck; it is quite distinct in spite of the small number of specimens recorded.

Comparable Senonian Spores

Pyrobolospora lobata (Dijkstra)

1949. *Triletes lobatus* Dijkstra. *Meded. geol. Sticht.*, N.S. 3, 25.

Dijkstra describes this as having " three ? or more ? lobes " to the neck with no tri-radiate ridges or contact faces ; he also mentions a " canule " within the neck. Vangerow (1954) confirms but gives very much smaller dimensions for the neck. This Senonian spore has sufficient characters to place it (unexamined) in *Pyrobolospora* rather than *Triletes*.

No attempt has been made to study the remaining *Lagenicula* species of *Triletes*, the first two of which may prove to belong to the new group, and the third to *Aphanozonati*. *T. dictyotus* Dijkstra 1949 is described as having tri-radiate ridges and contact faces ; Vangerow (1954) confirms. *T. capulifer* Dijkstra 1952 is apparently trilete from the figure although the description does not state this clearly. *T. costatus* Dijkstra 1952 is described as having a short neck but the figure is not clear on this point ; *T. lanaris* Vangerow 1954 is almost certainly a synonym but again the figure is not convincing.

Further Consideration of the Organ-genus Pyrobolospora

Dijkstra (1951) thought his species *T. hexapartitus* might be separated from *Triletes* but had insufficient specimens. The three species of *Pyrobolospora* now described are believed to be representatives of a new group distinguishable from *Lagenicula* (sensu Potonié and Kremp), chiefly by their formation of a neck from the outer exoexine and not from a raised part of the whole spore-wall with trilete scar clearly visible. The shape of each of the three neck leaves in *Lagenicula crassiaculeata* (Zerndt) is closely similar to that of each of the six leaves of *P. hexapartita* but the former spore has well developed contact faces.

Contact faces are presumably only present when the separation of the spores of the tetrad is delayed, as in many lycopods. In *Pyrobolospora* the observed sculptine detail can only have been formed on separated spores, and the observed tri-radiate scar (*P. vectis* and *P. pyriformis*) is not so small as to suggest unequal development or abortion of the other spores ; the various terata described support this view. *Pyrobolospora* is not placed in the Division *Triletes*, as this can be done if necessary later, after re-organization (see also next section of paper).

Kryshtofovichia africana Nikitin comes at first sight nearer to *Pyrobolospora* than *Lagenicula* as it lacks contact faces and bears complex (hooked) appendages. In section it appears to have an inner entire spore-wall but also in section each individual neck leaf has a definite central space suggesting complexity of structure. There are (only) three neck leaves and Nikitin describes the enclosed space as the androcamera, figuring the appropriate microspores within it. Mrs. R. Scott (1906) figures a curious apparently solid neck-like organ on her " Triletes diabolicus " which is perhaps a true *Lagenicula* spore, from the cone *Lepidostrobus foliaceus*. This appears to be similar to the solid " spongy " neck in some specimens of *Pyrobolospora pyriformis*. As suggested by the description of the latter, the varied forms of neck may represent different stages of maturity and function. The genus *Lagenicula* erected apparently on external form is in need of further investigation which should be possible now that its provenance is established.

Chronologically *Pyrobolospora pyriformis* is the earliest species of the new group, being present in the lower part of the Dutch Wealden (Lower Neocomian) and continuing through the English series ; no spore of this type has been recorded from the Jurassic. *P. vectis* and *P. hexapartita* are both from the Upper Wealden Marls of the Isle of Wight and their probable equivalents. Two or three other types have been seen in the English Wealden but so far in too small numbers to describe. Some Dutch Senonian spores are mentioned below but nothing of this type is yet known from later beds. It will be seen from the descriptions that *P. vectis* has been the most fully studied.

C. Triletes Reinsch emend. schopf : Section Aphanozonati Schopf

Two species of megaspore described below are now interpreted as belonging to this group ; the separated perispore of one has been described by Horst (1954) under the name *Dictyothylakos* and presumed to be algal. The old nomenclature is continued for the present, as although Potonié and Kremp (1954) have elevated *Triletes* from a genus to a group in the Carboniferous, no consquent Mesozoic reorganization has yet been attempted.

Triletes galericulatus Dijkstra 1951 emend.

Plate XII, figs. 1, 2

Abstract diagnosis.—Smooth black megaspore about 350μ diameter ; tri-radiate ridges and pointed appendages together often form an apparent low neck.

Reference Preparations.—S.M. K2269-71.

Locality and Horizon.—Cliff 200 yards N.E. of Cliff End Point, near Fairlight, Sussex ; Lower Ashdown Sand, locality 48 B. (Osborne White, 1928, p. 33).

Emended Description.—Spore-body has rounded equatorial outline and an average diameter 300–400μ (mean 345μ, 30 specimens). Tri-radiate ridges prominent 200μ long, 50μ high, and 20μ broad ; the upper part of the contact face in each inter-radius bears three or more upwardly directed spines 100μ long, 20μ diam. at their bases and tapering to a point (Plate XII, figs. 1, 2) ; the arrows on the two figures indicate the same point on the periphery. Spore coat black or dark brown, smooth, 20μ thick. In certain lateral compressions (Dijkstra, Plate ii, fig. 16) there may be an elliptical outline but the spore is believed to have been essentially spherical.

Comment.—In nearly all specimens the space between the spines and tri-radiate ridges is packed with sediment and the structure is only revealed after cleaning with hydrofluoric acid. Although Dijkstra's description leads to the classification of the spore in the section *Lagenicula*, his second figure (Plate iii, fig. 5) appears to show one of the spines separated. Arcuate ridges mentioned by Dijkstra have not been seen but the present illustration suggests that the thickness of the wall may give the appearance of such a structure.

Comparison and Discussion.—No other spore of this type has been found in the Wealden. The change in classification of this pore removes the last member of the section *Lagenicula* in the Wealden (Dijkstra, 1951).

Occurrence.—*T. galericulatus* also occurs rarely in the Upper Fairlight Clay and never more than three specimens have yet been recorded from any one sample.

Triletes retiarius sp. nov.

Plate XI, figs. 3, 4

1954 *Dictyothylakos pesslerae* U. Horst (pars) Geologie, Berlin, 3, 610–613.

Type Preparations.—S.M. K2271–5.

Locality and Horizon.—Author's locality S47 (see under *Pyrobolospora hexapartita)* ; Wealden Marls, Isle of Wight.

Description.—Trilete megaspore, probably spherical but the whole upper (apical) surface may be preserved infolded (Plate XI, fig. 3). Diameter 520–700μ (mean 620μ, 6 specimens). Tri-radiate ridges 200μ long, 10μ broad, and 30μ high ; contact faces smooth except in one specimen where the perispore covers the whole spore surface. Distal hemisphere entirely covered (Plate XI, fig. 4) with a mesh-like (net) perispore sometimes in different layers of different mesh size. The chief mesh strands are 15–35μ wide and 10μ thick (Plate XII, fig. 9) but may also be circular in cross-section exactly as described and figured by Horst (1954) and figured by Michael (1936). Spore coat is brown but shows a red luminous effect in places : thickness of exoexine 15μ, intexine very thin 2–3μ and continuous with tri-radiate lips.

Occurrence.—Pieces of detached mesh have already been seen in samples from seven quite separate Wealden localities in England, in addition to the German localities.

Comment.—This does not necessarily account for some of the larger pieces of mesh (up to 4·5 mm. long) described by Horst, although they could perhaps be unrolled perispore from larger specimens. Two further specimens found here consist of large pieces (5–6 mm. long) of the cutinized coat of an unknown body of about 1 cm. diam. and perhaps spherical. The coat is 50μ thick with apparently also a very thin inner layer ; there are patches of irregular mesh perispore of the type described above, attached to the outside surface. No suggestion can yet be put forward about the identity of these parts of larger objects, but Horst's attribution to the *Algae* is considered unlikely ; Seward (1913, p. 103) describes and figures scales of *Eury-Cycadolepis* which bear a raised reticulum although it is on a larger scale than *Dictyothylakos* in the specimens concerned. A microspore, *Dictyotriletes* Naumova with this type of perispore was described in 1937.

In addition to other bodies it seems possible that with further specimens there may prove to be more than one ordinary megaspore like *T. retiarius* with variants of this type of perispore.

Derivation of Name.—Lat. *retiarius*, the net-fighter.

V. Notes on the Descriptions

The recommendations of the Palynology Conference at the Third Heerlen Carboniferous Stratigraphy Congress (Dijkstra in Erdtman, 1952) are followed, except that greater magnification ($\times$ 100) is used for most figures ; this is considered justifiable if it is clearly stated and if a simple multiple of 50 is used. The magnification ($\times$ 50) was obviously chosen for the much larger Carboniferous spores. The various measurements given for spores are known to be imperfectly expressed by a simple mean value ; the use of space necessary, however, for a fuller record is not considered justified as size alone is of little value with megaspores. Colour of spore-coats is given as it may be helpful but the state of preservation and the degree of treatment can obviously alter it.

Adequate photography has not yet been possible owing to depth of focus required on such small objects, and the different lighting frequently required for appendages and spore-body. Author's locality numbers given will be amplified where necessary in a subsequent paper.

VI. Botanical Attribution of Dispersed Megaspores

There is little to add to Dijkstra's discussion (1951) of the possible heterosporous plants which could have borne these spores. The various members of the *Lycopodiales* are probably represented by the *Aphonozonati* and *Zonales* sections of *Triletes* in the sense of Dijkstra ; his reference (1951, p. 17) to " possible *Hydropterideae* " (presumably *Marsiliaceae*) in the German Wealden is clarified by the kindness of Dr. Hiltermann who sent me photographs of the spores ; they are almost certainly *Triletes pseudotenellus* Dijkstra of the section *Zonales*, and thus probably from lycopods.

The *Pyrobolospora* group of spores appear to be quite separate, although it is as well to record that they occur in the same samples as *Aphonozanati* and *Zonales*. The elaborate and delicate appendages of *P. vectis* and *P. medusa* suggest aquatic habit ; on the other hand authors including Vakhrameiev (1947) stress that one of the major factors in the northern hemisphere Lower Cretaceous world was rapidly increasing aridity, and this group is so far confined to the Cretaceous. The function of the various types of neck cannot be guessed until more material has been studied.

Although records of derived spores are becoming more numerous (Teichmüller, 1953), that complication is considered very unlikely at least in the case of *Pyrobolospora vectis* and *P. hexapartita*, as relatively few specimens are even damaged.

The possibility of discovery of new heterosporous groups is emphasized in an article on the *Noeggerathiinae* by Halle (1954), although that group is not involved here. Even the use of the term " megaspore " where microspores are not seen should really be open to criticism ; the arbitrary nature of the size limit of 200μ will be seen from the size-ranges given in this paper.

VII. General Remarks

Papers quoted and unpublished work separate the Middle Jurassic, Wealden, and Senonian megaspore floras in one province (Britain and the Netherlands) which are quite distinct from each other ; in my opinion there is reasonable prospect of continuing the process to finer sub-divisions. Gilbert and Harris (1953) and Jongmans (1952, p. 12) both also take an optimistic view of this problem, the plant microfossils not being confined to megaspores. From the point of view of evolution of floras it is perhaps worth noting a rather greater similarity (in this small megaspore section) between Wealden and Senonian than between Middle Jurassic and Wealden.

Acknowledgments

Dr. H. Hamshaw Thomas, Professor T. M. Harris, Professor P. Allen, and Dr. S. J. Dijkstra have frequently helped me with generous advice. The drawings have been made with great patience by my wife, Pamela Hughes. Mr. W. Stigwood has very kindly assisted me in making sections of some of the spores.

All the specimens are from the author's collection and the preparations numbered S.M. are deposited in the Sedgwick Museum, Cambridge.

REFERENCES

ALLEN, P., 1954. Geology, etc., of the London-North Sea Uplands in Wealden times. *Geol. Mag.*, XCI, 498–508.
ANDERSON, F. W., 1939. Wealden and Purbeck Ostracoda. *Ann. Mag. nat. Hist.*, Ser. 11, iii, 291–310.
—— 1940. Ostracod Zones of Purbeck and Wealden (Abstract). *Advancement of Science*, I, 259.
CHALONER, W. G., 1953. The Megaspores of Sigillaria. *Ann. Mag. nat. Hist.*, vi, 881–897.
—— 1954. Mississippian Megaspores from Michigan, etc. *Contr. Mus. Palaeont., Univ. Michigan*, XII, 22–35.
DIJKSTRA, S. J., 1949. Megaspores, etc., from the Senonian of South Limburg. *Meded. geol. Sticht., S'Gravenhage*, N.S. 3, 19–32.
—— 1951. Wealden Megaspores and their Stratigraphical value. *Ibid.*, N.S. 5, 7–21.
—— 1952. The Stratigraphical value of Megaspores. *C.R. III Congr. Strat. Carb. Heerlen*, 163–168.
DOYLE, J., 1953. Gynospore or Megaspore—a restatement. *Ann. Bot.*, N.S. xvii, 67, 465, etc.
ERDTMAN, G., (Ed.) 1952a. Palynolygy, Aspects and Prospects, III. *Svensk. bot. Tidskr.*, 46, 153–172.
—— 1952b. On Pollen and Spore Terminology. *The Palaeobotanist, Lucknow*, I, 169–176.
GILBERT, J. L., and T. M. HARRIS, 1953. The Occurrence of determinable plants, etc. *Geol. Mag.*, xc, 219.
HALLE, T. G., 1954. Notes on the Noeggerathiinae. *Svensk. bot. Tidskr.*, 48, 368–380.
HORST, U., 1954. Merkwürdige Gebilde in Kohlen aus dem Wealden. *Geologie*, Berlin, 3, 610–612.
JONGMANS, W. J., 1952. Coal Research in Europe. II *Conference on Origin and Constitution of Coal*, Crystal Cliffs, Nova Scotia.
MICHAEL, F., 1936. Paläobotanische Studien in der Nordwestdeutschen Wealdenformation. *Abh. preuss. geol. Landesanst.*, N.F., H.166.
NIKITIN, P., 1934. Fossil plants of the Devonian of Voronezh. *Bull. Acad. Sci. de Russie, Classe Sci. Math. et Naturelles*, Sér. VII, i, 1091–2.
OSBORNE WHITE, H. J., 1921. Geology of the Isle of Wight. *Mem. Geol. Surv. Gt. Brit.*
—— 1928. Geology of the country near Hastings, etc. *Ibid.*
POTONIÉ, R., 1953. Zur Paläontologie der Sporites. *Paläont. Z.*, 27, 32–36.
—— 1954. Stellung der Paläozoischen Sporengattungen im naturlischen System. *Ibid.*, 28, 103–139.
POTONIÉ, R., and G. KREMP, 1954. Die Gattungen der paläozoischen Sporae dispersae und ihre Stratigraphie. *Geol. Jb., Hannover*, 69, 111–194.
SEWARD, A. C., 1913. Contributions to our knowledge of Wealden Floras. *Quart. Journ. Geol. Soc.*, lxix, 85–116.
SCOTT, R., 1906. On the Megaspore of Lepidostrobus foliaceus. *New Phytol.*, v, 116–119.
TEICHMÜLLER, M. and R., 1953. Karbonkohle-Gerölle in der Unterkreide des Niedersächsischen Beckens. *Z. deutsch. geol. Ges.*, Hannover, civ, 459–473.
THIERGART, F., 1949. Die stratigraphische wert Mesozoischer Pollen und Sporen. *Paläontographica*, lxxxix B, 1–34.
VAKHRAMEIEV, V. A., 1947. Le rôle des conditions géologiques dans le developpement et la diffusion des Angiospermes au Crétacé. *Bull.*

Soc. nat. Moscou, Sér. géol., 22, 6, 3–17. (Traduction : Centre d'Etudes et de Documentation Palaeontologiques, Paris.)

VANGEROW, E. F., 1954. Megasporen usw. aus der Aachener Kreide. *Paläontographica*, 96 B, 24–38.

ZERNDT, J., 1934. Les Megaspores du Bassin Houiller Polonais, Partie I. *Trav. géol., Kraków*, 1, 1–55.

—— 1937. Partie II, *Ibid.*, 3, 1–78.

SEDGWICK MUSEUM,
CAMBRIDGE.

EXPLANATION OF PLATE X

FIGS. 1–3.—*Pyrobolospora vectis* gen. et sp. nov. Wealden Marls (S18), Sandown, Isle of Wight. × 100.
  1. Spore with some appendages removed.
  2. Broken spore showing neck chamber.
  3. Large spore with two necks.

FIGS. 4–5.—*Pyrobolospora hexapartita* (Dijkstra). Wealden Marls (S47), Compton Bay, Isle of Wight. × 100.
  4. Typical spore.
  5. Specimen consisting of two spores with necks united.

EXPLANATION OF PLATE XI

FIGS. 1–2.—*Pyrobolospora pyriformis* (Dijkstra). Fairlight Clay (80A), Covehurst Bay, near Hastings. × 100.
  1. Spore showing body-fold (left centre) running up into neck, and neck lobe (right centre) projecting down.
  2. Small specimen with complete neck, showing three body-folds and two lobes.

FIGS. 3–4.—*Triletes retiarius* sp. nov. Wealden Marls (S47), Compton Bay, Isle of Wight. × 100.
  3. Spore with proximal hemisphere infolded.
  4. Spore with perispore partly detached.

FIG. 5.—*Dictyothylakos* sp. Wadhurst Clay (113A), Cooden, Sussex. × 100. Detached perispore fragment. (Phot.)

FIG. 6.—*Pyrobolospora vectis* sp. nov. Wealden Marls (S40B), Brixton Bay, Isle of Wight. × 100. Specimen with two necks.

EXPLANATION OF PLATE XII

FIGS. 1–2.—*Triletes galericulatus* Dijkstra. Ashdown Sands (48B), Cliff End, near Hastings. × 100. Two views of same spore.

FIGS. 3–8.—*Pyrobolospora vectis* sp. nov. Wealden Marls (S18), Sandown, Isle of Wight. × 100. Serial sections (see Text-fig. 2). (Phot.) 3, zero ; 4, + 7µ ; 5, + 14µ ; 6, + 21µ, apex of spore ; 7, + 35µ, neck chamber ; 8, + 63µ, neck.

FIG. 9.—*Triletes retiarius* sp. nov. Wealden Marls (S47), Isle of Wight. × 100. Detached perispore from type locality.

FIGS. 10–11.—*Pyrobolospora vectis* sp. nov. Same locality as above (S 18). × 100. (Phot.)
  10. Dissection of spore-apex showing tri-radiate laesurae from below. See also Text-fig. 1.
  11. Spore showing two translucent appendages.

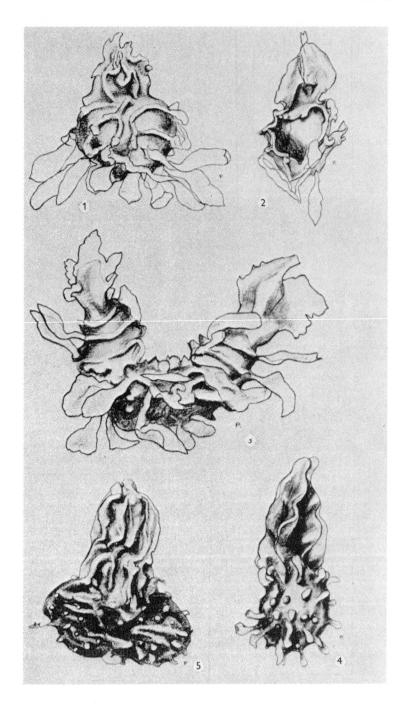

WEALDEN MEGASPORES.

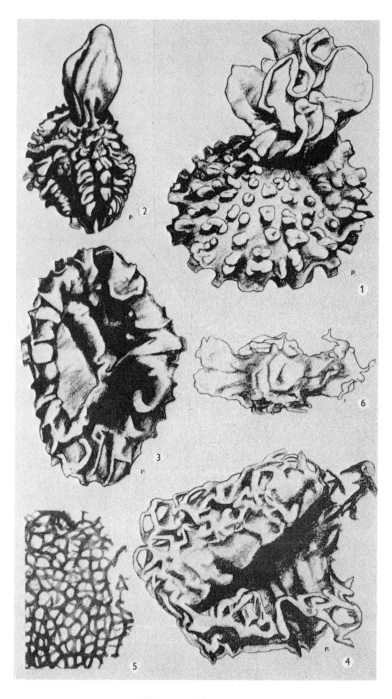

WEALDEN MEGASPORES.

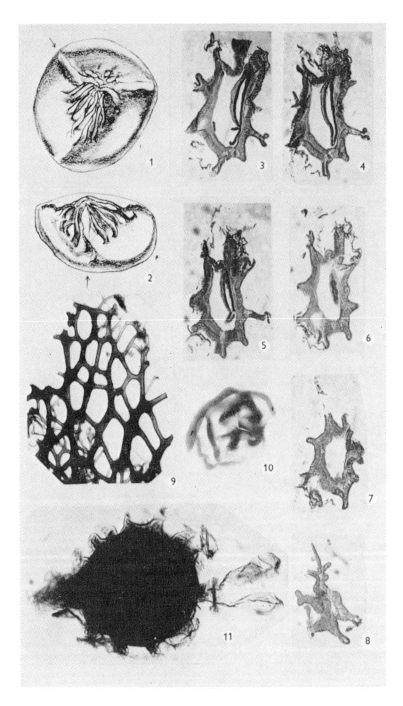

WEALDEN MEGASPORES.

Part V

DISTRIBUTION OF SPORES AND POLLEN IN SEDIMENTS

Editors' Comments
on Papers 16 Through 19

16   MULLER
     Palynology of Recent Orinoco Delta and Shelf Sediments: Reports of the Orinoco Shelf Expedition; Volume 5

17   NEVES
     Excerpts from *Upper Carboniferous Plant Spore Assemblages from the* Gastrioceras subcrenatum *Horizon, North Staffordshire*

18   CHALONER
     The Carboniferous Upland Flora

19   SMITH
     Excerpts from *The Palaeoecology of Carboniferous Peats Based on the Miospores and Petrography of Bituminous Coals*

Although many early palynological papers stressed the fact that miospores can be found in sedimentary rocks of all types, continental as well as marine, it was early realized that they are by no means totally independent of facies (Kuyl et al., Paper 6). Muller's account (Paper 16) of the distribution of spores and pollen in the modern Orinoco deltaic complex, and in the associated marine sediments of the Gulf of Paria and to the north of Trinidad, showed that different assemblages were found in different facies. Some offshore assemblages were clearly affected by the direction of the prevailing wind, whereas water currents controlled the distribution of other assemblages. Water transport was the main factor in bringing some species into the area of study, and reworking being commonest in the delta itself. Spores most affected by wind transport were produced mainly by trees growing in exposed areas at high altitudes, whereas spores of low shrubby plants growing close to the water's edge tended to be principally water distributed.

A study to determine lithological dependence of spore assemblages was carried out by Neves (Paper 17). Assemblages from a coal, a nonmarine shale, and a marine shale of very similar age (Carboniferous) were compared, with interesting results. The coal and nonmarine shale assemblages were very similar, the dominant spore genus being *Lyco-*

spora (a spore known to be produced by *Lepidodendron*, a major arborescent constitutent of the Coal Measures forest). The marine shale assemblage, however, was quite different and was dominated by *Florinites*, a Cordaite pollen grain. It was evident that the *Lycospora* assemblage was produced by the plants which grew in the coal swamp and its immediate environs; and Neves suggested that the *Florinites* assemblage was produced by the plants which inhabited the marginal marine areas. Support for this view can be adduced from the work of Cridland (1964) on the anatomy of a Cordaite stem; this shows that it may have had aerial roots similar to those found in the modern mangrove, which lives in the marginal marine/salt marsh environment.

This interpretation was challenged by Chaloner (Paper 18), who pointed out that Kuyl et al. (Paper 6) had attributed heavy fluctuations in mangrove pollen to the rise and fall of sea level with consequent decrease and increase in the area of their habitat, the coastal plain. Since Northern Hemisphere Carboniferous Coal Measures are known to be paralic (i.e., marginal marine), Chaloner suggested that it was the Coal Measures plants which were playing the role of the mangrove and inhabiting the coastal plain environment. When the sea level rose, they would be almost completely wiped out; the only plants that survived to produce pollen must have lived in an environment protected from flooding, on higher ground. Thus the flora of the marine shale, far from representing the marginal marine vegetation, actually gave a unique glimpse into the flora living on the Carboniferous Uplands. Rocks which form uplands are generally destroyed by erosion and leave no trace of their former presence, far less of their former vegetation. Support for Chaloner's view comes from work carried out by Muir (reported in Chaloner and Muir 1968), and by Muller (1966).

For many years, coals have been classified as different lithotypes—vitrain, clarain, fusain, and durain. These lithotypes can be further subdivided; and, indeed, most coal seams are made up of repeated successions of microlithotypes. Smith (Paper 19) physically separated the coal lithotypes and then looked at the spore assemblages in each individual lithotype. He found that distinct associations were to be observed and that these played a consistent role in the history of the coal. At the base of the seam, there was a Lycospore phase representing the climax forest vegetation of the time. This was followed by a transition phase, which in turn was succeeded by a Densospore phase. There is some evidence that *Densosporites* is produced by small shrubby herbaceous Lycopods; and, with the additional evidence that the Densospore phase is found in dull coal (duro-clarain) which contains a great deal of ash or inorganic matter, it appears likely that the Densospore phase represents an open peat moor situation. Incursion phases, characterized by algal coals and/or influx of detritus, can occur at any stage.

Smith relates the phase to changes in the water table, which is low during the Lycospore phase and high in the Densospore phase. The incursion phase represents flooding. Thus, for any coal seam, it is possible to determine an entire geological history of the coal swamp. The different phases can occur many times in a single seam; they represent colonization and recolonization of the swamp by the forest plants and their repeated decline.

Spore assemblages are, therefore, very dependent on facies. For serious stratigraphic work, it is essential that similar environments should be compared; otherwise, the results obtained may well be misleading.

REFERENCES

Chaloner, W. G., and Muir, M. D. (1968). Spores and Floras. In: D. C. Murchison and T. S. Westoll (eds.), *Coals and Coal-Bearing Strata.* Edinburgh: Oliver & Boyd, pp. 127-146.

Cridland, A. A. (1964). *Amyelon* in American Coal Balls. *Palaeontology* 7:186-193.

Muller, J. (1966). Montane pollen from the Tertiary of northeastern Borneo. *Blumea* 14:231-235.

16

Copyright © 1959 by Micropaleontology Press, The American Museum of Natural History

Reprinted from *Micropaleontology* 5(1):1-32 (1959)

Palynology of Recent Orinoco delta and shelf sediments:
Reports of the Orinoco Shelf Expedition; Volume 5

JAN MULLER
Brunei Shell Petroleum Company, Ltd.
Seria, State of Brunei,
British Borneo

ABSTRACT: *Palynological study of Recent Orinoco sediments shows the relative influence of location of source area and transport by air and water currents on pollen dispersal. Distribution of fungi, cuticles, reworked material, Hystrichosphaeridae, and foraminifera have also been studied. Variations in microfossil content with depth in core samples are found to be related to depositional history. The significance of the results in palynological and paleogeographical studies of older sediments is discussed.*

[*Editors' Note:* Table 1 and diagram 1 are not reproduced here.]

INTRODUCTION

General remarks

Experience with palynological correlation in clastic sediments of Tertiary age has led to the conviction that water-transport of pollen grains and spores is more important than wind-transport in explaining certain distributional phenomena. The present study has, consequently, been undertaken with the primary purpose of investigating transport and deposition of pollen in an environment of predominantly clastic sedimentation. The distribution of other microfossil groups has also been studied. In the following discussion, the term "pollen" will be used as comprising both pollen grains (derived from Spermatophyta) and spores (derived from Pteridophyta).

The samples taken by the Orinoco Shelf Expedition from the Recent sediments of the Gulf of Paria and on the outer shelf north and east of Trinidad proved to be well suited to this purpose. Many samples were available for study, but because of lack of time the examination had to be restricted to a limited number only. In making the selection, the guiding principle was to obtain a regional picture. For the Gulf of Paria and the outer shelf this was possible without much difficulty, but the equally interesting Orinoco delta could only be sampled to a very limited extent, due to poor accessibility. In January, 1956, the author was able to accompany a surveying party to the eastern Orinoco delta for a period of four weeks. Transportation was by flat-bottomed launch and canoe along the main tributaries, and there were short side-trips into the back-swamp area also. Samples were collected from the river bottom, levee, and back-swamp sediments, but because of the great variety in local conditions of vegetation and sedimentation, this sampling is still very incomplete. Valuable information on the distribution of vegetation was obtained, however. The main emphasis is therefore laid on the study of the offshore sediments. Three cores, at suitably located spots on the shelf, were selected for a closer study of the vertical variations in microfossil content. Some samples were examined especially at the request of the sedimentologists, in order to check the supposed sub-Recent age. The location of all samples is plotted in text-figure 1.

In interpreting his results, the author was greatly aided by the very detailed hydrographic and sedimentological results discussed by van Andel and Postma (1954) and by Koldewijn (1958) and Nota (1958). Gratitude is further expressed to the Cía. Shell de Venezuela for permission to publish the results of the present study. Thanks are also due to Professor P. H. Kuenen, Dr. A. M. Oosterbaan, Dr. A. Ford, Dr. B. L. Meyer, and Dr. B. W. Koldewijn for valuable suggestions and for reviewing the manuscript, and to Dr. T. van Andel for encouragement and stimulating discussions during the course of the work.

291

Preparation of samples

In the laboratory the samples were subjected to standard chemical treatment. After drying, crushing and mixing to a uniform powder, an accurately weighed amount between 1 and 5 grams, depending upon the richness of the sample in pollen, was boiled for 20 minutes with concentrated hydrofluoric acid. When the samples contained a large amount of calcium carbonate they were pre-treated with 10 per cent hydrochloric acid. After centrifuging off the remaining hydrofluoric acid they were washed one or more times, if necessary, with hot 10 per cent hydrochloric acid to remove silica gels formed during the reaction with hydrofluoric acid. This chemical treatment is quantitative, and the next step is to measure the concentration of microfossils in the residue. For this purpose the residue was transferred to a graduated centrifuge tube and a measured amount of a mixture of glycerine and alcohol was added, varying between 1 and 10 cc., again depending upon the concentration of microfossils to be expected. The mixture was then shaken very thoroughly to obtain an even suspension. Next, two drops were taken at random with a standard pipette from the suspension, transferred to a slide, and covered with a standard-sized cover glass. The number of microfossils could then be counted, and, as the number of drops from the pipette per cubic centimeter of liquid is known, the number of microfossils per gram of dried sediment could easily be calculated. This method produced consistent results with duplicate measurements; three slides were generally counted for each sample and the average value determined.

At the same time, the percentage ratios in which pollen types and spores occur were determined. For this purpose it was often necessary to count more slides in the case of poor samples. It was difficult to concentrate pollen in these samples because, here especially, large amounts of silica gels were often formed, which could not be entirely removed by hydrochloric acid treatment, thereby greatly hampering counting.

Identification of microfossils

The residue contained a large variety of microfossils. It was not possible to study all groups equally well, and in the following discussion the main emphasis will he placed on the pollen, which served the primary aim of the study. In addition, those microfossils that have proved to be of practical interest in the study of fossil sediments have also been studied.

In the first place, the total quantity of pollen per gram of sediment was measured. Next, the percentage ratio in which the various pollen types occurred was determined. These types were differentiated at first on a morphologic basis, and were then compared with a collection of Recent acetolysed pollen slides that had been prepared from identified fresh or herbarium material. Most of the more common and interesting types could thus be identified.

The next group of microfossils recorded were the fungal spores. These were merely counted per gram, and no differentiation between types was attempted. Algal remains were very scarce. Only *Phycopeltis* was seen regularly in the delta deposits. *Pediastrum* was extremely rare, and *Botryococcus* completely absent. Hystrichosphaerids were determined, but no attempt at a detailed study was made.

The abundance of cuticles and their average sizes were estimated, but a further analysis was not made, although identification is probably possible and would have some importance for facies interpretation. Tracheid remains, reworked from older deposits, were also noted. Among the animal microfossils, only the remains of the smaller foraminifera that are left after the chemical preparation were recorded. No attempt at specific or generic identification was made, however.

ENVIRONMENT

Topography

A detailed description of the area has been given by van Andel (1954). In text-figure 1 the main features are shown, such as the llanos plains, where older Quaternary sediments form the surface; the vast alluvial plain of the Orinoco delta (also called the Delta Amacuro); the mountainous peninsula of Paria; and the islands of Trinidad and Tobago, which are also, in part, mountainous. The shelf edge is indicated at the 100-fathom depth contour. The topographic features are of importance insofar as they influence vegetation and movement of sediments. Pollen transport and deposition is, of course, partly dependent on the location and altitude of source areas in relation to areas of deposition.

Climate

Every pollen grain and spore released by a flower or sporangium is transported for some distance through air before settling on a land or water surface. The principal climatic factors that determine the extent of this transport phase are wind strength and direction, and rainfall pattern.

According to Beard (1946), the prevailing winds on Trinidad are easterly, varying from northeast in the dry season to southeast in the rainy season. The average velocity at sea level in Port of Spain was 4 m.p.h., and the highest velocity recorded 25 m.p.h. Conditions are typically breezy. Probably this pattern also holds for the remaining part of the area under investigation. In any case, the coast lines lie under the general influence of the Northeast Trade Winds. Unfortunately, not much is known about the air circulation at higher altitudes, but it is unlikely that many pollen grains manage to reach these layers.

Rainfall also influences pollen distribution. The heavy, and often daily, tropical showers that are common in a large part of our area tend to wash the air clear of dust and pollen. Thus transport facilities are restricted.

An attempt was made on the first Gulf of Paria survey, from April to June, 1952, to sample the atmosphere for pollen. A vacuum-cleaner was rigged up on one of the mastheads of the survey vessel. The apparatus had been tested in Holland, where it worked efficiently, but the air samples obtained from the eastern Gulf of Paria failed to produce a single pollen grain. This may be a direct indication that airborne transport of pollen in our area is limited.

Noticeable quantities of airborne pollen can be expected only in the Gulf of Paria and on the northern Trinidad and Paria shelf, close to the shore lines. It appears extremely unlikely that airborne pollen reaches the shelf area east of Trinidad and east of the Orinoco delta.

Hydrography

When pollen settles upon a moving water surface it floats for some time and then slowly descends to the bottom. The actual speed of this process is difficult to measure directly, as it is influenced by many factors of unknown magnitude. The floating time depends further on the specific weight of a particular grain, the surface properties of the wall, and the speed with which the protoplasm decays. Settling in water is retarded by turbulence, especially in moving water masses. In addition, the slight density difference between fresh and salt water will have an influence. Finally, the strength of the local tidal currents is an important but hardly measurable factor.

Therefore, only the residual water movements will be described. In this respect we are fortunate in having at our disposal the results of the detailed hydrographic survey made in the area of investigation by Postma (van Andel and Postma, 1954), and also in Koldewijn's study (1958).

The main features of the water circulation pattern are summarized in text-figure 2, and can be described as follows:

1) Orinoco River system: The drainage area is formed by the Llanos depression, which is bordered by the Cordillera de la Costa, the Andes, and the Guayana highlands. Pollen may thus be gathered from mountain and plains vegetation. Reworked pollen can be expected to be freed by the erosion of exposed Tertiary and Cretaceous strata in the Andean foothills. At Barrancas the Orinoco distributaries enter the delta plain and are divided into several branches. Residual current velocities decrease, and sedimentation of suspended material begins. In the lower delta there is a gradual transition to the tidal reaches of the estuaries.

The delta watercourses can be divided into the muddy Orinoco distributaries, through which the main Orinoco discharge is carried seaward, and the back-swamp rivers, which are recognizable by their clear "black" water. The latter drain excess rain-water from the peaty back-swamps. The delta is largely flooded in the rainy season, whereas in the dry season the water level falls, and the levees are exposed. The back-swamps remain waterlogged, however, for the greater part of the year. Pollen may be gathered by the moving Orinoco water masses at any stage during their journey through the delta, but at the same time, some pollen is already being deposited on the levees and in the back-swamps.

Erosion of sub-Recent delta deposits takes place in the meander belts of the tidal channels in the outer delta, especially in the eastern part. Considerable amounts of pollen that has been deposited earlier in a back-swamp area may therefore be liberated again and carried off seaward.

The net result of these processes is that the water masses discharged from the estuaries into the sea contain pollen of various origins in suspension. The pollen may have been derived directly from the coastal vegetation bordering the estuaries, or it may have had a shorter or longer history of air transport, water transport, or reworking in one or more cycles. Turbulence in the moving water masses, which is especially strong in the tidal region, will tend to mix the suspended pollen thoroughly. The degree of turbulence cannot be measured directly, but the effect can be judged by the degree of homogeneity with which sediment and pollen are deposited. In the back-swamp area, water movements are much more restricted, and lateral transport of pollen in shallow water can presumably take place only during flood stages.

2) Northern llanos river system: The sediments of the northern part of the delta plain were not accumulated by the Orinoco River but by several smaller rivers, such as the Rio Grande, San Juan, and Guanipa, which descend from the Serrania del Interior and drain the northeastern part of the llanos. These rivers transport much less water and carry little sediment. They reach the sea in wide estuaries without bars and of simple morphology. As a pollen-gathering area they differ slightly from the main Orinoco estuaries.

3) Gulf of Paria and the open sea: When the river discharge from the estuaries comes into contact with moving marine water masses, the nature and direction of the water circulation pattern changes considerably. Marine water movement in this area is dominated by a strong westward current, which forms part of the Atlantic equatorial current system. The main features of water movement that are of especial interest here are summarized below, following the detailed descriptions given by Postma (1954):

a) The residual current pattern is shown in text-figure 2. Note here especially the eddy-like current present in the eastern part of the Gulf of Paria.

b) Estuarine circulation existing in the rainy season: Oceanic water enters the Gulf of Paria and the large eastern estuaries in subsurface layers, as far as the western Gulf, while the lighter, fresh river discharge flows seaward on top.

c) Existence of rather strong tidal currents, especially in the estuaries.

These data are invaluable for an understanding of the water-borne phase of pollen transport. In general, it appears possible for pollen escaping in suspension from the estuaries to be transported over large distances by marine currents. The estuarine circulation causes pollen that settles on the water surface in the estuaries to be transported seaward farther and in larger quantities than pollen that is already in suspension in the river water flowing through the delta.

Locally, marine currents may be so strong that active scour of the sea bottom takes place, as, for example, in the Serpent's Mouth. This phenomenon, together with wave erosion of exposed cliffs composed of soft Tertiary formations, such as those along the southwestern point of Trinidad, and erosion of shallow banks, indicates possible sources of reworked pollen and other microfossils.

Phytogeography

The demarcation of pollen source-areas is a prerequisite for full understanding of pollen dissemination patterns. For this purpose a botanical inventory of the terrestrial and floating vegetation of a large region surrounding and partly covering the area of deposition is necessary.

Unfortunately, only the floras of Trinidad and Tobago can be considered as being adequately known. Only scattered information is available from the Peninsula of Paria, the Orinoco delta, and the Venezuelan llanos. The author was able to make a short botanical reconnaissance in the Orinoco delta in January, 1956, and made first-hand observations on the distribution of plant communities and of palynologically important plant species. Reference is made to the detailed studies of Beard (1946, 1949) for Trinidad and Tobago, and some remarks on the plant communities encountered in the Orinoco delta, and their relations with the soil and water conditions governing their distribution, will be made here, as these large swampy areas form one of the main sources of pollen. The phytogeography of the remaining area is of lesser importance and will not be further discussed. The author was able to distinguish seven different vegetation types in the Orinoco delta, based on his own field observations, a study of aerial photographs, and a comparison with the literature (Beard, 1946; Lindeman, 1953). They are the following:

a) Mangrove forest: A very uniform vegetation type, virtually consisting of only two species, *Rhizophora mangle* and *Avicennia nitida*, in varying proportions. *Laguncularia racemosa* occurs locally. It forms belts along the coasts and for some distance upstream along the major tributaries, approximately to the point where the brackish tidal influence ceases. The typical soil of this vegetation type is a soft bluish-gray clay, strongly penetrated by a dense root felt. The absence of *Avicennia* in the strongly dissected southeastern part of the delta is striking. Picas B, D and F (see diagram 1) traverse this vegetation type.

b) Mixed swamp forest: A tall, mixed forest with a rather restricted number of dominant trees, among which *Symphonia globulifera*, *Pterocarpus officinalis*, *Bombax aquaticum* and the palms *Euterpe* sp. and *Manicaria sacchifera* could be recognized. It forms extensive forests in the lower delta, often directly behind the mangrove belt, especially in

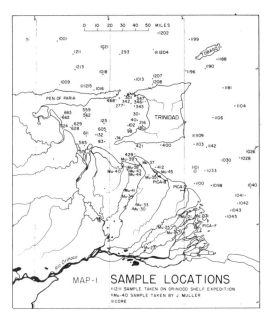

TEXT-FIGURE 1

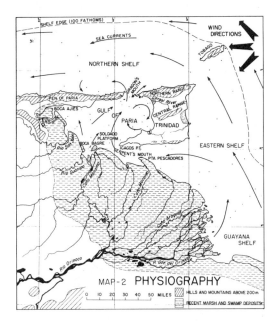

TEXT-FIGURE 2

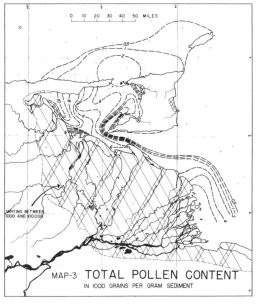

TEXT-FIGURE 3

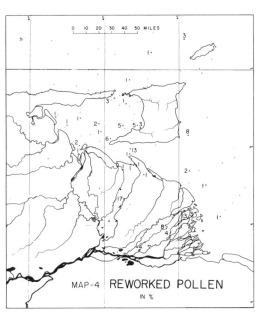

TEXT-FIGURE 4

the southeastern part of the delta. The grayish clay soil is covered with a layer, 30–50 cm. thick, of very soft humic clay with abundant decaying plant material. Tidal influence is restricted, but seasonal variations in water level are probably also quite small. Picas E and F (see diagram 1) traverse this vegetation type.

c) Erythrina swamp forest: This is a rather open mixed forest type, with heavy undergrowth. *Erythrina glauca* is the dominant tree, but some palms (*Mauritia* and *Euterpe*) may also occur. This forest type is typical more of the upper delta, where seasonal variations in water level predominate, than of the lower delta, which lies more under the influence of tidal fluctuations and was found on the levees, traversed by Pica E. A stiff clay soil is present here (see diagram 1).

d) Palm swamp: An open, bush-like vegetation with emerging clusters of palms, mainly *Mauritia* sp., *Manicaria sacchifera*, and *Euterpe* sp. Heavy undergrowth and soft peaty soil deposits make this an exceptionally difficult vegetation type to traverse. Pica D penetrates into this type, and pure peat deposits are formed here (see diagram 1). The soil is permanently waterlogged, and fluctuations in water level, if any, are seasonal. It covers extensive areas in the central delta, locally known as "Morichales."

e) Herbaceous swamp: Low vegetation with strongly varying local dominance of ferns, sedges, low shrubs, *Montrichardia* sp., and, locally, floating grass. Stunted trees and scattered *Mauritia* palms may occur. The vegetation remains more or less permanently inundated, with only seasonal variations in water level. Peat deposits are formed here. Unfortunately, no penetration into this vegetation type was possible, and consequently no data on the pollen deposition could be gathered here. It covers great expanses in the central delta.

f) Dune and beach ridge vegetation: This type of vegetation, which consists of small trees and shrubs, occurs locally on sandy beach ridges in the coastal area, but is believed to be of little importance as a pollen source.

g) Rain forest: The higher levees in the upper delta, which are flooded only seasonally, at very high river level, were originally covered by rain forest, which is now largely cleared for agricultural purposes. Remnants of this forest are still visible here and there in the shape of very large trees with a dome-like crown emerging above a several-storied forest.

The horizontal distribution of these vegetation types follows certain ecological rules. This makes it possible to recognize several landscape regions in the delta. Firstly, the upper delta can be recognized, where seasonal fluctuations in water level predominate and well developed levees are present. The original vegetation of these levees was rain forest. Depending on the water level, the back-swamp area is covered with *Erythrina* swamp forest, palm swamp, or herbaceous swamp, or open water is present where drainage is blocked by high levees. Levee building is much less pronounced in the central delta, and although the caños here show distinct tidal fluctuations, the latter do not penetrate far into the back-swamps. This is the domain of the *Erythrina* swamp forest, palm swamp, and herbaceous swamp, and it is probable that these vegetation types here represent a serial succession, controlled by the decreasing supply of mineral sediment.

The outer delta, where the influence of tidal fluctuations is strongly felt and where an adequate supply of mineral sediment is present, is characterized by the dominance of mixed swamp forest in the inner belt and of mangrove forest in the outermost belt. The southeastern part of the outer delta is strongly dissected by a network of meandering tidal channels, and differs in this respect from the central sector, where a few large, straight Orinoco branches traverse a very extensive but largely undrained back-swamp area. The former area is covered by mixed swamp forest, growing on a clayey soil, the latter by palm swamp and herbaceous swamp on a peaty soil. The coast of the southeastern part also has a different character, as indicated by the absence of *Avicennia*, which is common along the northern and northwestern delta shores, where mud-flats are more extensive.

All of these variations in delta vegetation and sediments are of direct interest in a palynological study, mainly because they indicate the locations of potential source areas for important pollen types. In the separate discussion of the distribution of these types, further details will be given on this subject. The approximate location of the source areas is given in the accompanying maps and graphs.

DISTRIBUTION OF MICROFOSSILS
Introduction

In this section the distribution of the principal microfossil types will be discussed, with the aid of a number of maps and graphs. The complete documentation is given in the distribution chart (Table 1).

The surface data will be discussed first (text-figs. 3-23). The microfossil content of the samples taken in the delta picas is given in a separate chart (diagram 1), because the details could not possibly be shown to scale on the other maps. Next, the variations in the microfossil content of the cores will be considered (diagram 2), and finally some data on sub-Recent or Pleistocene samples will be given.

Surface samples

1) *Total pollen content:*

Text-figure 3 shows the variations in total pollen content of the samples examined. On this map the land areas that are not cross-hatched represent outcropping Pleistocene or older sediments. On the eastern shelf, stations 1026 and 1181 also show older sediments exposed on the sea floor.

In the delta the pollen content of scattered samples is indicated. They show a high but variable pollen content. The nature of this variation can be studied better in the chart (diagram 1), where the total pollen content of the surface samples from the picas is given.

The location of these four picas is shown in text-figure 1. The topography is indicated on an exaggerated vertical scale. Pronounced natural levees are found along Caño Sacupano (Pica E) and Caños Jefe-Cubaca and Guayo (Pica F). In these two picas the average mineral content of the samples is also highest. These levees have an elevation of not more than 1 meter above mean tide level, and their deposits have a low organic content. Picas D and B are situated in areas where less mineral sediment penetrates. Levees are therefore less pronounced here. In the back-swamp areas, peat formations are found that are practically free of mineral admixture. In Pica E the total pollen content fluctuates rather strongly but reaches a minimum in the levee deposits, where slightly coarser mineral sediments were laid down.

The local presence of prolific pollen producers such as *Terminalia* and *Rhizophora* further influences the total pollen content, because high percentages of these types go together with peaks in total pollen content. In Pica F the total pollen content is also lowest in the levee deposits. Pica D penetrates far into a peat area, and here the total pollen contents are lowest on the levee, because of the relatively high mineral content. In the peat deposits they are also rather low, but this is due to the large amount of organic accumulation here. Pica B shows similar relationships between total pollen content and content of organic matter.

In general, it can be concluded that the total pollen content of the delta samples is at its maximum where both mineral supply and organic accumulation are at their minima, which is in sediments containing 20-50 per cent organic matter. It varies further in relation to the local presence of prolific pollen producers.

It would not be possible, even if the samples were available, to show this variation in pollen content on the large-scale map of text-figure 3. The pollen content is therefore shown only as a cross-hatched pattern, which indicates a range of variation between 1000 grains/gram on natural levees and the maximum of 100,000 grains/gram in back-swamp deposits. The landward boundary of this cross-hatched area represents the limit of Holocene swamp deposition. Outside of this limit no Recent pollen is preserved.

The sediments that accumulate in open water offshore are much more evenly sampled, and show a different, but also quite regular, variation in pollen content. It was possible here to construct lines of equal pollen content ("isopollens") at intervals of 1000 grains/gram (text-fig. 3). The overall picture thus obtained shows, first of all, a consistent decrease seaward, with the heaviest pollen concentrations opposite the larger Orinoco distributaries. This seaward decrease is most clearly visible off the eastern delta. Here, at first, a gradual decrease down to 3000 grains/gram is visible. Then a rather steep decline to 500 grains/gram is shown, and farther seaward, the pollen contents remain more or less constant.

Considering the predominant easterly wind, it must be concluded that practically all of the pollen contained in the sediments of this area is derived by water transport from the large eastern estuaries. This is further confirmed by the fact that the highest concentrations are in front of the Rio Grande – Caño Araguao group of estuaries, whereas farther along the coast in a north–northwesterly direction, where large estuaries are absent, the pollen contents decrease until the Caño Macareo is reached. It is also clear that the strong Equatorial Current deflects the pollen-bearing estuarine discharge in a north-westerly direction, the pronounced drop from 3000 grains/gram to 500 grains/gram marking the zone of maximum deflection.

Settling of the suspended pollen and coarser sedimentary particles reaches a maximum opposite the estuaries because of decreasing current velocities. However, not all of the pollen settles on the outer shelf, because appreciable quantities are still found here. Some of these pollen grains are derived, of course, from the Guayana shelf, following the current direction.

Approaching the Serpent's Mouth channel from the southeast, close to the delta coastline, very low pollen contents are observed, because of the absence of large estuarine outlets. The wind directions here still impede direct airborne transport of delta pollen seaward. However, along the southern coast of Trinidad, evidence of a slight increase in pollen content is seen, and this increase must be due to an airborne supply from Trinidad with northeasterly winds, as the rocky coast here, without major rivers or swamps, does not favour water transport of pollen from the island into the sea.

Near the narrowest portion of the Serpent's Mouth channel, a new supply of pollen becomes evident in the concentration opposite the Caño Macareo estuary. The Macareo discharge, which is rich in pollen, is carried with the westward-flowing marine current alongside of oceanic waters with a low pollen content. Evidently very little mixing occurs, which results in a very steep decline in the pollen content of bottom sediments between Punta Pescadores and the southwestern point of Trinidad. The same pattern is visible on Postma's map of the concentration of suspended matter at the surface (van Andel, Postma and Kruit, 1954, text-fig. 23) and the distribution of Secchi disc visibility (ibid., tex-fig. 25) during the early rainy season, which confirms the idea that the pollen distribution mainly reflects waterborne transport.

The effect of the entrance of oceanic water, which is poor in pollen, through the Serpent's Mouth channel into the Gulf of Paria alongside of water that is rich in pollen derived from the Caño Macareo remains visible in the pollen content of the bottom sediments in the Eastern Gulf as an elongated patch of low pollen contents, which is bordered by the isopollen line of 3000 grains/gram. This is accentuated by the existence of the eddy-like current here. Toward the Trinidad coast, the sediments again show an increase in pollen content, which is most probably due to airborne supply from Trinidad. Opposite the mouth of the Caroni, which is the largest river on Trinidad, a very pronounced concentration of pollen is visible, which is probably caused by pollen that is produced locally in the small Caroni delta and carried seaward with the river discharge.

The pollen distribution in the Eastern Gulf shows a different pattern. In the large funnel-shaped Boca Vagre estuary, low pollen contents are shown, which increase seaward rather rapidly only to show, farther to the northeast, the normal gradual decrease which is also found opposite the other Orinoco estuaries. This anomaly is caused by the coarse-grained material that is being deposited in the Boca Vagre by the Caño Manamo, while the sedimentation of the finer fractions, together with the pollen grains, finds its maximum somewhat farther offshore in quieter waters. It should be recalled here that the bottom sediments of all the large Orinoco distributaries consist largely of rather coarse sand, which does not contain any pollen grains at all. The track of the Boca Vagre discharge is indicated farther along in the Central Gulf by the northeasterly-trending tongue of higher pollen concentrations near the Dragon's Mouth.

Still farther westward along the delta front, in the Rio San Juan estuary and in the Boca Ajies, far higher pollen concentrations are found than in the Boca Vagre. It has already been mentioned that these rivers carry very little sediment. This factor, together with the narrowing shape of the Gulf here, which increases the number of wind-transported pollen grains caught per unit of water surface as compared with the Central Gulf, easily explains these very high pollen concentrations.

In the area of the Dragon's Mouth the isopollen lines show clearly the discharge of Gulf water, which is rich in pollen, into the ocean, in which the pollen content of the bottom sediments is lower. There is, however, one slight anomaly in the Northern Gulf, where a tongue of sediments poor in pollen is noticeable near the southern Paria coast. This very probably reflects the entrance of an undercurrent of ocean water into the Gulf through the Dragon's Mouth, the existence of which has been determined hydrographically by Postma (van Andel, Postma and Kruit, 1954). This ocean water is poor in pollen and dilutes the pollen-rich Gulf water, as a consequence of which the bottom sediments also show a lower pollen content.

From the Dragon's Mouth westward, the westerly deflection of the Gulf discharge along the northern Paria coast is indicated by the 1000 and 500 isopollen lines. In the northwesternmost corner of the shelf, where the lowest pollen contents occur, the supply has become almost negligible.

Northeastward from the Dragon's Mouth, a concentration of pollen can be observed north of Trinidad and west of Tobago, whereas farther eastward, south of Tobago, a decrease is visible. This pattern indicates a noticeable supply of primarily airborne pollen from Trinidad and Tobago. Also, there is probably some supply of pollen derived from the eastern Orinoco delta and carried in suspension over the shallow area of non-deposition present east of Trinidad. The bottom sediments of this area consist mainly of coarser-grained material and calcarenites, and Pleistocene deposits occur locally at or near the surface, which makes them rather unsuitable for palynological study. The few Recent samples examined from the eastern shelf, however, have the low pollen content typical of outer shelf sediments.

The main conclusions are:

1) The Orinoco delta has been the source area for most of the pollen contained in the offshore sediments. This can be deduced from the distinct decrease in pollen content from the estuaries seaward.

2) Marine currents are the main medium of transport. This is shown by the correspondence between the pattern of decrease in pollen content with the current and the sediment-transport pattern.

3) The effect of wind transport is small and is noticeable only in pollen concentrations to leeward of Trinidad and Tobago. This is corroborated by the negative results of air sampling in the eastern Gulf of Paria.

The total pollen content is further influenced by the following factors: 1) Distance from source; 2) rate of sedimentation; 3) coarseness of sediments; and 4) presence of carbonates and organic matter.

From a palynological point of view it would be interesting if the total pollen content of the sediments could be converted into the amount of pollen deposited on a given surface (1 cm².) in a given length of time (one year). In this way the so-called "absolute pollen frequency" could be obtained. The surface data alone do not permit this, however mainly because the sedimentation rate is too imperfectly known. It appears that no simple relationship exists between van Andel's figures for the rate of deposition during the last 700 years and the total pollen content at the surface. In the Western Gulf, low values of 30 cm. per century coincide with very high pollen contents because of the proximity of the pollen source area. In the Boca Vagre, a high rate of 100 cm. per century produces a low total pollen content; and along the southwestern coast of Trinidad, low rates of 12–20 cm. per century are associated with low pollen contents due to the small supply of pollen. Assuming constant conditions during the last 700 years, it is possible to derive from these figures an estimate of the absolute pollen frequency expressed in grains/cm²./year. The values obtained would vary from 3600 in the Western Gulf, to 2400 in the Boca Vagre, and 168 in the Eastern Gulf (assuming an average water content of the fresh sediment of 60 percent (van Andel), and a specific weight of the clay material of 2.0).

It is, of course, also possible to calculate the total pollen content using the clay fraction of the sediments only. In this way a more regular decrease in total pollen content seaward would be obtained. However, the straightforward determination of total pollen content per gram of dry sediment was preferred, because this is virtually the only way in which this value can be determined in older, indurated sediments, in which clay content and rate of sedimentation are usually difficult to measure.

2) *Percentage distribution of pollen:*

The frequency with which each type occurs has been calculated in percentages, and these values have been plotted separately on distribution maps and the source areas determined as accurately as possible. It then appeared possible to recognize various groups of types on the basis of similarities in their distribution patterns. From each group, one or more examples will be discussed in detail below.

REWORKED POLLEN

First of all, it appeared necessary to separate reworked older pollen from Recent grains. Tertiary pollen reworked by the Orinoco headwaters in the Andean foothills and deposited in the recent Orinoco sediments is rather easily distinguishable from Recent pollen by its flattened shape and poor preservation. It is often corroded and stains differently. Pleistocene pollen is more difficult to recognize as such on the basis of preservation, and the separation of reworked older Holocene pollen from Recent pollen is next to impossible.

In text-figure 4 the distribution of obviously reworked grains is plotted as a percentage of total pollen content. This map shows that reworked grains are concentrated in certain areas. In the first place, they are rather abundant in the levee deposits in the delta, where values up to 17 per cent may occur. These grains have evidently been eroded and redeposited by the Orinoco river system. In the outer delta and in the offshore zone near the delta front, the percentage of reworked pollen drops because the supply of locally produced Recent pollen outnumbers the reworked material.

Near Trinidad a second source of reworked grains is visible. The high percentages in the Gulf of Paria around the southwestern point of Trinidad are evidently related to the Tertiary sediments, which are locally eroded by wave abrasion on shoals and cliffs and which also yield fossil foraminifera and Radiolaria (see Kruit, in van Andel, Postma, and Kruit, 1954, text-fig. 60). South of Trinidad, in the Serpent's Mouth channel, reworked material also occurs. The reworked grains found east and north of Trinidad on the shelf have probably been derived from eroded Pleistocene deposits. They may have been originally Pleistocene in age, or may have been reworked at that time from the Tertiary.

In the following paragraphs, the reworked pollen has not been included in the pollen total on which the percentages of the Recent pollen types have been calculated, as their occurrence bears no relation to Recent plant distribution and pollen dissemination, and their inclusion would only tend to obscure the Recent distribution patterns.

Mauritia sp. (PALMAE)
Plate 1, figure 1

The typical echinate, monocolpate pollen of *Mauritia* cannot easily be mistaken. The peculiar foundation of the short, rounded spines, and the nearly spherical shape, are distinct characteristics. The grains are medium heavy.

Mauritia is very common in large expanses of swamp and marsh vegetation in the central and western Orinoco delta. It occurs in varying abundance from scattered occurrences to dense stands, the so-called "Morichales." It is largely absent in the brackish-water belt along the delta front, nor does it occur in the closed mixed swamp forest found in the eastern delta. It also occurs locally on Trinidad in the swamp and marsh environments (Beard, 1956). In the Venezuelan llanos it is especially common along the river courses. In text-figure 5 the occurrence is indicated by cross-hatching. The relative abundance could not be mapped accurately.

The distribution of *Mauritia* pollen as shown in text-figure 5 and diagram 1 shows, in general, strongly fluctuating percentages in the central delta. These percentages vary in direct proportion to the local abundance of the mother plant. In the mixed swamp forest of the eastern delta, *Mauritia* is not abundant, and here the percentages are low, as shown most clearly in Picas E and F. In Pica D, however, the peat area is covered by a Morichal vegetation, and this is reflected in the relatively high pollen percentages here. In Pica B, a high *Mauritia* percentage in one peat sample indicates the abundance of the *Mauritia* palm. It is evident from these data that atmospheric mixing is unable to equalize this locally strongly variable pollen production to any appreciable degree.

Offshore, where lower but more regularly distributed percentages are found, concentrations are visible in the Gulf of Paria and, less clearly, in a small area opposite the eastern delta. Apparently the western estuaries, which traverse the areas where *Mauritia* is most abundant, carry a proportionally larger amount of *Mauritia* pollen into the Gulf than is possible for the eastern estuaries. Most of these grains apparently settle in the Gulf, and very few escape through the Dragon's Mouth to the northern shelf.

Avicennia nitida Jacq. (VERBENACEAE)
Plate 1, figure 2

The tricolporate *Avicennia* pollen is easily recognized by its reticulate sculpture, rather thick wall, subprolate form, and medium size. The mother plant forms one of the main constituents of the mangrove forest. The areal distribution is indicated in text-figure 6, and is restricted to the outer brackish-water belt of the delta, but the species penetrates inland along the river courses as far as the brackish influence is felt. Its almost complete absence in the strongly dissected eastern delta is remarkable, although it is abundant along the straight northern and northwestern delta edge. *Avicennia* has not been observed in the outermost vegetation belt, but generally occupies the inner parts of the mangrove forest. It also occurs locally in the Trinidad swamps.

The general distribution in the delta shows that in the eastern delta, where *Avicennia* is scarce, no pollen was found. In the central delta, *Avicennia* pollen is frequent in the source area, but apparently only very few grains are carried inland by wind transport.

In Picas E, F, and D, *Avicennia* is also virtually absent. Noticeable percentages are reached only in Pica B, where *Avicennia* is about equally as abundant as *Rhizophora*. Nevertheless, the highest percentages of *Avicennia* pollen in the delta are much lower than those reached by *Rhizophora*, which indicates that *Avicennia* is comparatively under-represented in relation to the abundance of the mother plant in the vegetation, which equals that of *Rhizophora*.

Offshore, an area with a higher average percentage can be separated from one with a lower average. The area with higher percentages, restricted to the western Gulf of Paria, is closest to the main source area. In the eastern part of the Gulf and on the northern shelf, values are low, and apparently the rather heavy *Avicennia* pollen is not carried over great distances in suspension but settles comparatively soon. The pollen distribution is closely related to the source area, and water appears to have been the main medium of transport.

Rhizophora mangle L. (RHIZOPHORACEAE)
Plate 1, figures 3, 3a

Rhizophora pollen is, in general, fairly easy to recognize, although rather strong variation in size and in the appearance of pores has been observed, which is probably due to differences in preservation. The pollen is small, spherical, nearly psilate, and the most typical characteristics are the equatorially elongated pores, which are combined with three rather weakly pronounced colpi.

Rhizophora is the second main mangrove species, and generally forms the outer vegetation belt along the delta front and, together with *Avicennia*, penetrates some distance inland along the estuaries. The width of the *Rhizophora* belt varies with the amount of mud accumulation. In the eastern delta, the belt is never very broad

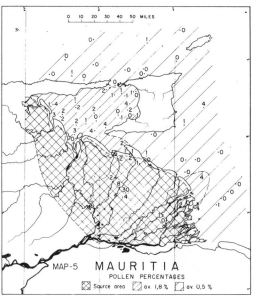

TEXT-FIGURE 5

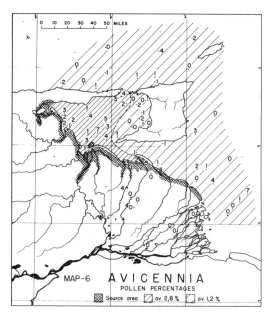

TEXT-FIGURE 6

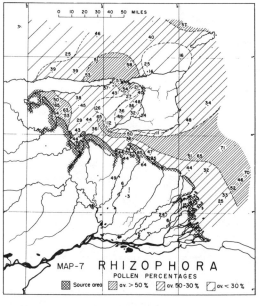

TEXT-FIGURE 7

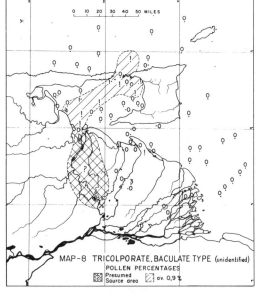

TEXT-FIGURE 8

or continuous because of the numerous meandering estuary branches, where strong tidal currents are eroding concave banks, and mud accumulates only on convex banks, where point-bar deposits are laid down. On these point bars, *Rhizophora* grows in dense stands, whereas the back-swamp vegetation is exposed on the eroded banks. On the shores of the Gulf of Paria, the situation is different; the coastal mud-flats are more extensive, and the *Rhizophora* belt is more continuous and wider. The species occurs locally in Trinidad in the small swamps on the western shore (Beard, 1946). The approximate distribution of the *Rhizophora* belt is indicated in text-figure 7.

The map of pollen distribution (text-fig. 7) shows low values in the central delta and high values within the source area. In Pica E (diagram 1), which is situated at the upstream limit of the *Rhizophora* vegetation, the pollen is still regularly present but in low percentages. Apparently these grains are windblown from the outer vegetation belt. In Pica F, a well developed *Rhizophora* belt 30–40 meters in width is found on the weakly defined levees, and *Rhizophora* pollen is dominant in the samples there.

In relation to the occurrence of the mother plant, *Rhizophora* pollen is heavily over-represented in the pollen total. In Pica D, the *Rhizophora* belt is clearly expressed in the pollen percentages, but it appears that the wind has distributed the pollen in low quantities over the entire back-swamp area. Pica B traverses a well developed *Rhizophora* belt, and accordingly, percentages of *Rhizophora* pollen are very high. This distribution in the delta samples indicates that *Rhizophora* is a very prolific pollen-producer, and it is due to this fact that the pollen grains are found regularly in the central delta outside the source area, having been transported through the air by the predominantly northeasterly to southeasterly winds.

Rhizophora can be classified as a wind pollinated species, and in this respect it is well adapted to its special environment, where, due to the strong sea breeze, its flowers are seldom visited by insects, as observed by the author. Offshore, a rather complicated-looking pattern appears, characterized mainly by a number of areas with higher concentration alternating with areas of lower concentration. These areas with higher concentration are not always in contact with the source area. The key to this anomalous situation lies in the following facts:

1) *Rhizophora* pollen is very small, and is therefore easily transported over a large distance by wind.

2) The exposed location of the *Rhizophora* belt and the high pollen production favour offshore transport because *Rhizophora* pollen is the last to settle on the outflowing surface water.

These facts, in combination with wind directions and current pattern, offer a logical explanation for the offshore percentage distribution pattern. The author's hypothesis is set forth below:

In the western corner of the Gulf of Paria, the high values opposite the Rio San Juan and Boca Ajies estuaries are caused by a low supply of other pollen, due to the small extent of the inner swamp belt and the restricted drainage. The extensive *Rhizophora* vegetation, on the other hand, is very favorably located for its pollen grains to be trapped on the water surface, because the estuaries are at right angles to the prevailing wind direction.

Opposite the Boca Vagre, the *Rhizophora* pollen percentage is lower because the large discharge of the Caño Manamo has been collecting pollen from a much larger area. There, *Rhizophora* pollen does not outnumber the other types so markedly. The same holds true for the Macareo estuary and for the area opposite the eastern delta, where *Rhizophora* also grows less abundantly. Seaward from the eastern delta, a distinct and gradual increase in percentage is observed. In the table below, these figures are analyzed in relation to distance offshore and total pollen content per gram:

Sample number	Distance offshore (in miles)	*Rhizophora* (%)	Total pollen/g.	Total *Rhizophora*/g.
1045	20	25	3827	958
1043	30	33	3033	1011
1042	38	52	383	199
1041	47	46	271	225
1040	55	70	294	206

It must be remarked that the distance offshore has been measured in a straight line following the traverse of the samples, whereas it should actually have been measured following the current pattern. In addition, the pollen content of samples 1041 and 1040, in particular, has probably been largely derived from the Guayana shore. Nevertheless, the same trend is visible in the samples taken somewhat farther north. These figures demonstrate clearly that the *Rhizophora* percentage is inversely proportionate to the total pollen content. If the *Rhizophora* content per gram is calculated, a normal decrease is found due to the progressive settling of *Rhizophora* pollen.

The offshore increase of *Rhizophora* observed in this area is therefore relative only, and is probably due mainly to the relatively small size of the *Rhizophora* pollen in comparison with the bulk of the other pollen types discharged from the estuaries. *Rhizophora* pollen is settling farthest seaward and is thus increasing in percentage. This is thought to be a distinct case of progressive sorting of pollen grains, resulting in a relative quantitative increase in the smallest grains seaward. The increase is augmented by estuarine circulation, because *Rhizophora* also happens to occupy the outermost vegetation belt and is thus more easily transported in the seaward-flowing upper water layers.

On the basis of these results, obtained in an area where transport must have taken place almost exclusively by

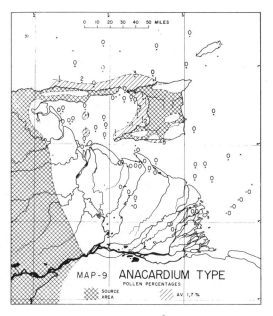

TEXT-FIGURE 9

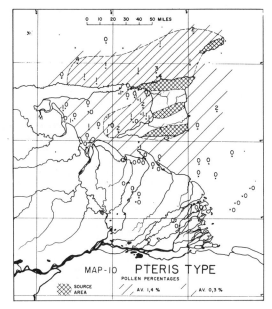

TEXT-FIGURE 10

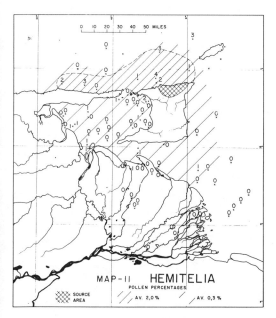

TEXT-FIGURE 11

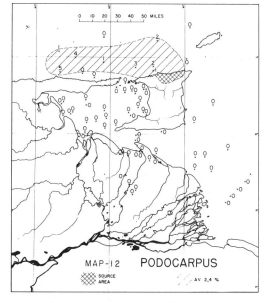

TEXT-FIGURE 12

water, it is now possible to proceed to the further analysis of the *Rhizophora* pollen percentages in the Gulf of Paria and on the northern shelf, where the possible presence of windblown pollen and the more complicated marine current pattern are confusing factors. The continuation of the patch with high *Rhizophora* values on the outer shelf is clearly visible in the Serpent's Mouth and can also be seen penetrating into the Gulf of Paria up to a short distance past Icacos Point. Then, following the marine current pattern, an area with rather low *Rhizophora* values is found in the eastern Gulf, more or less coinciding with a patch of low total pollen content. The decrease in *Rhizophora* percentages here may be caused by a large supply of other types from the Caño Macareo and Boca Vagre and by wind supply from Trinidad. The *Rhizophora* percentages again increase in the Dragon's Mouth and on the northern shelf, in a pattern similar to that shown by the total pollen content. As virtually no *Rhizophora* vegetation exists along the rocky northern shore of Paria and Trinidad, the only explanation for this increase in *Rhizophora* percentages is selective sorting in the northward-moving suspended pollen, which has its main supply from the western delta. If distance between main source and zone of maximum *Rhizophora* deposition is estimated, taking into account marine-current velocity and direction, values are obtained that are similar to those of the area opposite the eastern delta.

Both on the eastern and northern shelves, *Rhizophora* percentages again show a tendency to decrease away from the source of delta supply. This indicates a supply of other pollen types from other sources, and on the northern shelf it is probable that windblown pollen from Trinidad, Tobago, and the Paria peninsula tend to decrease the percentages of *Rhizophora* grains reaching the outer shelf area. For the eastern Trinidad shelf, however, this explanation is unlikely, in view of the prevailing northeasterly to southeasterly winds, and here it is probable that a supply of reworked Pleistocene pollen may have lowered the *Rhizophora* percentages.

Tricolporate, baculate type (unidentified)

This pollen type is characterized principally by its spherical form and baculate sculpture. It resembles *Ilex* pollen, but at present no certain determination can be given. Because the mother plant is uncertain, the source area can only be inferred indirectly from the pollen distribution (as happens only too often in fossil sediments). As shown in text-figure 8, this type occurs in the central delta and less frequently in the eastern delta.

This distribution, together with the remarkable distribution offshore, which is restricted to a narrow tongue traversing the central Gulf of Paria, starting in the Boca Vagre and ending with branches just outside the Dragon's Mouth and in the eastern Gulf, strongly suggests that this type of pollen is exclusively discharged by the Boca Vagre estuary. Its distribution in the Gulf sediments then indicates the stream along which the main water-mass leaving the Boca Vagre is moving through the Gulf. This stream more or less coincides with the tongue of higher total pollen content shown in text-figure 3, which was also believed to indicate the travelling stream of the Boca Vagre discharge.

The preceding pollen types can be grouped together as delta-pollen because their main source lies in the vast swamps and marshes covering the Orinoco delta. Various other pollen types belong to this group but will not be discussed because they are found less frequently and their distribution patterns are therefore less informative. The next group of pollen types that will be discussed are those of Trinidad, under which heading are grouped all types having their main origin on the islands of Trinidad and Tobago and on the peninsula of Paria.

Anacardium type (ANACARDIACEAE)
Plate 1, figure 4

The typical striate-rugulate, tricolporate *Anacardium* type of pollen is rather distinctive and is probably derived mainly from the genus *Anacardium*. The two most important species represented in our area are *Anacardium occidentale*, a fruit tree commonly planted in all cultivated areas outside the swamps and also occurring spontaneously in the llanos, and *Anacardium excelsum*, a large dry-land forest tree, which may occur locally.

Some other members of the Anacardiaceae, such as *Spondias*, which have similar pollen, may also be represented. In general, however, Anacardiaceae do not occur in the swamps and marshes. Lindeman (1953) mentions only *Spondias mombin* occurring on beach ridges on the Guayana coast. Accordingly, the *Anacardium* type of pollen was found only once in a levee deposit in the delta, and this single grain may easily have been derived from the cultivated levees in the upper delta or from the llanos area farther upstream.

In text-figure 9, a distinct concentration is shown in the offshore bottom sediments along the southwestern and northern coasts of Trinidad and along the northern coast of the Paria peninsula. This distribution pattern differs completely from those of the delta pollen, and it is evidently controlled by wind transport. The grains must have been subjected to some water transport while settling to the sea bottom, but the *Anacardium* type may be considered as exhibiting a typically windblown pattern of distribution. The scattered occurrences of *Anacardium* pollen north of Boca Vagre can be attributed to transport via the Caño Manamo from cultivated areas upstream.

Pteris type (cf. PTERIDACEAE)

This type-group comprises all trilete spores that possess a distinct, smooth equatorial flange, a heavily verrucate-rugulate or reticulate sculpture on the distal side, and a smooth or nearly smooth sculpture on the proximal side. Their size falls generally between 40 and 50µ. These

ORINOCO DELTA PALYNOLOGY

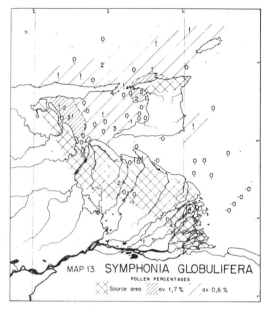

TEXT-FIGURE 13

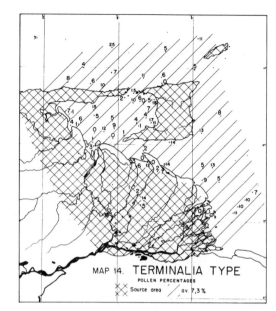

TEXT-FIGURE 14

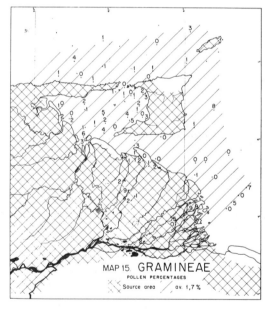

TEXT-FIGURE 15

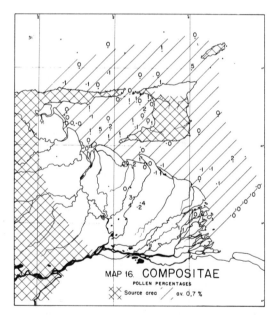

TEXT-FIGURE 16

spores are most probably derived from ferns belonging to the genus *Pteris*, but they may occur in other genera and even in other families. The author's reference collection was unfortunately not extensive enough to furnish a certain generic or specific determination. Lindeman (1953) does not mention the genus in his extensive study of the Surinam swamp vegetation. Posthumus (1928) lists several species of *Pteris* from Surinam, but mostly from the upstream mountainous areas. Beard (1946) unfortunately gives no data on the distribution of the genus *Pteris*, except one reference to *Pteris multiserialis*, which is a tree-fern growing in montane rain forest above 2500 feet altitude. In view of this scanty information and the uncertainties in determination, the source area cannot be accurately defined.

It seems improbable, however, that the spores of this group were derived from the swamp and marsh environment. Their most likely source area is the hilly-montane forests of Trinidad, Tobago, and, possibly, the Paria peninsula. This is clearly confirmed by the spore distribution shown in text-figure 10. Only one grain was found in the delta deposits, whereas offshore they are distinctly concentrated in the eastern Gulf of Paria, north of the Paria peninsula and north of Trinidad. This again is a predominantly windblown pattern of distribution. In comparison with *Anacardium*, *Pteris* has a wider distribution. This may be due to a difference in the source areas; that of *Pteris* probably lies at a higher altitude than that of *Anacardium*, thus giving the *Pteris* spores a better chance of reaching higher atmospheric levels because of the strong upward movement of air masses, especially on the mountain slopes, and consequently longer transport. It may also be partly caused by *Pteris* spores being liberated more abundantly than *Anacardium* pollen.

Hemitelia sp. (CYATHEACEAE)
Plate 1, figure 5

Hemitelia spores are easily recognizable by their perforated outer layer. These perforations are generally three in number and are situated in the equatorial zone between the arms of the trilete scar (see Erdtman, 1943, pl. 27 figs. 450–451). This type of spore may also be produced by other genera of Cyatheaceae.

Unfortunalely, the source area of these typical spores cannot be accurately defined. Beard (1946) does not mention *Hemitelia* from Trinidad, but, as his attention was mainly focussed on the tree vegetation, it is possible that his information was not complete as far as ferns are concerned. He does mention *Hemitelia grandifolia* from the montane rain forest of the Lesser Antilles north of Trinidad (Beard, 1949); the genus, or a related one, could therefore well be expected on Trinidad. This is at least strongly suggested by the spore distribution pattern shown in text-figure 11, which shows *Hemitelia* spores as absent in the delta and opposite the eastern delta. Southwest and north of Trinidad and north of Paria, they are found scattered over a rather wide area, but with concentrations close to the northern coasts of Trinidad and Paria. It seems most probable, therefore, that the spores are derived from the northern range and that wind transport has been dominant in determining the distribution pattern. It is possible that the faintly indicated concentration opposite the Boca Vagre points to water transport via the Orinoco from the Andes and the coastal range, as *Hemitelia* is known to occur there locally.

Podocarpus sp. (CONIFERAE)
Plate 1, figur 6

The saccate grains of *Podocarpus* are unmistakable and cannot be confused with the pollen of any other plant occurring in the area of investigation. *Podocarpus* can safely be assumed to be absent from the Orinoco delta, although Beard mentions localized occurrences in marsh forest on alluvial terraces in Trinidad. He further classifies the occurrence of *Podocarpus* as follows: "Very scarce in evergreen seasonal forest, regularly present in lower montane forest, and abundant in montane rain forest." Its present-day source area is therefore almost entirely restricted to the northern range of Trinidad.

This localized source area is clearly reflected in the pollen distribution pattern shown in text-figure 12. The restricted concentration north of Trinidad and Paria indicates wind transport, mainly during the season of southeasterly winds. Of course, sea currents must have modified the pattern by secondary transport westward. The similarity of distribution to that of *Hemitelia* is striking, and suggests a common main source.

The foregoing discussion has dealt with the distribution patterns of pollen types with only one main source area. A few patterns that have originated by supply from multiple source areas will next be shown. In extreme cases this leads to a diffuse distribution which is shown especially by pollen from ubiquitous and anemophilous species.

Symphonia globulifera (GUTTIFERAE)
Plate 1, figure 7

The pollen of this tree is easily recognizable by its large size, slightly angular outline interrupted by five irregular pores, and smooth, thick cavate wall, which has a typical corroded appearance. *Symphonia globulifera* is especially abundant in the extensive swamp forest covering the easternmost part of the Orinoco delta, but it also occurs regularly in the remaining swamp and marsh area and in the marginal forests along the lower river courses. The plant association in which it occurs here so abundantly is comparable with Lindeman's *Symphonia globulifera* marsh forest. In Trinidad, according to Beard, the tree occurs locally in a similar vegetation, but is also rather frequent in evergreen seasonal forest, *Mora* forest, and lower montane and montane rain forest. The occurrence of a single species in such widely different environments as tidal swamp on the

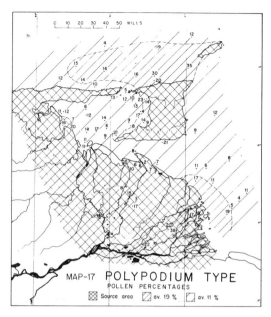

TEXT-FIGURE 17

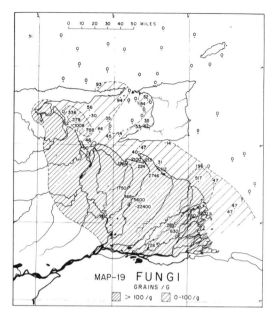

TEXT-FIGURE 19

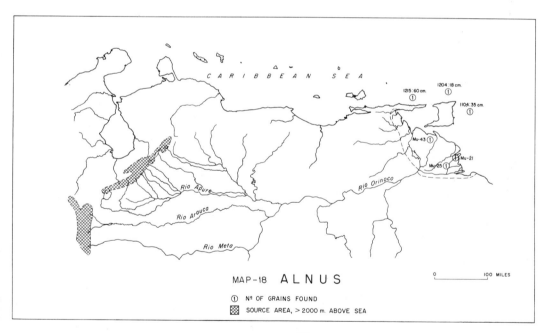

TEXT-FIGURE 18

one hand and mountain forest on the other is very remarkable, and diminishes its value as a climatic and ecological indicator in fossil sediments.

In its distribution, shown in text-figure 13, *Symphonia* pollen shows the combined effect of multiple source areas and limited transport due to the relative heaviness of the grain. In the delta it reaches its highest percentages in the eastern part, especially in Pica E, where the tree is very abundant. In the central part, where the tree is decidedly less frequent, only scattered pollen is found. Offshore, opposite the eastern delta, the grains are conspicuously absent, although the tree is common up to the outermost vegetation belt. This is probably due to size selection, as a result of which the heavy *Symphonia* grains settle close to the coast line. Unfortunately, no samples were available from this area.

In the western Gulf of Paria, a local concentration can be observed. These grains are probably derived from the coastal swamp vegetation. In the central Gulf the *Symphonia* pollen is very scarce, which proves that the marine currents were able to transport the pollen over a short distance only. Around Trinidad another concentration appears, and it is possible that these grains may have been carried by wind transport from the forested areas on Trinidad. Finally, the rather high percentages found at stations 74 and 468 are probably indicative of reworking. This cannot be confirmed from their state of preservation, because reworked older *Symphonia* grains have almost the same appearance as Recent ones.

Terminalia type (cf. COMBRETACEAE)
Plate 1, figures 8, 8a

The identification of this pollen type is certain when it is well preserved. It is a very small, smooth-walled prolate tricolporate grain with narrow colpi and inconspicuous pores, and, as its most characteristic feature in equatorial outline, it possesses a depression between the colpi that suggests the presence of three more colpi without pores. This pollen type occurs in the genera *Buceras*, *Conocarpus*, and *Terminalia*, all of which belong to the Combretaceae. *Buceras* is not mentioned by Beard or Lindeman, but enough information is available on the other two genera to make it possible to outline the potential source area of this pollen type.

The genus *Conocarpus* is represented by a single species, *Conocarpus erectus*, which is one of the rarer mangrove plants. It could be expected in the delta, although it has not been observed by the author. In Surinam (see Lindeman) and on Trinidad (see Beard), it is rare. It is therefore likely that the bulk of the pollen grains under discussion have been derived from the genus *Terminalia*. This genus has very widespread distribution in different environments. In Trinidad, *Terminalia dichotoma* is a characteristic swamp species, and *Terminalia amazonia* is frequent in the marsh forest. The latter species, however, also occurs predominantly in evergreen, semi-evergreen, and deciduous seasonal forest, as well as in lower montane and seasonal montane rain forest. *Terminalia lucida* and *Terminalia catappa* are found in littoral woodland (see Beard). It is obvious from this list that the potential source area of this pollen type is large and widely distributed. If the fact that this pollen type is small and easily transported is also taken into account, the wide distribution shown in text-figure 14 is largely explained.

In the delta it is locally dominant in the eastern part (Picas E and F). The mother plant, unfortunately, could not be identified here. In the central delta it seems to be less abundant (Picas D and B). Offshore, it is distributed in a typically diffuse pattern, without distinct concentrations. It is therefore impossible to evaluate the influence of the various source areas and the nature of the transport medium.

GRAMINEAE

The large family Gramineae has one basic and easily recognizable pollen type, characterized by its single bordered pore and virtually smooth wall. Although a smaller and a larger type were observed, no attempt at further identification was made.

Gramineae are dominant in the semi-cultivated Venezuelan llanos and in other cultivated areas on Trinidad and Paria. In the delta they occur locally along river banks, where floating grass mats of *Leersia hexandra* are a common sight. In the upper part of the delta, *Gynerium sagittatum* borders the natural levees. They also occur in special types of herbaceous swamps. They are scarce in swamp forests and in upland forests.

The potential source areas are thus widely scattered, and this is clearly reflected in the pollen distribution shown in text-figure 15. In the delta, samples from the levees show a somewhat higher percentage than the backswamp samples, such as those taken in Pica E in swamp forest. Offshore, a very scattered distribution pattern emerges. Weak concentrations are found, first along the western coast of Trinidad, indicating some wind transport from the island, and secondly in the southeastern Gulf of Paria and in the Boca Vagre, indicating waterborne supply, presumably from cultivated areas upstream. Finally, the scattered high values found on the eastern shelf should be mentioned. These high values may indicate waterborne supply from the Guayana coast, where large swamp-savannah areas have been described.

COMPOSITAE

Here, the Compositae pollen comprises only typical tubiflorae and liguliflorae types. The tubiflorae type was by far the more common of the two. No attempt at further detailed identification could be made.

Compositae as a group scarcely occur in the typical swamp and marsh vegetation, nor are they abundant in the upland forests of the area under consideration. Most of the pollen found has probably been derived from the savannah and cultivated areas on Trinidad, Tobago, and Paria, and in the Venezuelan llanos.

The pollen distribution shown in text-figure 16 indicates low values in the delta, especially in the back-swamps, but a significantly high value of 11 per cent in a levee deposit in the upper delta, in the cultivated area. Offshore, a vague concentration is visible in the central and eastern Gulf of Paria and along the northern coast of Trinidad and Paria, although Compositae pollen is notably absent in the western Gulf. It is also absent opposite the eastern delta except for a weak concentration in the northeast, which probably indicates some supply from the Orinoco tributaries.

Polypodium type (cf. POLYPODIACEAE)

The *Polypodium* type includes all monolete bean-shaped spores with either a smooth or a verrucate sculpture. It is a heterogeneous group, as this morphological type occurs in the Polypodiaceae as well as in some other fern families. The source area cannot be accurately defined; these spores may have been derived from nearly all the environments present in this area, except possibly the llanos and the cultivated areas.

Extensive fern swamps occur in the delta. Accordingly, the general distribution of the spores shows a wide range of variation related to the local frequency of the mother plants. In Picas E and F they are abundant, and in Pica D the transition between the *Rhizophora* belt and the morichal vegetation is marked by a dominance of *Polypodium* spores in the sediment (see diagram 1). In the coastal belt the percentages are lower (Pica B). Offshore they were found in all samples, often in considerable percentages (see text-fig. 17). Opposite the eastern delta seaward decrease due to size selection can be observed. Early settling of the rather large and sometimes heavily sculptured *Polypodium* type of spore is favoured. In the Gulf of Paria, there are only slight concentrations in the Rio San Juan estuary and along the western coast of Trinidad. On the northern shelf, rather high values north-northwest of Trinidad and west of Tobago indicate airborne supply from these islands. Thus, the influences of multiple source areas and mass production have resulted in wide and diffuse distribution.

A similar distribution pattern is shown by various other pollen types, such as *Acrostichum*, Amaranthaceae, Cyperaceae, *Ilex*, *Virola*, and psilate Palmae. Other types are less common and have more scattered distribution; these include *Bombax aquaticum*, *Caesalpinia*, *Caryocar*, *Ceiba*, *Ceratopteris*, *Croton*, *Erythrina*, Malpighiaceae, Malvaceae, Meliaceae, Mimosaceae, Myrtaceae, *Polygonum*, *Pterocarpus*, and Sapotaceae. For *Erythrina* and *Pterocarpus* it was possible to establish that the scarcity of their pollen, which is in strong contrast to the abundance of the mother plants, is due to the small amount of pollen liberated by the papilionaceous flower types of both genera.

Alnus sp. (BETULACEAE)
Plate 1, figure 9

Very characteristic *Alnus* pollen was found as single grains in six samples, the locations of which are indicated in text-figure 18. In the delta they were recovered from levee deposits, and on the shelf they were found in cores at depths of 35, 60 and 180 cm.

The nearest occurrence of the species *Alnus Mirbelii* and *Alnus jorullensis* is in the Andes, at altitudes between 2000 and 3000 meters, in the Venezuelan states of Trujillo, Mérida, and Táchira, and in Colombia (see text-fig. 18). Within this area, *Alnus* grows preferably along the banks of mountain streams and brooks. It is quite evident that these grains cannot have been wind-transported to the area of deposition, but must have been carried by the Orinoco River system from the Andean sources, over a distance of more than 500 miles, to the delta and shelf. Of course, this is an exceptional case, and *Alnus* has been especially favoured because it is a very prolific pollen producer and grows in a favourable habitat beside running water, but nevertheless its average quantity in the area of deposition scarcely amounts to 0.5 per mille. Some of these grains, however, may have been reworked from Pleistocene deposits.

3) *Distribution of other microfossils:*

FUNGI
Plate 1, figure 10

Spores and hyphae of fungi are extremely abundant in most delta deposits, where these micro-organisms play an important part in the decaying processes which liquidate most of the plant litter produced by the exuberant swamp vegetation. In text-figure 19 the amount of fungal spores per gram of sediment is plotted. Their abundance in the delta is obviously. Offshore concentrations are found opposite the northeastern corner of the delta and in the western part of the Gulf of Paria. The smaller amounts in the Boca Vagre and San Juan estuaries can be explained in the same way as the lower pollen content, that is, by the comparatively larger amounts of coarser mineral fractions. A continuous belt of medium values is found opposite the entire delta, along both the southwestern and western Trinidad coasts, and along part of the northern coast of Paria. In the central and eastern Gulf and on the outer shelf, no fungal spores were found.

It may be assumed that fungal spores, in comparison with pollen, are less subject to initial wind transport because they are generally produced in the thick mat of decaying plant material at the soil surface of a generally very dense plant cover. They are probably liberated mainly from the back-swamp soils during heavy rainfall or by erosion of river banks. Once freed, they may be transported over fairly large distances because of their generally small size ($10-30\mu$).

The offshore distribution pattern indicates multiple source areas and moderate distances of transport. Apparently, they are liberated not only from delta soils but also from upland Trinidad soils. The lack of initial wind transport causes less wide distribution in comparison with *Rhizophora* pollen. In the delta they are largely liberated by erosion, and thus tend to be concentrated in the lower water levels in the estuaries. Because of the estuarine circulation pattern, the particles from these levels have a smaller chance of escaping offshore than pollen, which first settles on the surface and which is more easily carried off seaward. The local concentrations opposite a few delta estuaries may indicate areas where deposits particularly rich in fungi are being eroded.

Cuticles

In text-figure 20, the sizes and abundances of plant cuticles are plotted, expressed on a scale of five points. Size and abundance are generally correlated. In delta sediments, cuticles are very abundant and of large average size. The cuticles found offshore, concentrated in a belt opposite the large delta estuaries, have been derived by stream erosion from these delta sediments. Both the abundance and the size of the cuticle fragments decrease rapidly offshore, and only the smallest fragments escape to the outer shelf. Some secondary concentrations at stations 74, 180, and 468 may indicate the presence of locally reworked material. The similarity between the cuticle and the fungus distribution is evident.

Tracheids
Plate 1, figure 11

The distribution of tracheid remains is indicated in text-figure 21. These peculiar tissue fragments have all been reworked from pre-Eocene deposits, of which they are characteristic. They are mainly eroded by the Orinoco headwaters in the Andes. The plants from which they originated are not known with certainty, but presumably are gymnosperms. In the levee deposits they are frequent, but offshore they are mainly concentrated in a belt opposite the larger estuaries. Some secondary concentrations along the southwestern Trinidad coast and locally in the northern Gulf of Paria indicate reworking from other local sources. Reworking is also indicated here by the distribution of reworked pollen. On the outer shelf, tracheid remains are virtually absent.

Hystrichosphaeridae
Plate 1, figure 12

The Hystrichosphaeridae are a group of planktonic organisms of little-understood systematic position. Most probably they represent the cyst-like resting stage of certain unicellular algae of the dinoflagellate group. Morphologically they are characterized by their small size (40–60μ) and smooth wall, which is covered with numerous hooked, hollow spines. It is possible to distinguish various morphological types, but as the systematic relationships between these types is not yet known, they are here placed in a single group and referred to as "*Hystrix*."

The distribution of *Hystrix* per gram of sediment is shown in text-figure 22. It is absent in the littoral zone opposite the Orinoco delta, and is concentrated mainly in the eastern Gulf of Paria, close to the Trinidad shore. This distribution pattern suggests that *Hystrix* may be a marine organism, which probably cannot flourish below a certain salinity, but the turbidity in the littoral zone may also have been a limiting factor.

The concentration in the eastern Gulf suggests that optimum conditions for the development of *Hystrix* prevail there. Hydrographic data for this area indicate the presence of clear water with a chlorine content of 18–22 parts per thousand and with temperatures up to 29° C., which is slightly higher than the temperature of normal ocean water as it is carried into the Gulf by the Atlantic equatorial current through the Serpent's Mouth. It is possible that the relatively high temperature is favourable for the development of *Hystrix*. It is also worth mentioning in this connection that the *Hystrix* specimens from this part of the Gulf are larger and better developed than those from the outer shelf, suggesting some direct environmental influence on the growth and morphological appearance of *Hystrix*. On the other hand, the abundance of *Hystrix* is also related to the rate of sedimentation, and the influence of this factor is hard to isolate without further detailed ecologic study of the living population. For the time being, *Hystrix* can be considered only as a general facies indicator for a marine environment.

Foraminifera
Plate 1, figure 13

In many samples, the remains of small foraminiferal tests were found, which had survived the chemical treatment. The most common type found was the probably chitinous inner lining of tests, left after solution of the calcareous outer layers. In a few cases, fluoride replacement of the original calcareous test was observed, as described by Grayson (1956). The distribution of the first-mentioned group is shown in text-figure 23.

Since there was at first some doubt whether or not these remains, which occur either in a rotaloid or bolivinoid form, were actually derived from foraminifera, because of their small size and different appearance, it was decided to dissolve some known foraminifera under the microscope in order to find out whether any remains were left. These experiments were made by B. B. 't Hart, palaeontologist with Shell Condor (Colombia). The sample taken at station 1041 was washed in the ordinary way for foraminiferal study, with floating elements preserved as is normally done for ostracode study. The foraminifera of the finest fraction (200 mesh) were separately transferred to a glass slide and treated with a 10 per cent HCl solution. The average diameter of this foraminiferal fraction proved to be 0.16 mm., and all specimens tested were megalospheric forms belonging to

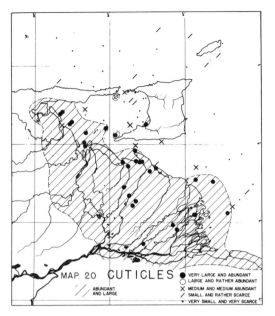

TEXT-FIGURE 20

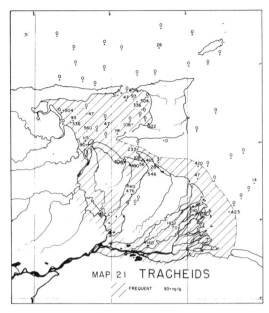

TEXT-FIGURE 21

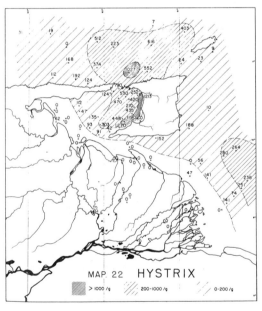

TEXT-FIGURE 22

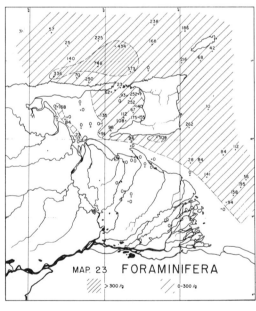

TEXT-FIGURE 23

the genera *Bolivina, Cassidulina, Cibicides, Eponides, Globigerina, Globorotalia, Nonion, Planulina, Quinqueloculina,* and *Siphonina*. Of these, only specimens of *Cibicides* and *Planulina* left a chitinous residue similar to those observed in the palynological preparations. This is in agreement with data given by Piveteau (1952), who states that only the Rotaliidea, to which *Cibicides* and *Planulina* belong, possess a chitinous inner test in the juvenile stage. The other genera apparently do not possess this property, with the possible exception of *Quinqueloculina*, which may develop a chitinous test under certain conditions. Grayson's statement that he was not able to detect these chitinous inner tests proves, therefore, only that the specimens he dissolved happened to belong to genera that do not have this property. His subsequent conclusion that these inner coats do not exist at all is therefore an invalid generalization. Wilson and Hoffmeister (1952), and later van Veen (1957), have already drawn attention to the phenomenon of microforaminifera and have arrived at the correct conclusions concerning their origin. In this connection it is worth noting that Faegri and Iversen (1950) mentioned and correctly interpreted the occurrence of chitinous remains of marine foraminifera in their pollen residues. The foraminiferal nature of the microfossils under discussion may therefore be taken as proven, although a more detailed study of these small but sometimes fully grown representatives of a restricted group of genera is evidently desirable.

Returning now to their distribution pattern, which is given without any attempt at specific determination but which mainly covers remains of Rotaliidea, it appears that they are widely distributed except in the delta and in the major part of the pro-deltaic-littoral zone. A notable exception to this pattern is the small elongate patch in which they occur in the latter zone opposite the western estuaries. It is of interest to note that this occurrence was not detected by Kruit in his study of the Recent foraminifera in the Gulf of Paria (van Andel, Postma and Kruit, 1954). Generally speaking, it may be concluded that foraminiferal remains of the type under discussion are more abundant in the marine than in the littoral and estuarine environments.

Investigation of cores

Although the present study is essentially concerned with microfossil distribution in Recent sediments only, it was felt desirable to investigate a few core samples in order to test the homogeneity of microfossil distribution in time. Three cores were selected, taken at stations 1101, 1204 and 1215. These showed the deepest penetration in areas which lie comparatively far from local deltaic influence, and their lithology consists of a homogeneous clay sediment, which was also a favourable factor. The cores were examined at 10–20 cm. intervals, but unfortunately the top and bottom layers of the cores were no longer available, as the material had been used for other purposes. The results are shown in diagram 2, where the variations in abundance are indicated for the most important microfossil groups.

CORE 1101

From 70–195 cm., the total pollen content is higher than at the surface, but below 195 cm. a sharp decrease downward is observed to values that are lower than those found at the surface. This indicates that, in the period represented, there was a considerable shift in the pollen distribution pattern, which in itself is not surprising, as the core was taken rather close to the large eastern estuaries. There is little variation in the per cent of reworked pollen, and the low values found off the eastern delta are typical both of the surface and of deeper layers. *Mauritia* percentages are uniformly very low. *Avicennia* percentages average 3 per cent both at the surface and in the core. *Rhizophora* percentages fluctuate in the core at approximately 50 per cent, which is normal in the surface area. If the fluctuations are compared with the total pollen curve, it appears that a decrease in the total pollen content is always accompanied by an increase in *Rhizophora* percentages and vice versa. This relationship in a vertical direction is identical with the one observed in the surface sediments in a section perpendicular to the coast, and the explanation given previously holds here too: Progressive sorting tends to increase the proportion of small *Rhizophora* grains seaward. In the core, this relationship is not absolute because the samples at 225 cm. and 255 cm., which have the lowest pollen content per gram, do not show the highest percentage of *Rhizophora*. This indicates that in this period the *Rhizophora* vegetation suffered a setback in areal extent as compared with the higher part of the core, where *Rhizophora* pollen was supplied in larger quantities. In agreement with surface observations in this area, the mountain element is almost completely unrepresented in the core. Two single *Hemitelia* grains recovered may have been carried by long-distance transport from an area other than Trinidad. Only scattered grains of *Symphonia globulifera* were found in the core, which is in line with the virtual absence of these heavy grains at the surface in this area. An average of 9 percent of the *Terminalia* type in the upper six samples checks very closely with the surface average. The lowermost sample shows a significantly lower value. Gramineae show very low frequencies in the upper part of the core, in accordance with surface values, but the lowermost sample shows a significant increase. Compositae show a similar pattern: Low frequencies in the upper part, in accordance with surface distribution, but a significant increase in

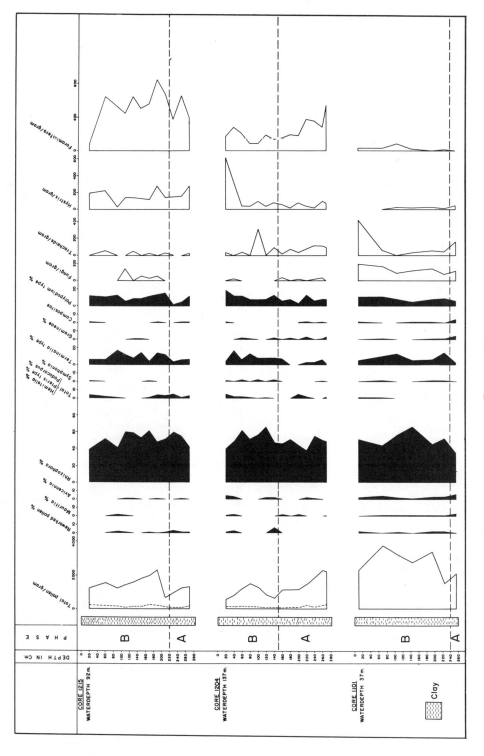

DIAGRAM 2

MICROFOSSIL DISTRIBUTION IN CORES

percentage in the lowermost sample. The Polypodiaceae type of spore averages about 9 per cent in the core, which checks with surface values. The fungi concentration shown in the surface distribution at stations 1100 and 1101 appears to have been present during the entire time-span covered by the core, although there is a slight decrease in the lower part. The high tracheid content found at the surface is evidently an exceptional occurrence. *Hystrix* shows low values throughout the core, in agreement with surface values. Foraminifera values are slightly higher in the upper part of the core, and conform to surface values.

Summary of development: The variations in microfossil content in the upper 2 meters of the core are rather small, and indicate fluctuations in the hydrographic regime such as could have been caused by variations in the Orinoco discharge or by slight shifts in the positions of the eastern estuaries. In the two lowest samples, however, a more marked change in microfossil content appears. The total pollen content drops sharply, and a decrease in *Rhizophora* percentage is coupled with an increase in Gramineae and Compositae. This probably reflects a somewhat more profound change in pollen supply and sedimentation, which may have been caused partly by a climatic change.

Core 1204

From the surface down to 80 cm., the total pollen content shows a very regular and significant increase, which is then followed by a similar decrease. A small increase is seen next, and then, from 160 cm. to 200 cm., the values remain constant. Finally, down to 270 cm., a gradual and significant increase occurs, attaining values that are four times higher than those found at the surface. These fluctuations are indicative of shifts in pollen supply, which, because of the distant location of the core, far from the delta source, must have some regional significance. Reworked pollen is observed only occasionally, in line with surface observations in this area. *Mauritia* occurs scattered in the same frequency as at the surface. The rather high percentage of *Avicennia* found at the surface persists down to 40 cm., then scattered occurrences are found at lower levels. It is hard to explain these high *Avicennia* values other than by assuming that they represent reworked grains, an assumption which is supported by the coincidence of the higher *Avicennia* percentages with the two maxima of reworked grains.

The *Rhizophora* percentage is lower in the surface layer than it is between 60 cm. and 120 cm. Below 120 cm., values show a tendency to decrease again.

There is a certain agreement between the curves for total pollen and for *Rhizophora*, especially in the upper part of the core. Here, the total pollen maxima between 40 cm. and 120 cm. also show high *Rhizophora* percentages. The minima in total pollen content at the surface and between 140 cm. and 220 cm. also show a lower average *Rhizophora* percentage. The strong increase in total pollen below 220 cm. is, however, not entirely matched by a comparable *Rhizophora* increase. This parallel trend is the reverse of what has been described for core 1101, and can be explained by different supply conditions in the area of core 1204. Here we are in an area where the *Rhizophora* share in the waterborne-pollen mass is decreasing westward because the largest part has already settled out and the airborne supply of other types is becoming more prominent. This wind-supplied share may be assumed to be constant, so that any drop in the delta supply, which here consists of more than 80 per cent *Rhizophora*, will cause a relative increase in the airborne pollen. In core 1101, this compensating influence was absent, which explains the difference. As was to be expected from the surface data, the mountain element in the pollen flora is found regularly, and decreases only with high *Rhizophora* values. This confirms the idea that the supply of this element remained essentially constant during the time of deposition of the core. There appears to have existed a rather persistent supply of the heavy *Symphonia globulifera* pollen in the area north of Trinidad, at least for the period covered by the upper part of the core. These grains must either have been carried by wind from Trinidad or have been reworked, as the probability that they were carried by water from the eastern Orinoco delta is negligible. It is, at present, difficult to decide between these two possibilities. *Terminalia* values are higher in the upper part of the core than in the lower part. At the surface, the general trend in the area again appears to be toward slightly lower percentages. Gramineae are slightly more frequent in the lower than in the upper part of the core. Compositae are very scarce throughout. *Polypodium* forms a mainly windblown element, derived from Trinidad and Tobago, and consequently shows a trend in the percentage fluctuations which is the reverse of the *Rhizophora* curve. Fungi are more frequent in the lower half of the core, and their decrease upward is in line with their almost complete absence in the surface layers of the area. Especially in the lower part of the core, below 100 cm., tracheids seem to have been supplied regularly and sometimes abundantly, indicating reworking from close by. The source of this material probably must be sought in Pleistocene and older Holocene sediments, which have been locally attacked by wave erosion. In the youngest layers and at the surface, this supply has almost ceased. The rather high *Hystrix* values found at the surface appear from the core record to be a relatively recent feature, for which it is hard to find a satisfactory explanation at present. Foraminiferal content fluctuates rather strongly, but shows, nevertheless, an increasing trend in the lower 40 cm. of the core.

Summary of development: The rather well-marked variations in microfossil content in core 1204 center around the marked but regular variations in total pollen content and the alternation between the curves for *Rhizophora* and the windblown elements. If the total values for the windblown pollen per gram of sediment are computed, a nearly straight line is obtained, with only a minor increase near the bottom, as can be seen from the chart, where it is plotted as a dotted line. Furthermore, the increase in *Terminalia* percentage in the upper part seems significant.

The constancy of the windblown supply indicates that no major changes in shoreline position took place during the period covered by the core sediments. The variations in total pollen content are therefore most probably attributable to the same causes as were indicated for core 1101, that is, changes in the hydrographic system caused by variations in Orinoco discharge. The location of the core makes it improbable that these were caused only by local shifting of estuaries, but suggests instead a more widespread cause.

Core 1215

From the surface down to 120 cm., the total pollen content remains almost constant, but farther down a gradual increase is noted, reaching a maximum at 190 cm., after which a sharp drop takes place. This is followed by a renewed increase which levels off farther down. Reworked pollen is scarce throughout, in accordance with the general surface distribution, but seems to be slightly concentrated in the lower part. *Mauritia* is almost completely absent, in accordance with the general surface distribution. *Avicennia* occurs scattered, with a frequency similar to that found at the surface. In this core also, a parallel trend is observed in both the curve for total pollen and the curve for *Rhizophora* per cent. The main supply of *Rhizophora* at this location must have been through the Dragon's Mouth, but some contribution from the eastern shelf is also to be considered. In relation to the two supply directions, the core is located in the zone of decreasing *Rhizophora* supply and increasing windblown share in the pollen total. The explanation of the parallel trends between total pollen content and *Rhizophora* per cent given for core 1204 therefore holds here also. As in core 1204, we see here also an alternation between the trends for the mountain-pollen curve and the *Rhizophora* curve. The same explanation can be given as for core 1204. *Symphonia globulifera* is very scarce, in accordance with the general surface distribution. From 0–210 cm., rather high and strongly fluctuating *Terminalia* percentages are found, whereas in the lower part, percentages tend to decrease somewhat. Gramineae and Compositae are both very scarce, in accordance with the general surface distribution. The *Polypodium* curve shows a trend similar to that in core 1204, with alternations of its peaks with *Rhizophora*. The explanation is the same: a constant supply of windblown *Polypodium* spores and a variable supply of waterborne *Rhizophora* pollen grains. The curve for fungi content shows peculiar maxima between 110 cm. and 200 cm. In the upper part, fungi are absent, in accordance with the general surface distribution. Tracheids are rather regularly present in low numbers, which contrasts with their absence at the surface. This is similar to what was observed in core 1204, although the values reached there are higher. Values for *Hystrix* fluctuate rather strongly but are, on the average, in line with those found in the general surface area. The surface values for foraminifera are considerably lower than those from 60 cm. downward. This suggests that the area with a high foraminiferal content at the surface north and northeast of station 1215 may formerly have had a wider distribution.

Summary of development: The most striking variations in microfossil content are visible in the curves for total pollen, *Rhizophora*, the mountain element, and *Terminalia*. If the curve for total windblown pollen per gram is computed, slightly more variation is observed than in core 1204. Changes in supply from the Orinoco delta dominate the pollen diagram here also. The high values for *Hystrix* and foraminifera found throughout the core emphasize the depositional environment, which differs from that of the other cores.

Discussion of Cores

From the preceding detailed descriptions it will be clear that the average microfossil content in the upper part of the cores is typical of their respective environments, with respect to both pollen supply and biofacies characteristics. It is only in the lower part that marked deviations from the present-day situation are indicated. The relations between the windblown element and the delta-supplied element in the pollen content of the sediments accumulating on the northern shelf, as deduced from the surface distribution, are confirmed by the results of the core investigation. The seaward increase of *Rhizophora* at the surface due to progressive sorting is also confirmed by the results of the core study.

A closer study of the variations in pollen content suggests that they may have a regional cause. The evidence for this lies, first of all, in the fluctuating total pollen content. From the surface downward, all three cores show a gradual increase in total pollen content, followed by a more sudden decrease to values that are equal to or even lower than those found at the surface. This minimum, which is found at 225 cm., 140 cm., and 210 cm., respectively, in cores 1101, 1204, and 1215, is followed by a renewed gradual increase downward. This earliest phase is also characterized by decreasing *Rizophora* and *Terminalia* percentages, while grasses and Compositae increase in frequency. In agreement with their more comparable locations in relation to supply cores 1204 and 1205 show a closer similarity to each other in the trends of their pollen curves than they do to core 1101.

The fact that the total pollen content rises only in the lower part of core 1204 up to values more than four times as high as those found at the surface suggests that this core penetrated the oldest sediments. On the basis of this evidence, two developmental phases are tentatively distinguished. The older, Phase A, appears to be characterized by a regularly decreasing total pollen content, an average *Terminalia* percentage of 4 per cent, and frequencies (per cent of samples in which they occur) of 55 per cent each for Compositae and Gramineae. Phase B is marked first by a rather sudden increase in total pollen content, followed by a decrease; high *Rhizophora* values; an average *Terminalia* percentage of 10 per cent, and frequencies for Compositae and Gramineae of 13 per cent and 18 per cent, respectively.

Although a full evaluation of the significance of this regional development must await a more detailed investigation of additional core material, it seems worthwhile to offer a few suggestions as to its possible ultimate cause. The main factors operating in the area that may have shown their influences in the pollen diagram are: 1) The local shifting of Orinoco estuaries due to normal sedimentary processes; 2) sea level changes; and 3) climatic changes. Shifting the position of a main estuary changes the total pollen distribution and the percentage ratios of the various types. Van Andel has found evidence that a shift of the main Orinoco discharge from the eastern delta toward the Macareo and Boca Vagre estuaries took place about 700 years ago. It appears possible that the sharp decrease in core 1101 from 70 cm. upward, and in core 1204 from 80 cm. upward, may reflect this change, as this shift would tend to reduce the delta-pollen supply at these locations. There is not such a marked decrease visible from 100 cm. upward in core 1215, probably because the delta supply reaches this location via the Dragon's Mouth from the western estuaries, which increased their discharge simultaneously. Further study of cores, especially from the Gulf of Paria, is required in order to test this hypothesis.

Sea-level changes may cause general shifts in the positions of coast lines, especially of the flat delta coast. Van Andel cites evidence of a regression of sea level toward the 6–7 meter depth line, which he thinks took place from 1000 to 1500 years ago, and which was followed by a transgression about 1000 years ago. The sharp increase in total pollen content found in core 1101 between 230 cm. and 190 cm., and in core 1215 between 210 cm. and 190 cm., as well as the more gradual increase in core 1204 between 140 cm. and 80 cm., may be due to this lowering of sea level, resulting in a seaward movement of the delta front. The extent to which the lines of equal pollen content migrated seaward in this case can be deduced from the depth contours, and is found to be roughly proportionate to the increase in total pollen content registered in the cores at the depths indicated.

Climatic change may be expressed both by changes in the composition of the vegetation and by changes in the amount of Orinoco discharge as a result of changes in rainfall in the drainage area. The only indication that suggests a climatic change is found in the increase of *Terminalia* and decrease of Compositae and Gramineae at the boundary between phase A and phase B, but the nature of this change cannot yet be indicated with the scarce data available.

Notwithstanding the provisional nature of these reflections on the causal relations of the changes observed in the diagrams, it is tentatively concluded that these changes can be used for time-stratigraphic correlation, as they appear to be of a regional nature. The boundary between phases A and B in particular is regarded as closely approaching a time line. Unfortunately, no radiocarbon dating is possible in the pure clay cores that are most suitable for palynological study, and direct confirmation of this hypothesis must await correlation with nearby cores in which dated mollusk beds are present.

An estimate of sedimentation rates at each location can be given, if the correlation suggested above proves to be correct. Core 1101 shows the fastest rate of sedimentation, in accordance with its position in the fluvio-marine environment, close to the large eastern estuaries. Core 1204 has a much slower rate of sedimentation, which is evidently related to its distance from the source of sediment in the eastern estuaries. Core 1215 shows a rate of sedimentation that is only slightly lower than that found in core 1101. In this connection it is of interest to note that apparently no direct relationship between rate of sedimentation and total pollen content exists, as the average pollen content in phase B of core 1215 is roughly intermediate between those of cores 1101 and 1204, although the thickness of phase B is the same as in core 1101. This is not surprising, as the sources and supply conditions of clay and pollen are naturally different.

It is also of interest to investigate to what extent *Hystrix* and foraminiferal content are related to the rate of sedimentation. There is, again, no direct relationship, but the increase in both *Hystrix* and foraminiferal content in the upper part of core 1204, as well as the decrease in foraminiferal content in the upper part of core 1215, may tentatively be assumed to have been connected with the above-mentioned shift in the supply of sediment about 700 years ago.

If the tentative absolute dating proves to be correct, it will be possible to calculate both the rate of sedimentation for phase B, and the average total amount of pollen settling per unit of time/surface, the so-called "absolute pollen frequency." Calculated for phase B, the average rate of sedimentation/100 years is 14 cm. in cores 1101 and 1215, and 9 cm. in core 1204. These values are somewhat lower than the estimates given by van Andel for the central Gulf of Paria. In cores 1204 and 1215, this difference is in line with their greater distance from the sediment source. These values check surprisingly well with estimates made indepentently by Koldewijn (1958) for the northern shelf. His estimate for core 1204, based on the depth of the top of the Pleistocene, which can be traced on the echo-sounding graph, is 8 cm./century, which is virtually identical with the palynological estimate of 9 cm./century. His estimate for core 1215 is also 8 cm./century, which is markedly lower than the palynological estimate of 14 cm., but here it must be borne in mind that the latter value refers to the top layer of Holocene sediments, whereas the former value is based on the thickness of the entire post-Pleistocene sediment layer. This difference indicates an increased rate of sedimentation in the youngest Holocene phase, which may have been due to an increased supply of sediment to this area during the past 700 years, caused by the shift in the Orinoco discharge toward the Macareo and Boca Vagre estuaries referred to previously. Thus the palynological and sedimentological data are in essential agreement, and this constitutes a strong argument for the correctness of the palynological correlation and the subsequent stratigraphic interpretation.

The absolute pollen frequency can next be calculated in the same way as was already done for the Gulf of Paria, by assuming an average water content of 60 per cent for the fresh sediment (van Andel), and a specific weight of 2.0 for the clay material. The following values are then obtained: Core 1101, 315 grains/cm^2/year; core 1204, 69 grains/cm^2/year; core 1215, 171 grains/cm^2/year.

These values, and the figures given earlier for the Gulf of Paria, can be compared with the values obtained from direct measurements by Hesselman for the Gulf of Bothnia and by Faegri for the North Sea (Faegri and Iversen, 1950). These figures, however, deal with windblown pollen only, and a direct comparison can be made only for cores 1204 and 1215, in which it has been estimated that a minimum of 10 per cent is windblown. This gives minimum values of 7 and 17 windblown grains/cm^2/year for cores 1204 and 1215, respectively. These figures are much lower than those found in the Gulf of Bothnia, but are quite comparable to the value of 17 grains/cm^2/year found in the Norwegian Sea 450 km. from the Norwegian coast. The much smaller distance off the coast of cores 1204 and 1215 in comparison to the latter locality is counterbalanced by the preponderance of adverse northeasterly trade winds north of Trinidad.

Thus it appears that the results of the investigation of the three cores not only are of interest as a background for understanding the Recent distribution patterns of pollen and other microfossil groups, but also that they reveal important features which can be utilized in the study of Holocene depositional history and in the interpretation of palynological changes in general.

Recognition of sub-Recent sediments

On the southeastern Trinidad shelf, sediments that are of proven or supposed Pleistocene age lie at or near the surface, pointing to reduced sedimentation due to wave action in the shallow areas there. Several samples were investigated palynologically in order to confirm, if possible, the supposed Pleistocene age.

The salient features in the microfossil content of these samples are given in the table below.

Station	Depth	Total pollen/g.	Total Hystrix/g.	Per cent Rhizophora	Radiocarbon age
1026	0–10 cm.	1568	0	6	
1103	195 cm.	2610	0	41	
1104	35 cm.	3696	0	8	
1109	100 cm.	12,500	0	27	17,820 ± 600 yrs.
1142	192 cm.	1453	0	11	
1181	15 cm.	1250	0	1	

If these figures are compared with the values typical of the Recent sediments in this area, it will be clear at once that these are sediments that have been deposited in an environment differing strongly from the one prevalent today. Pollen contents are much higher than normal in the Recent surface layer, pointing, in the case of sample 1109 at 100 cm. depth, even to deltaic or very near-shore conditions. In general, all samples seem to have been deposited during a period in which the coast line had migrated far seaward, which would have been possible only with glacial eustatic lowering of the sea level, thus making a Pleistocene age highly probable. In sample 1109 at 100 cm. depth, this has actually been proved by radiocarbon dating. *Rhizophora* percentages are much lower than those found today in the surface sediments, which points to a different paleogeography. The remaining pollen types also occur in different ratios. Finally, Hystrichosphaeridae are completely absent, in strong contrast to their regular occurrence in the Recent muds. Although direct proof of a Pleistocene age cannot be provided by palynological examination without further study of proven Pleistocene sections, the assumptions of the sedimentologists appear to be well justified, and a Pleistocene to early Holocene age can be safely assigned to the samples indicated.

DISCUSSION OF RESULTS

Transport and sedimentation of pollen

The preceding analysis of variations in the pollen content of sediments at the surface and in cores has made it possible to follow the processes of transportation and sedimentation of pollen. The relative importance of wind versus water transport has been largely evaluated from the foregoing.

The dominating factor is the presence of a large delta which acts as a major pollen source and transport agency by its discharge of water. Overrepresentation of pollen produced by local swamp vegetation is the rule in delta sediments because of restricted transport facilities, but offshore the pollen suspension in water is thoroughly mixed by turbulent flow and widely distributed by marine currents.

Pollen settles gradually during transport, which results in a decrease in the total pollen content away from the source area. Settling is accompanied by a certain degree of size selection, the smallest grains being relatively concentrated the farthest seaward. In addition, the effect of the mainly airborne supply from Trinidad and Tobago is apparent, but quantitatively it remains of only minor importance.

The airborne phase of pollen transport, which in general is of short duration for delta-produced pollen and much longer for pollen from Trinidad and Tobago, was not extensively investigated. Pollen sampling of the atmosphere above the Gulf of Paria produced negative results. However, a tentative estimate of the absolute frequency of windblown pollen in the two cores from the north Trinidad shelf proved possible, and is in line with observations made by other authors. The waterborne phase of pollen transport could, on the contrary, be further analyzed in relation to the large amount of hydrographic and sedimentological data discussed by Van Andel and Postma and by Koldewijn.

For a direct comparison of the distribution of pollen and sedimentary particles, the settling velocities of pollen grains in water will have to be determined. The first difficulty is the determination of the specific weight of pollen, both in the fresh state and as an empty exine. Erdtman (1943) indicates a range of 0.4–1.2 for fresh pollen. This suggests that many pollen grains will float for a time before settling, a fact that is well known, especially of coniferous pollen. The specific weight will probably increase when the living content dies and is replaced by water. Wolfram (1954) indicates a range of 1.1–1.2 for the specific weight of sub-Recent pollen. The settling velocity of these empty pollen exines is also dependent on their size, shape and ornamentation, all of which are extremely variable. According to Stoke's law, the settling velocity of a pollen grain with a diameter of 0.03 mm. and a specific weight of 1.2 would be 4 cm./hour. Settling experiments carried out by the author with pollen derived from Recent Orinoco sediments, however, gave higher values. For *Rhizophora* pollen (20μ, spherical, smooth), a settling velocity of 7 cm./hour was found, and for *Symphonia* pollen (40μ, suboblate, smooth), 17 cm./hour. On the basis of an average settling speed of 12 cm./hour, pollen grains would then bear comparison with mineral particles of 0.004–0.008 mm. diameter (fine silt). A direct comparison is now possible between the distribution of water-transported pollen from the delta and the distribution of other suspended matter which is also mainly derived from the delta estuaries. According to Postma, the average grain-size of the suspended floccules in the Gulf of Paria is 0.015 mm., and the material suspended at the estuarine surface is even coarser and is estimated at 0.03 mm. As the settling velocities of these particles are of the order of 80–100 cm./hour, they must have been kept in suspension by considerable turbulence, otherwise they could not have been distributed over the entire Gulf of Paria. Thus, for pollen grains, with their considerably lower settling velocity, transport in suspension over the Gulf of Paria is theoretically possible. This is in agreement with the distribution pattern actually found. The heavier grains are mainly restricted to the coastal belt opposite the estuaries, where the largest amounts of pollen also settle, but the light *Rhizophora* pollen escapes in large quantities, together with the suspension of the finest clay particles, through the Dragon's Mouth onto the northern shelf. Even on the outer shelf edge, the dispersal limit has not yet been reached.

Palynological provinces

It is possible to distinguish palynological provinces on the basis of the source areas and transport and distribution patterns of pollen. These provinces, which have a certain resemblance to mineral provinces as distinguished in sedimentary petrology, would then be characterized by depositional associations of specific pollen types.

Thus, in the delta region, which forms one large tract characterized mainly by the confluence of source and sedimentation areas, it is possible to differentiate between the levee province and the back-swamp province in that the deposits of the former carry additionally reworked pollen and pollen derived from the upstream drainage basin. In the back-swamp province, a large number of sub-provinces can be distinguished, which reflect the mosaic-like distribution of plant communities in relation to sediment supply.

The offshore provinces are generally characterized by a uniformly mixed supply. Opposite the eastern Orinoco estuaries, the eastern Orinoco shelf provinces can be distinguished. Here, supply from the forested eastern delta is dominant. An inner belt, with higher total pollen content, and an outer belt, with lower total pollen content and progressive sorting effects, can be separated. East of Trinidad, the supply of reworked Pleistocene material, coupled with a very low content of delta pollen, indicates another province. Airborne supply is absent over the eastern shelf.

In the Gulf of Paria, conditions are more complex, and more provinces can be recognized in this smaller area than on the eastern shelf. Although insufficient data preclude the delineation of provincial boundaries on a quantitative basis, a provisional subdivision can be made, as follows:

The western Gulf province is characterized by a high total-pollen content, with a predominantly local supply from the extensive mangrove belts along the estuaries and shores. The central Gulf province is characterized by a rather high total-pollen content and waterborne supply mainly from the Boca Vagre and Caño Macareo. The eastern Gulf province is a typically mixed one because of waterborne supply through the Serpent's Mouth, causing low total pollen content together with a qualitatively important airborne supply from Trinidad *(Anacardium)*. On the northern shelf, a *Podocarpus-Hemitelia* province can be recognized, characterized by a notable windblown supply from the mountains of Trinidad and Tobago.

Facies recognition

The data on the distribution of pollen and other microfossil groups can also be utilized in studies of depositional environment, and therefore are valuable in facies recognition and in the paleogeographic reconstruction of ancient sedimentary basins. The recognition of palynological provinces makes it possible to distinguish rather sharply the characteristics of each of several major types of depositional environment. The data on the distribution of microfossils originating from the delta, such as cuticles, fungi, and reworked tracheids, on the one hand, and on the distribution of the planktonic Hystrichosphaeridae on the other, add information on the location of estuaries and on salinity. The significance of the foraminiferal remains described should be evaluated in connection with conventional micropaleontological studies of the sediments.

The consistent decrease seaward of the total pollen content of the sediments, and the subsequent modifications of this general trend by local differences in transport facilities, further indicate the possibility of reconstructing main sea-current patterns and the positions of large estuaries and of shore lines in general. In this respect, palynology, in conjunction with sedimentological studies, can make a valuable contribution to paleogeographic interpretation. In view of the practical interest in these problems, a more detailed discussion will be given here.

The positions of the coast lines in the area under study is not invariably reflected in the pollen content of the sediments. The steep, rocky coast of the narrow Paria peninsula shows its presence only by a slight concentration of windblown pollen on the northern side. The northwestern and southern Trinidad coasts are more clearly expressed in the pollen concentrations because of the larger source area inland and the more favourable wind direction. The eastern Trinidad coast, on the other hand, is not expressed in the palynological data because of adverse wind directions and the absence of an appreciable river supply.

The positions of the delta coast line and the mouths of the large estuaries in particular are indicated clearly by the increase in total pollen content toward the main pollen sources. In addition, the disappearance in this direction of *Hystrix* and foraminifera, and the increase in cuticle and fungi content, can be used as guides in detecting the location of the littoral zone. The crossing of the delta shore line, which is taken to be at the outer limit of swamp vegetation and which is not expressed by any distinct lithologic change in the sediments, is registered by an abrupt increase in the total pollen content and a shift in percentage ratios toward local dominance, first of the mangrove association, and farther inland of the other swamp associations. Local variations, such as the lower pollen content in the coarser-grained Boca Vagre deposits and on the levees, of course, disturb this general picture to a certain extent. In addition, the situation opposite the northeastern delta shore line between Caño Macareo and Caño Araguao is worth noting. Here, no major estuaries open into the sea, and wind directions are shoreward. Pollen concentrations in the littoral zone here decrease from east to west parallel to the shore line, because of the predominant east-west current. From the foregoing it can be concluded that in making paleogeographic interpretations, the general direction of a delta shore line can sometimes be detected, but that the actual location is difficult to predict.

The interest in the general problem of shore-line detection is indicated by the "Symposium on finding of ancient shore lines," the abstracts of which were published in the Journal of Sedimentary Petrology (vol. 23, pp. 125–128, 1953), as well as by a United States Patent granted in 1954 to W. S. Hoffmeister, of the Carter Oil Company, entitled "Microfossil prospecting for petroleum." In this patent it is claimed that the positions of ancient shore lines can be determined on the basis of total pollen per gram, the ratio of large to small pollen, and the abundance of *Hystrix* and foraminifera. Hoffmeister's methods are evidently based on data that are closely comparable with the evidence obtained by the present study. For example, his figure of 7500 pollen grains per gram of sediment under nearshore conditions corresponds closely to values determined in the Orinoco sediments.

GENERAL CONCLUSIONS AND PRACTICAL APPLICATIONS

The distribution of pollen, spores, and other microfossils in the Recent Orinoco sediments, as discussed in detail in the preceding chapters is shown to be related principally to the locations of source areas and to the transporting air and water currents. In the delta, transport of pollen grains is restricted, and the local swamp flora is dominant in the sediments. Offshore, the pollen is better mixed. Pollen produced by the delta flora is carried seaward with the Orinoco discharge and incorporated in the offshore sediments over a wide area. In most areas it far outnumbers windblown pollen derived from Trinidad, Tobago, and the Paria peninsula. In the process of gradual settling of the pollen load transported by marine currents, some size selection of grains takes place. Waterborne pollen grains compare in settling speed with fine silt particles, and there are similarities in distribution between these particles and pollen. The total pollen content offshore is shown to reflect the hydrographic current pattern.

Pollen grains reworked from older sediments were recognized, and are most abundant in levee deposits in the delta. They are probably derived from Terti-

---

EXPLANATION OF PLATE 1

1, pollen grain of *Mauritia* sp., × 1000, Sample Mu-326; 2, pollen grain of *Avicennia nitida*, × 1000, Sample Mu-347; 3, pollen grains of *Rhizophora mangle*, Sample Mu-152: a, polar view, × 1000; b, equatorial view, × 1000 (oil immersion); 4, pollen grain of *Anacardium* type, × 1000, Sta. 1013; 5, spore of *Hemitelia* sp., × 1000, Sta. 346; 6, pollen grain of *Podocarpus* sp., × 1000, Sta. 1009; 7, pollen grain of *Symphonia globulifera*, × 1000, Sample Mu-3; 8, pollen grains of *Terminalia* type, Sample Mu-3: a, polar view, × 1000 (oil immersion); b, equatorial view, × 1000; 9, pollen grain of *Alnus* sp., × 1000, Sample Mu-43; 10, fungal spore, × 1000, Sample Mu-70; 11, tracheid, × 250, Sta. 180; 12, hystrichosphaerid, × 700, Sta. 1207; 13, foraminifer, × 650, Sta. 1202.

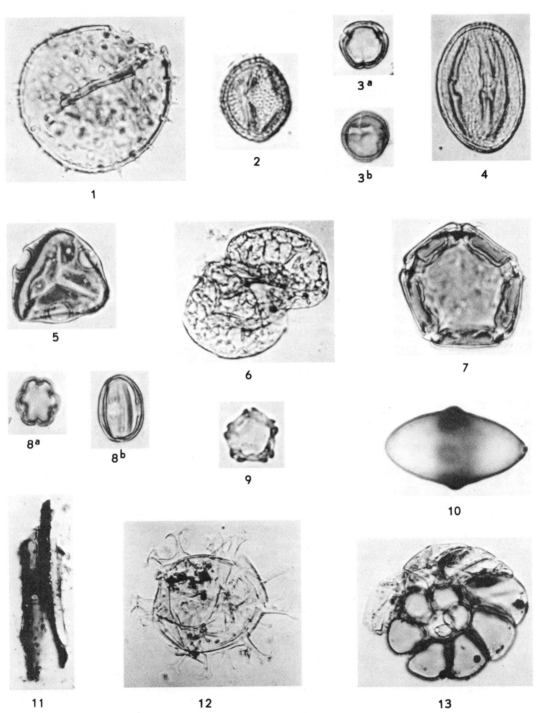

ary strata in the Andean foothills, which are being eroded by the Orinoco headwaters. They are also found locally concentrated near outcrops of Tertiary and Pleistocene strata, which are being eroded by wave action.

The distribution of fungal spores, leaf cuticles, and reworked tracheid fragments is mainly restricted to the delta and the littoral belt. The distribution of remains of Hystrichosphaeridae and foraminifera shows the influence of salinity in the offshore area.

These results derived from the surface layer of sediments are supplemented by investigations of three core samples, in which it was found that the microfossil content exhibits variations with depth. These variations can be related to changes in hydrography, sea level, and climate, and appear to have been contemporaneous over the area of investigation. Marked differences from present-day conditions were found only in some of the older Holocene and Pleistocene samples from the eastern shelf area, collected at or below the surface of the sea-floor.

The broader significance of the results obtained lies in their value in refining and perfecting palynology as an aid in geologic exploration. In summary, it can be said that they demonstrate the high potential sensitivity of the palynological method and the degree of integration of this method with sedimentology and paleogeography. The subject of floral change in time could only be touched upon, but the extent to which a detailed palynological study of otherwise homogeneous shale sections might reveal slight but synchronous and causally interrelated changes in climate, sediment supply, and facies is demonstrated in the results of the core investigations.

The results can be applied in the evaluation of local versus regional floral changes and of the reliability of correlation within a single facies province as compared with correlation between different facies provinces within one or more sedimentary basins. This, in practice, will to some degree eliminate empirical procedures in discriminating between regional markers and local ones, especially when working with small-scale quantitative floral changes, and will therefore restrict the number of type-sections that must be examined before reliable zonation can be established. Whenever fossil pollen-types can be identified botanically, the possibility of interpreting paleoecologic change will be increased and it will generally be possible to separate, in a more direct way, the influence of local facies shifts in the pollen diagram from that of regional, climatically controlled changes in vegetation. For example, the fact that it is possible to recognize *Rhizophora* pollen in the Oligo-Miocene sediments of Venezuela, in combination with the knowledge obtained in the present study of its peculiar distribution pattern, has led to its exclusion from the restricted set of regional zone markers, and at the same time has established this type as a highly useful environmental indicator.

However, it is realized that the results of the present study should not be generalized indiscriminately. Strictly speaking, they apply only to the area of Recent Orinoco sedimentation, and many basins may be found in which different conditions of pollen supply and sedimentation exist and in which the rules formulated above may not hold.

BIBLIOGRAPHY

ANDEL, T. VAN, POSTMA, H., AND KRUIT, C.
1954 – *Recent sediments of the Gulf of Paria (Reports of the Orinoco Shelf Expedition; Vol. I)*. K. Nederl. Akad. Wetensch., Verh., ser. 1, vol. 20, no. 5.

BEARD, J. S.
1946 – *The natural vegetation of Trinidad*. Oxford.
1949 – *The natural vegetation of the Windward and Leeward Islands*. Oxford.

ERDTMAN, G.
1943 – *An introduction to pollen analysis*. Waltham.

FAEGRI, K., AND IVERSEN, J.
1950 – *Text-book of modern pollen analysis*. Copenhagen.

GRAYSON, J. F.
1956 – *The conversion of calcite to fluorite*. Micropaleontology, vol. 2, no. 1.

HOFFMEISTER, W. S.
1954 – *Microfossil prospecting for petroleum*. United States Patent no. 2.686.108.

KOLDEWIJN, B. W.
1958 – *Sediments of the Paria–Trinidad Shelf (Reports of the Orinoco Shelf Expedition; Vol. III)*. Thesis, Amsterdam.

LINDEMAN, J. C.
1953 – *The vegetation of the coastal region of Surinam*. Thesis, Utrecht.

NOTA, D. J. G.
1958 – *Sediments of the Western Guyana Shelf (Reports of the Orinoco Shelf Expedition; Vol. II)*. Wageningen, Landbouwhogeschool, Meded., vol. 58, no. 2.

PIVETEAU, J.
1952 – *Traité de paléontologie*. Paris.

POSTHUMUS, O.
1928 – *The ferns of Surinam and of French and British Guiana*. Malang, Java: 196 pp.

VEEN, F. R. VAN
1957 – *Microforaminifera*. Micropaleontology, vol. 3, no. 1.

WILSON, L. R., AND HOFFMEISTER, W. S.
1952 – *Small foraminifera*. Micropaleontologist, vol. 6, no. 2.

WOLFRAM, A.
1954 – *Versuche zur Trennung der Sporomorphen von organischen und anorganischen Beimengungen unter Berücksichtigung der Wirkung des Ultraschalls auf Kohlenmazerate*. Geologie, vol. 3, no. 5.

17

Copyright © 1958 by Cambridge University Press

Reprinted from pp. 1-3 and 12-18 of *Geol. Mag.* **95**(1):1-18 (1958)

Upper Carboniferous Plant Spore Assemblages from the *Gastrioceras subcrenatum* Horizon, North Staffordshire

By R. NEVES

ABSTRACT

From a locality in North Staffordshire. Microspore assemblages are described from the Six Inch Mine coal seam, the immediately overlying carbonaceous shales, and the marine shale bed. Marked differences occur within this thin sequence in the proportions of certain genera present; in particular, the dominance of *Florinites* in the marine shales is discussed. Whilst the proportionate numbers of genera vary, the actual species present are more or less the same, though a greater variety exists in the shales. Three new genera and seven new species are described.

INTRODUCTION

THE aim of the work has been to confirm the presence of microspores in carbonaceous shales of Upper Carboniferous age from the North of England, by the use of maceration techniques, and to compare the variety and distribution of spores in shales with assemblages from coals. Workers in palynology concerned with the description of Carboniferous spores have confined their attention mainly to coal seam assemblages. These assemblages, as several authors have observed (Hoffmeister, Staplin, and Malloy, 1955, p. 372), represent a flora of limited ecological association. Consequently in order to determine the overall distribution and range of the various spore types it is necessary to examine a wider range of ecologies, such as may be represented by the spores contained in sediments of all types.

Maceration techniques developed allow the microspores, contained in many carbonaceous shales, to be isolated in a condition that enables description and comparison with those obtained from coals. The following account compares qualitatively and quantitatively the assemblages of spores from a coal seam, a " non-marine " shale, and a marine shale. The three rock types are in a continuous and restricted vertical sequence and were selected from the same locality in order to reduce the effects of evolutionary changes and to minimize the possibilities of regional variation. The selection of the actual horizon was conditioned by the fact that this work forms part of a broader study of the Namurian and Lower Westphalian spores from North Derbyshire and Staffordshire. The locality chosen was an exposure of the *Gastrioceras subcrenatum* horizon in a stream gully near The Wash (G.R. SK 014662), Quarnford, Staffordshire. The sequence consists of a thick seat earth overlain by a thin coal seam, $1\frac{1}{2}$ to 2 inches; the grey roof shales (18 to 24 inches) yield plant fragments and are succeeded by approximately 6 inches of black carbonaceous shales containing goniatites. The goniatite bearing shales were taken as the true marine deposits whilst the underlying grey shale contained no fossils and is subsequently referred to as " non-marine ". F. W. Cope (1946) correlated the coal seam with the Six Inch Mine of Lancashire and the Pot Clay Coal of Yorkshire. The horizon lies, therefore, at the junction of the Namurian and Westphalian measures in North Staffordshire.

Channel samples were taken from the three rock types and the microspore assemblages obtained were used as the basis for this study.

Previous Work

Relatively few authors have published work on the Namurian plant spores and in particular on spores of this age from argillaceous sediments.

Raistrick (1937), Millott (1939), and Knox (1947) studied microspores of Namurian age in Great Britain during their analysis of assemblages from coal seams only. Full studies were rendered somewhat difficult particularly in the Central English area since the coal-seams there are spasmodic in occurrence and as Millott (1939, p. 320) pointed out, the physical character of the coals creates problems of maceration leading frequently to a poor microspore yield. More recent work by Horst (1955) is contained in an account of the Namurian " Dispersed Spores " from Mahrish-Ostrau and West Upper Silesia, the result of sampling from coals alone. The results from this work were incorporated in the monograph of Potonie and Kremp (1955 and 1956). Hoffmeister, Staplin, and Malloy (1955) published the results of an investigation into plant spores derived both from coals and shales contained within the Mississippian of Illinois and Kentucky. These authors described several new genera and species and discussed the relative abundance of spore types in the various sediments. They noted a great variation in the relative abundance of genera within short vertical ranges though the same genera and species were generally in common throughout. No particular mention was made of *Florinites* which shows such a marked variation in numbers in the present study.

MACERATION TECHNIQUES

The treatment of both shale and coal was essentially similar in that the microspores were released from the humic material by oxidation with Schulze solution and subsequent washing with potassium hydroxide solution.

A channel sample of coal was mechanically crushed and macerated with Schulze solution for a period of 12 hours. The residue was washed with 10 per cent potassium hydroxide and distilled water in a sinter glass funnel until the filtrate was colourless. The spore concentrate remaining was pipetted off and mounted in glycerine jelly.

The shales were at first treated in order to remove argillaceous and other mineral matter and to obtain a concentrate of organic residues. For this purpose, shale fragments (approximately $\frac{1}{4}$ inch square) were immersed in bromine for a period of up to 48 hours, which effectively broke up the shale into a fine aggregate. The bromine was then washed off with distilled water in a pressure-assisted filtration system. The bromine-free material was transferred to a polythene flask and heated in 40 per cent hydrofluoric acid on a water bath for 16 to 20 hours, a time sufficient for the removal of the mineral matter. The acid was decanted off and the mineral-free organic concentrate was then oxidized by Schulze solution for a period of 4 to 6 hours. Thereafter the method described for coal preparation was followed.

The bromine plays an important part in this procedure, which apparently results in a mechanical disintegration of the shale possibly through the accompaniment of mild oxidation of organic and some inorganic constituents. Bromine treatment increased the ease and efficiency with which the hydrofluoric acid attacked the mineral residues to such an extent that slides prepared from the shales so treated contained little or no argillaceous residue. The use of bromine and hydrofluoric acid in this technique has no deleterious action upon the spores themselves. A check was carried out on a sample of coal which had been treated with bromine and hydrofluoric acid prior to oxidation with Schulze solution. Slides prepared from this maceration showed the microspores to be comparable in size and relative abundance to those obtained from the same coal but prepared by the normal method of coal maceration.

[*Editors' Note:* Material has been omitted at this point.]

Microspore Assemblages

This detailed study of the microspore assemblages from a coal seam, marine and " non-marine " shale, which together constitute a relatively thin sequence, has revealed significant differences in both types of spores present and their relative numerical proportions.

For comparison, the relative abundance of each genus in the three assemblages are listed in Table 1 as percentages. The three right-hand columns show the number of species of each genus occurring at each horizon. The generic frequencies shown in the left-hand columns are also plotted as a histogram (Text-fig. 4).

In all cases the spore counts are based on 750 specimens; however, in the case of the coal and " non-marine " shale counts the first 250 specimens observed were used to determine the proportion of *Lycospora* to other genera. The subsequent 500 specimens were

TABLE 1.—LISTS OF GENERIC PERCENTAGES AND SPECIES TOTALS

	Genera	Percentage of genera in assemblage			Number of species observed		
		Coal	" Non-marine "	Marine	Coal	" Non-marine "	Marine
1	*Leiotriletes*	0·2	1·3	3·8	1	4	3
2	*Punctatisporites*	1·4	0·8	12·4	3	1	3
3	*Calamospora*	5·3	3·0	4·8	2	4	4
4	*Granulatisporites*	0·2	0·4	1·2	2	2	5
5	*Cyclogranisporites*	0·5	1·2	1·4	1	3	2
6	*Planisporites*	1·2	0·2	2·2	2	1	1
7	*Apiculatisporites*	8·3	5·4	2·4	5	5	9
8	*Lophotriletes*	0·1	S	2·0	1	1	2
9	*Acanthotriletes*	0·2	S	0·4	1	1	3
10	*Raistrickia*	S	0·1	2·4	1	1	3
11	*Verrucosisporites*	0·5	0·2	3·0	2	1	2
12	*Convolutispora*	S	S	1·0	1	1	2
13	*Camptotriletes*	1·3	0·7	0·3	1	1	3
14	*Microreticulatisp.*	S	0·1	1·7	1	1	2
15	*Dictyotriletes*	1·3	2·0	S	2	1	2
16	*Reticulatisporites*	NS	NS	0·2	0	0	2
17	*Knoxisporites*	S	S	S	1	1	1
18	*Triquitrites*	S	0·1	0·2	1	2	1
19	*Ahrensisporites*	0·1	0·6	S	1	1	1
20	*Simozonotriletes*	NS	NS	S	0	0	1
21	*Lycospora*	76·0	75·0	S	3	3	1
22	*Anulatisporites* / *Densosporites*	0·3	0·6	0·7	{1 / 1}	1 / 2	1 / 1
23	*Cirratriradites*	S	0·2	NS	1	1	0
24	*Laevigatosporites*	0·1	0·1	NS	1	2	0
25	*Latosporites*	S	NS	NS	1	0	0
26	*Punctatosporites*	NS	0·1	NS	0	1	0
27	*Alatisporites*	NS	NS	0·3	0	0	2
28	*Endosporites*	0·2	0·1	0·3	1	2	2
29	*Schulzospora*	1·4	3·3	NS	1	1	0
30	*Florinites*	1·0	2·6	58·3	3	3	5
31	*Schopfipollenites*	S	0·4	NS	1	1	0
32	*Mooreisporites*	NS	NS	S	0	0	1
33	*Discernisporites*	NS	S	S	0	1	2
34	*Proprisporites*	NS	NS	S	0	0	1
35	Spore Type A	NS	NS	S	0	0	1
36	Spore Type B	NS	NS	S	0	0	1
37	Spore Type C	NS	NS	S	0	0	1
	Incertae Sedis.	0·4	1·5	1·0			

S = Frequency less than 0·1 per cent.   NS = Not observed.

considered disregarding *Lycospora* in order to find the relative abundance of the remaining genera more accurately.

From the distribution of genera through the sediments it is clear in the first instance that the coal and " non-marine " shale are closely comparable in the predominance of *Lycospora*, *Apiculatisporites* and, *Calamospora* in their assemblages; the dominance of the latter two genera is not so marked in the shale, whilst *Schulzospora*, *Florinites*, and *Dictyotriletes* correspondingly increase in numbers. The assemblage from the marine shale is very different from the others in the occurrence of genera. The reduction in numbers of *Apiculatisporites* continues and *Lycospora* becomes extremely rare; *Florinites* now assumes the

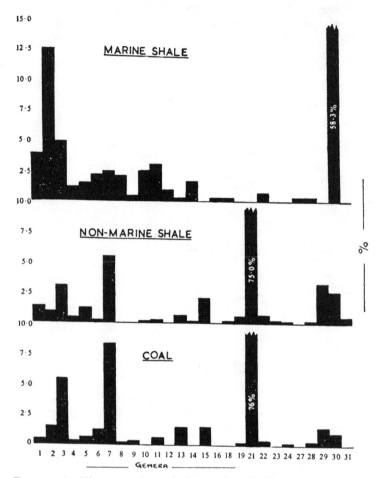

TEXT-FIG. 4.—Histogram of generic frequencies. Numbers below histogram refer to genera numbered in Table 1.

dominant role, constituting 58·3 per cent of the total assemblage. This predominance of *Florinites* is not restricted to this horizon, for in spore assemblages from marine shales at various horizons in the Millstone Grit and lower Coal Measures, *Florinites* has been observed by the author to be the dominant constituent. On the other hand, *Schulzospora* does not maintain the increase in numbers it showed in the " non-marine " shale and is rare at the marine horizon. *Punctatisporites* is present in its greatest proportion in the marine shale; *Schopfipollenites* reaches a maximum in the " non-marine " shale, but forms less than 0·1 per cent of the coal assemblage and appears to be absent from the marine shale macerations. The *Densosporites-Anulatisporites* group is not common in any of the sediments examined but shows a slight increase in the marine shale assemblage. *Raistrickia* is rare in the coal and " non-marine " separations but is a fairly common constituent (2·4 per cent) of the marine shale assemblage. The new genera *Discernisporites, Mooreisporites,* and *Proprisporites* occur as " Accessory Spores " (in the sense used by Raistrick, 1934) in the marine shale microflora. Only isolated examples of *Discernisporites* have been observed from the " non-marine " shale. Complete species lists have been compiled and these show that the total number of species investigated reaches 108, of which seventy-two are present in the marine shale, fifty in the " non-marine " shale, and forty-three in the coal seam. It is clear that whilst many species are common to all three horizons, additional species are present in the shales and the widest variety is present in the marine shales.

The marked contrasts in totals of species and relative abundance of different genera in the spore assemblages from the three sedimentary types calls for some explanation; at this stage of the work such explanations must, however, be regarded as tentative.

Since the coal seam is *in situ*, as indicated by its associated rootlet bed, it may be expected that the microspore assemblage will be derived from a local, specialized swamp type flora. This seems to be the case, for relatively few genera, namely *Lycospora, Apiculatisporites,* and *Calamospora*, form the bulk of the assemblage.

The similarity of the " non-marine " shale and coal seam assemblages is marked by the dominance of the two genera *Lycospora* and *Apiculatisporites*. At the same time increases in several genera, including *Dictyotriletes, Florinites,* and *Schulzospora,* and slight decreases in others, i.e. *Apiculatisporites, Calamospora,* and *Planisporites*, could be interpreted as a broadening, however slight, of the source of derivation of the " non-marine " shale microflora.

The dominance of *Florinites* in the marine shale is striking, but in addition the other genera are also fairly well represented, which when considered in conjunction with the greater variety of forms at the specific level seems to indicate an even more varied parent flora.

However, the abundance of *Florinites* raises a further problem which cannot be fully explained on the evidence of this work alone. Nevertheless, there are three possible reasons for its abundance: (1) Its large air sac would assist widespread dispersal by wind and possibly rivers; (2) It may have had greater resistance to chemical and vegetable decay; (3) A locally concentrated parent flora marginal to the area of deposition.

Potonié and Kremp (1955, p. 22), in a discussion of the ecology of Carboniferous plant spores, thought some structural features found in mega- and microspores assisted their dispersal. For instance, the large air sac, which is a diagnostic character of the genus *Florinites*, would obviously be of assistance, particularly with regard to wind dispersal, and may account for the abundance of this genus in the marine shale. However, spores of the *Endosporites* type show very little increase in the marine shale (0·3 per cent compared to 0·2 per cent in the coal). This would indicate that the possession of a large air sac may not be the only factor, or even the most important one in accounting for the distribution of *Florinites*. Furthermore, the relatively large proportion in the marine shale of spores of the *Punctatisporites* type, which are essentially smooth, spherical spores, apparently possessing no real dispersal advantages, would tend to detract from theories depending on structural control of dispersal.

It is also very difficult to assess the importance of degradation of the microspores or the relative resistance of various spore types to the prevalent forms of decay. That microspores are subjected to various types of decay or destructive agencies was observed at an early stage by Reinsch (1884, vol. i, p. 4, when he attributed some destructive elements on spores to " minute plant parasites "). A complete discussion of the forms of attack that have been observed is beyond the scope of the present paper, but is worthy of note that differing spore types frequently show a characteristic pattern of destructive attack. The general preservation of the microspores, however, in all the separations from the three horizons here described, allows the majority of specimens to be identified at the specific level. For example, the few specimens of *Lycospora* seen in the marine shale separations can be recognized as species occurring abundantly in the separations obtained from the two other horizons and there appears to be little difference in their general condition.

The overall appearance of the material, including spores, in slides prepared from the maceration of the two shale types and the coal seam, do, however, exhibit subtle differences in preservation of the organic remains. In preparations of marine shale, the spores have a waxy yellow appearance and the wood tissues are typically very dark brown or black, whilst from the coal seam the spores are golden brown and the wood and cuticle tissues are shades of dark and light brown. Although maceration affects to some extent the intensity of colouration in the organic matter, the quality of preservation remains different. Further work is necessary on the subject of preservation under various environments, but the mode of preservation in these sediments is sufficiently different to allow marine and non-marine organic remains to be distinguished.

The suggestion that the abundance of *Florinites* in the marine shales is due to the peculiar distribution of the parent flora seems possible. The spores or pollen grains attributed to the genus *Florinites* are believed to be derived from plants with Cordaitalean (Florin, 1936) and Coniferalean (Potonié and Kremp, 1955, p. 22) affinities. Hence, a flora in which these plants abounded, marginal to the marine depositional area, could be envisaged as providing this high concentration of *Florinites* in the marine sediment.

Acknowledgments

The author wishes to express his gratitude to Professor L. R. Moore for his advice and encouragement. Thanks are also extended to Dr. C. Downie and Mr. H. J. Sullivan who have read through the manuscript and offered many useful suggestions; to Mr. G. S. Bryant for his assistance in preparing the plates and to Miss T. McCarthy for her help in producing the manuscript. The work was carried out by the author in the Department of Geology, University of Sheffield, whilst he was in receipt of a Robert Styring Research Grant.

REFERENCES

COPE, F. W., 1945. The Contorted Rocks of the Southern Pennines. *Quart. Journ. Geol. Soc.*, ci,

FLORIN, R., 1936. On the structure of the pollen grains in the Cordaitales. *Sörtryk ur Svensk Bot. Tidskr.*, xxx, 624–650.

HOFFMEISTER, W. S., F. L. STAPLIN, and R. E. MALLOY, 1955. Mississippian plant spores from the Hardinsburg Formation of Illinois and Kentucky. *Journ. Pal.*, xxix, 372–399.

HORST, U., 1955. Die Sporae Dispersae des Namurs von Westoberschlesien und Mährisch-Ostrau. *Palaeontographica*, xcviii, 137–226.

KNOX, E. M., 1942. The microspores of some coals of the Productive Coal Measures of Fife. *Trans. Inst. Min. Eng.*, ci, 98–112.

—— 1947. The microspores in coals of the Limestone Coal Group in Scotland. ibid., cvii, 155–163.

KOSANKE, R. M., 1950. Pennsylvanian spores of Illinois and their use in correlation. *Illinois State Geol. Surv., Bull.* 74.

LUBER, A. A., and I. E. WALTZ, 1938. Classification and Stratigraphic value of spores of some Carboniferous coal deposits in the U.S.S.R. *Trans. Cent. Geol. Prosp. Inst.*, Fasc. 105, 43 pp.

MILLOTT, J. O'N., 1939. The microspores in the coal seams of North Staffordshire. Pt. I, The Millstone Grit—Ten Foot Coals. *Trans. Inst. Min. Eng.*, xcvi, 317–353.

NAUMOVA, S. N., 1937. The spores and pollen of the coals of the U.S.S.R. *Abstracts Papers, XVII Internat. Geol. Congr.*, U.S.S.R., 1937.

POTONIÉ, R., and G. KREMP, 1954. Die gattungen der palaozischen Sporae Dispersae und ihre Stratigraphie. *Geol. Jahrb.*, lxix, 111–194.

—— —— 1955. Die Sporae Dispersae des Ruhrkarbon ihre Monographie und Stratigraphie mit Ausblicken auf Arten andere Gebiete und Zeitabscnitte, Teil I. *Palaeontographica*, xcviii, 1–136.

—— —— 1956a. Teil II. ibid., xcix, 85–191.

—— —— 1956b. Teil III. ibid., c, 65–121.

RAISTRICK, A., 1934. The correlation of coal seams by microspore content. Pt. I—The seams of Northumberland. *Trans. Inst. Min. Eng.*, lxxxviii, 142–153.

—— 1937. The microspore content of some Lower Carboniferous coals. *Trans. Leeds Geol. Assoc.*, v, 221–6.

REINSCH, P. F., 1884. *Micro-Palaephytologia Formationis Carboniferae.* Vol. I, *Continens Trileteas et Stelideas.* Theo Krische, Erlangen and London.

WILSON, L. R., and E. A. COE, 1940. Description of some unassigned plant microfossils from the Des Moines Series of Iowa. *Amer. Mid. Nat.*, xxx, 182–6.

THE CARBONIFEROUS UPLAND FLORA

W. G. Chaloner

University College, London

Sir,—Mr. R. Neves is to be congratulated on his paper (1958) on the spores of an Upper Carboniferous coal and its associated shales, which throws new light on a relatively unknown aspect of the Carboniferous flora. Mr. Neves shows that there is a striking abundance of pollen of the type produced by cordaites and conifers in his marine shale, in contrast to the associated coal and non-marine shale, where this pollen is a relatively trivial element. He also suggests that this is the case in other Carboniferous marine deposits. I believe that the implications of this are more far reaching than Mr. Neves suggests.

Spores in Carboniferous marine shale have a peculiar ecological significance in that they represent mainly the "spore rain" from what may be regarded as the permanent land of that time, in contradistinction to the spores of the swamp vegetation abundantly represented in coal. That cordaite and conifer pollen proves to be very abundant in such a marine shale is of particular interest in this context. Mr. Neves suggests that "a flora in which these plants (cordaites and conifers) abounded marginal to the marine depositional area could be envisaged as providing the high concentration of *Florinites* in the marine sediment". But the flora immediately adjacent to the marine depositional area would presumably be of the typical coal swamp type; and in this, as he shows from the coal itself, pollen of the cordaite and conifer type is a relatively trivial component.

A much simpler, and I believe more attractive, hypothesis is that the cordaite and conifer dominated vegetation was present throughout the depositional cycle in a situation practically unaffected by it—namely in the areas of higher land adjacent to the coal swamp. Before a marine invasion the spores of the lycopods, calamites, pteridosperms and ferns in the swamp in the central part of the basin would far outweigh locally the pollen rain coming from the relatively remote (if extensive) upland cordaite and conifer vegetation. When a marine inundation flooded the basin, the swamp vegetation would be temporarily displaced to a marginal belt, while that of the upland areas would be virtually unaffected. The marine mud deposited in the centre of the basin would continue to receive the same pollen rain from the uplands as before, but that from the restricted marginal swamp would now be in relatively lesser proportion. This hypothesis would explain Neves' observed abundance of *Florinites* as being a direct result of the marine phase.

An analogous situation to that postulated here occurs in the Tertiary pollen sequence in Venezuela (Kuyl et al., 1955) in which changes in the pollen from the upland vegetation can be discerned against the background of locally dominant mangrove swamp vegetation which responds rapidly to minor changes in water level.

Whether this explanation of Neves' observations is valid or not, the spore contents of these marine shales are worth serious study, for these may be the only source of evidence that we shall ever have on the upland vegetation of that time. It is tempting to speculate that the true conifers, as distinct from the cordaites, may have been an important component of this upland flora during Carboniferous time long before they became a conspicuous element in the (macro-)fossil flora in the Permain. If we can learn to recognise the various Palaeozoic plants from their spores and pollen this type of palaeo-ecology holds interesting possibilities.

REFERENCES

Kuyl, O. S., J. Muller, and H. T., Waterbolk, 1955. The application of palynology to oil geology with special reference to western Venezuela. *Geol. en Mijnbouw*, N. S. 17, 49–75.

Neves, R. 1958. Upper Carboniferous Plant Spore Assemblages from the *Gastrioceras subcrenatum* horizon, North Staffordshire. *Geol. Mag.*, xcv, 1–19.

THE PALAEOECOLOGY OF CARBONIFEROUS PEATS BASED ON THE MIOSPORES AND PETROGRAPHY OF BITUMINOUS COALS

BY A. H. V. SMITH

(*Read at Sheffield, 13th January, 1962*)

CONTENTS

		page
I.	INTRODUCTION	424
II.	PREVIOUS RESEARCH	425
III.	SELECTION OF COALS AND INVESTIGATION PROCEDURES	426
	(a) Petrographic examination and analysis of coal	427
	(b) Isolation and analysis of miospores	428
IV.	MIOSPORE ASSEMBLAGES OF COAL SEAMS	428
	(a) Characterization of assemblages in terms of "dominant species"	430
	(b) Concept of miospore phases and their sequential relationship	432
	(c) The miospore associations of the different phases	435
V.	PETROGRAPHY OF THE COAL SAMPLES	440
VI.	RELATIONSHIP BETWEEN THE MIOSPORE PHASES AND THE PETROGRAPHY	443
VII.	THE FORMATION OF CARBONIFEROUS PEATS	448
	(a) The origin of petrographic types of coal	450
	(b) Possible influence of water levels	452
	(c) Possible influence of climate	456
VIII.	CONCLUSIONS	460
IX.	ACKNOWLEDGEMENTS	461
X.	REFERENCES	462
	APPENDICES	466
	DISCUSSION	470

SUMMARY

The lack of ecological data concerning the vegetation of the peat deposits and adjoining areas in the Carboniferous period is a limiting factor in the use of spores for stratigraphical purposes. For this reason, a petrological and palynological investigation of a number of coal seams containing crassidurain, in the Yorkshire Coalfield, was undertaken. The results show four distinct assemblages of miospores, each assemblage being more or less associated with coal of a distinctive petrographic type. Their vertical sequence is similar in the different seams. The succession culminates in the crassidurain, above which, under favourable conditions, the sequence is reversed. The peats producing this sequence are considered to be autochthonous in origin although a partly allochthonous coal type is also recognized.

Existing theories, which do not take account of palynological evidence, attribute the petrographic differences in the humic coals to varying degrees of aerobic decomposition controlled by environmental factors such as degree of drainage, or depth of water covering the peat surface at the time of deposition. They do not satisfactorily explain all the new evidence. The possibility that at least part of the sequence, and particularly the crassidurain, was the result of climatic factors is suggested.

[*Editors' Note:* Material has been omitted at this point.]

IV. Miospore Assemblages of Coal Seams

The large number of miospore species recorded by Potonié and Kremp (1955, 1956) in their synopsis of Carboniferous spores of the Ruhr is a reflection of the varied composition of the vegetation of the period, whose members largely reproduced by means of spores. The extent to which all the spore-bearing plants growing on the swamps and on the dry land are represented in the spore and pollen rain is unknown. It is possible that the spores and pollen grains of the Carboniferous plants are more representative of the floras than the pollen grains of later times, which may be under-represented in the pollen rain due to the evolution of floral morphology (Faegri and Iversen, 1950, p. 31).

Many of the genera and species of small spores recorded by Potonié and Kremp from the coal seams of continental Europe and North America occur in Britain but, in any one seam, the

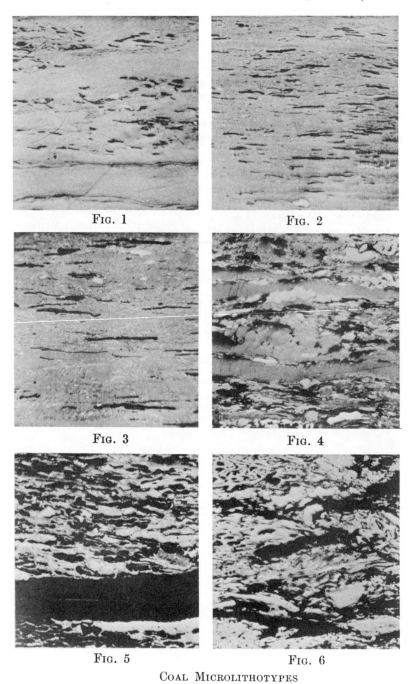

Fig. 1

Fig. 2

Fig. 3

Fig. 4

Fig. 5

Fig. 6

Coal Microlithotypes

number of species encountered is relatively small. The assemblages from the thin coal samples used in this investigation generally comprise between twenty and forty species. There is no evidence that thick seams yield a greater variety of spores than thin seams, the richness of the total assemblage probably depending more on the variety of environments occurring during the accumulation of the deposits than the length of time taken to accumulate. In many instances a particular type of vegetation or sequence of vegetation types must have persisted over considerable areas since the qualitative and quantitative composition of the miospore assemblages remain more or less unchanged for many miles (Raistrick, 1934; Hacquebard, 1952a; Guennel, 1952; Wilson and Hoffmeister, 1956; Dybova and Jachowicz, 1957; Tomlinson, 1957); sometimes only the qualitative features remain constant (Kosanke, 1954). However, changes do occur in the composition of the total assemblage of the same seam at different localities, presumably owing to variations in the sequence of environmental conditions associated with peat formation. Differences between seams may occur for the same reason.

It is obvious that the Carboniferous peat environment of the northern hemisphere provided conditions which were on the whole suitable for the preservation of spores. Certain petrographic coal types, such as some durains, yield spores which exhibit various amounts of weathering but there are few horizons in bituminous coal seams in this country which fail to yield any spores. The changes which have been shown to occur in the frequency and vertical distribution of the spores in coal seams are considered to be due primarily to changes in the composition of the vegetation in response to changing environmental conditions, and not to preferential decomposition of some of the spore exines in certain environments.

EXPLANATION OF PLATE 29

Illustrations of coal microlithotypes, cut perpendicular to the bedding, polished, and viewed by reflected light, using oil immersion objectives.

FIG. 1.—Vitrite and clarite. Distinct micro-banding, spores mainly *Lycospora* spp., × 140.

FIG. 2.—Canneloid clarite. Lacking distinct micro-banding, spores mainly *Laevigatosporites desmoinensis*, × 140.

FIG. 3.—Part of the above at higher magnification to show finely divided vitrinite and micrinite grains, × 340.

FIG. 4.—Clarodurite. Mixture of vitrinite, exinite (crassi- and tenui-) and inertinite (mainly massive micrinite), × 140.

FIG. 5.—Durite. Inertinite, mainly massive micrinite. Megaspore exine in lower part of photograph, × 140.

FIG. 6.—Durite. Inertinite, mainly semifusinite, × 140.

(a) Characterization of assemblages in terms of "dominant species"

It is reasonably certain that by selecting thin layers of coal with particular regard to the uniformity of petrographic composition, each assemblage of miospores will be mainly derived from a vegetation growing under reasonably constant environmental conditions. However, there will be some contamination from contemporary vegetation growing in other habitats.

The species lists, from the 179 samples selected in this way, show that in any one sample only a small number, and sometimes only one species, is recorded in any abundance. These abundant species do not necessarily represent the dominant plants in the vegetation but they may nevertheless be used to characterize a particular vegetation type. In this paper the abundant spores are referred to as the "dominant species" of the assemblage and are arbitrarily defined as the minimum number of species constituting fifty per cent of the total number of spores counted. The choice of fifty per cent gives a reasonable number of species for analysis; twenty-five per cent results in too few, whilst a value of seventy-five per cent results in too many species, some of which are not sufficiently abundant to justify inclusion in the dominant species. In the samples investigated the dominant species vary in number from one to five. Miospore assemblages which are dominated by one, two or three species are about equally represented and account for about ninety per cent of the samples. Assemblages in which there are four or five dominant species are only recorded from seven and one per cent of the samples respectively. The dominant species comprise a total of nineteen species, nine of which are recorded from fewer than five per cent of the

TABLE 1

The occurrence of ten of the dominant species, being those which occur in five per cent or more of the samples examined

Dominant species	Number of occurences in samples examined	Percentage of total samples examined	Number of seams from which recorded
Laevigatosporites desmoinensis	88	49	9
Lycospora pusilla	63	35	11
L. brevijuga	54	30	10
Punctatosporites minutus	52	29	8
Densosporites sphaerotriangularis	35	20	6
Crassispora kosankei	27	15	7
Lycospora granulata	14	8	5
Densosporites duriti	14	8	3
D. loricatus	11	6	5
Dictyotriletes bireticulatus	9	5	1

samples. In Table 1 are listed the ten dominant species which occur most often (see also Plate 30) and Table 2 shows the frequency with which each of the six most often recorded dominant species occurs in association with the other dominant species.

TABLE 2

The relationship of occurrence of dominant species

Dominant species	*Lycospora brevijuga* or *L. pusilla*	*Laevigatosporites desmoinensis*	*Densosporites sphaerotriangularis*	*Punctatosporites minutus*	*Crassispora kosankei*
Lycospora brevijuga or *L. pusilla*	—	35	2	39	13
Laevigatosporites desmoinensis	35	—	15	30	12
Densosporites sphaerotriangularis	2	15	—	Nil	5
Punctatosporites minutus	39	30	Nil	—	12
Crassispora kosankei	13	12	5	12	—
Densosporites duriti	2	11	2	3	Nil
D. loricatus	2	10	Nil	3	Nil
Lycospora granulata	2	10	Nil	1	1
Dictyotriletes bireticulatus	1	8	4	Nil	2

Three features which probably have ecological significance, emerge from this analysis:

1. The first concerns the association of certain species of *Lycospora*, *Laevigatosporites* and *Densosporites*. Whilst *Laevigatosporites desmoinensis* is frequently found associated with *Densosporites sphaerotriangularis* or with either *Lycospora brevijuga* or *L. pusilla*, the occurrence of *D. sphaerotriangularis* and either of these species of *Lycospora* in the dominant species was only recorded twice.

EXPLANATION OF PLATE 30

Illustrations of coal macerals.

FIG. 1.—Resinite. Resin bodies embedded in vitrinite, × 140.

FIG. 2.—Semifusinite. Fragments of semifusinite embedded in matrix of mineral sediment, × 340.

FIG. 3.—Massive micrinite, × 340.

FIG. 4.—Granular micrinite, × 340.

Some miospores diagnostic of the different phases, × 480.

FIG. 5.—a. *Lycospora pusilla* (Ibr.) S.W. & B.
  b. *Densosporites sphaerotriangularis* Kos.
  c. *Laevigatosporites desmoinensis* (Wils. & Coe) S.W. & B.
  d. *Punctatosporites minutus* Ibr.
  e. *Crassispora kosankei* Pot. & Kr.

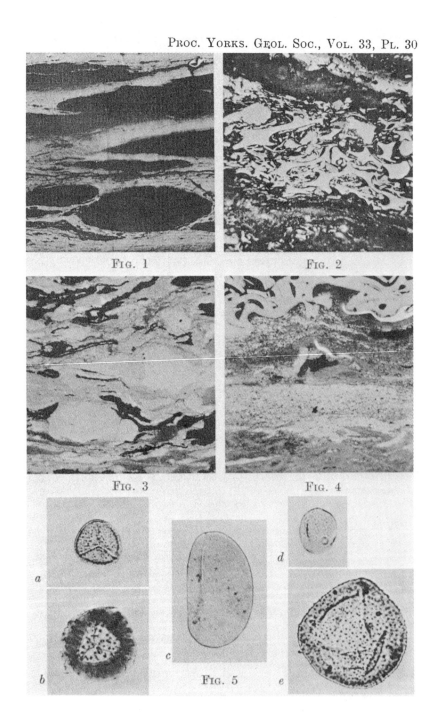

Coal Macerals and Miospores

2. *Punctatosporites minutus* occurs with similar frequency in association with the two species of *Lycospora* and with *Laevigatosporites desmoinensis* but is not recorded as occurring with *Densosporites sphaerotriangularis* in the dominant species. *Crassispora kosankei*, often in association with *Punctatosporites minutus*, also occurs with similar frequency with the species of *Lycospora* and with *Laevigatosporites desmoinensis*. *Crassispora kosankei* does, however, occur with *Densosporites sphaerotriangularis* although less often than with any other species.

3. Four species, *Densosporites duriti*, *D. loricatus*, *Lycospora granulata* and *Dictyotriletes bireticulatus* are recorded more often in association with *Laevigatosporites desmoinensis* in the dominant species than with any other species.

This analysis may provide a useful basis for the further study of the plant communities and the environments in which they grew, even allowing for the fact that the spores in the coal may not be present in the same proportions as their parent plants in the vegetation. It is assumed from the recurrence of the same abundant miospores in the seams investigated that the plant communities and the same environmental conditions occurred fairly often and were widespread.

(b) Concept of miospore phases and their sequential relationship

It has long been known that the assemblages from bituminous coal seams tend to be dominated by species of *Lycospora* and *Densosporites*. The present investigation has shown that other species, among them *Laevigatosporites desmoinensis*, *Crassispora kosankei* and *Punctatosporites minutus* are numerically important elements of the assemblages from restricted portions of seams. Each dominant, or group of dominant, species probably represents a distinct type of vegetation growing under a particular set of edaphic conditions, that is, conditions relating to the soil environment. The evidence of the miospores clearly supports this suggestion. The assemblages examined by the author from suites of samples from coal seams show not only that the different horizons within a seam have different dominant species but also that there is a regular pattern in the sequence of changes.

In order to demonstrate these features clearly, and to avoid a multiplicity of spore names, it is necessary to recognize and name the distinctive associations of spores which occur within the Carboniferous coal seams and which are believed to be derived from the different plant communities. The concept of the miospore phase was introduced for this purpose (Smith, 1957, p. 349). A miospore phase is defined as that part of the seam profile which is characterized by the occurrence of a particular association of species comprising a dominant, or group of dominant, species together with certain less abundant species. The term "association" is used in preference to "assemblage" to denote a group of species which generally occur together and may be related in the ecological sense. The dominant species of a phase are, therefore, diagnostic of the phase and may be made the basis for its recognition. The diagnostic species which have been used to determine the phases are listed in Table 3.

TABLE 3
Species used to determine the different phases

Lycospore	Transition	Densospore	Incursion
Lycospora brevijuga	*Leavigatosporites desmoinensis*	*Densosporites sphaerotriangularis*	*Crassispora kosankei*
L. pusilla	*Densosporites duriti*		*Punctatosporites minutus*
	D. loricatus		
	D. striatus		
	Dictyotriletes bireticulatus		
	Lycospora granulata		

The earlier analysis (p. 431) suggested that assemblages can be grouped on a basis of their dominant species into four categories and these form the basis of four phases—Lycospore, Transition, Densospore and Incursion. The Lycospore and Densospore phases are named after their dominant genera; "Transition" is so named because of its position in the phase-sequence, and "Incursion" is probably associated with periods of flooding. The last two are discussed further below.

Each of the 179 assemblages studied was allocated to one of these four, which, with the exception of the Incursion phase, are determined by the dominance of certain species, as given in Table 3. Those appearing with *Laevigatosporites desmoinensis* in the Transition phase are those that occur as dominant most frequently with that species (see Table 2). The numbers of *Crassispora kosankei* and *Punctatosporites minutus* generally fluctuate together, as can be shown statistically by correlating their abundance at different levels; the Incursion phase was thus recorded wherever the combined total of these two species exceeded fifteen per cent, for although they are rarely numerous enough to dominate their associated assemblage, the palynological and petrological evidence suggests that the phase so defined does represent a distinctive type of swamp environment.

An examination of the sequences of phases, excluding for a moment the Incursion phase, has shown that the Densospore is preceded and followed (when not in juxtaposition with the roof) by the Transition phase. The Lycospore phase (when not next to the floor) is likewise always preceded and followed by the Transition phase. The simplest sequence is thus:

Lycospore – Transition – Densospore – Transition – Lycospore.

The above sequence may be modified in various ways; it may be protracted by the repeated alternation of any two phases before the third makes its appearance, or peat formation may have been terminated at any stage in the sequence. The thickness of coal associated with the different phases may vary from less than one inch to many feet. In no case have the Densospore and Lycospore phases been found in juxtaposition.

In contrast to the others the Incursion phase does not appear to occupy a unique position in this order. Analytical study shows that there are as many instances where its occurrence is unaccompanied by any change in the dominant species of the neighbouring miospore assemblages as there are records of a change. Where a change of phase is associated with the occurrence of the Incursion phase there are about equal numbers of instances of the change being between Lycospore and Transition as between Transition and Densospore phases. It is probably significant that the Incursion phase has not so far been recorded from within the Densospore phase. Where the same phase occurs above and below the Incursion phase, the species associated with *Crassispora kosankei* and *Punctatosporites minutus* are, with one exception, those of the accompanying phase; where the phases above and below are different the species associated with *C. kosankei* and *P. minutus* may either change within the Incursion phase or immediately before or after it. There is one record of the Incursion phase occurring at the bottom of the seam and another one of it at the top. In both these instances the spores associated with the Incursion phase species were mainly those characterizing the Lycospore phase. The sequence of phases occurring in the Hazel, Swallow Wood and Parkgate seams are shown in Plates 31, 32 and 33 respectively. That recorded from the Hazel seam differs slightly from the one previously published (Smith, 1957) owing to the different criteria now employed to determine certain of the phases.

The Parkgate seam at Grange colliery contains, at about the middle of the seam, a massive horizon of crassidurain approximately eleven inches in thickness, and separated from two thin horizons of similar petrographic type on either side by bright coal. The quantitative examination has been confined to the lower durain band and the bright coal immediately below it and to part of the upper durain band and the bright coal between it and the roof of the seam. The coal intervening between the two thin bands of crassidurain was sub-divided into twenty samples and the spores isolated from these coals were examined qualitatively. It is reasonably certain that only two phases occurred in these samples, the Densospore phase which occurred through the greater part of the massive dull coal, and the Transition phase which occurred through the bright coal above and below.

The repetition of phases at different levels in the profiles of the Parkgate and Hazel seams suggests that the changes in the vegetation were not the result of the increasing depth of peat upon which the plants were growing.

(c) The miospore associations of the different phases

Broadly speaking it is true to say that individual miospore species occur most often and in their highest numbers in one or two phases and decline, sometimes markedly, in frequency and numbers in the others. The phases in which a given species is a characteristic component are those which occur in vertical juxtaposition in the ecological succession. Thus a species which occurs often in the Lycospore phase may also occur fairly frequently in the Transition phase but will probably be rare in the Densospore phase. Those species which occur at least in one phase in forty-five per cent of its samples are listed in Appendix C; with two exceptions all these species are found in twenty-five per cent of the total samples. Species excluded are relatively rare. The percentage occurrence and the maximum recorded range of abundance are given for each species, and each species is allocated to the phase in which it occurs most often in the maximum numbers; it may thus be considered diagnostic of that phase. In general the values for occurrence and abundance attain their maxima in the same phase.

Among the less common species omitted from this table are some which, because they are largely restricted in their occurrence to certain phases, are equally diagnostic. Examples are *Alatisporites pustulatus* and *Verrucosisporites papillatus* which characterize the Transition and Densospore phases as well as the Incursion phase, where they are associated with Transition phase species; these two species however have not been recorded from the Lycospore phase. *Triquitrites triturgidus* is of similar occurrence but it has once been recorded from the Lycospore phase. *Schopfipollenites* has a relatively high incidence in the Lycospore and Incursion phases but has not been recorded from the Densospore and only rarely from the Transition phase. *Ahrensisporites* sp. has been recorded from all phases but from a higher proportion of the samples of the Densospore than of any other phase.

The data provide some evidence of an allochthonous element in the miospore assemblages. Species such as *Granulatisporites microgranifer* and *Planisporites spinulistratus* which occur in small numbers in most samples irrespective of the phase may represent such an element. A number of species which may be of allochthonous origin are more common in the assemblages of the Incursion phase than elsewhere—a point discussed below; these include *Pityosporites westphalensis* which Williams (1955, p. 470) has suggested might have come from plants growing on the higher ground surrounding the swamps.

A feature of ecological significance which emerges from this analysis is that in some instances different species of the same genus have their maximum abundance in different phases. Examples are cited below from four genera and in each case the morphological differences between the allied species are slight.

Species	Phase of maximum numbers
Densosporites duriti	Transition
D. sphaerotriangularis	Densospore
Endosporites globiformis	Lycospore
E. zonalis	Transition
Calamospora pallida	Incursion
C. parva	Transition
Lycospora brevijuga and *L. pusilla*	Lycospore
L. granulata	Transition

Also of ecological significance is the number of species comprising the miospore assemblages assigned to the different phases; these are analysed in Table 4, based on counts of five hundred spores. It can be seen that the richest miospore assemblages occur in the Incursion phase whilst the poorest are in the Lycospore and Densospore phases.

Further analysis shows that each phase is characterized not only by one or a small number of abundant species but also by associations comprising a small number of the less common species. These less common species and the phases which they

TABLE 4

Percentage occurrence of the numbers of species comprising the miospore assemblages of the different phases

Phase	Percentage of counts of 500 spores			
	with less than 20 species	with 20 to 29 species	with 30 to 39 species	with 40 or more species
Lycospore	% 54	% 31	% 15	% —
Transition	6	65	27	2
Densospore	64	28	4	4
Incursion with Lycospore phase species	14	52	25	9
Incursion with Transition phase species	—	40	48	12

TABLE 5

Some less common species diagnostic of the miospore phases

Lycospore	Incursion associated with Lycospore phase species	Incursion associated with Transition phase species	Transition	Densospore
Apiculatisporis spinosaetosus	*Calamospora pallida*	*Cirratriradites saturni*	*Anapiculatisporites spinosus*	*Leiotriletes adnatoides*
Granulatisporites parvus	*Planisporites spinulistratus*	*Granulatisporites granulatus*	*Calamospora parva*	*Verrucosisporites microtuberosus*
Granasporites cf. *irregularis*	*Raistrickia fibrata*	*Raistrickia saetosa*	*Convolutispora* spp.	*V. sifati*
		Reticulatisporites reticulatus	*Dictyotriletes bireticulatus*	
			Lycospora granulata	

characterize are listed in Table 5. The extent to which these species characterize the phases is illustrated diagramatically in Fig. 1 where each histogram records the percentage, in the total samples of each phase, of assemblages containing all the species of the particular association under consideration. The figure also shows the values of average percentage abundance, in assemblages of each phase, of the more abundant species which typify the different phases. Thus in the case of the Lycospore phase the histogram shows that forty-four per cent of all the assemblages assigned to this phase (first column) contain the species *Apiculatisporis spinosus*, *Granulatisporites parvus* and *Granasporites* cf. *irregularis*. On the other hand these species are only recorded from fourteen per cent of the samples of the Transition phase (fourth column) and from none of the samples of the Densospore phase (fifth column). The pecked line on the same histogram shows that the average abundance of *Lycospora brevijuga* and *L. pusilla* falls from sixty-five per cent in samples of the Lycospore phase to ten per cent in the Transition phase and to less than four per cent in the Densospore phase.

The data shown in Fig. 1 clearly suggest that the vegetation of the Carboniferous swamps comprised different associations of plants at different stages in the growth of the swamps. The plants, or plant groups, associated with the different phases are known to some extent from the palaeobotanical affinities of certain Carboniferous miospores (Table 6). Only evidence concerning miospore genera referred to in this paper is considered. Further work may show that certain miospore genera have

438 A. H. V. SMITH

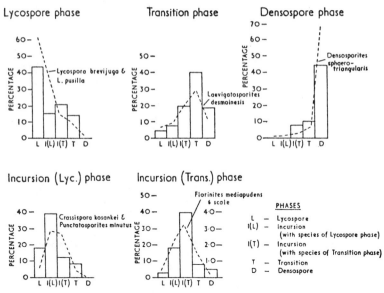

FIG. 1.—Composition of miospore phases. In each diagram the pecked line represents the average percentage of the most abundant diagnostic species of the phase. The histograms show for all phases the occurrence of assemblages containing all the less abundant species characteristic of a particular phase (see Table 5).

TABLE 6

Palaeobotanical affinities of certain miospore genera

Miospore genus	Phases characterized by genus*	Palaeobotanical affinities	Authority
Calamospora	Transition, Incursion	Calamite Sphenophyll	Hartung 1933 Remy and Remy 1955
Cirratriradites cf. *saturni*	Incursion	Lycopod (herbaceous)	Chaloner 1954
Convolutispora	Transition	Fern	Remy and Remy 1955
Cyclogranisporites	Lycospore, Incursion	Fern	Remy and Remy 1955
Densosporites cf. *loricatus* type	Transition	Lycopod (herbaceous)	Chaloner 1958a
cf. *sphaerotriangularis* type	Densospore	Lycopod	Bharadwaj 1958
Dictyotriletes	Transition	Pteridosperm	Benson 1904
Endosporites	Lycospore, Transition	Lycopod	Chaloner 1958b
Florinites	Incursion	Cordaites, Conifer	Florin 1936, 1938-40

Miospore genus	Phases characterized by genus*	Palaeobotanical affinities	Authority
Granasporites	Lycospore	? Lycopod (arborescent *Sigillaria*?)	Moore 1946, Chaloner 1953a
Granulatisporites	Lycospore, Incursion	Fern (?Pteridosperm)	Knox 1938
Laevigatosporites spp. greater than 40 microns	Transition	Calamite	Reed 1938, Andrews and Mamay 1951, Remy 1960
Leiotriletes	Densospore	Fern	Knox 1938, Mamay 1950, Remy and Remy 1957
Lophotriletes	Incursion	Fern	Remy and Remy 1957
Lycospora	Lycospore, Transition	Lycopod (arborescent *Lepidodendron*) (?*Lepidophloios*)	Chaloner 1953b, Felix 1954 Andrews and Pannel 1942
Planisporites	Incursion	Pteridosperm	Kidston 1923-25
Punctatosporites recorded as indeterminate thin exines	Lycospore	?Lycopod (arborescent *Sigillaria*?)	Moore 1946, Chaloner 1953a
Punctatosporites	Incursion	Fern	Mamay 1950, Remy and Remy 1957
Raistrickia	Incursion	Fern	Radforth 1938, Mamay 1950, Remy and Remy 1955
Spencerisporites	Transition	Lycopod	Chaloner 1951
Verrucosisporites	Densospore	Fern	Remy and Remy 1955, 1956, 1957, Bougnères and Remy 1957

*Where two phases are recorded they are characterized by different species.

affinities with plant groups other than those indicated in this table. Features of particular interest suggested by the table are that the arborescent Lycopods probably played a significant role in the vegetation of the Lycospore phase, whereas Sphenopsids became important in the Transition phase where ferns, pteridosperms and herbaceous Lycopods are also represented. Little can be deduced about the vegetation of the Densospore phase. In contrast considerable information is available concerning the affinities of the species of the Incursion phase; apart from the species of *Florinites*, which may be allochthonous in origin, it may be significant that herbaceous forms predominate.

[*Editors' Note:* Material has been omitted at this point and Plates 31, 32, and 33 have not been reproduced.]

It is also useful to study the relationship between microlithotypes and miospore phases in the context of the seam profiles, as summarized in Plates 31, 32 and 33. From these the following generalizations arise:

1. There is a relationship between the amounts of vitrite and clarite and the numbers of *Lycospora pusilla* and *L. brevijuga* (direct relation) and Incursion phase species *Crassispora kosankei* and *Punctatosporites minutus* (inverse relation) throughout those parts of the profiles assigned to the Lycospore phase or to the Incursion phase containing species of the Lycospore phase. The relationship is most striking in the Parkgate seam. In those parts of the profile where the two species of Lycospore are not common, the numbers of *C. kosankei* and *P. minutus* appear to be independent of the numbers of *Laevigatosporites desmoinensis* and *Densosporites sphaerotriangularis*.

2. Samples of the Transition phase in proximity to the Lycospore phase tend to have a high content of vitrite and clarite, whilst those in proximity to the Densospore phase tend to contain durite; intermediate samples quite often contain clarodurite. The progressive increase or decrease in the amount of clarodurite in passing through the Transition phase towards or away from the Densospore phase which these observations suggest, is not invariable and fluctuations in the petrography within the Transition phase are not uncommon. With only two exceptions all samples from the profiles recorded as having, at least in part, canneloid appearance (that is, lacking distinct microbanding and with relatively finely divided vitrinite and inertinite, Plate 29, figs. 2, 3) belong to the Transition phase. Nearly all these samples had a high content of vitrite and clarite. The exceptions belong to the Incursion phase.

3. Samples at the base of an Incursion phase nearly always contain substantial amounts of one or more of the microlithotypes clarodurite, durite and fusite. These amounts vary considerably within, and may decline towards the end of, the phase. The change from the Lycospore to the Incursion phase will generally, therefore, be marked by a change in the petrography (Fig. 2), which reflects a change in the environment. However the change in the petrography towards the end of the Incursion phase in some instances suggests that the characteristic species may have persisted for a time after the environment had reverted to its earlier condition. When the Incursion phase is preceded or followed by the Transition or Densospore phase there may be no apparent change in the microlithotype content of the samples

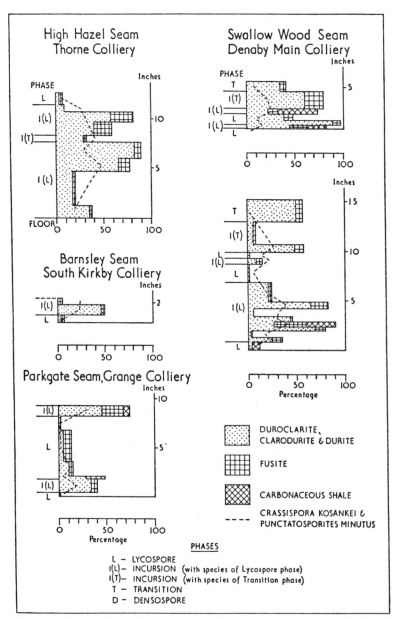

Fig. 2.—Petrography of the Incursion phase and the occurrence of *Crassispora kosankei* and *Punctatosporites minutus* (includes only instances where the Incursion phase is preceded by vitrinite-rich coal of Lycospore phase).

although there may be differences in the nature of the inertinite content.

The preceding analysis in this section shows the range of petrography associated with the different miospore phases. Both the Lycospore and Densospore phases are each more or less restricted to coals with a particular type of petrography, whereas coals containing Transition and Incursion phase assemblages are more variable in their petrography. A study of profiles shows that the petrography of a sample is determined to some extent by its position within the phase. The petrography of samples at the base and towards the top of the Transition phase tends to be similar to that of the adjacent phases. In contrast the junction between Lycospore and Incursion phases may be marked by an abrupt change, while the end of the Incursion phase may be preceded by a gradual change in the petrography.

VII. The Formation of Carboniferous Peats

A discussion of the palaeoecology of these peat swamps may usefully be preceded by a summary, under three headings, of the main facts that have emerged from this and other recent investigations on miospores in coal seams:

1. The existence of a petrographic and palynological sequence implies a progressive change in the environment which affected both the vegetation and the mode of its decomposition. The sequence, not always complete, occurred in many of the successive swamps, now represented by coal seams of the Carboniferous period. The gradual change between the Lycospore and Transition phases and the more abrupt change between the Transition and Densospore phases may imply different causative factors.

2. The association of distinctive spore assemblages with the two principal types of durain. The durains associated with the Densospore phase will be considered first. In this country they occur in coals of Upper and Lower Carboniferous age although their occurrence in the Lower Carboniferous is less well documented. In the Upper Carboniferous they have a discontinuous stratigraphical range, being recorded from a number of coals of the Lower and Middle Coal Measures (Westphalian A and B), but only from seams high in the sequence of the Upper Coal Measures, for instance in the Farrington and Radstock series of the Bristol and Somerset coalfield (Westphalian D). These durains are usually found in the middle and upper portions of the coal seams in which they occur, the lower portions being mainly bright coal (Hoffman, 1933). They usually persist more or less unchanged in thickness for considerable distances but are rarely found in thin coals. In France this type of durain occurs in coal of limnic origin (Alpern, 1959). The spores are usually

well preserved but the assemblage is restricted in species; the maceral accompanying the spores is massive micrinite. The amount of mineral sediment and pyrites identified microscopically is generally less in these durains than in those associated with the Incursion phase. The chemical data of Wandless and Macrae (1934, p. 6) on black and grey durains, which in general correspond to crassi- and tenuidurains respectively, confirm this distribution.

In contrast the durains associated with assemblages of the Incursion phase are found in most coals of Coal Measures age and can occur at any horizon in the seam, although they have not so far been found between durains associated with the Densospore phase. The extent of their lateral persistence is not generally known but in the Swallow Wood seam of Yorkshire, the band locally known as "Jabez" is of this type. The predominant maceral is semifusinite, often in association with mineral sediment. Material believed to be of fungal origin is characteristic (Stach and Pickhardt, 1957, p. 157). The assemblage of spores is rich in species but they are generally poorly preserved.

3. Results of regional studies of miospore distribution and frequency in the different coalfields of Britain. Seams, or groups of seams, deposited more or less contemporaneously (assumed from stratigraphical relationship to a marine band for instance) in the different coalfields, have been found to possess similar assemblages which may differ significantly from those of other seams above and below. Some common environmental factor must have been operating in areas as far apart as, for example, South Wales, Yorkshire and Durham.

Theories of peat formation in Carboniferous times have, until recent years, largely been the concern of geologists and have been incidental to the larger problems of cyclic sedimentation. Subsidence and water levels are, therefore, the factors usually considered although the parts played by isostatic and eustatic movements are uncertain.

Coal petrologists mostly accept the view that the type of peat was determined by the depth of water this being determined by purely geological factors. It is suggested here that these views do not satisfactorily explain the new evidence, and that the possible influence of climate on peat formation in Carboniferous times should be considered. There is a close parallel between the sequence of miospore phases demonstrated in certain coal seams and the general pattern of tree pollen distribution, induced by climatic changes, in the Pleistocene interglacial peats. The species towards the top and bottom of these are the same, and differ from those occurring in between. A knowledge of the plants concerned shows that this pattern reflects a temporary climatic amelioration, preceded and succeeded by a colder climate (Godwin, 1956, p. 298). Snyman (1961) has not demonstrated this type of pattern in the Gondwana coals, where it might have been expected, however, nor is there any evidence that temperature changes affected the formation of Carboniferous peats in the northern hemisphere. It is therefore necessary to consider other climatic factors, and in order to compare the various views of coal petrologists with those presented here, the following summary is given.

(a) The origin of petrographic types of coal

In general the views of coal petrologists on the environments associated with the formation of the different coal types are based on the state of preservation of the original plant tissues and their present chemical composition. Thus vitrain and clarain, which contain abundant evidence of their botanical origin, are generally considered by coal petrologists and coal chemists to have formed under anaerobic conditions with the minimum cover of surface water. Van Krevelen and Schuyer (1957, p. 66) even refer to "dry" conditions. The fine-grained micrinite and syngenetic pyrite which characterize the clarains formed from fern peats described by Teichmüller (1952, p. 611) are both considered to indicate a strong reducing environment. Teichmüller, however, suggests an environmental distinction between spore-rich and cuticle clarains formed subaquatically, and the spore-poor clarains resulting from wood peats. It is unlikely that the clarain lithotypes originated under saline conditions; according to Spackman (1958, p. 422) peats encountered today in marine and strongly saline water, whether under temperate, subtropical or tropical climates, are of an amorphous or finely fibrous type.

There is less agreement about the conditions under which durain formed and usually no distinction is made between the two petrographic types, and Teichmüller's work on brown coals (1950) does not provide any evidence, nor did she find anything resembling massive micrinite or a coal type analagous to crassidurain.

Several authorities regard durain as a subaquatic deposit but having an allochthonous origin (Raistrick and Marshall, 1939, p. 195; Hacquebard, 1943; Maurenbrecker, 1944). A distinction between allochthonous spore-rich durain formed under open water and autochthonous spore-poor durain formed under initially aerobic conditions is made by Teichmüller (1952, p. 608) and accepted by Mackowsky (1955). The present author, however (1957), considers the spore-rich durain (the crassidurain of the Densospore phase) to be autochthonous. Snyman (1961) also considers durains of South African coals to be autochthonous and of subaquatic origin.

Massive micrinite has been cited as evidence of the subaquatic origin of durain. Teichmüller (1952, p. 612) concluded from a study of coal balls that this maceral originated from decomposing organic muds and not from peat. The intimate association of both forms of micrinite with boghead algae impressed Schopf (1952, p. 318) as evidence that at least in some instances this maceral must imply extremely wet conditions. Mackowsky (1953) was unable to decide whether massive micrinite should be regarded as a substance of strictly anaerobic origin or as the end product of very short aerobic and intensive anaerobic disintegration. The same author (1955) indicated that the inertinite content, including micrinite, of subaquatic, spore-rich durain is derived hypauthochthonously from the durains formed in the relatively dry environment. The distinction between these two types of durain is considered to be one of particle size. The transition between crassidurain and cannel coal, which has been studied by the author in the Silkstone seam of Yorkshire (Fenton and Leighton, p. 893), might be taken as supporting evidence for the aquatic origin of this type of durain, as cannels are generally agreed to have formed as humic mud on lake floors. However, if cannel coal and crassidurain formed in similar environments it is strange that vitrinite should often be the main petrographic constituent of the former whilst massive micrinite should characterize the latter. A possible explanation has been suggested by van Krevelen and Schuyer (1957, p. 66) that the water over durain peat was less acid, but if this were so the lack of fish remains in this type of coal is surprising, as they are locally abundant in cannels.

Other workers attribute the origin of durain to an autochthonous deposit formed from peats subjected to a slight degree of drainage and partial access to air (Timofeyev, 1955, and other Russian workers). Stach (1955, p. 107) suggests from the infrequent occurrence of pyrite in crassidurain that the formation of this microlithotype took place essentially under aerobic conditions although this need not imply exposure to air for long periods. Such an environment is required by those workers who attribute a partially aerobic origin to massive micrinite. Karmasin (1952, p. 86) influenced by the work of Thomson, considers that massive micrinite results from peat formed in open, tree-less moors which, although normally wetter than forest moors, allowed the peat to be subject to brief exposures and aerobic conditions. Svoboda (1955) considers massive micrinite formed by an initial oxidation process in those parts of the seam where the original peat was associated with drier environmental conditions. On the basis of chemical studies Dormans *et al.* (1957, p. 338) suggest that the precursor of micrinite differed chemically from that of vitrinite only in having been subject to dehydrogenation.

The duroclarains and claroduraains are generally considered to have formed in environments which are intermediate in character between those which gave rise to the vitrainous and the durainous coal lithotypes.

The numerous authors who have contributed opinions as to

the origins of fusain are reviewed by Stutzer *et al.* (1929), Stach (1952b) and Francis (1961, p. 623). The generally held view is that the occurrence of fusinite and to a lesser extent semifusinite in coals is the direct result of forest fires. This theory has recently received fresh support at least for fusites occurring in the northern hemisphere, in the work of Terres *et al.* (1956) based on measurements of specific heats, and the work of Harris (1958, p. 452) which gives figures for the incidence of forest fires from natural causes. However, many authors while accepting the forest fire theory of origin, recognize that fusinite and semifusinite can also arise in other ways (Hacquebard, 1952b, p. 503; Marshall, 1955, p. 775; Stach, 1956; van Krevelen and Schuyer, 1957, p. 63).

(b) Possible influence of water levels

Little is known for certain about the conditions under which peat formation began in the Carboniferous period although it is generally accepted that it formed in fresh water swamps and land-locked coastal tracts at or near sea level. The initial stages of peat formation are linked with the problem of seat-earth formation. The evidence on the origin of Carboniferous seat-earths is considered by Huddle and Patterson (1961, p. 1645). There are two opposing theories concerning the prevailing soil environment. The first, accepted by Taylor and Warner (1960, p. 81) is that they formed in well-drained areas and represent a soil on which generations of plants grew before peat deposition began. This theory is not favoured by Huddle and Patterson who state that "most seat rocks formed in a water-logged environment and their development was generally followed immediately by the accumulation of peat. Long continued subaerial weathering in a well-drained area under stable conditions is not necessary for formation of all the kaolinite found in seat rock". On this view it can be assumed that the plants growing on the seat-earth were adapted for living in a semi-aquatic environment. Such an environment could arise in the basin of deposition when sedimentation had reduced the depth of water sufficiently to permit the colonization by plants. One might expect, therefore, a distinctive spore assemblage in the seat-earths and in the coals at the base of seams derived from the semi-aquatic vegetation. For instance, coal formed from reed peat, or more often peat rich in wood remains and containing abundant pollen of *Taxodiaceae* (swamp cypress), is found above the clay gyttja (organic mud) at the base of the Tertiary brown coal deposit of the Rhineland. There is no evidence, however, from the present investigation or from present experience to show the existence of a distinctive assemblage of spores or of peat type at this stage in the formation of Carboniferous coal seams. The bright coals of the Lycospore phase at the base of seams do not differ in petrological or palynological composition from bright coals which occur at other horizons in the seams. It is possible that the arborescent Lycopods were more adaptable than some modern plants to a large range of water depth. An examination of the spore assemblages in seat-earths is at present being undertaken for further evidence on this stage of seam formation.

If we accept the view that the peat at the base of seams formed in shallow water, some inches only in depth, the sequence culminating in the Densospore phase could be due to progressive changes in the water levels with respect to the peat surface, although there is no clear evidence of the direction of the change. If the sequence corresponded to an increasing depth of water cover, the Densospore phase would represent an environment like that of the present day forests of swamp cypress. This would accord with the many authorities who believe that durain is a subaquatic deposit.

There are, however, difficulties in accepting the view that crassidurain was formed in this way. First, there is no reason why a depth of water favourable to the vegetation of the Densospore phase should not have occurred somewhere and at some time during the Westphalian C in Britain which is conspicuously low in crassidurain. Second, plants in general are very sensitive to their environment and can only tolerate small changes in the situation in which they are growing. Vegetation adapted for life in a semi-aquatic environment will require a certain specified depth of water to flourish. Assuming that the vegetation of the Densospore phase required this condition, it is difficult to imagine that the depth of water could have remained within the required limits over areas equal in extent to the known deposits of crassidurain. This difficulty can, to some extent, be met by postulating that conditions associated with the Densospore phase were confined to a belt which gradually transgressed the existing vegetation as the depth of water became suitable during a period of subsidence. However, it is difficult to accept this explanation since the Densospore phase does not occur in all seams and does not often occur in those positions in seams where subsidence might be thought to have taken place, namely preceding dirt partings and at the top of seams.

For these reasons it is difficult to accept the subaquatic theory of origin of crassidurain. It is equally difficult to imagine that the sequence of phases was the result of a gradual drying out of the bog surface as postulated by Boddy (1938, p. 123) for the origin of durain in general. The beds of crassidurain are sometimes more than a foot in thickness. The formation of a considerably greater thickness of peat is considered unlikely to have taken place in a relatively dry environment.

Let us now consider the durains usually associated with part of the Incursion phase, which essentially occurs at random in the seam profile and contains the durains enriched in fusinite and containing variable quantities of mineral sediment. If these durains were the result of a drying out of the swamp surface, this would probably bring about changes of miospore assemblage preceding and succeeding the event, but there is no evidence of this.

On the other hand a flooding of the bog surface by oxygenated water could satisfactorily account for the observed facts. Flooding would occur at irregular intervals, the flood waters carrying mud, charred wood and spores in suspension and providing environments in which the processes of decay were largely aerobic. The rise in water level might also prove fatal to some part or all of the existing vegetation in the affected area. Thus the profile of the Hazel seam shows that where the Transition phase was interrupted by the Incursion phase *Lycospora granulata* and *Dictyotriletes bireticulatus* show a reduction in numbers, as presumably did their parent plants; in contrast the numbers of *Laevigatosporites desmoinensis* remain at about the same level of abundance. This partial elimination of species might permit colonization by new plants or exploitation of the area by plants already present in the vegetation and capable of spreading rapidly. The latter supposition is the more likely since the two key miospore species of the Incursion phase, *Crassispora kosankei* and *Punctatosporites minutus*, occur in small numbers in other phases. The alteration of the water level may only have been of comparatively short duration but in some cases the effect on the vegetation was probably more lasting judging by the thickness of coal associated with the phase. In some instances the Incursion phase may have had a more profound effect on the vegetation and the subsequent history of the deposit in that the Incursion phase in some instances follows the Densospore before the Transition phase appears.

The semifusinite and fusinite of the Incursion phase might well have been formed by forest fires. It is not considered that the phase owes its occurrence entirely to forest fires sweeping through the swamp vegetation, since although their effects might have been sufficiently devastating to permit the establishment of a temporary vegetation comprised of relatively fast growing species, they would not account for the weathered state of the spore exines. It would also be difficult to attribute the occurrence of mineral sediment and massive micrinite to forest fires. Massive micrinite, although not the dominant form of inertinite in these coals, does occur in relatively large amounts and its origin is more satisfactorily explained by the decay of plant debris in oxygenated flood waters.

Much of the semifusinite is considered to have been derived, like the mineral sediment, from outside the margins of the swamp, as postulated by Kendal (1923, p. 7). However, it is difficult to account for the widespread occurrence of semifusinite if it was entirely allochthonous. If the vegetation growing near the point of entry of the semifusinite was dense it might be expected to act as an effective filter. On the other hand fires, sweeping periodically through the vegetation of the swamp might provide a regular supply of charcoal, which was mechanically eroded by the rising flood waters. This process of attrition need not necessarily have persisted throughout the accumulation of the miospore assemblages of the Incursion phase, which in some instances was protracted, because the petrography of successive samples within this phase show that high fusite values occur only intermittently (Fig. 2).

If the Incursion phase is in fact symptomatic of an increase in the volume of water entering the swamp from outside its margin, there is a strong possibility that some of the characteristic spores of the phase may be allochthonous. Muller (1959, p. 9) has shown that water can be an important agent in disseminating spore exines and that currents can counter the effect of the prevailing wind in introducing species to the area of deposition. It is perhaps relevant that among the species that attain their greatest numbers in the Incursion phase are three of *Florinites* and one of *Pityosporites*. Neves (1958) has shown that the former form a much larger proportion of the assemblage in marine beds of the Millstone Grit and Lower Coal Measures than they do in coal seams of the same age; he deduces from this that *Florinites* perhaps represents an element in the marginal swamp flora, but Chaloner (1958c) suggests a more probable origin from an upland flora. The pollen grains of *Florinites* in coal of the Incursion phase may, therefore, very well be allochthonous, having been carried by water to the area of deposition. It is interesting to note that the pollen of *Pinus*, which somewhat resembles that of *Pityosporites* and *Florinites*, is more resistant to decomposition than that of certain other types (Sangster and Dale, 1961).

The picture that emerges is of a mixed vegetation composed of arborescent and herbaceous forms. Some part of the Incursion phase assemblages is composed of species characterizing either the Lycospore or the Transition phase depending on which of these phases precedes or follows. The proportion in which elements of these phases persist in the Incursion phase assemblages depends on the extent to which they survived the changes taking place in the environment.

The high incidence of material considered to represent the living and resting stages of fungal organisms in the Incursion phase is worth noting and may provide further evidence of an allochthonous element in this phase. Muller (1959, p. 19) records that spores and hyphae of fungi are extremely abundant in most delta deposits, having been transported by the rivers following their liberation from the back-swamp soils during heavy rainfall or by erosion of the river banks.

Although fluctuating water levels are considered to be a major factor in determining the occurrence of the Incursion phase, they are not thought to be the primary cause of the sequence of phases, beginning with the Lycospore and terminating with the Densospore phase, but they may have a secondary influence which is discussed below.

(c) Possible influence of climate

It is desirable to preface this theme by seeing first what conclusions recent authors have reached regarding climate and climatic changes in the Carboniferous period and second, the effect of climate on the formation and development of certain geologically recent peats. Geological and palaeobotanical evidence of past climates is summarized by Nairn (1961), and Blackett (1961), in a discussion of ancient climates and rock magnetism, has provided further support for the view that the coals of North America and Europe were deposited in or near tropical latitudes in Permo-Carboniferous times. Schwarzbach (1961) in considering the evidence of the Carboniferous flora points out that thick coal formation is not valid climatic evidence for the determination of temperatures since peat forms at the present time in temperate as well as tropical latitudes; it does however indicate high humidity. Particularly relevant to the present discussion is his statement that "the vast extent of the European and North American coal regions shows that one must not think of 'oases' with locally high water levels, but instead a really humid climate (particularly in the Upper Carboniferous) must be presumed"

It is often assumed that the climate in the northern hemisphere in the Carboniferous period was fairly uniform. Whether or not one accepts the view of Schwarzbach (1961, p. 263) that the absence of growth rings in Carboniferous woods may only be due to the primitive state of development of the plants, the important question for the present discussion concerns the existence of relatively long term climatic rhythms. Direct evidence is lacking but climatic cycles have been postulated to account for cyclic deposition in Carboniferous times (Wanless and Shepherd, 1936; Wheeler and Murray, 1957). The views of these authors are not universally accepted, but Schwarzbach (1961) has pointed out that a less marked climatic rhythm might accompany the sedimentary rhythm arising from tectonic causes because of the influence of the changing tectonic-orographical conditions on the climate.

Hollingworth (1962) favours the "glacial control" hypothesis originally put forward by Wanless and Shepherd. He states "there can be little doubt that cyclic sedimentation associated with oscillation of sea-level reached its highest degree of perfection at a time broadly coincident with that of the Gondwanaland glaciations. It now appears to be widely accepted that the time-span of the glaciations in the southern continents ranges from the Lower Carboniferous into the Lower and probably the Middle Permian and that during this time . . . there were many phases of glaciation". Thus it is not unreasonable to believe that climatic changes of sufficient magnitude to cause the growth and decay of these ice-sheets would be world-wide in their effect and might bring about repeated fluctuations in temperature and humidity even in tropical latitudes.

At the present time peat is being formed under a variety of climates over areas comparable in size to the Carboniferous peat deposits. For instance, vast areas of tundra are covered with muskeg (Sjörs, 1961) and in warmer climates thick peat deposits are forming in suitable environments such as the well-known Dismal Swamp of Virginia and North and South Carolina, estimated by Osbon (1919) at originally 2200 square miles. Other examples are referred to later.

It is generally known that peat will form wherever the water level is permanently or for long periods at, or just above, the surface of the ground irrespective of climate, but it is less often appreciated that peat which begins to form under such conditions may, if the climate is wet and the atmosphere very humid, continue to accumulate to levels well above the original ground surface. Raised bogs (the German Hochmoor) are of this type, deriving their name from their slight convex surface. The raised bogs of temperate climates have been studied in detail by plant ecologists and their development is known to be the result of a succession of vegetation cycles. The changes in the vegetation and the peat type are largely due to the control of the edaphic conditions by the plants themselves, although the rate of peat formation depends on the climate.

The extensive peat swamps of Malaya may be of this type and they lie at low altitudes, mainly near the coast and always in regions of high rainfall with no well-marked dry season. Polak (1933) described similar tree-covered bogs from Sumatra, and recently Richards[1] gave an account of those of North West Borneo:

> "The peat swamps have a convex surface and are analogous to the raised bogs of temperate climates. The plant communities form more or less concentric zones and are dominated by trees. In some zones the Dipterocarp *Shorea albida* is dominant to the exclusion of almost all other tall-growing species. The zonal communities form a series with respect to number of species (which diminishes towards the centre), abundance of *Nepenthes* and other features.
> "A boring by Muller and Anderson at Marudi in Sarawak showed that the peat may be as much as 13 metres thick and the pollen profile suggests a succession beginning with mangrove and ending with the stunted *Combretocarpus-Dactylocladus* community."

[1] In report of meeting of the British Ecological Society. *J. Ecol.*, 1960, **48**, p. 764.

The sequence of peat types referred to invites comparison with the sequence of phases found in the present investigation. It is also interesting to note that these tropical peats are tree-covered as were those of the Carboniferous period.

Raised bogs are usually of limited areal extent although a number of such bogs may form in close proximity. However, if precipitation and atmospheric humidity are very high peat formation may take place almost independently of the surface topography. The "blanket" bogs of oceanic cool temperate climates are of this type. Blanket bogs, as their name implies, form an extensive and thick peat layer covering the general surface of the country except for the steeper slopes and rocky outcrops; small pools or considerable areas of water may occur within the confines of the bog. This type of peat occurs in Western Ireland over extensive areas of West Galway and Mayo; from the continuous layers of pine stumps which occur at the base of these Irish peats in many parts it is probable that the soil was at one time relatively dry and that peat formation began when the climate became sufficiently wet. Both raised bogs and blanket bogs are acid and poor in plant food.

It is worth noting that changes in climate which are known to have occurred during the development of these temperate peat deposits can be correlated with changes in physical appearance. The "Grenzhorizont" marked by a sudden change in the colour and texture of the peat has been recorded from many bogs in north-west Europe and is considered to mark the onset of increased precipitation (Godwin, 1956, p. 33). There is thus the possibility that changes in the character of the peat of older geological deposits may be due to climatic factors. Although this has not been demonstrated, for instance, in the case of the Rhineland brown coal, it is not clear how far the changes recorded in the vegetation and peat types were the result of climatic or edaphic conditions.

In considering the relationship between climate and peat formation in Carboniferous times there is at present no evidence to show what part climate played in the initial stages of peat formation. This difficulty arises in part from uncertainty about the edaphic conditions associated with the formation of the seat-earths. If at that time the water table was below the soil surface an increase in precipitation might easily have led to water-logging and peat formation. However, it seems more reasonable to suppose that the primary cause of peat formation was the existence of a semi-aquatic environment as previously discussed. It is suggested that after the initial phase, peat accumulation took place primarily as a result of very humid conditions.

The sequence of phases is interpreted either as the succession of vegetation types resulting from the formation of a peat structure analogous to the raised bogs or as a result of climatic changes taking place during the history of the deposit. In some instances both factors may have been operative.

The first hypothesis requires a relatively humid climate throughout the period of peat formation. The sequence of phases follows the general rise in the bog surface and is associated with changing edaphic conditions in the direction of increasing acidity and impoverishment of plant foods. The culmination of these changes is marked by the occurrence of the Densospore phase which therefore indicates a specialized environment generally requiring the prior accumulation of a considerable thickness of peat. This may explain the occurrence of this phase, usually at only one level, in the middle or upper portions of the thicker coal seams. The phase does not imply any drying out of the bog surface, but because the surface was not actually covered by water, plant decay took place mainly under aerobic conditions, although subsequently decomposition was anaerobic. These are the conditions suggested by Mackowsky (1953) for the formation of massive micrinite.

The repeated alternation of Lycospore and Transition phases, which have been described from some seams, and the changes taking place after the Densospore phase, suggest that the succession was deflected, often only temporarily, by flooding of the bog surface either following a period of exceptionally heavy rain (the presence of the Incursion phase is evidence of this) or by an increase in the rate of subsidence. The reversion to the Lycospore phase following the Densospore and Transition phases points to a return to the more favourable conditions for plant growth prevailing at an early stage of peat formation. The fact that the Densospore phase was often terminated by the Incursion phase was possibly due to the influx of oxygenated water rich in plant foods, so altering the edaphic conditions that the vegetation of the Densospore phase was no longer able to survive in competition with species able to grow in a less extreme environment. The chief objection to this hypothesis is that peat development would tend to occur at different rates depending on local conditions thus leading to deposits with individual beds of limited extent and variable thickness whereas beds of crassidurain often persist more or less unchanged in thickness over wide areas.

The second hypothesis, which relates the sequence of phases to climatic changes, attributes the formation of crassidurain to an increase in precipitation and atmospheric humidity ultimately resulting in a deposit of the blanket bog type. Temperature changes were possibly also involved.

On this hypothesis it is assumed that the petrography of the crassidurain is attributable indirectly to the climate. The factors affecting the processes of decomposition and the ultimate petrography of the deposit are complex; climate, vegetation and edaphic conditions all play a part but climate has an important influence over the other factors. It is also possible that the effects of climate on the vegetation must be considered in relation to other factors such as the level of the ground water. For instance, if the initial change from Lycospore to Transition phase was induced by a slight change in climate the subsequent repeated alternation with the Lycospore phase, already mentioned, may be due to the influence of ground water levels rather than to frequent changes of climate. Similarly the changes following the Densospore phase, when this is near the roof of the seam, are likely to be due to a rise in ground water level due to subsidence, rather than to a return to the climate prevailing at an early stage in the development of the deposit.

This second hypothesis provides a possible explanation for the absence of crassidurain in coals of Westphalian C age in Britain, namely the lack of a climate favouring the establishment of the vegetation of the Densospore phase. The incidence of crassidurain is not so common that frequent changes in climate need be postulated. Not all seams contain horizons of crassidurain but those that do appear from the spore evidence to occur at approximately the same stratigraphical horizons in the different coalfields. As a corollary to this theory it is suggested that where a land surface became exposed during that period when the climate favoured the growth of the vegetation of the Densospore phase, crassidurain would be found at the base of the resulting seam providing other factors were suitable for its formation.

It is possible that both hypotheses account for the formation of seams containing horizons of crassidurain. In this case two conditions must have been satisfied before the vegetation of the Densospore phase became established, namely, the suitability of the edaphic environment, determined by the state of bog development, and the climate.

VIII. Conclusions

In the preceding discussion an attempt has been made to establish certain facts concerning the edaphic conditions and the nature of the vegetation associated with areas of peat deposition in the Westphalian in Britain. These facts are now summarized.

The miospores provide evidence for the existence of four distinct vegetation types or plant "associations" corresponding to the four miospore phases recognized. The coals associated with each phase are generally petrographically distinct. The peats corresponding to these petrographic types are considered to have been subject to different degrees of aerobic decomposition.

The initial stage of peat formation is assumed to have taken place under a shallow covering of more or less stagnant water in which decomposition was anaerobic. Such conditions might have persisted for unlimited periods providing that the balance between peat formation and subsidence was maintained. The vegetation at this stage was forest and the miospore assemblages belong to those of the Lycospore phase.

The progressive withdrawal of the ground water cover, either as a result of eustatic changes in sea-level or by the elevation of the bog surface through the accumulation of peat in a humid climate, resulted in an increasing tendency for decomposition to be aerobic at least in the initial stages. The changes in the edaphic conditions brought about by one or other or both of these factors, together with possible changes in climate, resulted in a gradual replacement of the forest by a more open vegetation, richer in species. The miospore assemblages of the Transition phase belong here.

There is a reasonable possibility that coals assigned to the Densospore phase may be attributable to climatic factors such as high precipitation and humidity and not to increasing depth of water cover. Changes of temperature may also be involved. It is not clear how far the vegetation was also dependent on edaphic factors, for instance absence of surface water and high acidity. Decomposition is assumed to have been initially aerobic; the processes of decay proceeded further than in other environments, only spore exines retaining their identity. The vegetation of the phase comprises very few species.

A return to climatic conditions prevailing at an earlier stage might have brought about a change in the vegetation without influencing the mode of decomposition. This would give rise to assemblages of the Transition phase in association with massive micrinite (tenuidurain). The increase in the amount of vitrinite in the profiles following a Densospore phase is most easily ascribed to the presence of ground water cover brought about by subsidence.

The course of the succession just outlined was often interrupted or deflected by flood waters carrying varying amounts of mineral sediment and semifusinite in suspension. These flood waters had a more or less catastrophic effect on the existing vegetation. The occurrence of this phase is marked not only by a distinctive petrography but by a miospore assemblage particularly rich in species.

IX. Acknowledgements

I wish to thank Mr. H. W. Pearson for reading this paper and for his helpful criticism, Dr. M. Teichmüller for information on the Rhineland brown coals, Mr. J. W. Fowler for considerable help in the preparation and polishing of the coal samples, and the National Coal Board for permission to publish this paper.

X. REFERENCES

ALPERN, B., 1959. Contribution à l'étude palynologique et pétrographique des charbons français. 314 pp. Thesis privately published, Paris.

ANDREWS, H. N. and S. H. MAMAY, 1951. A new American species of *Bowmanites*. *Bot. Gaz.*, **113**, 158-65.

—— and E. PANNELL, 1942. Contributions to our knowledge of American Carboniferous floras, II. *Lepidocarpon*. *Ann.Mo. Bot. Gdn.*, **29**, 19-35.

BALME, B. E., 1956. Inorganic sulphur in some Australian coals. *J. Inst. Fuel.*, **29**, 21-2.

BENSON, M., 1904. *Telangium Scotti*, a new species of *Telangium* (*Calymmatotheca*) showing structure. *Ann. Bot. Lond.*, **18**, 161-77.

BHARADWAJ, D. C., 1958. On *Porostrobus zeilleri* Nathorst and its spores with remarks on the systematic position of *P. bennholdi* Bode and the phylogeny of *Densosporites* Berry. *Palaeobot. India*, **7**, 67-75.

BLACKETT, P. M. S., 1961. Comparison of ancient climates with the ancient latitudes deduced from rock magnetic measurements. *Proc. Roy. Soc.* (A), **263**, 1-30.

BODDY, R. G. H. B., 1938. The mode of deposition of coal seams. A microscopic study. *Trans. Instn. Min. Engrs. Lond.*, **96**, 100-8, 121-3.

BOUGNÈRES, L. and W. REMY, 1957. Mitteilung über die Sporen von *Zygopteris*. *Abh. Dtsch. Akad. Wiss. Berl.*, **4**, 1-4.

CHALONER, W. G., 1951. On *Spencerisporites* gen. nov., and *S. karczewskii* (Zerndt), the isolated spores of *Spencerites insignis* Scott. *Ann. Mag. Nat. Hist.*, **4**, 861-73.

—— 1953a. On the megaspores of *Sigillaria*. *Ann. Mag. Nat. Hist.*, **6**, 881-97.

—— 1953b. On the megaspores of four species of *Lepidostrobus*. *Ann. Bot. Lond.*, **17**, 263-93.

—— 1954. Notes on the spores of two British Carboniferous lycopods. *Ann. Mag. Nat. Hist.*, **7**, 81-91.

—— 1958a. A Carboniferous *Selaginellites* with *Densosporites* microspores. *Palaeontology*, **1**, 245-53.

—— 1958b. *Polysporia mirabilis* Newberry, a fossil lycopod cone. *J. Paleont.*, **32**, 199-209.

—— 1958c. The Carboniferous upland flora. *Geol. Mag.*, **95**, 261-2.

CROSS, A. T., and M. P. SCHEMEL, 1952. Representative microfossil floras of some Appalachian coals. *C. R. Congr. Avanc. Ét. Stratigr. Carbonif.*, Heerlen, Vol. 1, 123-30.

DORMANS, H. N. M., F. G. HUNTJENS and D. W. VAN KREVELEN, 1957. Chemical structure and properties of coal. XX. Composition of individual macerals. *Fuel Lond.*, **36**, 321-39.

DYBOVA, S. and A. JACHOWICZ, 1957. Microspores of the Upper Silesian Coal Measures. *Prace Inst. Geol.*, Vol. 23, 328 pp., Warsaw.

EDWARDS, W., J. J. WALKER and A. M. WANDLESS, 1938. A new interpretation of the Pollington boring, Yorkshire. *Trans. Instn. Min. Engrs. Lond.*, **95**, 147-72, 321-4.

FAEGRI, K. and J. IVERSEN, 1950. *Text-book of modern pollen analysis*. 168 pp., Ejnav Munksgaard, Copenhagen.

FELIX, C. J., 1954. Some American arborescent lycopod fructifications. *Ann. Mo. Bot. Gdn.*, **41**, 351-94.

FENTON, G. W. and L. H. LEIGHTON, 1956. The coal survey in Yorkshire—a review. *Trans. Instn. Min. Engrs. Lond.*, **116**, 887-922.

FLORIN, R., 1936. On the structure of the pollen-grains in the Cordaitales. *Svensk. Bot. Tidskr.*, **30**, 624-51.

—— 1938-40. Die Konifern des Oberkarbons und des Unteren Perms. *Palaeontographica*, **85**, 1-263.

FRANCIS, W., 1961. *Coal, its formation and composition*. 2nd edit., 623 pp., Edward Arnold, London.

GLAGOLEV, A. A., 1934. Quantitative analysis with the microscope by the "Point" method. *Engng. Min. J.*, **135**, 399-400.

GODWIN, H., 1956. *The history of the British Flora.* 383 pp., Cambridge.

GREBE, H., 1953. Beziehungen zwischen Fusitlagen und Pollenführung in der Rheinischen Braunkohle. *Paläont. Z.*, **27**, 12-5.

GUENNEL, G. K., 1952. Fossil spores of the Alleghenian coals in Indiana. *Rep. Ind. Dep. Geol.*, No. 4, 40 pp.

HACQUEBARD, P. A., 1943. Kohlenpetrographische Studiën. *Meded. Geol. Sticht.* Serie C-III-2, **1**, 1-129.

—— 1952a. A petrographic investigation of the Tracey seam of the Sydney coalfield, Nova Scotia. *2nd Conf. Origin and Const. Coal,* Nova Scotia, 298-318.

—— 1952b. Opaque matter in coal. *Econ. Geol.*, **47**, 484-516.

HARRIS, T. M., 1958. Forest fire in the Mesozoic. *J. Ecol.*, **46**, 447-53.

HARTUNG, W., 1933. Die Sporenverhältnisse der Calamariaceen. *Arb. Inst. Paläobot. Berl.*, **3**, 95-149.

HOFFMAN, E., 1933. Neue Erkenntnisse über die Vorgange der Flözbildung. *Bergbau*, **46**, 89-94.

HOLLINGWORTH, S. E., 1962. The climatic factor in the geological record. *Quart. J. Geol. Soc. Lond.*, **118**, 1-21.

HUDDLE, J. W. and S. H. PATTERSON, 1961. Origin of Pennsylvanian underclay and related seat rocks. *Bull. Geol. Soc. Amer.*, **72**, 1643-60.

INTERNATIONAL COMMITTEE FOR COAL PETROLOGY, 1957. *Glossary of terms,* C.R.N.S., Paris.

KARMASIN, R. W., 1952. Deutung des Fazieswechsels in den Flözen Erda und Ägir auf Grund mikropetrographischer Schlitzprobenuntersuchungen. *Bergb-Arch.*, **13**, 74-99.

KENDALL, P. F., 1923. The physiography of the coal swamps. *Rep. Brit. Ass.* (for 1922), 49-78.

KIDSTON, R., 1923-25. Fossil plants of the Carboniferous rocks of Great Britain. *Mem. Geol. Surv. Gt. Brit. Palaeont.* Vol. 2, 670 pp.

KNOX, E. M., 1938. The spores of Pteridophyta with observations on microspores in coals of Carboniferous age. *Trans. Bot. Soc. Edinb.*, **32**, 438-66.

—— 1942. The microspores in some coals of the Productive Coal Measures in Fife. *Trans. Instn. Min. Engrs. Lond.*, **101**, 98-112.

KOSANKE, R. M., 1950. Pennsylvanian spores of Illinois and their use in correlation. *Bull. Ill. Geol. Surv.*, **74**, 1-126.

—— 1954. Correlation of coals and spore analysis. *Bull. Missouri School Mines*, Tech. Ser., No. 85, 11-6.

KREMP, G. O. W., 1952. Sporen-Vergesellschaftungen und Mikrofaunen-Horizonte in Ruhrkarbon. *C. R. Congr. Avanc. Ét. Stratigr. Carbonif.*, Heerlen, Vol. 1, 347-57.

—— and N. O. FREDERIKSEN, 1960. Recognition of coal lithotypes by palynological methods. *Proc. Int. Cttee. Coal Petrology*, No. 3, 75-83.

MACKOWSKY, M. T., 1953. Probleme der Inkohlung. *BrennstChemie*, **34**, 182-5.

—— 1955. Der Sedimentions Rhythmus der Kohlenflöze. *Neues Jb. Geol., Min. Pal.* Monatshafte (B), Heft 10, 438-49.

MAMAY, H., 1950. Some American Carboniferous fern fructifications. *Ann. Mo. Bot. Gdn.*, **37**, 409-76.

MARSHALL, C. E., 1955. Coal Petrology. *Econ. Geol.*, 50th Anniv. Vol., 757-834.

MAURENBRECKER, A. L. F. J., 1944. Kohlenpetrographische Studiën. *Meded. Geol. Sticht.* Serie C-III-2, 1-108.

MILLOTT, J. O'N., 1939. The microspores in the coal seams of north Staffordshire Pt. I—The Millstone Grit—Ten Foot coals. *Trans. Instn. Min. Engrs. Lond.*, **96**, 317-53.

MOORE, L. R., 1946. On the spores of some Carboniferous plants; their development. *Quart. J. Geol. Soc. Lond.*, **102**, 251-98.

MULLER, J., 1959. Palynology of recent Orinoco delta and shelf sediments. *Micropalaeontology*, **5**, 1-32.

NAIRN, A. E. M., 1961. *Descriptive palaeoclimatology.* 380 pp., Inter. Sc. Publ. Ltd., London.

NEVES, R., 1958. Upper Carboniferous plant spore assemblages from the *Gastrioceras subcrenatum* horizon, north Staffordshire. *Geol. Mag.*, **95**, 1-19.

OSBON, C. C., 1919. Peat in the Dismal Swamp, Virginia and North Carolina. *Bull. U.S. Geol. Surv.*, 711—C, pp. 41-59.

POLAK, E., 1933. Uber Torf und Moor in Niederlandisch Indien. *Verh. Akad. Wet. Amst.*, **30**, 84 pp.

POTONIÉ, R., 1952. Die Bedeutung der Sporomorphen für die Gesellschaftsgeschichte. *C. R. Congr. Avanc. Ét. Stratigr. Carbonif.*, Heerlen, Vol. 2, 501-6.

—— and G. KREMP, 1955-56. Die Sporae dispersae des Ruhrkarbons. Ihre Morphographie, und Stratigraphie mit Ausblicken auf Arten anderer Gebiete und Zeitabschnitte. *Palaeontographica*, Abt. B, 1955, Teil I, **98**, 1-136; 1956, Teil II, **99**, 95-191; 1956, Teil III, **100**, 65-121.

RADFORTH, N. W., 1938. An analysis and comparison of the structural features of *Dactylotheca plumosa* Artis sp. and *Senftenbergia ophiodermatica* Göppert sp. *Trans. Roy. Soc. Edinb.*, **59**, 385-96.

RAISTRICK, A., 1934. The correlation of coal seams by microspore content. I. The seams of Northumberland. *Trans. Instn. Min. Engrs. Lond.*, **88**, 142-53.

—— and C. E. MARSHALL, 1939. *The nature and origin of coal and coal seams.* 282 pp., English Univ. Press.

REED, F. D., 1938. Notes on some plant remains from the Carboniferous of Illinois. *Bot. Gaz.*, **100**, 324-35.

REMY, R., 1960. *Bowmanites nindeli* Remy. *Monatsber. Dtsch. Akad. Wiss. Berl.* (2), **2**, 122-4.

REMY, W. and R. REMY, 1955. Mitteilungen über Sporen, die aus inkohlten Fruktifikationen von echten Farnen des Karbon gewonnen wurden. *Abh. Dtsch. Akad. Wiss. Berl.*, **1**, 41-7.

—— 1956. *Noeggerathiostrobus vicinalis* E. Weiss und Bemerkungen zu ähnlichen Fruktifikationen. *Abh. Dtsch. Akad. Wiss. Berl.*, **2**, 1-11.

—— 1957. Durch Mazeration fertiler Farne des Paläozoikums gewonnene Sporen. *Paläont. Z.*, **31**, 55-65.

SANGSTER, A. G. and H. M. DALE, 1961. A preliminary study of differential pollen grain preservation. *Canad. J. Bot.*, **39**, 35-43.

SCHOPF, J. M., 1952. In discussion of Hacquebard, 1952a, *q.v.*

SCHWARZBACH, M., 1961. The climatic history of Europe and N. America. pp. 255-91 in Nairn, 1961, *q.v.*

SJÖRS, H., 1961. Surface patterns in boreal peatland. *Endeavour*, **20**, 217-24.

SLATER, L., M. M. EVANS and E. EDDY, 1930. The significance of spores in the correlation of coal seams, Pt. I. The Parkgate seam—south Yorkshire area. *Fuel Research Board, Survey paper*, No. 17, pp. 28.

SMITH, A. H. V., 1957. The sequence of microspore assemblages associated with the occurrence of crassidurite in coal seams of Yorkshire. *Geol. Mag.*, **94**, 345-63.

—— 1958. The methods and application of coal microscopy. *J. Quekett Micr. Cl.*, **5**, 76-94, 131-43.

SNYMAN, C. P., 1961. *Der Petrographie Südafrikanischer Gondwanakohlen. Ein Vergleich zwischen Gondwana- und Euramerischen Kohlen.* 137 pp., Bonn.

SOMERS, G., 1952. *Fossil spore content of the Lower Jubilee seam.* 30 pp., Nova Scotia Res. Foundation.

SPACKMAN, W., 1958. The maceral concept and the study of modern environments as a means of understanding the nature of coal. *Trans. N.Y. Acad. Sci.* (2), **20**, 411-23.

Stach, E., 1952a. Die Vitrit-Durit Mischungen in der petrographischen Kohlenanalyse. *BrennstChemie*, **33**, 361-70.
— 1952b. Geschichte der Kohlenmikroskopie in "Mikroskopie in der Technik", Edit. Hugo Freund, 2, 1-64, Umschau Verlag, Frankfurt a.M.
— 1955. Crassidurain—a means of seam correlation in the Carboniferous coal measures of the Ruhr. *Fuel*, **34**, 95-118.
— 1956. La sclerotinite et son importance pour l'origine de la durite. *Ann. Min. Belg.*, 1re livraison, 73-89.
— and W. Pickhardt, 1957. Pilzreste (Sklerotinit) in paläozoischen Steinkohlen. *Paläont. Z.*, **31**, 139-62.
Stutzer, O., K. A. Jurasky, A. Duparque and Th. Lange, 1929. Fusit, Vorkommen, Entstehung und praktische Bedeutung der Faserkohle. *Schr. Brennstoffgeol.*, 139 pp.
Svoboda, J., 1955. Mikrinit a jeho genese. *Uhli*, **5**, 211-3.
Taylor, G. H. and S. St. J. Warne, 1960. Some Australian coal petrological studies and their geological implications. *Proc. Int. Cttee. Coal Petrology*, No. 3, 75-83.
Teichmüller, M., 1950. Zum petrographischen Aufbau und Werdegang der Weichbraunkohle. *Geol. Jb.*; **64**, 429-88.
— 1952. Vergleichende mikroskopische Untersuchungen versteinerten Torfe des Ruhrkarbons und der daraus entstandenen Steinkohlen. *C. R. Congr. Avanc. Ét. Stratigr. Carbonif.*, Heerlen, Vol. 2, 607-13.
Terres, E., H. Dahne, B. Nandi, C. Scheidel and K. Trappe, 1956. Die Entscheidung der Frage der Entstehung von Faserkohle auf Grund ihrer spezifischen Wärmen. *BrennstChemie*, **37**, 269-77, 342-7, 366-70.
Thomson, P. W., 1950. Grundsätzliches zur tertiären Pollen- und Sporenmikrostratigraphie auf Grund einer Untersuchung des Hauptflözes der rheinischen Braunkohle in Liblar, Neurath, Fortuna und Brühl. *Geol. Jb.*, **65**, 113-26.
Timofeyev, P. P., 1955. Conditions of formation of genetic types of coals and their connection with cycles of sedimentational environment in the Donbass. *Doklady Akademii Nauk S.S.S.R.*, **102**, 809-12 (in Russian).
Tomlinson, R. C., 1957. Coal Measures microspore analysis: a statistical investigation into sampling procedures and some other factors. *Bull. Geol. Surv. G.B.*, No. 12, 18-26.
Van Krevelen, D. W. and J. Schuyer, 1957. *Coal Science*, 352 pp., Elsevier, Amsterdam.
Wandless, A. M. and J. C. Macrae, 1934. The banded constituents of coal. *Fuel*, **13**, 4-15.
Wanless, H. R. and F. P. Shepherd, 1936. Sea level and climatic changes related to late Paleozoic cycles. *Bull. Geol. Soc. Amer.*, **47**, 1177-206.
Wheeler, H. E. and H. H. Murray, 1957. Base-level control patterns in cyclothemic sedimentation. *Bull. Amer. Assoc. Petrol. Geol.*, **41**, 1985-2011.
Williams, R. W., 1955. *Pityosporites westphalensis*, sp. nov. an Abietineous type pollen grain from the Coal Measures of Britain. *Ann. Mag. Nat. Hist.*, **8**, 465-73.
Wilson, L. R., 1961. Palynology as a tool in economic Geology. *Micropaleontology*, **7**, 472-4.
— and W. S. Hoffmeister, 1956. Plant microfossils of the Croweburg coal. *Circ. Okla. Geol. Surv.*, **32**, 58 pp.
Zetsche, F. and O. Kälin, 1932. Eine Methode zur Isolierung des Polymerbitumens (Sporenmembranen, Kutikulen, usw.) aus Kohlen. *Braunkohle*, **31**, 345-63.

[*Editors' Note:* Appendix A on the source of coal samples; Appendix B, petrographic terms of the Stopes—Heerlen nomenclature; Appendix C on the occurrence and abundance of miospores in phases; and the discussion have been omitted.]

AUTHOR CITATION INDEX

Aario, L., 90
Abelson, P. H., 90
Adamson, C. L., 244
Afzelius, B. M., 16, 24, 90
Allen, K. C., 101
Allen, P., 282
Alpern, B., 365
American Commission on Stratigraphic Nomenclature, 218
American Geological Institute, 50
Ammosov, I. I., 50
Ananova, Y. N., 234
Andel, T. van, 322
Anderegg, H., 219
Anderson, F. W., 282
Anderson, R. Y., 218
Andrews, H. N., 365
Andrusov, D., 240
Angold, R. E., 24
Anisgard, H. W., 219
Argue, C. L., 24
Arkhipov, S. A., 90
Armstrong, T. A., 124
Arnold, C. A., 182
Asbeck, F., 90

Backus, R. C., 93
Baher, C. L., 50
Bair, C. R., 50
Balme, B. E., 365
Barghoorn, E. S., 182
Bartlett, A., 219
Bartlett, H. H., 266
Barton, D. C., 50
Beard, J. S., 322
Belsky, C. V., 218
Bennett, A. J. R., 90
Bennie, J., 164, 266
Benson, M., 365
Bentall, R., 124, 165
Berger, F., 90

Berger, I. A., 90
Bergius, G., 50
Bharadwaj, D. C., 365
Blackett, P. M. S., 365
Bless, M. J. M., 223
Boddy, R. G. H. B., 365
Bode, H., 50
Bogolyubova, L. I., 50
Bolli, H. M., 218
Boltenhagen, E., 218
Bosshard, U., 92
Boswell, P. G. H., 266
Büttcher, H., 50
Böttcher, H., 50
Bowler, J. W., 124
Bown, C. C., 24
Bradley, W. H., 182
Bratzeva, G. M., 218
Brauns, F. E., 90
Bräutigam, F., 90
Brelie, G. von der, 90
Brooks, J., 7, 24
Brosius, M., 90
Brouwer, A., 90
Brown, C. A., 90
Brown, H. R., 244
Bukatsch, F., 90

Campbell, M. R., 50
Campo-Duplan, M. van, 164
Caribbean Petroleum Company, 124
Cepek, A. G., 90
Chaloner, W. G., 101, 164, 282, 290, 365
Chandler, M. E. J., 266
Chapman, B., 24
Chiguryayeva, A. A., 90
Chitaley, S. D., 164
Christ, H., 218
Christensen, J. E., 24
Cizancourt, M. de, 124
Clapp, F. G., 51

Author Citation Index

Clark, G. L., 90
Clarke, R. T., 235
Cocking E. C., 25
Coe, E. A., 332
Colombo, U., 90
Cook, A. C., 244
Cookson, I. C., 101, 164, 182, 218
Cope, F. W., 331
Correia, M., 7
Couper, R. A., 101, 182, 218
Covas, G., 218
Cridland, A. A., 290
Croneis, C., 50
Cross, A. T., 218, 365
Cushing, E. J., 90

Dahne, H., 368
Dale, H. M., 92, 367
Dalziel, J. M., 219
David, T. W. E., 244
Davis, M. B., 90, 235
Denizot, J., 24
Dettmann, M. E., 101
Dewers, F., 91
Dickinson, H. G., 24
Diener, S., 90
Dijkstra, S. J., 266, 282
Dobbin, C. E., 50
Dormans, H. N. M., 365
Dorsey, G. E., 50
Doyle, J. A., 101, 282
Doyle, W. T., 16
Drooger, C. W., 124
Drugg, W. S., 218
Dufour, J., 124
Dulhunty, J. A., 50
Dumait, P., 90
Dunbar, A. E., 24
Duparque, A., 368
Dybowa, S., 240, 365

Eberl, B., 90
Eby, J. B., 50
Echlin, P., 6, 24
Eddy, E., 367
Edelman, C. H., 90
Eder, H., 90
Edwards, K. L., 219
Edwards, W., 365
Ehrlich, H. G., 16
Elphick, J. O., 51
Elsik, W. C., 6
Emiliani, C., 90
Emmons, W. H., 50
Engler, A., 218
Erdtman, G., 16, 24, 90, 124, 164, 182, 218, 282, 322

Ergol'skaya, Z. V., 50
Evans, M. M., 367

Faegri, K., 90, 124, 322, 365
Fagerlind, F., 24
Feder, N., 24
Felix, C. J., 90, 365
Fenton, G. W., 365
Ferguson, I. K., 6
Fichter, H. J., 219
Florin, R., 331, 365
Florschütz, F., 90, 91, 92, 124
Flynn, J., 25
Francis, W., 365
Frank H., 25
Fritsche, H., 90
Fuchs, H. P., 218
Fuchs, W., 50
Fuller, M. L., 50
Funnel, B. M., 218

Garcia de Mutis, C., 220
Gelletich, J., 220
Gibson, M., 90
Gijzel, P. van, 6, 7, 90, 220
Gilbert, J. L., 282
Gischler, C. E., 90
Glagoley, A. A., 366
Gleason, S., 90
Godwin, H., 24, 101, 366
Gonzalez Guzman, E., 91, 218
Goodwin, R. H., 90
Gothan, W., 218
Gottschewski, G. H. M., 91
Grabner, A., 90
Gradzinski, R., 240
Grant, P. R., 7
Grayson, J. F., 322
Grebe, H., 366
Grichuk, V. P., 235
Gripp, K., 91
Groot, C. R., 91
Groot, J. J., 91, 235
Gropp, W., 50
Guennel, G. K., 164, 366
Gullvåg, B. M., 24
Gutjahr, C. C. M., 91

Haberlandt, H., 91
Hackner, B. L., 219
Hacquaert, N., 91
Hacquebard, P. A., 366
Haitinger, M., 91
Halle, T. G., 282
Hammen, Th. van der, 90, 91, 101, 220
Harris, T. M., 282, 366
Harris, W. K., 218

Härtel, O., 164
Hartung, W., 366
Havinga, A. J., 6, 91
Havlena, V., 240
Heck, E. T., 50
Heckman, C. A., 24
Hedberg, H. D., 124
Hedemann, H. A., 91
Heinen, W., 91
Heslop-Harrison, J., 6, 7, 12, 16, 24
Heuveln, B. van, 91
Hieronymus, C., 266
Hinder, H., 50
Hobson, G. D., 90
Hoffmann, E., 50, 366
Hoffmeister, W. S., 124, 322, 331, 368
Hollingsworth, S. E., 366
Holmes, A., 91
Hopping, C. A., 218
Hörhammer, L., 124
Horner, H. T., Jr., 24
Horst, U., 164, 282, 331
Hou-Liu, S. Y., 219
Hsü, K. J., 223
Hubman, C. W., 124
Huck, G., 50
Huddle, J. W., 366
Huggler, K., 93
Hughes, N. F., 219
Hull, H. M., 92
Hultzsch, A., 91
Hume, G. S., 50
Hutchinson, H., 219
Hutjens, F. G., 365

Ibrahim, A. C., 135, 164, 165
International Committee for Coal Petrology, 50, 91, 366
Ivanov, M. V., 91
Iversen, J., 51, 90, 91, 101, 124, 164, 235, 322, 365

Jachowicz, A., 240, 241, 365
Jacob, H., 91
Jansonius, J., 92
Jardiné, S., 219
Jones, D. H., 92
Jones, D. J., 235
Jones, I. W., 50
Jongmans, W. J., 282
Jonker, F. P., 90, 124
Jurasky, K. A., 368
Jux, U., 91

Kälin, O., 93, 368
Kalt, P., 12
Karmasin, R. W., 366

Karrer, P., 12
Karweil, J., 50, 91
Katchalsky, A., 25
Kedves, M., 101, 219
Kempf, E. K., 247
Kendall, P. F., 366
Kidston, R., 164, 266, 366
Kirchheimer, F., 50, 91
Kisch, H. J., 91
Klaus, W., 124, 165
Knox, E. M., 135, 266, 331, 366
Koldewijn, B. W., 322
Korczynski, M. S., 25
Kosanke, R. M., 124, 164, 331, 366
Koslow, W. P., 50
Kötter, K., 91
Kremp, G., 165, 282, 332, 366
Kreulen, J. W., 50
Krevelen, D. W. van, 50, 91, 365
Krieg, A., 91
Kröger, C., 91
Kruit, C., 322
Krutzsch, W., 219
Ksiazkiewicz, M., 241
Kuyl, O. S., 51, 219, 333
Kuznetsov, S. I., 91

Lahee, 51
Lange, Th., 368
Langenheim, J. H., 219
Larson, D. A., 16, 25, 91
Lata, P., 219
Leidelmeyer, P., 219
Leighton, L. H., 365
Lersten, N. R., 24
Lesley, J. P., 51
Le Thomas, 25
Levenshteyn, M. L., 51
Lewis, C. W., 16, 25, 91
Lewis, E., 51
Liddle, R. A., 124
Liechti, J., 12, 93
Liechti, P., 219
Lilley, E. R., 51
Lindeman, J. C., 219, 322
Linskens, H. F., 91
Loeblich, A. R., 219
Loose, F., 165
Luber, A. A., 135, 164, 331
Lugardon, B., 25
Lundgren, D. G., 25
Lyalikova, N. N., 91

Maarleveld, G. C., 92
McClymont, J. W., 16
McCoy, 51
McFarlan, 51

McIntyre, D. J., 6
Mackenzie, A., 12
Mackowsky, M. T., 366
Macrae, J. C., 91, 368
Mädler, K., 164
Magloire, L., 219
Maier, D., 91
Malloy, R. E., 331
Mamay, S. H., 365, 366
Manten, A. A., 219
Marshall, C. E., 51, 366
Martin, H. H., 25
Martin, P. S., 92
Martin, R., 219
Massalski, A., 24
Matveyeva, O. V., 90
Maurenbrecker, A. L. F. J., 366
Mayr, E., 219
Mehner, W., 91
Mencher, F., 124, 219
Menéndez Amor, J., 91
Menten, A. A., 101
Meuter, F. de, 92
Michael, F., 282
Miller, J. B., 219
Millott, J. O'N., 331, 366
Miner, E. L., 182
Molchanov, I. I., 51
Mollenhauer, H. H., 25
Moody-Stuart, J., 219
Moore, L. R., 366
Moret, L., 124
Moulton, G. F., 51
Muir, M. D., 7, 223, 290
Mukherjee, A. N., 51
Muller, A., 51, 92
Müller, H., 219
Muller, J., 6, 101, 219, 290, 333, 367
Murray, H. H., 368

Nabli, M. A., 25
Nairn, A. E. M., 367
Nandi, B., 368
Naumova, S. N., 135, 164, 331
Nayar, B. K., 219
Neves, R., 333, 367
Niggli, P., 164
Nikitin, P., 135, 282
Nota, D. J. G., 92, 322
Notestein, F. B., 124
Nowak, J., 241

O'Brien, T. P., 24
Ode, W. H., 92
Ohrbohm, R., 223
Oliphant, E. M., 92

Oltmann, O., 25
Osborn, C. C., 367
Osborne White, H. J., 282
Oswald, A., 12
Overbeck, F., 91, 92

Pannell, E., 365
Pant, D. D., 164, 182
Patijn, R. J. H., 51
Patteisky, K., 51
Patterson, S. H., 366
Pearse, A. E. G., 25
Pease, D. C., 16
Penrose, R., 50
Petijn, R. J. H., 51
Petrascheck, W., 51
Petris, S. de, 25
Petteisky, K., 50, 92
Pettitt, J. M., 25, 101
Pflug, H. D., 92, 101, 124, 129, 164, 166, 182
Phillips, L., 223
Pike, K. M., 182, 218
Piveteau, J., 322
Pocock, S. J., 92
Pojnar, E., 25
Polak, B., 92, 367
Posthumus, O., 322
Postma, H., 322
Potonié, H., 164, 165
Potonié, R., 51, 101, 102, 124, 165, 182, 183, 218, 219, 220, 282, 332, 367
Power, S., 51
Przibram, K., 92

Raadshooven, B. van, 124
Radforth, N. W., 367
Radomski, A., 240
Raistrick, A., 51, 101, 124, 183, 332, 367
Ralska-Jasiewiczowa, M., 92
Ramanujam, C. G. K., 220
Rao, A. R., 183
Rawat, M. S., 220
Reed, F. D., 367
Reeves, F., 51
Reger, D. B., 51
Reid, C., 266
Reid, E. M., 266
Rein, U., 124
Reinsch, P. F., 165, 332
Reissinger, A., 124
Remy, R., 367
Remy, W., 365, 367
Renz, H. H., 219
Reyment, R. A., 220
Reynolds, E. S., 16

Rogers, H. D., 51
Rogers, J., 51
Ross, J. B., 51
Rowley, J. R., 16, 25, 92
Ruch, F., 92
Russel, W. L., 51
Russell, L. S., 93

Saad, S. I., 220
Sacchi Vialli, G., 92
Sadebeck, R., 266
Sangster, A. G., 92, 367
Sass, L. C., 124
Schärer, G., 93
Schaub, H. P., 124, 220
Scheidel, C., 368
Schemel, M. P., 365
Schmitz, H., 92
Schnack, B., 218
Schneider, W., 92
Schopf, J. M., 51, 124, 165, 367
Schucht, F., 92
Schultze, R., 51
Schulze, G. M., 165
Schuyer, J., 50, 91, 368
Schwarck Anglade, A., 220
Schwartz, W., 51, 92
Schwarzbach, M., 367
Scott, R. A., 102, 282
Selvig, W. A., 92
Semmers, D. R., 51
Sengupta, S., 25
Senn, A., 124
Seward, A. C., 166, 282
Seyler, C. A., 51
Shaw, G., 6, 7, 12, 24, 92
Shellhorn, S. J., 92
Shepherd, F. P., 368
Short, K. C., 220
Siedlecki, S., 241
Šilar, J., 241
Simonicsics, P., 101
Simpson, G. G., 220
Simpson, J., 101, 183
Sitte, P., 92
Sjörs, H., 367
Sjöstrand, F. S., 16, 24
Skok, V. I., 51
Skvarla, J. J., 16, 91
Slater, L., 367
Smith, A. H. V., 367
Snyman, C. P., 367
Somers, G., 367
Southworth, D., 25
Spackman, W., 367

Spinner, E., 247
Sprunk, G. C., 92
Stach, E., 368
Stach, H., 51
Stadnichenko, T., 51
Stamp, L. D., 266
Stanley, E. A., 92
Staplin, F. L., 6, 92, 331
Stäuble, A. J., 220
Steel, M., 92, 223
Stevens, C. H., 92
Stevens, K. R., 52
Stolk, J., 220
Storm, W., 51
Stuart, M., 52
Stutzer, O., 368
Suggate, R. P., 51
Sutton, F. A., 129
Svoboda, J., 368

Tappan, H., 219
Tarr, R. S., 52
Taylor, G. H., 244, 368
Teichmüller, M., 50, 51, 52, 92, 129, 282, 368
Teichmüller, R., 50, 52, 91, 92, 129, 282
Terhune, R. W., 52
Terres, E., 368
Teunissen, D., 92
Thiergart, F., 129, 165, 183, 282
Thiery, J.-P., 25
Thiessen, R., 183
Thom, W. T., 52
Thompson, A. B., 52
Thomson, P. W., 92, 101, 129, 165, 166, 183, 368
Timofeyev, P. P., 368
Tiwari, L. P., 219
Todd, R. G., 92
Tomlinson, R. C., 368
Toschew, G., 52
Totten, C., 25
Tralau, H., 220
Trappe, K., 368
Traverse, A., 223
Trifonow, I., 52
Troel-Smith, J., 51, 124, 164
Tschudy, R. H., 102
Tyzhnov, A. V., 51

Unrug, R., 240
Urban, J. B., 92

Vakhrameiev, V. A., 282
Val'ts, I. F., 52

Author Citation Index

Van Hoeken-Klinkenberg, P. M. J., 220
Van Krevelen, D. W., 368
Van Raadshooven, B., 220
Van Steenis, C. G. G. J., 220
Varma, C. P., 220
Veen, F. R. van, 322
Veenstra, H. J., 92, 93
Venitz, H., 165, 183
Vicari, H., 93
Vierssen Trip, P. H., 266
Viete, G., 92
Vimal, K. P., 183
Vronina, K. V., 90
Vyshermirskii, V. S., 52

Wagenbreth, O., 92
Waksman, S. A., 52
Walker, J. J., 365
Wallis, W. E., 219
Waltz, I. E., 164, 331
Wandless, A. M., 365, 368
Wang, W. S., 25
Wanless, H. R., 368
Warne, S. St. J., 368
Waterbolk, H. T., 219, 333
Waters, P. L., 244
Webster, R. M., 183
Wee, W. M. ter, 92
Weidel, W., 25
Wettstein, D. von, 16
Wetzel, W., 91, 92, 93

Weyland, H., 218
Wheeler, H. E., 368
White, D., 52, 93, 183
Wijmstra, T. A., 90
Williams, R. C., 93
Williams, R. W., 368
Willison, J. H. M., 25
Wilshire, H. G., 244
Wilson, L. R., 52, 93, 124, 129, 165, 183, 235, 322, 332, 368
Wilson, J. T., 220
Winchester, D. E., 183
Wodehouse, R. P., 129, 183
Wolcott, P. P., 219
Woldstedt, P., 93
Wolf, M., 93
Wolfram, A., 322
Woolnough, W. C., 52
Wyatt Durham, J., 129
Wymstra, T. A., 220

Yeadon, A., 12, 92
Young, M. R., 93

Zagwijn, W. H., 93
Zeiller, M. R., 266
Zerndt, J., 166, 241, 266
Zetzsche, F., 12, 52, 93, 368
Ziegler, E., 12
Zoldani, Z., 241
Zonneveld, J. I. S., 93

SUBJECT INDEX

Abies, 63, 109
Acacia octosporites, 175
Acanthotriletes sp., 244
Acetolysis, 1
Acmopyle, 15
Aetheotesta elliptica, 155
Agglutination, 23
Ahrensisporites sp., 344
Ahrensisporites angulatus, 237, 239
Alangium barghoornianum, 181
Alatisporites, 105, 157, 237, 239
Alatisporites pustulatus, 344
Alatisporites trialatus, 239
Aletes, 161
Aletes, 105
Algae, 111
Alisporites, 157
Alisporites spp., 243, 244
Alnipollenites, 158
Alnipollenites verus, 203, 210, 216
Alnus, 63, 113, 116, 190
2-aminoethanol, 17, 19, 26
Anacardium-type, 304, 319
Anacolosa, 113
Angiosperm
   aperture types, 108
   pollen, taxonomy of, 106
Anteturma (=oberabteilung), 98
Anthers, 8
Anthracite, 32
Anthurium sp., 13, 18
Anulatisporites, 329
Anulatisporites coronarius, 237
Aphanozonati, 267
Apiculati, 136, 160
Apiculatisporites, 327, 328, 329
Apiculatisporites spinosus, 346
Aratrisporites spp., 243
Archaeopteris latifolia, 159
Archaeotriletes, 227
Artemisia, 24
Asteraceae, 189, 190

Auriculati, 160
Auritotriletes, 160
Avicennia nitida, 294, 300, 312, 314
Azolla, 155
Azonaletes, 161
Azonomonoletes, 161
Azonotriletes, 136

β-carotene, 5, 11
β-carotene polymer, 11
Bacterial cell walls, 18
Betula, 63
Betulaceaepollenites, 181
Bituminous coal, 32
Blechnum, 18
Bombacaceae, 113
Bombacacidites annae, 201, 202
Bombax aquaticum, 294
Borneo, Tertiary palynology of, 185ff.
Botryococcus, 292
Brandon lignite of Vermont, 96, 167ff.
Buttinia andreevi, 201

Cadargasporites senectus, 243
Calamospora, 105, 138, 139, 327ff.
Calamospora breviradiata, 238
Calamospora flexilis, 239
Cancellatisporites cancellatus, 238
Carbohydrate, 16
Carbonization, 4, 26, 27ff.
   in coal samples, 40, 229ff.
   data, recording of, 38ff.
   effects of pressure on, 33ff.
   effects of temperature on, 32ff.
   effects of time on, 32ff.
   in sand-shale samples, 40, 41, 43, 48
Carbonization studies
   applications of, 48, 69
   preparation techniques for, 36ff.
   selection of standard miospores for, 36ff.
   techniques for, 37, 38

Subject Index

Carbon ratio, 27
  theory, 27
Caribbean stratigraphic data, 184ff.
Carotenoid esters, 5, 11, 12
Carotenoids, 4, 5, 8, 11, 12
Carpinus, 63
Carya, 176
Caryapollenites, 162
Caving, 27, 223
Ceratopteris, 196
Cedroidites, 141
Cedrus, 141
Cheirolepidaceae, 113
Chenopodiaceae, 63, 106
Chlamydospermae, 113
Cicatricosisporites, 177
Cicatricosisporites dorogensis, 190, 196, 200, 201, 207, 208, 210, 214
Cingulati, 161
Cingulizonates asteroides, 237, 239
Cingulizonates tuberosus, 237
Circulisporites parvus, 244
Circumpollis, 113
Cirratriradites, 105, 150, 157
Cirratriradites punctatus, 238
Classification
  biological relationships in, 97
  morphographic systems of, 97, 98, 137ff., 158ff.
  natural system of, 98
  Raistrick's system of, 92
  seminatural, 98
  systematic, 97
Classopollis, 113
Coal
  assemblages, influence of climate on, 359ff.
  brown, 32
  formation, influence of water level on, 355ff.
  measures correlation, 2, 96, 97
  rank, 27, 28, 53ff.
  recycled fragments, 222, 236ff.
  related to fluorescence, 80ff.
  studies related to carbonization, 29ff.
  types, 290ff.
    origin of, 353ff.
Coalification, 4, 53ff., 242
  and fluorescence, 77ff.
Compositae, 106, 113, 116, 189, 190, 308
Concentration of pollen in fluviomarine sediments, 191
Conocephalum, 18
Converrucosisporites cameroni, 243
Cordaitales, 113
Cornaceoidea, 176
Correlation of coal measures. See Coal

Corrosion of miospores, 55
Coryli? -pollenites coryphaeus, 142
Corylus, 63
Corylus-Typ, 176
Crassidurain, 1, 334
Crassispora kosankei, 341, 342, 343, 349, 357
Crassoretitriletes vanraadishooveni, 200, 203, 209, 210
Ctenolophon, 116
Ctenolophonidites lisamae, 201, 202, 211, 212, 213
Cuticles, 310
Cyatheaceae (*hemitelia*-type), 116
Cycadopites sp., 244
Cycadopites nitidus, 243, 244
Cyclogranisporites leopoldii, 240
Cyperaceae, 63
Cystites, 161
Cystosporites, 138, 157
Cystosporites giganteus, 156
Cystosporites verrucosus forma *abortivus*, 250, 255

Dacrydium, 175
Dacrydiumites mawsoni, 175
Daughter cell, 5
Densospora, 150
Densospore phase, 345ff.
Densosporites, 105, 113, 157, 289ff., 339ff.
Densosporites decorus, 237, 239
Densosporites duriti, 341
Densosporites loricatus, 341
Densosporites sphaerotriangulatus, 339ff., 369
Densosporites spinosus, 237, 238, 239
Desmochitina, 227
Dicarboxylic acid, 8
Dicolpatae, 104
Dicolporatae, 104
Dictyophyllidites mortoni, 243, 244
Dictyotriletes, 327, 329
Dictyotriletes bireticulatus, 341, 357
Diporatae, 104
Disaccites, 161
Discernisporites, 329
Dispersal of pollen grains and spores, 191ff., 288ff.
Duplexisporites gyratus, 243
Duplospores, 108
Dyadeae, 104

Echitricolporites mcneillyi, 203, 210
Echitricolporites spinosus, 187, 188, 200, 203, 209
Echitriporites trianguliformis, 193, 201
Ektexine, 5

376

Subject Index

Endexine, 5
Endoplasmic reticulum, 14
Endosporites, 105, 113, 157, 330
Environments of deposition, 211ff., 292ff.
Ephedra, 108, 113
Ephedra distachya, 116
Ephedra strobilacea, 116
Epilobium, 23
Ericaceae, 63, 106, 176
Ericaceae-pollenites roboreus, 175
Ericacipites longisulcatus, 173
Ericales, 109
Erythrina glauca, 296
Euterpe sp., 294, 296
Exine, 4, 5, 8, 17, 19
  color change in, 6, 27
  degradation of, 5, 19, 31ff.
  layers, 5, 150ff.
    terminology of, 154
Extratriporopollenites spp., 106

Fagidites, 176
Fagus, 63, 174
Fagus silvaticoide, 176
Fatty acids
  branched chain, 11
  straight chain, 11
Fixed carbon, 27, 53ff.
Florinites, 105, 113, 157, 289ff., 325ff., 333, 348ff., 358
Florinites antiquus, 237, 239
Florinites ovatus, 238
Florinites similis, 237
Florschuetzia levipoli, 200, 203, 208, 209
Florscheutzia meridionalis, 203
Florschuetzia semilobata, 203
Florschuetzia trilobata, 203, 208
Fluorescence
  color determinations, 61ff.
  data
    presentation of, 66ff.
    recording of, 65ff.
  and geological age, 65
  related to fossilization, 85ff.
  secondary, by staining, 57
  techniques
    in geology, 58
    in palynology, 6, 55, 59ff.
Fluorescence studies, preparation techniques for, 55, 59ff.
Flysch, Polish (Carpathian), 22, 236ff.
Foraminifera, 310
Formalin, 13
Fossil spores and pollen grains. *See* Miospores
Foveotricolpites irregularis, 201

Foveotricolpites perforatus, 202, 211, 212, 213
Foveotriletes margaritae, 201, 202, 211
Franciscan melange, 222
Fungi, 111, 309

Gemmastephanocolpites gemmatus, 202
Geochemistry, organic. *See* Organic geochemistry
Geothallus, 14
Germinal apertures, 14
Germination
  distal, 4
  equatorial, 4
  proximal, 4
Gnetopsis, 157
Gramineae, 106, 113, 189, 308
Granasporites cf. *irregularis*, 346
Granulatisporites, 158
Granulatisporites microgranifer, 344
Granulatisporites minor, 243
Granulatisporites parvus, 346
Grimsdalea magnaclavata, 202, 209, 216
Gymnospermous pollen classes, 105
Gymnosperm pollen, taxonomy of, 108

Haloragaceae, 180
Haloragacidites, 180
Hay fever, fluorescence studies, 56, 103
Hemitellia sp., 306, 310, 319
Heterocolpatae, 104
Heterospory, 6
Hexaporites, 176
Holocene palynology, 1
Homosporous plants, 4
Horniella, 18
Horniella clavaticosts (sic), 181
Hydrocarbon potential. *See* Miospores, used to assess hydrocarbon potential
Hystrichospherideae (Hystrix), 111, 291, 310, 314, 315, 317, 318, 320

Ilex-pollenites iliacus, 175
Illinites, 105, 157
Inaperturatae, 105
Inaperaturates, 161
Incursion phase, 345ff.
Index fossils, 99
Intine, 8
Intratriporopollenites instructus, 179
Ipomoea, 17
Isoetes, 248
Isospores, 4, 138

Jandufouria seamrogiformis, 202, 208, 215, 216

Subject Index

Kraeuselisporites differens, 243, 244
Kraeuselisporites pallidus, 243
Kryshtofovichia africana, 272, 277

Laevigati, 136, 160
Laevigatisporites glabratus, 148, 157
Laevigatosporites, 105, 156, 158, 162, 339
Laevigatosporites desmoinensis, 339, 341, 342, 349, 357
Lagenicula, 155, 156, 254, 269ff.
Lagenoisporites, 155, 156
Lagenotriletes, 136, 160, 269
Laguncularia racemosa, 294
Lamellae of unit membrane dimensions, 4, 5, 13, 14, 19. See also Tapes
Lead citrate, 13
Leguminosae? -pollenites quisqualis, 175
Leitfossilien. See Index fossils
Lemna minor, 18
Lepidodendron, 289
Lepidostrobus foliaceus, 155
Lepidostrobus major, 156
Liguliflorae, 116
Lilium, 5, 8
Lilium henryii (L. henryii), 8, 11, 12
Lipopolysaccharides, 24
Liquidambar, 179
Longapertites vaneendenburgi, 202
Lycopodium clavatum, 4, 8, 11, 15, 17, 18, 19, 21, 23
Lycopodiumsporites sp., 243
Lycospora, 105, 113, 150, 288ff., 339ff.
Lycospora brevituga, 339, 346, 349
Lycospora pusilla, 339, 346, 349
Lycospora granulata, 357
Lycospore phase, 345ff.
Lycostrobus scotti, 156

Macrostachya, 139
Magnastriatites howardi, 189, 196, 200, 201, 208
Magnoliaceae, 113
Manicaria sacchifera, 294, 296
Marine fossils in fresh water sediments, 231
Marker species
  intracontinental, 184
  pantropical, 184
  transatlantic, 184
Mauritia sp., 296, 300, 312, 314
Megaspores, 1, 4, 138, 246ff.
  carboniferous, 246ff.
  walls, ultrastructure of, 246
  Wealden (L. Cretaceous), 246ff.
Microreticulatisporites sp., 243
Microspores, 1, 4, 15, 138

Microsporites, 138, 157
Microsporites karczewskii, 237
Mimosaceae, 116
Miospore assemblages of coal seams, 335ff.
Miospores, 1, 3, 5, 96ff., 138
  age determination of, 4, 53ff., 65, 71ff.
  autofluorescence of, 53ff.
  distribution, factors influencing, 109
  experimental heating of, 6
  as paleotemperature indicators, 2, 27ff.
  as part of the plant life cycle, 4
  preservation of, in sediments, 109
  Tertiary, paleoecology of, 169
  used to assess hydrocarbon potential, 2, 27ff.
  used in stratigraphy, 99
Mohria, 177
Mohrioisporites, 177
Monocarboxylic acid, 8
Monocolpatae, 104, 105
Monocolpates, 161
Monoletes, 161
Monoporatae, 104
Monoporites annulatus, 193, 199, 200, 202, 206, 207
Monoporites (Graminidites) media, 174
Monosaccites, 161
Monosulcites sp., 244
Montrichardia sp., 296
Mooreisporites, 329
Morphology and taxonomy, 104
Mud, drilling, 27, 223
Multimarginites vanderhammeni, 202, 209, 215, 216
Murornati, 136, 160
Myrica, 63, 107
Myricaceae, 180
Myricales, 180
Myriophyllum, 176

Napites, 161
Neoraistrickia taylori, 243, 244
Nevesisporites limatulus, 243
Nexine, 5, 19. See also Endexine
Nigeria, Tertiary palynology of, 185ff.
Normapolles, 180
Nothofagus, 174
Nuphar, 15
Nymphaceae, 113
Nypa-type pollen, 189
Nyssa-phyllites, 177
Nyssa-pollenites (Nyssaceae-pollenites), 177

Olacaeae, 113

Oleaceen-typ, 176
Operculatae, 104
Ophioglossum, 18
Organic geochemistry of plant materials, 2, 6
Organic metamorphism (definition), 27. See also Thermal metamorphism
Orinoco delta palynology, 288
Osmium tetroxide, 19, 24
Osmunda, 18
Osmundacidites spp., 243, 244
Ozonolysis of sporopollenin, 5

Pachydermites diederixi, 216
Palaeogeographic reconstructions, 223
Palaeostachya, 139
Paleotemperature indicators. See Miospores, as paleotemperature indicators
Palmidites, 175
Palynological assemblages
  biological stain reactions of, 228ff.
  containing only older fossils, 230ff.
  containing only younger fossils, 231ff.
  contaminated, 53ff., 74ff., 223ff.
  with differential preservation, 230
  of mixed ages, 226ff.
Palynological data, limitations on, 198ff.
Palynological preparations, oxidation used in, 2
Palynological samples collection, 232
Palynological studies, techniques for, 110ff.
Palynological zones, controlling factors
  evolution, 113–116
  migration, 116–119
Palynology
  faulty techniques in, 222, 224ff., 232ff.
  holocene, 1
  in oil industry, 2, 27ff., 96ff., 103ff.
Palynomorphs, 224
Parkeriaceae *(ceratopteris)*, 116
Peat, 224
  carboniferous, formation of, 351ff.
Pediastrum, 292
Pericolpatae, 104, 112
Pericolporatae, 104, 113
Peridineae, 111
Periporatae, 104, 113
Periporopollenites stigmosus, 179
Perisyncolporites pokornyi, 214
Perotriletes, 175
Phenolic acids, 8, 12
Phycopeltis, 292
Physostoma, 156
Picea, 63, 171

Picea mariana, 233
Piceae? -pollenites alatus, 142
Pilasforites plurigens, 244
Pinus, 171, 191, 358
Pinus sylvestris, 8, 61, 63
Pityosporites, 157, 158, 358
Pityosporites westphalensis, 345
Planisporites, 329
Planisporites spinulistratus, 344
Plant evolution, evidence from palynology, 100
Plants fossils and geology, 268
Plasma membrane, 14
Pleistocene palynology, 1, 55ff.
Poa annua, 15
Poaceae, 63, 189
Podocarpidites, 181
Podocarpus, 119
Podocarpus sp., 306, 319
Pollen
  analysis, 1, 53ff., 96, 168
  classes
    dicotyledonous, 104
    monocotyledonous, 104
  diagrams, 54
  grain (definition of), 4
  morphology, 53
  ontogeny, 17, 23
  rain, 109
Pollenin, 5
Pollenites, 98, 136, 161
Pollenites granifer, 175, 176
Pollenites reclusus, 113
Polospores. See Miospores
Polyadeae, 104
Polyplicatae, 105
Polypodiisporites ipsviciensis, 243, 266
Polypodium sp., 309, 314
Polyporo-pollenites, 180
Polypor-poll. undulosus, 179
Polysaccharide nature of lamellae, 6, 23
Polysaccites, 161
Populis tremula, 15
Poroplanites, 108
Porostrobus (=porostrobosporites), 138
Precolpates, 161
Pre-quaternary palynology, 1, 103
Proprisporites, 329
Proteacidites dehaani, 193, 201, 205, 211
Proteacidites tubercalatus, 174
Protococcus, 233
Proxapertites cursus, 201
Proxapertites operculatus, 199, 201, 205, 206
Proximate analysis, 28
Psiladiporites minimus, 202, 208, 209, 215

Author Citation Index

Psilate (smooth), 28
Psilatricolpites operculatus, 202, 212
Psilatricolporites crassus, 201, 202, 207, 212, 213
Pteris-type, 304
Pterocarpus officinalis, 294
Punctatasporites, 159
Punctatisporites, 105, 159, 243, 244, 329, 330
Punctatisporites obliquus, 238, 240
Punctatisporites orbicularis, 238
Punctatosporites, 159
Punctatosporites minutus, 341, 342, 343, 349, 357
Punctatosporites walkomi, 243
Pyrobolospora, 267, 269ff.
Pyrobolospora hexapartita, 273ff.
Pyrobolospora lobata, 276
Pyrobolospora medusa, 276
Pyrobolospora pyriformis, 275ff.
Pyrobolospora vectis, 270ff.

"Q" values, 63
Quaternary palynology, 96
Quercoidites (Quercoipollenites), 176
Quercus, 63
Quercus-pollenites, 98
Quercus robur (heating experiments), 31ff.
Quisquilites, 227

Radiatisporites, 155
Raistrickia, 105, 329
Raistrickia pallida, 239
Recycling (=reworking), 222, 224ff.
Reditiporites magdalensis, 201, 202, 206
Reinschia, 150
Reinschospora, 105, 157, 162
Retibrevitricolpites triangulatus, 201, 206, 212
Reticulatisporites, 105
Reticulatisporites adhaerens, 238
Retitricolpites irregularis, 201
Retitricolporites guianensis, 190, 207, 212, 213, 214
Rhizophora, 109, 119, 187, 196, 297ff.
Rhizophoraceae, 106
Rhizophora mangle, 294ff., 300
Rotaspora, 150, 227
Rotatisporites, 155
Ruhrkarbons (Ruhr coals), 96, 136ff.

S. angulatus, 163
Scanning electron microscope (SEM), 18
Scapania nemorosa, 14
Schopfipollenites, 159, 344
Schulzospora, 327, 328

Schulzospora rara, 237
Sclerotites, 163
Seed (definition of), 4
Seeds of peas and beans, 18
Selaginella spp., 17, 56, 257, 258
Selagenellites (sic), 170
Sequoia, 15
Setosisporites hirsutus, 156
Sexine, 5. *See also* Ektexine
Silver granules on tapes, 21
Simozonotricetes intortus, 237
Sonneratia, 189
South America, Tertiary palynology of, 185ff.
Spectral ratios, 63
Spencerisporites, 138
Sphagnum, fluorescence of, 57, 61, 109
Spore wall, nature of, 150ff. *See also* Exine
Spores
  in coal, ecology of, 155
  dispersed, palaeontological studies of, 137ff.
  paleozoic, 96
    descriptive terminology for, 144ff.
  and pollen of living plants, 1, 4
  pteridophyte, taxonomy of, 108
  small. *See* Miospores
Sporites, 98, 136, 160
Sporites, used as generic name, 258
Sporomorphs (sporomorphae), 143, 173
Sporonin, 5
Sporonites, 98, 136, 160, 163
Sporopollenin, 2, 4, 5, 13, 14, 15, 21
  empirical formula, 5
  fluorescence and chemical nature of, 82ff.
  volatilisation of, 24
Stenotriletes, 227
Stephanocolpatae, 104
Stephanocolpites gemmatus, 202
Stephanocolporatae, 104
Stephanoporatae, 104
Stratigraphic control
  climate change as, 185ff.
  evolution as, 185ff.
  plant migration as, 185ff.
Stratigraphic correlation, using Tertiary miospores, 168, 210ff.
Stratigraphic leakage, 222, 223, 224ff.
Striaticolpites catatumbus, 201
Superbisporites, 150, 155
Subtriporopoll. simplex, 179
Symphonia, 190
Symphonia globulifera, 294, 306, 310, 315
Syncolpatae, 104
Synthetic carotenoid polymers, 4

Tapes, re-exposure of, 4, 17, 18, 19, 21, 23
Tapetal cells, 5, 16
Tasmanites, 227
Taxioidites, 142, 176
Tenuidurain, 1
Terminalia sp., 297ff., 308, 312, 315, 316
Tertiary assemblages
   interpretation of, 188ff.
   number of species in, 186
   statistical treatment of, 186ff.
Tertiary palynology, 96, 97, 185ff.
Tertiary pollen and spores, nomenclature of, 172
Tetracolpites, 175
Tetradeae, 104
Tetradopollenites laxus, 181
Tetrads, 4
Tetrad scar, 4
Thermal metamorphism, 6, 26, 27
Thiocarbohydrazide, 19
Thomsonia, 156
Thomson-Pflug nomenclature, discussion, 179
Thuja, 171
Tilia, 179, 180
Toluidine blue staining, 17, 21, 23
Tomato fruit protoplasts, 18
Tracheids, 310
Transition phase, 345ff.
Transmission electron microscope (TEM), 17, 21
Trapa natans, 155
Trichotomocolpatae, 104
Tricolpatae, 104
Tricolpites troedssoni, 113, 174
Tricolporatae, 104
Tricolporites, 175
Trilete mark, 18, 27. See also Tetrad scar
Triletes (as turma), 136, 160
Triletes, as generic name, 159, 160
Triletes auritus, 250, 251
Triletes brasserti, 250, 252, 253
Triletes cristatus, 156
Triletes galericulatus, 278
Triletes glabratus, 249, 250
Triletes hirsutus, 250, 252
Triletes horridus, 250, 253, 254
Triletes mamillarius, 249, 250, 251
Triletes praetextus, 250, 252, 253
Triletes retiarcus, 279
Triletes subpilosus, 254
Triletes superbus, 249, 250, 253
Tripartites, 156, 227, 228
Triplanosporites, 108

Triporatae, 104
Triporites, 175
Triquitrites, 156
Triquitrites auritus, 239
Triquitrites triturgidus, 344
Tsugapollenites igniculus, 154
Tuberculatisporites, 156
Tuberculatisporites gigantonodatus, 239
Tubuliflorae, 116
Turma (=abteilung), 98

Ubisch bodies, 15, 16
Ulmus, 63
Uranyl acetate, 13
UV irradiation. See Fluorescence, techniques

Valvisisporites, 155, 156
Verrucatosporites usmensis, 200, 207, 212
Verrucosisporites papillatus, 344
Verrucososporites sp., 243
Verrucososporites cf. *obscurus*, 240
Verrutricolporites rotundiporis, 201, 202, 208, 209, 210, 213, 215, 216
Vesicaspora, 157
Vitis forestdalensis, 181
Vitreisporites sp., 244
Vitreisporites pallidus, 243
Vitrinite, 28
Volatile matter, 28
Volcanic necks, coal fragments in, 222, 242ff.

Welwitschia, 113
Western Venezuela, 96, 103
   stratigraphic results, 119–124
White lines. See Tapes

"Xylan," 8

Zonales, 160
Zonaletes, 161
Zonati, 161
Zonation, using miospores, 108, 111ff., 199ff.
Zones
   intracontinental, 199ff.
   pantropical, 199ff.
   transatlantic, 199ff.
Zonocostites ramonae, 201, 202
Zonomonoletes, 161
Zonoratae, 104
Zonotriletes, 161

About the Editors

MARJORIE DOIG MUIR is presently on the staff of the Bureau of Mineral Resources, Canberra, Australia. She was born in Edinburgh, Scotland, on August 17, 1937, and took the B.Sc. degree at the University of Edinburgh. She subsequently graduated with a Postgraduate Diploma in Micropalaeontology and a Ph.D. in Palynology from University College, London. After this, she became Curator and then Lecturer in Geology at Imperial College, London, holding this appointment until immigration to Australia in 1977. During this time, she made several visits to palynological and palaeobotanical institutes in Poland, Czechoslavokia, and the Netherlands.

Dr. Muir has undertaken research in a number of fields, including the development and application of the scanning microscope and physical analytical methods to geology, Precambrian biostratigraphy, and the study of Mesozoic pollen and spores. She has written some fifty-seven papers and edited with others two books, *Sporopollenin* and *Quantitative Scanning Electron Microscopy* (both published by Academic Press). Dr. Muir has been a Council Member of the Palaeontological Association, and from 1971–1976 was Secretary to the International Commission for Palynology.

WILLIAM ANTONY SWITHIN SARJEANT is Professor in the Department of Geological Sciences, University of Saskatchewan. He was born in Sheffield, England, on July 15, 1935 and took the degrees of B.Sc. and Ph.D. at the University of Sheffield. He briefly held a Demonstratorship in Geology at the University College of North Staffordshire, subsequently becoming University Research Fellow at the University of Reading (1961–62), and was thereafter Lecturer in Geology at the University of Nottingham until immigration to Canada in 1972. During this time, he was for one year Visiting Professor at the University of Oklahoma.

Dr. Sarjeant is the author of some 160 scientific papers and books. He has undertaken research in a number of fields, from topographical mineralogy to geological bibliography and the study of vertebrate footprints, but his principal research has been on Mesozoic dinoflagellate cysts and acritarchs. He was a founder of the East Midlands Geological Society and first Editor of its journal, *The Mercian Geologist*, which now has an international circulation; his books include *Fossil and Living Dinoflagellates* (Academic Press, 1974). He was the first geologist to be honored by the award of a D.Sc. from the University of Nottingham.

DATE DUE